A. B. Semenov • S. K. Strizhakov • I. R. Suncheley

Structured Cable Systems

Springer

Berlin
Heidelberg
New York
Barcelona
Hong Kong
London
Milan
Paris
Tokyo

A. B. Semenov
S. K. Strizhakov
I. R. Suncheley

Structured Cable Systems

Translated by
N. Bolotnik and J. Bogatova

With 223 Figures

Springer

Andrey B. Semenov
E-mail: honor@it.ru

Stanislav K. Strizhakov

Igor R. Suncheley

I. T. Company
21A Krzhizhanovskogo Str.
Moscow, 117218
Russia

Translation from the Russian language edition:
Nikolay Bolotnik and Julia Bogatova

ISBN 3-540-43000-8 Springer-Verlag Berlin Heidelberg New York

Library of Congress Cataloging-in-Publication-Data applied for

Die Deutsche Bibliothek – Cip-Einheitsaufnahme
Semenov, Andrej B.: Structured cable systems / A. B. Semenov ; S. K. Strizhakov ; I. R. Suncheley.
Transl. by N. Bolotnik and J. Bogatova. - Berlin ; Heidelberg ; New York ;
Barcelona ; Hong Kong ; London ; Milan ; Paris ; Tokyo : Springer, 2002
(Signals and Communication Technology)
ISBN 3-540-43000-8

Springer-Verlag Berlin Heidelberg New York
a member of BertelsmannSpringer Science+Business Media GmbH
http://www.springer.de

Typesetting: data delived by authors
Cover design: design & production, Heidelberg
Printed on acid free paper SPIN: 10859689 62/3020/M - 5 4 3 2 1 0

Introduction

This study is devoted to the various aspects of cable system construction that are primarily focused on installation in office buildings. The term office building implies any building or its parts whose main areas are intended for personnel workplaces. A typical example of office buildings are business-centers, administrative buildings, financial institutions, ministries and other government bodies, buildings for design offices, education centers, and so on. The generic term “office” or “office building” will be used to designate this scope.

The present study pays a great deal of attention to the examination of cable systems used for the automation of workplaces for office personnel. At present, these are predominantly local area network (LAN) and public automatic branch exchange (PABX) cable systems. Examination of other telecommunications cable systems (fire and burglar alarms, access control, video monitoring, and installation of radio and loudspeaker communications) is beyond the scope of this study, however, these systems are partially discussed, if required. In order to provide standardization, it is recommended that they be constructed in compliance with the same approaches, components, and technologies.

In the mid-1980s, computer engineering as well as the LAN engineering was rapidly introduced into all parts of business and organization structures, which dramatically increased the information content transmitted within buildings (or a number of nearby buildings) without connection to the public switched telephone network (PSTN). The first generation of cable systems for information support was designed by computer engineers. The design efforts made by computer company specialists have focused on solving a rather narrow scope of issues in order to provide functioning of a specific and limited range of active network equipment supplied by only one manufacturer. It is obvious that such an approach has not enabled designers to pay special attention to the proper architecture and universality of the designed products. As a result, cabling installations were highly cusotmized, but due to restricted production were rather expensive. A change from one technology to another has always led to replacement of the cable system.

The changeover to new cabling is a very inconvenient process and requires considerable injections of capital and time, which hampers the efficiency of personnel. That is, it actually shuts down the work of the whole company or some of its departments for long periods of time. Even if the technology has not been changed (for example, when transitioning to the next generation of equipment made by the same manufacturer), the creation of new workplaces leads to certain difficulties for the maintenance service, since it requires routing of new cable segments.

Experience shows that during maintenance of cable systems for office buildings, it is not advisable to remove used cables from all types of cable pathways, as this may lead to damage of valid communications lines. Therefore new cables are routed directly above the existing ones in the course of transitioning to another cabling type. This results in the exhaustion of cable route capacity, which makes it impossible to create new wire communications lines.

The growth of subsystems created for building life support and for the support of its personnel leads to an increase in the number of services needing day-to-day maintenance. The services use the same cable routes, which often results in conflicts. In addition, specialists working there must have double functions, that is, it leads to an unreasonable distribution of human resources.

Table 1 demonstrates general statistical data on cost and service time of components for telecommunications infrastructure in office buildings. It implies that a properly designed cable system may potentially serve longer than the rest of the components and may therefore have a lower cost.

Table 1. Service time and capital injections made in various parts of building telecommunications structure [8]

	Software	Network equipment	Workstations and servers	Cable system
Service time, years	1.5–2	2.5–3	2–4	10–15
Capital injections, %	54	7	34	5

The solution of the problems which have been listed above is the creation of a cable system for an office building, which, at minimum, must possesses the following properties:

- is universal, that is, it can be used for transmission of signals from primary existing and prospective network equipment;
- enables quick and inexpensive creation of new workplaces and changes the topology of transmission channels without routing auxiliary cable lines;
- enables creation of a uniform maintenance service;
- built during building construction or remodeling of its premises into offices and has a maintenance warranty of 10 years or more.

A *structured cable system* (SCS) must meet the stated requirements. The SCS further defines a cable system whose construction approach meets three mandatory and several supplementary principles. The main principles of SCS are *structuring, universality,* and *redundancy.*

Structuring supposes the division of cabling and its accessories into separate parts or subsystems, each of which performs strictly defined functions and is supplied with a standardized interface for connection to other subsystems and network equipment. A part of any subsystem should include a number of termination facilities. These provide flexibility and enable the creation of structures with complex configurations, which can be easily and quickly changed and adapted to application-specific needs. When creating a system, a generic approach is used without association with any specific cable or termination hardware.

This enables designers, without any difficulty and at any level, to easily change both the optical and electrical signal transmission technologies, whose selection depends on the local conditions and the maximal technical and economical viability of the specific project.

The universality of a cable system means it is not adapted to a specific or very extended network technology, as it implies the construction principles of open architecture with the primary technical performance specified by standards. At the same time, normative documents specify parameters for electrical and optical cable pathways of separate subsystems and their interfaces. This enables the use of the cable system for signal transmission for various applications, and the reduction in the number of cable types to two: balanced (twisted pairs) and fiber-optic. The technical level of an element platform used for SCS development is specified by standards in order to provide for a minimum of 10 years of operation of telecommunications infrastructure.

Interconnection of SCS subsystems and their connection with active network equipment is made by means of a limited number of cords with universal connectors, which considerably simplifies the administration and adaptation of cable systems to various applications.

The ability of network equipment to use SCS cabling, which for various reasons fails to support transmission along balanced or fiber-optic cables, is ensured by the development of a range of adapters and device adapters. These elements are beyond the scope of the standards, however, designers create such products while taking SCS requirements into consideration.

Redundancy implies the introduction of auxiliary telecommunications outlets in SCS, whose number and location are defined by the work area topology and size. This allows the easy creation of new workplaces, as well as easy relocation of personnel and equipment. The use of the principle of redundancy enables designers to quickly adapt a cable system to specific technical needs and to avoid limiting the work of the office or its parts during technical or administration inputs. This principle is very important, as the lifetime of a SCS is several times longer than the lifetime of the rest of the components of a building telecommunications infrastructure.

Creation of a viable SCS and its operation are impossible without a number of supplementary conditions. A SCS must include:

- production catalogue;
- design norms and methods, which comply with the requirements of valid standards;
- ability to administrate it in compliance with standard procedures;
- personnel training system and manufacturer warranty.

A cable system that does not satisfy at least one of the supplementary or mandatory properties stated above is called proprietary [9].

Other SCS definitions (see example in [8]) are used in practice. Without going in a detailed analysis of these definitions, we would like to point out that, in the authors' opinions, all of them specify a technical object possessing the scope of the mandatory and supplementary properties given above. Based on that fact, one may confirm that all of them satisfy the described definition given here.

Application of SCS enables:

- a reduction in expenses because of its long lifetime and low operating costs, despite its rather high initial cost;
- an increase in and assurance of cable system reliability;
- the reconfigation and augmentation of telecommunications and computing systems of office buildings without changing the existing cabling;
- the simultaneous use of various network protocols and network architectures in one system;
- the avoidance of mess in cable pathways;
- the creation of a uniform maintenance service;
- the provision of the majority of existing and prospective network equipment of various classes with an information transmission platform because of the available standardized interface;
- quick troubleshooting and communications recovery or transition to reserve lines by using the principle of separate module construction.

This book is divided into 13 chapters. Chapter 1 highlights general information on SCS, including historical engineering background, structure of existing and prospective standards, primary subsystems, length restrictions of cable pathways, and schemes of subsystem construction that are allowable by standards.

Chapter 2 describes the physical processes of electrical signal transmission across balanced cables and optical signal transmission across fibers. It also reviews the system of parameters for electrical and optical cables, and various patching elements as well. Methods and techniques that enable the full use of the potential performance of electrical and optical SCS channels are given separately.

Chapter 3 is devoted to the element platforms for electrical SCS subsystems. It describes electrical cable products for various purposes as well as equipment cords, adapters, and auxiliary passive and active components, whose utilization considerably adds to SCS functions.

The element platform for SCS optical subsystems (cables, connectors, patching and termination equipment, preterminated assemblies, and adapters) is reviewed in Chapter 4.

Chapter 5 describes the primary auxiliary equipment that is used for construction of SCS in Russia and various options for 19-in mounting hardware and raceways, as well as numerous accessories. Special attention is paid to outlet facilities in standard office areas and halls.

Chapter 6 reviews problems that arise from SCS operation in small offices and industrial buildings, as well as in residential areas.

Chapter 7 examines active equipment that is included in the products delivered by some SCS manufacturers.

Chapter 8 describes some issues dealing with the development of protective and telecommunications grounding in office buildings.

Chapter 9 concerns some fire-safety issues for SCS cable products, including their standardization and selection of certification techniques.

Chapter 10 describes an approach to SCS design for architectural and telecommunications stages. The chapter contains administration information on industrial areas and horizontal and backbone cabling, and techniques to calculate the amount of equipment required for separate SCS subsystems. Chapter 10 includes an example of how to design a SCS in a four-story office building.

Chapter 11 provides information on installation procedures for various SCS subsystems at various organizations in telecommunications closets.

Chapter 12 highlights testing issues for electrical and optical SCS subsystems. It provides information on measurement equipment to determine the parameters of electrical and optical signal transmission channels and various measuring techniques.

Chapter 13 examines aspects of SCS administration and maintenance.

Chapter 14 contains some supplementary information.

The authors' opinions of future SCS development are given in the summary.

Contents

1 General Information on SCS

1.1 Origin of SCS and Development of Standards

The idea of creating a structured cabling system (SCS) to be the basis of the low-current cabling of a building was first proposed by designers of the AT&T Company (now Lucent Technologies) in 1983 [10]. The first successful attempt to create a unified cabling system for office information systems was undertaken by the IBM Corporation. During the 1980s, specialists of this company developed the IBM System that provided the functioning for Token Ring networks, AS/400 servers, terminals 3270, and other similar devices. This system was based on a two-pair balanced shielded cable with a wave resistance of 150 Ω. The operating potentials of this system were substantially extended by introducing components providing the transmission of telephone signals. The specification of the cabling part of the IBM System includes nine different types of cables (Table 1.1).

Table 1.1. Cable types according to the IBM specification

Cable type	Structure
Type 1	Two shielded twisted pairs of monolithic wires (22 AWG, 150 Ω) in a common braid-type shield
Type 2	Two shielded (22 AWG, 150 Ω) and four unshielded (22 AWG, up to 1 MHz) twisted pairs of monolithic wires in a common braid-type shield
Type 3	Four unshielded (22 or 24 AWG, up to 1 MHz) twisted pairs of monolithic wires
Type 4	Not specified
Type 5	Two multimode fiber-optic cables
Type 6	Plug wire; two shielded twisted pairs of monolithic wires (26 AWG) in a common shield
Type 7	Not specified
Type 8	Flat cable to be wired under carpet covering; two untwisted shielded pairs of monolithic wires (26 AWG)
Type 9	Two pairs of monolithic or bundled conductors (26 AWG)

It is interesting to note that the IBM Corporation has never produced the components of its own cabling system. These components are produced by other companies in accordance with IBM specifications. Of the nine types of cables, types 1 and 6 have become the most widespread. Cables of these types are still used in Token Ring networks, although for several recent years IBM has recommended the use of cables of types 3, 4, and 5 with eight-pole modular connectors. The functioning of the devices with coaxial and twin-axial interfaces is supported by a developed nomenclature of baluns included in the system.

Note also that now an improved modification of type 1 cable, known as type 1A cable, has been delivered to the market. In comparison with its predecessor, the new cable has improved transmission characteristics, achieved through appropriate changes in twist parameters of separate twisted pairs.

This cabling system has a number of drawbacks that prevent it from being widely used: it is rather expensive; its fitting possesses low technological efficiency; it is oriented primarily to products of IBM; and it is difficult to integrate into modern network systems[1].

In the late 1980s, the developers of technologies for data transmission along local networks applied much effort to increase the speed of data exchange, to improve the reliability of the equipment, and to reduce expenses on its production and operation. Twisted-pair cables turned out to be a suitable means for the implementation of communication channels in local networks; their production and fitting are technologically efficient. However, the absence of standards for this product hampered the development of advanced network technologies using balanced cables as a medium for transmitting information.

In 1985, Electronic Industries Association (EIA) started developing a standard for telecommunication cabling systems for buildings. The documentation was prepared by several working groups:

- TR-41.8.1 – working group on cabling systems for office and industrial buildings
- TR-41.8.2 – working group on cabling systems for residential constructions and office buildings with a low coefficient of utilization of usable area
- TR-41.8.3 – working group on cable channels for telecommunication cables
- TR-41.8.4 – working group on backbone cabling systems for residential constructions and office buildings with a low coefficient of utilization of usable area
- TR-41.8.5 – working group on standardization of terms and definitions
- TR-41.7.2 – working group on grounding and construction
- TR-41.7.3 – working group on the electromagnetic compatibility

In 1988, Telecommunications Industry Association (TIA) joined the work on standardization. In October of 1990, the first joint document prepared by these organizations – TIA/EIA-569 "Commercial Building Standard for Telecommunication Pathways and Spaces" – was approved. This document was prepared by working group TR-41.8.3. The approval of this document indicates the awareness of the fact that the construction of a high efficiency cabling system is impossible without imposing specific requirements on the architecture of the building in which the system should be setup.

---

[1] Not all manufacturers of active network equipment guarantee its normal operation in a medium with a wave resistance of 150 Ω. Therefore, the IBM cabling system is not completely unified and independent of applications.

In 1989, the well-known American research organization Underwriters Laboratories (UL) together with the Anixter Company developed a new classification for twisted-pair cables. This classification is based on the concept of *level.* The levels are defined in Table 1.2.

Table 1.2. Classification of twisted pairs in accordance with their levels

Cable type	Maximal frequency of signal	Typical applications
Level 1	Not specified	Power supply circuits and low-rate data exchange
Level 2	Up to 1 MHz	Voice communication channels and security systems
Level 3	Up to 16 MHz	Local networks Token Ring and Ethernet 10Base-T
Level 4	Up to 20 MHz	Local networks Token Ring and Ethernet 10Base-T
Level 5	Up to 100 MHz	Local networks with data transmission rate up to 100 Mbit/s

Working group TR-41.8.1 developed a commercial building standard for telecommunication cabling systems (document TIA/EIA-568), approved in July of 1991. This document defines the structure of cabling systems and the specifications of cables and connectors that may be used for the construction of such systems. It allows the use of unshielded twisted-pair cables with a wave resistance of 100 Ω and shielded twisted-pair cables with a resistance of 150 Ω, as well as coaxial cables with a resistance of 50 Ω and multimode fiber-optic cables. The document does not specify fiber-optic connectors.

In November of 1991, working group TR-41.8.1 issued additional specifications for balanced electric cables based on unshielded twisted pairs (Technical Bulletin TIA/EIA TSB-36). This document introduced the concept of *categories* for unshielded twisted-pair cables. The categories virtually completely coincide with levels in the UL&Anixter classification; see Table 1.3. In fact, only terms were changed, and the classification according to levels ceased to be used. The first two levels of twisted pairs for low-grade applications are not specified in the bulletin TSB-36.

In another appendix to Standard TIA/EIA-568 – Technical Bulletin TIA/EIA TSB-40 – additional specifications were given for connectors for unshielded twisted-pair cables. They also were classified into categories 3, 4, and 5. This bulletin requires utilizing connectors whose categories are not lower than those of the cables for which these connectors are designed.

In October of 1995, the second edition of Standard TIA/EIA-568 – TIA/EIA-568-A – appeared. This version included and refined all basic technical specifications of bulletins TSB-36 and TSB-40. The most important difference from the previous document was that coaxial cables were not recommended to be utilized when constructing new SCS. At the same time, the new document permitted the use of single-mode fiber-optic cables in backbone cablings.

Table 1.3. Correspondence between different standards

	International	American	European
General characteristics of the cabling system	ISO/IEC 11801	TIA/EIA-568-A	EN 50173
Design, installation, and administration	ISO/IEC 14763-1 ISO/IEC 14763-2	TIA/EIA-569 TIA/EIA-606 TIA/EIA-607	prEN 50174
Testing	CD 14763-3 CD 14763-4	TSB-67	
Cabling		TIA/EIA-569A	

In January of 1993, one more important document – TIA/EIA-606 Administration Standard for Telecommunication Structure of Commercial Buildings [138] – was approved. This document was prepared by working group TR-41.8.3. It defined the rules for keeping SCS documents during system operation (marking, record-keeping, schematic drawings, report compilation, etc.). The standard recommended electronic document keeping.

One more adjacent standard – TIA/EIA-607 – was adopted in August of 1994. It comprises specifications for grounding facilities used in buildings. The basic aim of a grounding system had been to prevent the personnel operating electrical equipment from receiving electrical shock. Standard TIA/EIA-607 specifies additional design requirements for grounding systems to provide efficient and reliable transmission of electrical signals through SCS. Documents TIA/EIA-568-A, TIA/EIA-569, TIA/EIA-606, and TIA/EIA-607 are USA national standards.

Rapid development of fiber-optic engineering, cost reduction of fiber-optic components, and their wide applications in office building cabling systems allowed using a centralized administration principle of SCS design. This principle substantially simplifies the SCS administration process. Possible designs of SCS with centralized administration and rules for their construction are specified in the technical bulletin TSB-72 issued in October of 1995.

In August of 1996, the technical bulletin TSB-75 appeared. This document substantially extended capabilities of designers, operators, and maintenance service of cabling systems for open offices.

In September of 1998, the technical bulletin TSB-95 was adopted which contained information on additional controllable parameters of the channel of category 5. To provide normal operation of the application Gigabit Ethernet, these parameters must correspond to standards.

In May of 1999, the subcommittee on standards TR.42.2 approved Standard TIA/EIA-570-A that specified optical connectors for subscribers sockets. This standard permitted one to use compact connectors of a new generation, together with SC-type connectors.

By 2000, the TIA subcommittee TR-42 had published a number of appendices to Standard TIA/EIA-568-A, which, in all probability, will enter the new edition of the American Standard without any substantial changes. (The draft title of this document is TIA/EIA-568-B.) For example, Appendix 1 specifies quantitative limitations for the parameters delay and skew, while Appendix 5 defines characteristics of the improved category 5e that surpass the standards of the aforementioned technical bulletin TSB-95.

Along with TIA/EIA, SCS standards were also being developed by International Organization for Standardization (ISO) and International Electrotechnical Commission (IEC). In 1995, these two organizations issued a joint document – Standard ISO/IEC 11801 "Information Technologies. Unified Cabling System for Customer's Buildings and Territory.". This document has unimportant differences from Standard TIA/EIA-568-A, related, basically, to the structure of the document, terminology, and the degree of development of some regulations. Note also that Standard ISO/IEC 11801 allows the use of twisted pairs with a wave resistance of 120 Ω and multimode optical cables with fibers 50/125 that are widely used in some European countries. The European Organization for Standardization CENELEC has prepared its own Standard EN50173[2], whose final edition appeared in August of 1995. The content of the English version of this document almost repeats that of the International Standard ISO/IEC 11801.

Standards of ISO/IEC and CENELEC are constantly being developed and extended. For example, these organizations adopted in January and February of 1999 documents that were similar to the aforementioned bulletin TSB-95 TIA/EIA.

In 1999, Standard ISO/IEC 14763-1 was adopted. This standard defines cabling system administration rules and is an analogue of the American standard TIA/EIA-606.

At the beginning of 2000, a completed edition of Standard ISO/IEC 11801 appeared, in which a number of new parameters of twisted-pair components and lines were introduced and the values of the traditional parameters were refined.

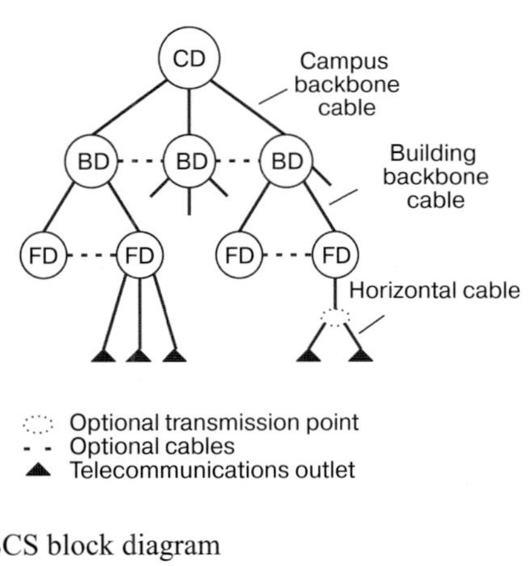

Fig. 1.1. SCS block diagram

All three standards are fairly close to each other and specify the basic issues related to the construction of SCS (Table 1.3). These standards have unimportant differences as regards the list of hardware components for the construction of SCS

[2] EN – Europa Norm

(Table 1.4), the ranges of parameters admitted for these components, the terminology, and the extent of specification of some issues. For this reason, in different cases one has to use both the international standard ISO/IEC 11801 and the American standard TIA/EIA-568-A, as well as TSB bulletins completing the latter document. Nevertheless, the differences occurring between early versions of SCS standards have been to a considerable extent overcome in the last ten years, and by the middle of 2000, the differences between the basic documents had become substantially less pronounced. SCS block diagram is shown in Fig. 1.1.

Table 1.4. Differences between the standards

Standard	ISO/IEC 11801	EN50173	TIA/EIA-568-A
Cable supported	UTP, FTP, STP	UTP, FTP, STP	UTP, STP
Cable with Z_B= 120 Ω	Allowed	Allowed	Not allowed
Diameter of the conductor, mm	0.40 – 0.65	0,40 – 0,6	0,511 - 0,643
Number of pairs in the horizontal cable	2 or 4	2 or 4	4
Category of the components	3, 4, and 5	3 and 5	3, 4 and 5
Attenuation of cables for cords	Greater by 50 %	Greater by 50 %	Greater by 20 %
Optical fiber 62.5/125	Main	Main	Main
Optical fiber 50/125	Alternative	Alternative	Not allowed
Shielded socket	Allowed	Allowed	Not allowed
Category of work place cables	5 + 3	5 + 5	5 + 3

Apart from international standards, some European countries have their own national standards. These standards take into account the requirements of national industries, historical traditions, legislative acts of allied areas, and other specific features. Such standards may be cited in the technical documents accompanying the SCS equipment delivered in the course of the implementation of joint projects. For example, the authors of this book have frequently encountered citations from DIN/VDE standards. This is explained by the fact that the cabling system ICCS was actively promoted on the Russian market by the German corporation Siemens for a long time until this business was sold in 2000 to the American company Corning. According to the server www.cabletesting.com, Australia and New Zealand also have national standards that are oriented mainly to American norms. It should be noted that the national standards familiar to the authors of this book do not essentially differ from the international, European, and American standards. These documents differ in terminology used and the extent of specification of some issues. For this reason, we will not consider these documents separately and will cite them only when necessary.

Unfortunately, by the middle of 2000, the work on the creation of the Russian national standards for telecommunication cabling systems was only at the initial stage. In addition, all cabling systems that are being constructed and operated in

Russia are analogues of some foreign systems. For this reason, the material presented in our book is mainly based on international and American standards. Additionally, when preparing the book, we used some Russian standardization documents, among which were "Regulations for Design of Electric Installations" and a number of Russian State Standards specifying design works, design documentation, and testing of cabling components.

1.2 SCS Structure

1.2.1 SCS Topology

Any structured cabling system is based on the tree-like topology which sometimes is also referred to as hierarchical star structure. The general structural diagram of a SCS is shown in Fig. 1.2. The nodes of the structure correspond to switch equipment of various types (distributor, in accordance with the terminology of Standard ISO/IES 11801). This equipment is usually arranged in technical rooms and its components are connected to each other and to information sockets at workplaces by means of electrical and optical cables.

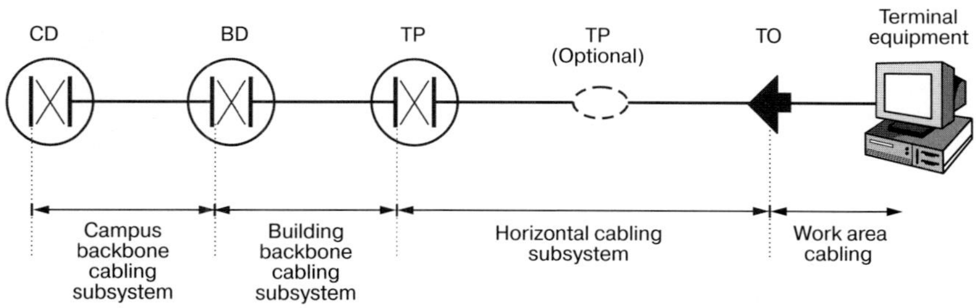

Fig. 1.2. SCS subsystems

The standards do not specify the types of switch equipment but only define its parameters. The fitting and operating of the switch equipment requires technical rooms. All input cables of these rooms must be connected to the switch equipment. All connections and switches during the construction and operation of a cabling system are performed on this equipment. This provides the SCS with flexibility, the possibility of easy reconfiguration, and adaptation to a specific application.

The use of the star topology is justified primarily by the fact that it can be used to support the operation of all basic network applications (Table 1.5). This table implies that the topology of such a type forms the platform that provides the functioning of all modern data transmission facilities.

Table 1.5. Logical and physical topologies of modern data transmission networks

Protocol	Logical topology	Physical topology
Token Ring	Ring	Ring, star
High-speed Token Ring	Ring	Ring, star
FDDI	Ring	Ring, star
Ethernet	Bus	Bus, star
Fast Ethernet	Bus	Star
Gigabit Ethernet	Bus	Star
ATM	Virtual channel	Ring, star

1.2.2 Technical Rooms

Technical rooms necessary for the construction of SCS and the information system of an entire enterprise are divided into instrument rooms and switch rooms.

In what follows, the term *instrument room* will be referred to as the technical room in which, apart from the SCS switch equipment, multiple-access network equipment (automatic telephone systems, servers, and concentrators) is placed. This room is sometimes called the server room, if the major engineering facilities placed in it belong to the local area network (LAN) equipment, or the communication center, if these facilities are related to the office automatic telephone system or the external communication system. The instrument rooms are equipped with raised floors, fire fighting, air conditioning, and access control facilities.

A Switch room is a room to place the SCS switch equipment, network equipment, and other sorts of auxiliary equipment. It is desirable that this room is located near the vertical post, so that it can have a telephone and the access control system. The overall level of equipment in switch rooms with engineering systems supporting its functioning is lower in comparison with that of the instrument rooms. The switch rooms are frequently called (store) technical rooms; sometimes they are called hub rooms.

The instrument room may be combined with the building switch room (BSR) In this case, the network equipment of the building may be connected directly to the SCS switch equipment. If the instrument room is separated from the switch room, its network equipment is connected to local switch equipment or to information sockets at workplaces. All campus backbone cables that are connected to building switch rooms converge in the campus cable switch room (CCSR). The building backbone cables that are connected to store switch rooms (SSR) converge in the BSR. The horizontal cables connected to information sockets at workplaces converge in the SSR. To increase the flexibility and reliability of the system, BSR can be connected by additional campus backbone cables and SSR by additional building backbone cables. This additional cabling is illustrated in Fig. 1.1.

The entire SCS may have only one CCSR and each building may not have more than one BSR. The CCSR and BSR are allowed to be combined, if these rooms are located in the same building. In a similar way, the BSR may be combined with SSR, if they are located on the same floor. If there are few workplaces on a floor or on a portion of a floor, then these workplaces are allowed (as an exception) to be connected to SSR of adjacent floors. The structure of the cabling system with connection to buildings is illustrated in Fig. 1.3.

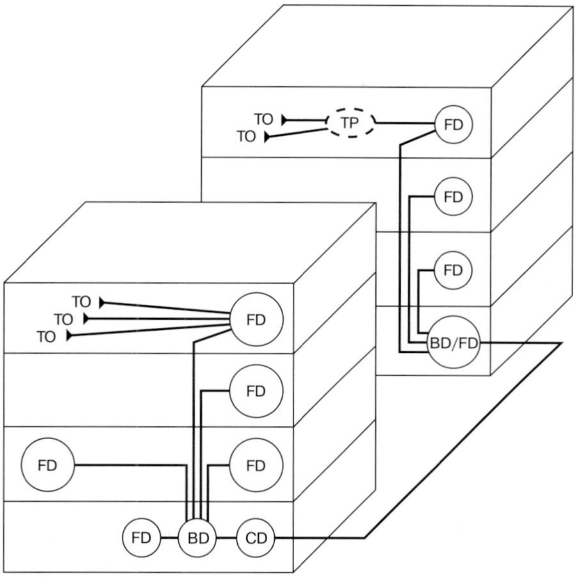

Fig. 1.3. The cabling system structure with connection to buildings, where *TO* – telecommunication outlet, *TP* – transition point, *FD* – floor distributor, *BD* – building distributor, *CD* – campus distributor

1.2.3 SCS Subsystems

In the general case, according to the international standard ISO/IES 11801, a SCS consists of three subsystems (Fig. 1.2)

- *Campus backbone system* (or primary system, according to the terminology of some European manufacturers) consists of campus backbone cables between CCSR and the BSR, switch equipment in the CCSR and the BSR to which the campus backbone cables are connected, and plug wires and breakdown cables in the CCSR. The campus backbone system is the basic subsystem for constructing a communication network between buildings compactly arranged on a common territory (campus). This subsystem frequently has a ring topology, which provides an additional increase in reliability due to the presence of doubling cables. For the same reason, the campus backbone cabling is sometimes designed in accordance with the double ring topology scheme. If the SCS is established autonomously only in one building, then the campus backbone cabling is absent.

- *Building backbone system*, which in some SCS is the vertical or secondary cabling, contains building backbone cables arranged between the BSR and FSR, the BSR and SSR switch equipment connected to these cables, as well as the switch cords and breakdown cables in the BSR. Cables of the building backbone system connect different floors of the building or separate rooms on one floor. If the SCS works only on one floor, the building backbone system may be absent.
- *Horizontal cabling system*, sometimes called the ternary subsystem, comprises building horizontal cables between the FSR and information sockets at workplaces, as well as between the sockets at work places, FSR switch equipment (to which the horizontal cables are connected), and switch cords and breakdown cables in the FSR. The horizontal cabling admits one change point at which the cable type may be changed (for example, for a flat cable with equivalent transmission characteristics to be arranged under a carpeting).

This classification of the SCS subsystems is used irrespective of the type of network or its implementation. Hence, this classification is the same, for example, for an office and industrial network. Sometimes, when convenient for the design or operational maintenance, a finer division of the SCS equipment into subsystems can be used. For example, the components for connection of the equipment to the SCS in the switch room may be combined into an isolated subsystem, whereas cords, adapters, and other components necessary at workplaces may form a workplace subsystem.

In the general case, in accordance with international documents on technical standards currently in force, an SCS consists of eight components:

- cabling equipment for the campus backbone system;
- switch equipment for the campus backbone system;
- cabling equipment for the building backbone system;
- switch equipment for the building backbone system;
- cabling equipment for the horizontal cabling system;
- switch equipment for the horizontal cabling system;
- change points;
- information sockets.

In most cases, the network equipment is connected to the SCS by a plug wire. In some cases, apart from a plug wire, an adapter may be needed to match signal and mechanical parameters of optical or electric interfaces (connectors) with those of the network equipment. For example, adapters are used to connect to SCS network equipment with interfaces V.24 (RS-232 C), cable TV devices, systems IBM AS/400 with terminals 5250, terminal controllers IBM 3724, and terminals 3270, as well as additional applications that have been designed for other types of cabling systems.

A workplace system provides the connection of network equipment at workplaces. The equipment needed for the implementation of this system completely depends on the specific characteristics of the application. This system is not a part of the SCS and is not specified by the standards ISO/IEC 11801 and TIA/EIA-568-A. However, these standards impose certain restrictions on the parameters of the workplace system. These restrictions will be discussed later.

1.2.4 Switching in SCS

An essential feature of any SCS is that its switching, unlike electronically controlled automatic telephone systems or computer network equipment, is always carried out manually by means of plug cables or jumpers. The most important implication of such an approach is that the functioning of the SCS is independent of the power source condition. Involvement in an SCS electronically controlled or electromechanical switching units requires use of an autonomous power source. At the contemporary stage of the development of cabling systems, this solution is unreasonable from both the economical and engineering points of view. A common number of switches in an operating system is several times per year, whereas the operating reliability of a power source is substantially lower than that of passive elements forming the cabling system. However, rejection of the use of an autonomous power supply has its disadvantages. For example, in this case,

- one has to use plug cables, which substantially increases the total mass and overall dimensions of the switching equipment and requires special measures to be undertaken to accomplish administration tasks;
- it is impossible to use standard switching units, controllers, transducers, and other equipment, which makes operating less convenient, increases the time required for fault tracing, and hampers the technical monitoring of the system.

Only separate designs that are aimed at the integration of active components into SCS subsystems and have been developed for repetition work are known. However, these designs have an auxiliary character; they cover port status polling, indication, and switching of signals of low-grade applications, but do not cover the process of transmission of information signals. None of these designs are specified by working standards or proposals for their perspective editions.

1.2.5 SCS Administration

The administration (management) scheme of an SCS is completely determined by its design. One distinguishes between single-point and multipoint administration. By *multipoint administration,* one understands the management of an SCS that has a classical hierarchical star design configuration. The main feature of this scheme is that it requires the switching of at least two cables to change the configuration in general. This provides the management with maximal flexibility and enables reconfiguration of the SCS to support new applications.

The *single-point administration* configuration is used in cases where the management of the cabling system should be as simple as possible. This type of administration is used solely for SCS that are established in one building and do not have a backbone subsystem. The basic feature of this management scheme is that all workplace information sockets are directly connected to a single technical room. The single-point administration can only be used for small networks. It implies that all cable switches are carried out at one place, which substantially simplifies the management of the cabling system.

1.2.6 SCS Cables

An effective way to increase the technical and economical efficiency of cabling systems for office buildings is to reduce the nomenclature of cables used for their construction. The international standard ISO/IES 11801 admits only the following types of cables to be used in SCS:

- twisted-pair balanced electric cables (shielded or unshielded) with a wave resistance of 100, 120, and 150 Ω;
- single-mode and multimode optical cables.

Electrical cables are used mostly for horizontal cabling. These cables can transmit telephone signals and low-grade data, as well as data of high-grade applications. At the present time, optical cables are rarely used in the horizontal subsystem, although their share rapidly increases ("fiber to the desk" solutions). In building backbone systems, electric and optical cables are utilized equally frequently. Electric cables are mostly used for transmitting telephone signals and data with clock frequencies not exceeding 1 MHz, while optical cables transmit data for high-grade applications. In campus backbone systems, optical cables dominate.

To allow switching from an electrical cable to an optical one when transmitting data with a speed of 10 Mbit/h or higher, the appropriate engineering equipment (medium converters or transceivers [6]) should be established in technical rooms. This equipment usually operates a certain group unit (e.g., a data concentrator, a remote unit of an automatic telephone system, or a controller of the building engineering system). Direct use of a fiber-optic cable for the transmission of telephone signals and low-grade data is economically unjustified. It is resorted to only in cases where either alternative solutions are impossible or an extremely high degree of protection of information from unauthorized access is required. Therefore, to improve the technical and economical efficiency of the entire network, one usually combines the process of conversion of electrical signals to optical ones with multiplexing.

For the construction of the horizontal subsystem, both shielded and unshielded cables are admitted by the standards. Shielded balanced cables potentially possess higher electrical (and in some cases also strength) characteristics in comparison with those of unshielded cables. However, shielded cables are extremely sensitive to the quality of wiring and grounding. In addition, these cables are substantially more expensive and have a lager mass and overall dimensions. For these reasons, unshielded twisted-pair cables are basically used for the transmission of electrical signals, at least in Russia. As has been mentioned above, the standards prescribe that one should use electrical cables with wave resistances of 100, 120, and 200 Ω when constructing SCS. The last two types of cables frequently demonstrate better characteristics as compared with those of the 100-Ω cable. However, for of a number of technical and economical reasons, these cables have not been widely used in Russia.

Multimode fiber-optic cables are used mostly in building backbone systems. Single-mode fiber optical cables should be used only for constructing long campus backbone systems.

The use of coaxial cables is not allowed by new standards and will be excluded from subsequent editions of the old standards. Networks based on coaxial cables have lower reliability, lower technological efficiency, and higher cost, in comparison with networks based on twisted pairs.

To provide network equipment with coaxial and triaxial interfaces to be utilized through an SCS, numerous adapters of various types are available.

1.3 Classes, Categories, and Their Relation to the Length of Cable Lines

1.3.1 Classes of Applications and Categories of SCS Cables and Connectors

The working version of the standard ISO/IEC 11801 divides all applications that can exchange data through twisted pairs into four classes – A, B, C, and D (Table 1.6). Class A is the lowest class and D is the highest class.

Table 1.6. Application classes according ISO/IEC 11801

Class of the line and the application	Definition
A	Telephone channels and low-frequency data exchange. Maximal signal frequency is 100 KHz
B	Medium exchange rate applications. Maximal signal frequency is 1 MHz
C	High exchange rate applications. Maximal frequency is 16 MHz
D	Superhigh exchange rate applications. Maximal signal frequency is 100 MHz
Optical	Applications using an optical cable as the transmission medium. Maximal frequencies are not lower than 10 MHz

To each class of applications there corresponds a class of communication lines that specifies the limiting electrical characteristics of the line to provide normal work for applications of this and lower classes (Table 1.7). The applications that use optical cables to transmit signals are referred to as the optical class of applications. At the moment of approval of the standard, the transmission bandwidth was not the limiting characteristic for such applications.

Table 1.7. Correspondance between application classes and cable categories

TIA/EIA-568-A	ISO/IEC 11801	EN 50173	ISO/IEC 11801
Cables and connectors		Applications	
-	-	-	A
-	-	-	B
Category 3	Category 3	Category 3	C
Category 4	Category 4	-	-
Category 5	Category 5	Category 5	D
-	Category 6	-	E
-	Category 7	-	F

It is also interesting to note that the standard ISO/IEC 11801 does not specify applications and lines with a maximum transmission frequency of 20 MHz corresponding to connectors and cables of category 4. This is due to the absence of widely used network applications with maximum transmission frequencies ranging from 16 to 20 MHz.

In a number of European countries, additional classes of applications are sometimes introduced. For example, in the German engineering literature, applications with an upper limiting frequency of 200 MHz are sometimes referred to as class D+, and applications with a limiting frequency of 300 MHz as class D++.

Standards ISO/IEC 11801 and TIA/EIA-568-A specify categories not only for cables but also for connectors. These categories are defined in accordance with the maximum frequency of the signal to which the corresponding cables and connectors are designed (Table 1.8). Cables and connectors of higher categories support all applications designed for cables of lower categories.

Table 1.8. Categories of cables and connectors

Category of a cable or a connector	Maximum frequency of a signal	Typical applications
Category 3	16 MHz	Local networks Token Ring and Ethernet 10Base-T, voice channels, and other low-frequency applications
Category 4	20 MHz	Local networks Token Ring and Ethernet 10Base-T
Category 5	100 MHz	Local networks with data transmission rate lower than 100 Mbit/s
Category 5e	100 MHz	Local networks with data transmission rate lower than 1000 Mbit/s
Category 6	250 MHz	Local networks with data transmission rate lower than 1000 Mbit/s
Category 7	600 MHz	Local networks with data transmission rate lower than 1000 Mbit/s

In September of 1997, IEC started its work on the standardization of two new classes of applications, E and F, as well as that of components for SCS of categories 6 and 7. Simultaneously, this organization is working on the specification of the so-called improved categories 5 (5+ and 5e) with an upper limiting parameter normalization frequency of 100 MHz. In fact, this specification fixes the engineering level achieved by the end of 1990 and at the same time normalizes a number of parameters to enable operation of the prospective application Gigabit Ethernet. The results of these developments are expected to enter the new editions of current working standards ISO/IEC 11801 and TIA/EIA-568A. The work in this direction is justified by the necessity to support operation of the prospective network equipment ATM (155 and 622 Mbit/s) and Gigabit Ethernet (1000 Mbit/s).

For applications of class E and SCS components of category 6, the upper limit frequency of parameters to be normalized was initially defined as 200 MHz. Then this frequency was increased to 250 MHz. This was justified by the requirement that the system should be able to support the functioning of two-pair interfaces of Gigabit Ethernet. Class F and the components of category 7 have been designed for frequencies not exceeding 600 MHz. The choice of this limit was justified to a considerable extent by the facts that the ATM equipment having a data transmission rate of 622 Mbit/s is widely used and that it is necessary to support the transmission of signals of multichannel analogue TV with an upper limiting frequency of 550 MHz. The basic distinctive features of communication lines of different categories are characterized in Table 1.9. A detailed discussion of the parameters given in this table is presented in Section 2.1.3. We only mention here that this table demonstrates that the higher the category, the higher the quality that is required of the system.

Table 1.9. Basic distinctive features of communication lines

Parameter	Category 5			Category 6	Category 7
	TIA/EIA-568-A	TSB-95	ISO/IEC 11801		
Frequency range, MHz	1-100	1-100	1-100	1-250	1-600
Attenuation, dB	24	24	24	21.7 (36)	20.8 (54.1)
NEXT, dB	27.1	27.1	27.1	39.9 (33.1)	62.1 (51)
PS-NEXT, dB	-	-	24.1	37.1 (30.2)	59.1 (48)
ACR, dB	3.1	3.1	3.1	18.2 (-2.9)	41.3 (-3.1)
PS-ACR, dB	-	-	0.1	15.4 (-5.8)	38.3 (-6.1)
EL-FEXT, dB	-	17	17.0	23.2 (15.3)	ffs
PS-FEXT, dB	-	14.4	14.4	20.2 (12.3)	ffs
Return losses, dB	-	8	10	12 (8)	14.1 (8.7)
Propagation delay, ns	-	548	548	548 (546)	504 (501)
skew, ns	-	50	50	50	20

ffs – for future study at the corresponding standardization committee.
Figures in parentheses indicate the values at the upper operation frequency.

For construction of lines of category 6, cables of all types (both shielded and unshielded) may be used. A modular connector is mostly used as a connecting unit. There are also systems designed for other types of connectors, of which most widespread are connectors of types 110 and 210. For the contemporary level of technological development, lines of category 7 require only shielded pair cables. The standard modular connectors that are currently available permit one to meet requirements of category 7 only for external contact pairs 1/2 and 1/8, which reduces the degree of unification. There are a number of experimental designs of connectors with improved parameters promising for use in lines of category 7. Nevertheless, some experts think that the improvement of the modular connector enabling it to be adapted to high frequencies does not make sense, and that it is necessary to proceed to a new design. By the middle of 1999, the final decision as regards the specification of types of connectors to be used in lines of category 7 had not been approved by the commissions on standardization. For more detailed information about the known developments in this area, see Section 3.2.6.

In what follows, we will not consider lines of classes A and B in detail. Hardware implementation of these lines on the basis of modern components causes no difficulties. SCS specialists are not very interested in these lines because they do not support operation of the Ethernet equipment, which is the most widespread at the present time and, hence, it is impossible to construct a modern SCS on these lines. We also will not discuss lines of categories 5e, 6, and 7 in detail, because they had not been specified by official standards by the middle of 1999.

The communication lines of a structured cabling system can be assembled The category of the cables and other components of the communication lines of a structured cabling system must not be lower than that of the entire system. For categories up to 5e inclusive, the inverse of this rule is also valid: a communication line assembled from the components of a certain category must support operation of all applications of its own and lower classes.

The standards ISO/IEC 11801 and TIA/EIA 568-A define that to be referred to a certain category specified by these standards, the communication line of a structured cabling system must satisfy the following three conditions

- technical characteristics of all cables, connectors, and connecting cords meet the requirements of the given category or exceed them;
- the communication line is designed with reference to the requirements of the standard (i.e., it satisfies all constraints imposed on the lengths of the cables, the number of switch points, and other parameters);
- the wiring completely corresponds to the requirements of the standards cited above.

1.3.2 Constraints on the Lengths of Cables and Cords in SCS

The standards ISO/IEC 11801 and TIA/EIA 568-A impose constraints on the maximal lengths of cables and connecting cords for the horizontal and backbone cabling subsystems. The lengths of the cables are shown in Fig. 1.4. and presented in Table 1.10. We emphasize once again that the maximal lengths of electric cable signal transmission lines are given for the lines constructed from a balanced cable and other components whose category is not lower than the class of the line.

The length of the cable of the horizontal subsystem is specified to be 90 m (plus 10 m for connecting cords). This length has been chosen based on the possibilities of a twisted pair to transmit signals of most widespread (by the time of approval of the standards) high-rate applications such as, for example, Fast Ethernet. Also, contemporary achievements in hardware components and the circuit design of transceivers for modern networks have been taken into account. Architectural features of standard office buildings have played more than a minor role in the choice of this specific value for the maximal length of the horizontal cabling.

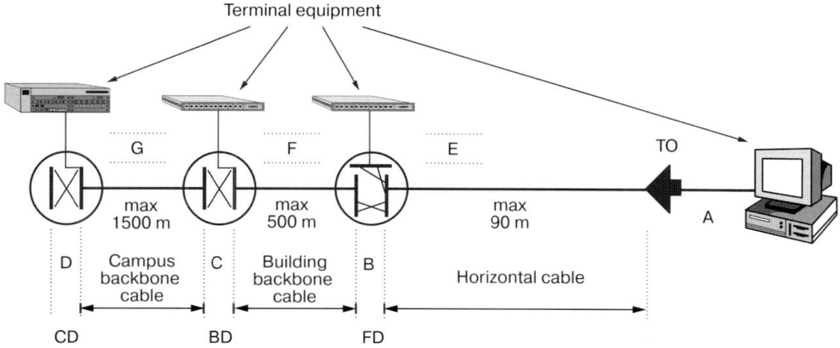

Fig. 1.4. Maximal cable lengths according ISO/IEC 11801

For horizontal cablings based on fiber-optic cables, the length of the line is constrained by the 90-m maximum because in this case all protocol restrictions on the maximal diameter of the collision domain for Fast Ethernet networks are guaranteed to be satisfied.

The basic intended use of a building cabling subsystem is the connection of technical rooms within one building. The maximal length specified by the standards for cables of such lines makes up 500 m.

Finally, in accordance with the standard ISO/IEC 1801, the campus cabling subsystem connecting separate buildings may include cables whose length does not exceed 1.5 km. It is also specified that the maximal length of backbone cables connecting the switch room of a store with that of the campus backbone system is not allowed to exceed 2000 m (500 m for the building cabling and 1500 m for the campus cabling), provided that plug wires and terminal cords have the standard length. For a single-mode cable, the limiting length can be increased to 3000 m. For modern fiber-optic cables used with the common standard equipment, this distance can be 100 km and more. However, the standards assume that to provide communication through such long distances one would use public communication lines and channels of different operators.

The lengths of switch and terminal cables depend on the selected scheme of connection of the network equipment, the type of the signal transmission medium, and the SCS subsystem to which a specific cable or a set of cables are related. In accordance with the standard ISO/IEC 11801 (2000 edition), the maximal total length of cables of cords in the horizontal cabling subsystem is specified:

- 9 m for crossconnect scheme, electric cable
- 10 m for interconnect scheme, electric cable
- 10 m for any scheme, optical cable

Note that in the 1995 edition of this standard, the maximal total length of cables was 10 m, irrespective of the connection scheme, i.e., the requirements of the standard have become more restrictive.

The maximal length of the plug wire used in switch rooms of backbone building and campus subsystems is specified by the standard ISO/IEC 11801 as 20 m. The length of terminal cords used for connecting the network equipment in these technical rooms should not exceed 30 m.

Table 1.10. Maximal lengths of cable lines depending on the cable type and the applications class

Signal transmission medium	Class of applications				Fiber
	A	B	C	D	
Signal transmission medium					
Balanced cable of category 3	2 km	200 m	100 m[a]		
Balanced cable of category 4	3 km	260 m	150 m		
Balanced cable of category 5	3 km	260 m	160 m	100 m	
Balanced cable (150 Ω)	3 km	400 m	250 m	150 m	
Multi-mode optical cable	-	-	-	-	2 km
Single mode optical cable	-	-	-	-	3 km[b]

[a]By a length of 100 m we understand the total length of the horizontal cable (must not exceed 90 m) and cords of all types.

[b]3 km is the limit specified by the standard. It does not correspond to the physical limit for single-mode optical fibers.

1.4 Additional Versions of SCS Topological Structure

In this section, we consider additional possible structures of the horizontal and building backbone cabling subsystems, part of which have not been specified by the working SCS standards. By the middle of 1999, these structures had been specified only by TIA/EIA technical bulletins and, in the opinion of most of the experts on SCS, were introduced to new editions of the standards without any essential changes. These new possible structures substantially enhance the designer's freedom of choice and permit one to considerably improve the technical and economical efficiency of a cabling system in a number of cases.

1.4.1 Possible Structures of the Horizontal Subsystem

The SCS horizontal subsystem, when based on twisted-pair cables, can be constructed in accordance with four different schematics, shown in Fig. 1.5. The first schematic is used the most frequently. This structure is formed by a continuous cable connecting the information socket with the switch board in the store switch room whose total length cannot exceed 90 m. In the second schematic, cables of two different types, which, however, have identical transmission characteristics, form the transmission line. These cables are connected to each other at the *transition point*. The international standard ISO/IEC 11801 allows two types of cable combinations: multipair + four pair and round + flat with the same number of pairs (four pairs in practice). The American standard TIA/EIA-508-A gives a more restrictive definition of the transition point. In accordance with this standard, at the transition point a flat cable is connected to a round cable.

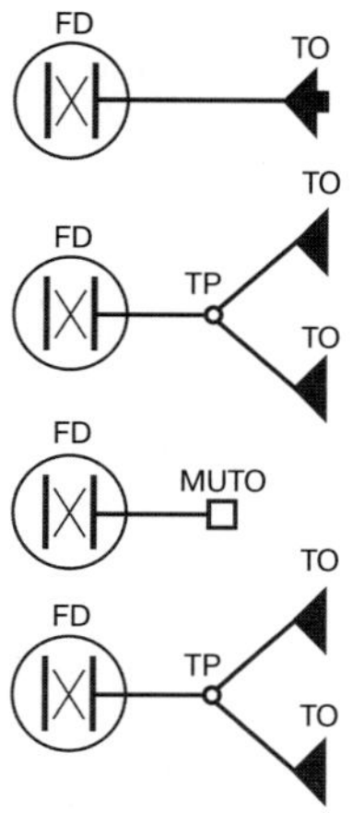

Fig. 1.5. SCS horizontal subsystem variations

The transition point is implemented on the common switch equipment; however, this equipment is not allowed to be used for cabling administration operations and for the connection of any active network units. Accordingly, switch and terminal cables must never be used at a transition point.

The remaining two schematics of the SCS horizontal subsystem are oriented primarily to the so-called *open offices*, i.e., to large working rooms divided into separate sections by either pieces of office furniture or light partitions. As a rule, people working in such offices frequently move between the working places, the working places are frequently reconfigured, and separate working places are grouped into zones. In open offices, one can use multiuser telecommunication outlets (MUTO) and consolidation points (CP). Both these schematics are specified by the technical bulletin TSB-75 [20] and adapt the solutions considered previously to a open office (Table 1.11)

Table 1.11. Comparison of different types of organization of the horizontal subsystem

Type of office	Direct connection	Multiuser connection
Usual office	Usual connection	Transition point
Open office	Multiuser outlet	Consolidation point

A MUTO is an outlet that serves several users. This unit is a separate component of equipment (see Section 3.3.4), which is mounted on columns and walls of buildings, under a raised floor, in floor boxes, and (quite seldom) in the space between the main and suspended ceilings. The maximal length, W, of the terminal cord connecting a MUTO with the network equipment at a workplace must not exceed 20 m and is defined by the relation

$$W = (102 - H) / 1.2 - 7 \text{ m}, \; W < 20 \text{ m} \tag{1.1}$$

where H is the length of the horizontal cable.

The coefficient 1.2 takes into account the elevated attenuation of a signal in a flexible bunched conductor (see Section 3.1.4). The constant coefficient 7 defines the maximal length for plug wires in the switch room. The length of the plug wire versus the length of the horizontal cable is plotted in Fig. 1.6. Relation (1.1) implies that for the length of a terminal cord not exceeding 20 m, the length of the horizontal cable does not exceed 70 m.

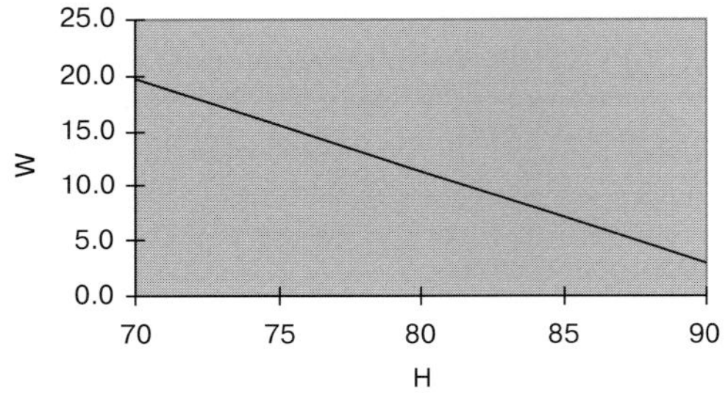

Fig. 1.6. Length of the plug wire versus the length of the horizontal cable

Thus, the total length of the terminal and switch cords in an open office may reach 27 m, in contrast to 10 m for a usual office, which leads to a substantial increase in the flexibility of the cabling system. However, due to an appropriate decrease in the length of the horizontal cable the overall attenuation of the signal transmission path turns out to be the same in both cases.

The consolidation point CP in an open office is a direct analogue of the transition point in the traditional topology. This point is connected with individual workplace outlets by means of short pieces of the horizontal cable that are continuations of the main cable segment. The designs with CP are recommended for cases where the personnel may change their workplaces but not as frequently as in cases where MUTOs should be preferred.

By analogy with traditional cabling, one must not use more than one transition point (either MUTO or CP) in any horizontal line of an open office. At the consolidation point, one is not allowed to connect active equipment or to perform administration operations.

1.4.2 Central Administration Topology

Systems with central administration are defined in the technical bulletin TSB-72 [21] and are related to the building cabling system based completely on optical cables. The definition of this document permits the designer in this case not to divide the cabling rigidly into the horizontal and building backbone subsystems. This makes it possible to use a simple one-level topology instead of the two-level star topology.

The principle of central administration allows one to

- considerably improve the controllability of the LAN, because of the possibility of forming any prescribed working groups on the physical level without using virtual connections;
- concentrate all active equipment at one place, which (i) increases the degree of protection of information from unauthorized access, (ii) reduces the demand of high-rate channels, and (iii) simplifies operational measurements;
- considerably reduce or (in some cases) reduce to zero the area allotted for store switch rooms.

The interest in using a central administration in practice has been dramatically increased due to mass engineering utilization of fiber-optic data transmission media that are not constrained by the 90-m physical limit characteristic of the twisted pair.

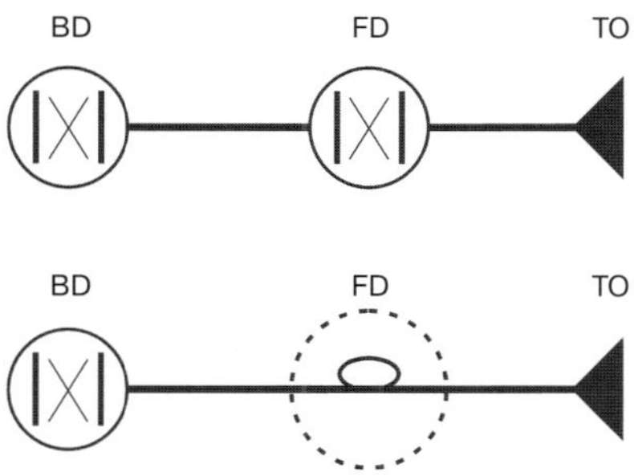

Fig. 1.7. System with central administration (with one intermediate connection and without intermediate connection)

According to the bulletin TSB-72, cabling systems of the type under consideration can be constructed with at most one intermediate connection. The design with one connection permits one to preserve the telecommunication structure of the building, since the switch equipment implementing this design is placed in the area allotted previously for store switch rooms. This design can be implemented in two modifications. The first modification may be called the *branching design* [22]. In this design, the backbone cable leads up to the switch room, and a service cable connected to the backbone cable by a permanent connector implements the subsequent cabling. The second modification is referred to as the *passive switchboard* [21]. It implies commutation by means of a common switch cord.

The maximal distance between a user outlet and the store switch room in this modification is 90 m. This is compatible with the standard TIA/EIA-568-A as regards the horizontal cabling and facilitates the return to the standard two-level topology. The maximal length of the channel with an intermediate connection has been chosen to equal 90 m to provide a cannel capacity of 1 Gbit/s for a link constructed on the optical fiber cable of type 62.5/125. This channel capacity permits the system to support applications such as Gigabit Ethernet or Fiber Channel, which have the maximal data transmission rate at the present time. By analogy with structures based on an electric cable, in which transition points of different types are used, no active equipment is arranged at the cross connectors.

The developers of the document TSB-72 consider that in this case, the protocol restrictions of the Fast Ethernet networks (mentioned in Section 1.3.2) are insignificant. This is the case, probably, because of the relatively small extent to which the fiber-optic equipment of standard 100Base-FX operating in the bandwidth share mode has spread.

For SCS without intermediate connections, the length of any channel is limited to 90 m, again in order to provide compatibility with the standard TIA/EIA-568-A. This, to a certain extent, restricts the possibilities for organizing systems with centralized administration in some types of office buildings but, on the other hand, permits one, in the extreme, to avoid using separate store switch rooms. If these rooms are given in the design, one can use the *through-type scheme* allotting special places in the switch rooms for the storage of coiled spare cables and arrangement of the switch equipment.

We list some additional restrictions and recommendations of the bulletin TSB-72:

- At the point of intermediate connection, the use of simultaneously permanent and plug-type connectors for optical fibers is not recommended.
- The basic type of a plug-type optical connector is SC in one-shot and duplex modifications.
- Permanent connections can be implemented either by means of welding or with metallic splices.
- In structures with one intermediate connection, one can use different types of cables for horizontal and backbone cablings, provided that the intermediate connections of fiber cables are permanent.
- Fibers and connectors should be identified and tagged in accordance with the requirements of the standard TIA/EIA-606.

1.5 Cable Sharing

The basic type of cable for horizontal subsystems of modern SCS is a four-pair balanced cable. This cable has four design modifications (see Section 3.1.2). Most of the widespread medium-rate (Ethernet 10Base-T, Token Ring) and high-rate (Fast Ethernet 100Base-TX, TP-PMD, ATM) applications require only two twisted pairs to operate. The other two pairs are not engaged and are grounded in some types of network interfaces. Hence, these pairs are useless for such interfaces. The electrical characteristics of horizontal cables required by the working standards permit one, in principle, to simultaneously transmit signals of several (from two to four) applications through such cables, with a negligibly small level of interference of these signals. This technical solution is an adaptation of methods of utilization of backbone cables to horizontal cablings and is referred to as *cable sharing* or *cable splitting*. This solution is permitted for practical utilization by the standards ISO/IEC 11801 and EN 50173.

For the practical implementation of cable sharing, a fairly large number of various appliances have been developed and produced. (These appliances will be considered in detail later.) They can be classified into the following groups:

- Y-adapters, as well as dual and three-throw balloons (see Section 3.4.2);
- Double adapter plugs (see Section 3.2.2);
- Branching cords (see Section 3.4);
- Connecting cords of special types (see Section 3.4.1);
- Dual and three-throw socket modules permitting one to arrange a single cable.

The listed appliances, apart from those of the last two groups, allow one to readily return (if necessary) to the standard four-pair organization of the horizontal segment of an electrical signal transmission line and, hence, do not violate the unification of the cabling system.

The standards do not impose any special requirements on the equipment used for cable sharing, apart from that of distinguishing the marking of sockets.

Note at the onset that the most suitable cables for simultaneously transmitting signals of two applications are multiple-twin and multiple-quadruple horizontal cables (see Section 3.1.2). These cables are formed by two identical elements enclosed in a common shell. However, for a number of reasons, these cables have not become widely used and are produced only by few manufacturers of SCS equipment.

The organization of an SCS in accordance with the cable sharing principle is beneficial, mainly for low-scale and medium-scale networks, mostly for the following two reasons:

- The expenditures on the horizontal cabling are relatively high and, hence, simultaneous transmission of signals from two applications through one cable provides a substantial savings in the money spent on the network organization.
- For networks of this type, the use of super-high-rate applications (for example, Gigabit Ethernet) that require four cables to be engaged simultaneously is substantially less important because a relatively small body of information is transmitted through such networks. In this case, the problem associated with the lack of transmission lines becomes remote.

The cable sharing principle is practiced on a large scale in some European countries. In these countries, this principle is used substantially more frequently than is the case for solutions based on two-pair cables. However, in Russia, by the middle of 1999, this solution had not been practiced on a large scale. This is for the following reasons:

- A considerable share of SCS in Russia are constructed in accordance with the requirements of the standard TIA/EIA-568A, which does not allow transmitting signals of two applications simultaneously through one horizontal cable.
- The cable sharing principle is especially effective in systems with individual shielding of separate pairs. The cost of hardware elements of such systems is relatively high, and the labor spent on the cabling is not compensated for by the savings from the reduction in the total length of the cables. For this reason, systems with individual shielding of separate pairs are utilized substantially less frequently than is the case for systems without such a shielding.
- The market of SOHO (small office and home office) and home networks, in which the transmission of different high-speed and wideband signals through one horizontal cable is widely used, is virtually absent from Russia.

The only solution that has been practiced in Russia on a relatively large scale is one based on the Y-adapter or the adapter insert that is functionally equivalent to the adapter. This solution is utilized in some structured cabling systems designed for the transmission of Ethernet 10Base-T signals and those of the analogue telephone through small-scale networks that are frequently not specified by standards.

1.6 Warranty Support of Modern SCS

A modern SCS is a complex high-technology product designed for long-term operation. In this connection, the system of warranties offered by SCS manufacturers on the components and the entire cabling is of great importance. The working standards do not prescribe any rigid rules in this regard. Only the standard ISO/IEC 11801 recommends the warranty term established for such products to be no less than 10 years. This term was justified to a considerable extent by the fact that the usual duration between redecorations in office buildings is about 9 years and that the rearrangement of cablings is usually combined with redecoration. It is from this standpoint that the principles and methods of warranty support de facto settled as standards will be considered in what follows.

At the present time, SCS manufacturers use various types of warranties. These can be classified into four basic groups (Fig. 1.8.).

The classical type of warranty is the *warranty on the components* or base warranty. It implies that all components of the cabling system do not have manufacturing defects and that, when operated in accordance with the intended use and technical conditions, they will preserve their operational integrity during some stipulated term from the time of purchase. The usual warranty term for the components is that of 5 years, however in recent years, the tendency of this term to be increased has been observed. For example, Lucent Technologies offers a 20-year warranty of this type on the products of the Gigaspeed series. The base warranty is provided only if the component has been purchased in accordance with the rules established by the SCS manufacturer.

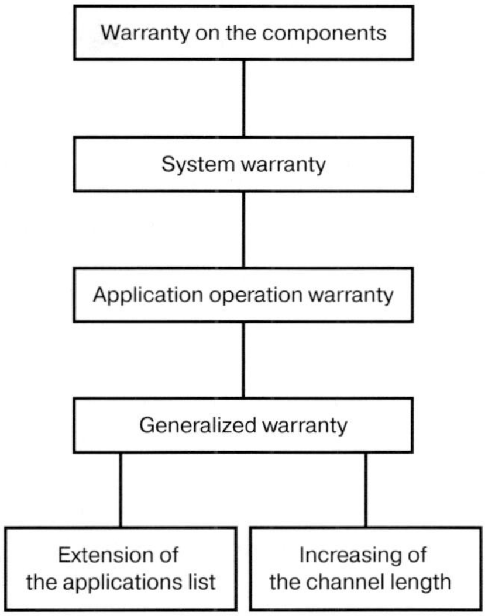

Fig. 1.8. Structure of SCS warranties

An extended warranty or *system warranty* is provided only for SCS that have been designed and established in accordance with all corresponding standards. This warranty implies the correspondence of the operating characteristics of the arranged cabling to ones specified by the standards. Most of the manufacturers define the extended warranty term for systems of category 5 from 15 to 16 years. The warranty term for systems with characteristics surpassing those of systems of category 5 are usually increased to 20 or even (by some manufacturers) 25 years. The basic principles of the system warranty policy can be formulated as follows:

- All components of the system must be officially allowed for use in the given SCS. Special permission of the manufacturer is required for use of components that are absent from the official list.
- The system must be arranged in complete agreement with the working standards. No exceeding, for example, of the length of cables and cords or the number of connectors in chains is allowed.
- The number of connection–disconnection cycles must satisfy the requirements of the standards.
- The system must be designed and constructed only by appropriately trained and authorized personnel. All changes and additions in the system must also be carried out by authorized personnel.

Some SCS manufacturers also put forward additional requirements obliging the users, for example, to use only the devices defined in a specific list for testing and to submit protocols of measurements.

The aforesaid implies that a system warranty includes the base warranty and even amplifies the latter as regards the warranty term. This apparent illogicality (the warranty term of the entire system exceeds that of any of its components) is accounted for by the fact that the cable in the system is not subjected to significant mechanical loads as is the case during its arrangement and, hence, operates in substantially less severe conditions.

Application operation warranty provides for the ability of an SCS that has been correctly designed and arranged (and, hence, has already had the system warranty) to support the operation of various applications. Since the late 1990s, a tendency has been clearly observed among SCS manufacturers to provide a new type of the application operation warranty that will be referred to as the *generalized warranty*. This type of warranty legally fixes the obligation of the manufacturer to improve certain parameters of the design above the level specified by the standards. This warranty has two varieties. The first variety is based on a list of applications that frequently also includes applications that are not supported officially by a standard SCS of a given category. Sometimes, it provides for the operational support of any application whose equipment has been originally designed for SCS of a given category. The second variety of the application operation warranty allows for the possibility of increasing the length of the line or channel (see Section 12.2) beyond the standard 100 m for some specific applications (companies BICC and ITT NS&S).

Thus, the application operation warranty, in the general case, indicates only the level of margins that the designer of a specific SCS assumed for his or her system. In other words, this warranty shows to what extent the requirements specified by the standard can be exceeded by applications in a given list. The declarations of some authors (see, e.g., [23]) that the warranty on the components and the system warranty do not make sense without the application operation warranty has no serious ground, since it is the modern network equipment that is designed "for the SCS", rather than vice versa.

The warranty of a certain type provided for the SCS is certified by a standard certificate of the manufacturer. The certificate can be issued either on the SCS established at a certain address (IT and Lucent) or on its owner (ITT NS&S, RIT Technologies, and Molex). In the latter case, the document has to be rewritten if the owner changes. The certificate is supplemented by the registration document containing a fairly complete description of the system. The structural schematic of the system and the results of its instrumental testing (if specified by the system establishment rules) may be attached to the registration document.

Warranty repair is usually performed by the company that has installed a specific SCS. Sometimes, this condition is stipulated by the partnership agreement between the manufacturer of the SCS and its system integrator. If this company is not able to perform this work for some reason, the manufacturer entrusts another partner with it or performs this work independently. The warranty repair is not performed if the system has failed due to improper operation, overloading, mechanical damages, damages caused by natural disasters, utilization of uncertified components, and in other similar cases.

For more detailed information about various aspects of providing and implementing warranty support, see [24].

1.7 Conclusions

A structured cabling system is a foundation of the information-calculating and telecommunication infrastructures of any modern enterprise, from a small business with several employees to a large corporation involving tens of thousands of people.

Modern SCS have a hierarchical star-like structure involving in the general case several subsystems, whose parameters are specified in detail by international standards, and an interface. The subsystems interact in accordance with certain rules. The integration of communication lines based on fiber-optic cables and electrical balanced cables provides the transmission medium for almost every class of contemporary and prospective network equipment. SCS constructed from commercially available components provide a maximal communication range of 3000 m and a maximal network throughput of no less than 1 Gbit/s.

The presence of a number of standardized structures of the SCS horizontal subsystem substantially extends the possibilities of the system to be adapted to specific local conditions. This makes it possible to obtain optimal (in terms of technical and economical efficiency) designs of networks for basic types of office rooms, both in old buildings and in prospective business centers.

The functional flexibility underlying the standards permits one to extend (if necessary) the SCS application area beyond the office building and construct cabling systems also in industrial and residential buildings.

Modern achievements in the technology of hardware components and the quality of design and construction provided by trained specialists permit the manufacturer to guarantee the reliable operation of the constructed cabling for 15 or more years. Accordingly, a properly installed SCS works for the entire term between major renovations of an office building.

2 Signal Transmission through SCS Channels

2.1 Transmission of Electrical Signals through Twisted Pairs

2.1.1 Methods of Transmission of Information through Twisted Pairs

There are two types of signal transmission through twisted pairs – unbalanced transmission (unbalanced circuits) and balanced transmission (balanced circuits). The twisted pair copper cables utilized in SCS permit one to construct both balanced and unbalanced circuits. All kinds of LANs use the balanced transmission of signals through twisted pairs. Unbalanced circuits are utilized for fire and security alarm warning systems, as well as for the transmission of DC power supply voltages, for example, from the automatic telephone system to telephones at workplaces.

For unbalanced transmissions, unbalanced circuits are used, i.e., one of the conductors is grounded at one or two ends. The signals are transmitted through the remaining wires and change relative to the ground. Unbalanced circuits are very sensitive to external electromagnetic radiation. It is apparent from Fig. 2.1 that the total voltage on the signal wire at the receiver inlet is the sum of the signal voltage, U_s, and the voltage induced by an external source of electromagnetic radiation, U_i. The induced currents at the grounding conductor flow down into the ground and, therefore, the voltage U_i on this conductor is zero. On the other hand, the signal conductor sends electromagnetic radiation into space. This leads to substantial attenuation of the signal during its propagation. One can somewhat improve characteristics of unbalanced circuits by using a common grounded shield. However, such a solution substantially complicates cabling and increases its cost.

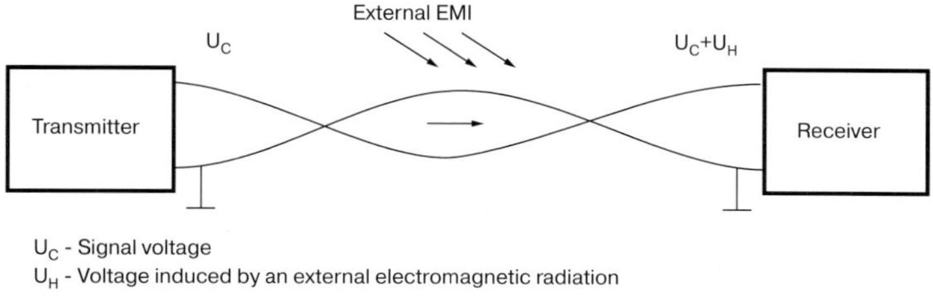

Fig. 2.1. Unbalanced circuit

Another disadvantage of unbalanced circuits is the galvanic coupling of the transmitter and the receiver. If the grounding system or the protective insulation is broken, high voltage can appear on either the grounding or the signal wire, which frequently leads to the failure of transmitting or receiving equipment on one or both terminals of the communication line. An advantage of unbalanced circuits is that only N+1 conductors (N signal conductors plus one common grounding conductor) are required to transmit N signals. Unbalanced circuits are used for transmitting low-frequency signals over short distances. A familiar example of such a circuit is the interface RS-232 (V.24).

A balanced circuit based on the balanced principle of data transmission is shown in Fig. 2.2. In balanced circuits, the transmitter and the receiver are galvanically decoupled by means of matching transformers. The secondary windings are fed only by the potential difference on the primary winding. It is apparent from Fig. 2.2 that induced currents in a completely balanced circuit lead to the antiphase change of the voltage U_i on the receiver's transformer primary winding and, hence, the resulting interference signal is not transmitted to the secondary winding. For this reason, the noise immunity of balanced circuits is substantially greater than that of unbalanced ones.

Accordingly, the vectors of electric and magnetic fields in the electromagnetic radiation of the conductors forming the pair are equal in magnitude and opposite in direction, since the currents in these conductors flow in opposite directions. Therefore, the resultant field is zero and there is no electromagnetic radiation.

Balanced circuits allow a substantially larger bandwidth for the signals to be transmitted as compared with unbalanced circuits. For this reason, balanced circuits are widely used in SCS. Balanced circuits with balanced transmission have two major disadvantages: first, 2N conductors are needed for the transmission of N signals (two wires for each signal) and second, it is impossible to transmit a constant component of a signal.

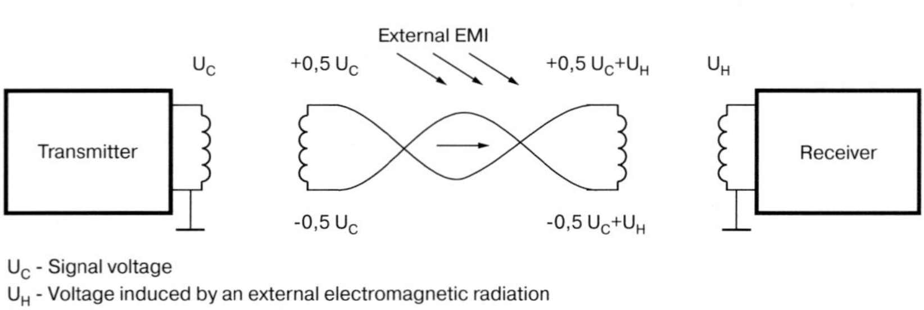

Fig. 2.2 Balanced circuit

2.1.2 Primary Electrical Parameters of a Twisted Pair

A twisted pair is a typical circuit with distributed parameters. Its equivalent circuit is shown in Fig. 2.3a. In some cases, one can use the simplified circuit of Fig. 2.3b. It simplifies some calculations while preserving acceptable accuracy. The electromagnetic properties of a twisted pair, as well as those of any other guide of electromagnetic oscillations, are completely determined by its primary parameters that include the (DC) resistance, *R*, and the inductance, *L*, of the conductors, as well as the capacitance, *C*, and the conductivity, *G*, of the insulator [25]. The parameters *R* and *G* characterize the energy loss; the resistance *R* characterizes the heat loss in the wire and in the shield (if any) and the conductivity *G* the loss in the insulator. The parameters *L* and *C* determine the reactivity of the twisted pair regarded as a waveguide and, hence, its frequency properties. Specific values of the primary parameters depend on the structure of the cable, for example, on the geometry of the individual components and their mutual arrangement, and on the materials of the conductors, insulators, and outer covers.

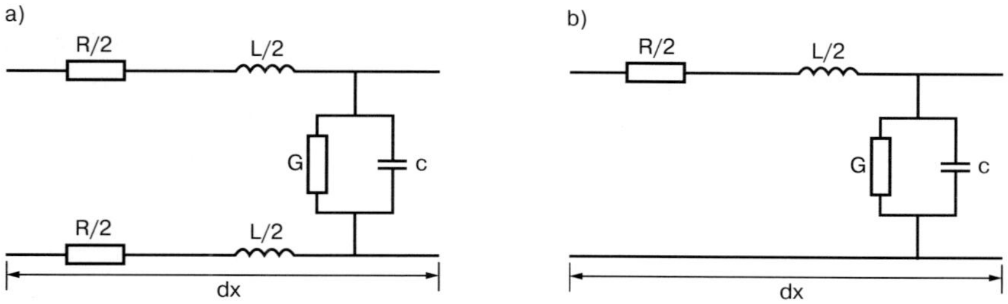

Fig. 2.3. Twisted pair equivalent circuit: *a)* general; *b)* simplified

Capacitance

A twisted pair is formed by two conductors separated by a solid insulator layer and an air clearance. Such a structure can be regarded as a capacitor in which the conductors play the role of plates and the insulator material and air play the role of a dielectric. This system has a noticeable capacitance that increases proportionally to the increase of the length. The electric capacitance of the twisted pair restricts the cabling bandwidth and leads to distortions of the high-frequency portion of the spectrum of the signal being transmitted.

The specific capacitance of an ideal twisted pair is defined by:

$$C = \frac{27.9\varepsilon}{\ln\frac{2D}{d}}, \text{ pF/m}, \tag{2.1}$$

where ε is the specific permittivity of the insulator material, *D* is the distance between the conductors of the pair, and *d* is the diameter of the conductor.

The capacitance is independent of the frequency. However, because of the specific features of the measuring techniques, the value of the capacitance is frequently indicated together with the value of the signal frequency at which the measurement has been performed. In accordance with the standard TIA/EIA-568-A, the capacitance per 100-m length should not exceed 6.6 nF for cables of category 3 and 5.6 nF for cables of categories 4 and 5. Some companies produce cables with substantially lower capacitance, taking into account their own technical conditions. For example, cable types 8606 and 8312 produced by the American company Ouabbin have a capacitance of 3.81 and 3.90 nF per 100 m, respectively

One should bear in mind that shielding increases the capacitance of a twisted pair by approximately 30% [26].

DC resistance

DC resistance (ohmic resistance) depends on the material of the conductor, its length, cross-sectional area, and temperature. The conductors of twisted pairs used in SCS are made of copper that has a low specific resistance. The smaller the cross-sectional area of the wire, the larger its length and the higher the temperature, the larger the DC resistance and, hence, the higher the signal attenuation in the twisted pair. According to the standard TIA/EIA-568-A, the DC resistance of any conductor of a twisted pair at a temperature of 20°C should not exceed 9.38 Ω per 100 m. To facilitate the experimental determination of the resistance in SCS technology, one usually measures it in accordance with the "loop scheme", i.e., from one end, with the far end of the twisted pair short circuited. The resistance measured in such a way is referred to as the DC loop resistance.

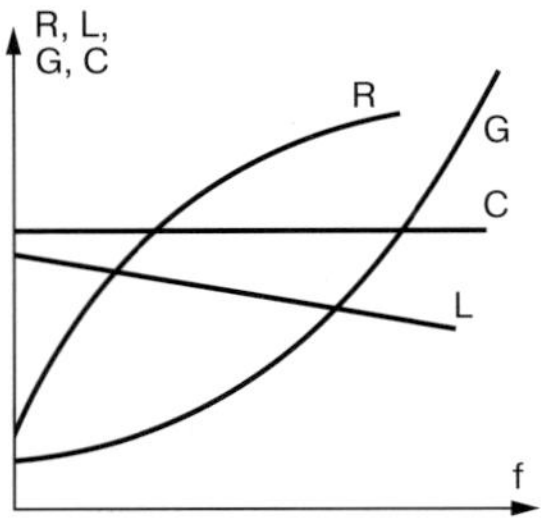

Fig. 2.4. Symmetrical cable parameters frequency characteristics

The DC resistance of a wire increases with the increase of the signal frequency. This is accounted for, first, by the displacement of the current toward the surface of the conductor (skin effect) and, second, by the fact that the current flows mainly along the surface turned to the second conductor (the proximity effect). As a result, the effective cross-section of the conductor is reduced and the resistance increases.

The quantitative measure of the decrease of the effective cross-sectional area of a conductor is given by the penetration of current. This quantity is numerically equal to the thickness of the skin layer concentrating 90% of the energy transferred by the electric current. For sufficiently high frequencies, the penetration of current is defined as follows [2]:

$$\Delta = 2.09/\sqrt{f}\text{, mm} \tag{2.2}$$

The frequency f is measured in kHz. The DC resistance R of a conductor is proportional to its cross-sectional area. For this reason, the expression (2.2) implies, in particular, that R is proportional to the frequency.

To minimize the deleterious influence of the skin and proximity effects, the twisted pair conductors for horizontal and backbone cables are constructed as a solid wire, rather than a bundle of several thin wires. One can use conductors formed by a bundle of thin wires only in cases where the primary requirement is to provide patch cords with a high flexibility and resistance to failure when repeatedly bent while the elevated attenuation of the signal is not strongly pronounced because of a relatively small total length.

The influence of the skin effect on the DC resistance of a conductor is taken into account when constructing cables for data transmission. The conductors of the cables designed for operation at high frequencies always have larger diameters and larger cross-sectional areas of the near-surface layers. Accordingly, the increase in the resistance with the growth of frequency is less pronounced in such cables. For example, the conductors of cables of category 5 operating at frequencies lower than 100 MHz typically have a diameter of 0.51–0.52 mm, whereas for 600-MHz cables this diameter is increased to approximately 0.6 mm.

Inductance

A twisted pair consists of two isolated conductors. Each of the conductors accumulates energy during the current flow through it and, hence, possesses the property of inductance. The inductance of a twisted pair has several components. *External inductance* L_1 depends on the geometry of the wire and the magnetic properties of the conductor material. Since copper is not a ferromagnetic material, this component also depends on the current flowing through the conductor. The definition of the external inductance implies its independence of the frequency. *Internal inductance* L_2 is accounted for by the magnetic field created by the current flowing through the conductor. As the frequency increases, the internal inductance decreases because of the skin effect. *Cover inductance* L_3 manifests itself only in shielded structures and is accounted for by a significant influence of metal shields on the magnetic field of the conductor. This component noticeably depends on frequency. The total inductance is defined as the sum of the separate components: $L = L_1 + L_2 + L_3$. It tends to decrease as the frequency increases.

Insulator conductivity

The insulator conductivity characterizes the quality of the materials and the technology of manufacturing insulating covers for separate conductors. The total insulator conductivity of a twisted pair can be represented as the sum of two components, $G = G_o + G_f$, where G_o takes into account leakage currents caused by imperfection in the dielectric and G_f the energy loss for the dielectric polarization.

By *polarization* one understands reorientation of bound dipoles in a dielectric under the action of a magnetic field. The alternating field causes vibration of the dipoles, which leads to an increase in the temperature of the dielectric. The heating of the dielectric, in turn, facilitates the vibration thereby causing the increase in the signal attenuation.

Specifically, polyvinylchloride, which is commonly used as the insulator material for cables of category 3, has many dipoles formed by the atoms of chlorine. Cables of category 5 utilize polyethylene or teflon possessing relatively low energy dissipation properties. Using foamed materials for insulator covers can provide even lower levels of energy loss. In these materials, the number of dipoles is considerably reduced because of a great number of air inclusions.

The insulator conductivity is measured in siemens. Also, one frequently uses the resistance of the insulator, which is the reciprocal of the conductivity, $R_{ins} = 1/G$. Typical values of this quantity range from hundreds of megaohms to several gigaohms. The voltage of the testing signal when determining the resistance or conductivity of the insulator may reach several thousand volts. Excessively low insulator resistance indicates a poor contact in connectors or damage in the cable.

2.1.3 Secondary Parameters of Cables and Lines Based on Twisted Pairs

The secondary parameters of a twisted pair are calculated on the basis of the primary parameters or determined experimentally, which happens more frequently. In technical documents, the secondary parameters are specified for a twisted pair, permitting one to design communication lines based on balanced cables using simple engineering techniques and to assess its suitability for transmitting signals of specific applications.

Impedance

In cabling engineering, the impedance (wave resistance) is defined as the resistance underwent by an electromagnetic wave propagating (without reflection) along any homogeneous waveguide, including a twisted pair. The impedance is a characteristic of a specific type of the cable; it depends only on the primary parameters of the cable and on frequency. The impedance and the primary parameters are related by the following simple expression:

$$z = \sqrt{\frac{R + j\omega L}{G + j\omega C}} \tag{2.3}$$

The impedance is equal to the input resistance of a line with infinite length and the terminating load coinciding with its own impedance. The impedance is measured in ohms and characterizes the quantitative relation between the electric and magnetic components of the wave propagating along the given line. It is apparent from the relation (2.3) and Fig. 2.3 that one can experimentally determine the impedance using the idling and short-circuit methods combined with the relation

$$z = \sqrt{z_o z_c} \tag{2.4}$$

where z_o and z_c are the resistances of the twisted pair for the idling mode and the short circuit at the far end, respectively. This is the method that is recommended by the Russian standard GOST 27893-88 to be used for measuring the impedance of cables [27].

The relation (2.3) implies that in the general case impedance is a complex quantity. The absolute value of this parameter, making up about 600 Ω at sound frequencies, decreases as the frequency increases and at high frequencies (over 1 MHz for modern cabling systems) tends to DC resistance.

$$z_\infty = \lim_{\omega \to \infty} \sqrt{\frac{R + j\omega L}{G + j\omega C}} = \sqrt{\frac{L}{C}} \tag{2.5}$$

According to the working standards, for a specific cable with an impedance of 100 Ω, the difference between the value given by the relation (2.5) and the nominal value should lie within an interval of ±15 Ω for frequencies higher than 1 MHz. The difference lying beyond this interval may indicate poor contacts in connectors or inadmissibly high mechanical loads during cabling (e.g., pressure, bending, tension, or torsion).

Attenuation

An electromagnetic signal propagating in a twisted pair gradually loses energy. This effect is referred to as attenuation or insertion loss. Attenuation is defined as the difference between the levels of the signal at the transmitter outlet and the receiver inlet and is usually measured in decibels (dB). An attenuation of 1 dB corresponds to a change in power by a factor of 1.26 or a change in voltage by a factor of 1.12. For more details, see Section 14.2.

In cabling engineering, one distinguishes between the proper attenuation and the working attenuation of the cable. The *proper attenuation* is the attenuation of the cable operating in ideal conditions. The ideal conditions imply the exact coincidence of the output resistance of the signal source and the load resistance with the impedance of the cable in the entire frequency range. In theory, the proper attenuation can be determined as the real part of the propagation factor γ, related to the primary parameters as follows:

$$\gamma = \sqrt{(R + j\omega L)(G + j\omega C)} \tag{2.6}$$

In experiment, one can determine the proper attenuation as the difference between the levels of the input and output signals in the case where the resistances of the signal source and the load are equal to each other and coincide with the impedance of the cable. When operating a real system, this condition may be violated, which leads to an increase in the attenuation. This attenuation is referred as the *working attenuation*. The above definitions imply a practical recommendation that to minimize the working attenuation and make it closer to the ideal attenuation the resistances of the signal source and the load should be equal to the impedance of the cable. Using the terminology of electrical engineering, one can say that the load of the signal source should be matched with that of the cable. The rigid norms specified by SCS standards for the admissible difference of the impedance from the nominal value in the entire range of operating frequencies, as well as a direct ban imposed on the utilization of components with different impedances in the same line, substantially help to fulfill this condition.

From relation (2.6), it follows that the attenuation depends on the frequency and also on the length, since all parameters occurring in this relation depend on the length. Analysis of formula (2.6) shows that the attenuation expressed in decibels is proportional to the length of a twisted pair for all frequencies. To simplify engineering calculations, it is convenient to use the attenuation factor or the attenuation per fixed length, α. For twisted pairs utilized in SCS, the fixed length in the definition of the attenuation factor is usually taken to be equal to 100 m, whereas in communication engineering the attenuation is commonly scaled to 1 km (thus the term kilometer attenuation). The attenuation factor, α, the length, L, and the attenuation, A, are related by

$$A\text{, dB} = \alpha\text{, dB/100 m} \times L\text{, m/100} \tag{2.7}$$

The lower the attenuation, the more powerful the signal at the receiver's input and the more reliable the communication (with other things being equal). The attenuation is primarily due to the DC resistance and the loss in the insulator. The attenuation is also influenced by the radiation of the electromagnetic energy and reflection of the waves.

Any conductor with alternating current flowing through it is a source of electromagnetic radiation. The radiation takes the energy away from the signal thereby increasing its attenuation. This effect becomes more pronounced as the signal frequency increases. If $\lambda < a$, where λ is the wavelength of an electromagnetic signal, and a is the space between cables, then the major portion of the energy is radiated into the surrounding space, which makes the transmission of the signal through an unshielded waveguide impossible. For a standard twisted pair, the value of the parameter a is about 2 mm, which corresponds to a critical frequency of about 15 GHz. This value exceeds the upper working frequency for typical unshielded balanced cables that are currently in use (~150 MHz) by approximately two orders of magnitude. The loss due to electromagnetic radiation increases with an increase in frequency. To minimize the radiation loss, balanced transmission is used in combination with twisting the conductors in pairs.

As has been mentioned previously, there is no electromagnetic radiation in an ideal balanced circuit. Such circuits do not exist in practice. The ideal circuit should be free from inevitable manufacturing allowances and defects and, moreover, the conductors in this circuit should lie infinitely close to one another and be tightened into an infinitely thin line, with the total current through this line being equal to zero. The smaller the diameter of the conductors and the thinner the insulation, the closer the conductors lie to each other. However, excessive reduction of the cross-sectional area of the conductor and thinning of the isolator is impermissible, since this leads to an increase in the attenuation due to the growth of DC resistance and the increase in the conductivity of insulator covers.

From the equivalent circuit shown in Fig. 2.3, one can conclude that attenuation tends to grow as the frequency increases. This is accounted for by the increase in the impedance of the longitudinal branch owing to the inductance L and the increase of the DC resistance R due to the skin effect and the proximity effect, as well as by the drop of the impedance of the transverse branch owing primarily to the capacitance *C*. According to the standard TIA/EIA-568-A, for a length of 100 m and a temperature of 20 C, the frequency dependence of the maximal admissible attenuation, $A(f)$, in the range beginning with 0.772 MHz for cables of categories 3, 4, and 5 is defined by

$$A(f) = k_1\sqrt{f} + k_2 f + k_3/\sqrt{f},\ \text{dB} \tag{2.8}$$

where A in dB is the maximal admissible attenuation; f in MHz is the signal frequency; and k_1, k_2, and k_3 are constant parameters to be determined from Table 2.1 depending on the length of the cable.

Table 2.1. Parameters for the relation of (2.8)

Cable category	k_1	k_2	k_3
3	2.320	0.238	0.000
4	2.050	0.043	0.057
5	1.967	0.023	0.050

Analysis of relation (2.8) shows that for high frequencies the attenuation of the cable is determined primarily by the first term and changes approximately proportionally to the square root of the frequency. Accordingly, the slope of the attenuation characteristic of the cable ranges from 5 to 15 dB per decade, depending on the frequency region, the value 15 dB/decade corresponding to frequencies equal to or higher than 15 MHz for cables of category 5.

Apart from the analytical definition of attenuation, the standard TIA/EIA-568-A also defines this parameter using tables, with an extension of the specified values to a lower frequency range. The tables can facilitate engineering calculations of communication lines intended for the support of some low-rate applications. In addition, they permit one to obtain the necessary information without any computations.

Unlike the American standard, the international standard ISO/IEC 11801 uses the table form for specification of the maximal attenuations allowed for cables of different categories as the main table. To determine the attenuation at intermediate frequencies, one applies linear interpolation using a semilogarithmic scale.

Figure 2.5 presents frequency characteristics of the maximal admissible attenuation calculated by relation (2.8) for cables of different categories. The approximation in accordance with relation (2.8) turns out to be very good and is frequently used by many manufacturers of cabling components to characterize their products. These manufacturers define their own values for the coefficients k_1–k_3 and extend the frequency range to 400 MHz and even to 550 MHz [49].

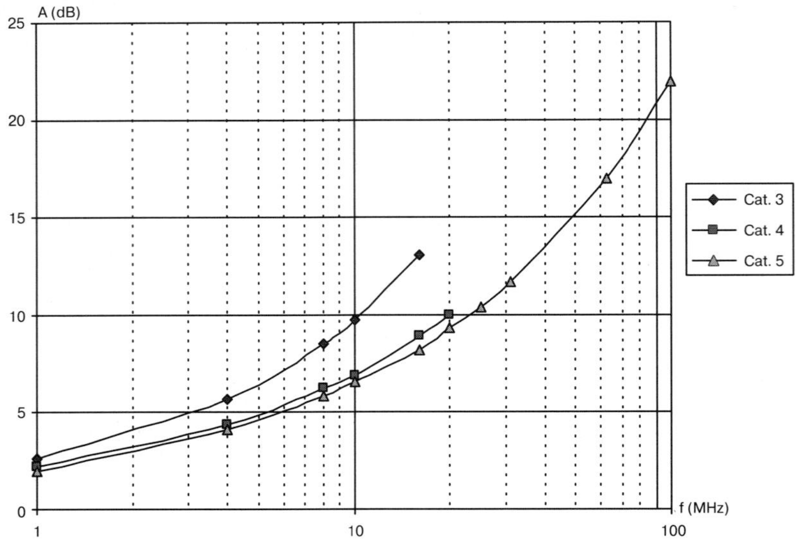

Fig. 2.5. Maximal admissible attenuation for cables of categories 3, 4, and 5 according TIA/EIA-568-A standart

Crosstalk attenuation

Since the balance of a twisted pair is not ideal, part of the energy of a signal being transmitted through this pair is converted into electromagnetic radiation that induces currents in neighboring pairs. This effect is called crosstalk. Crosstalk superimposed on a useful signal plays the role of noise. If the level of crosstalk is significant, it may distort the signal being received, which reduces the communication quality.

The difference between the level of the signal being transmitted and that of the crosstalk induced by this signal in a neighboring pair is called the *crosstalk attenuation*. The crosstalk attenuation can be classified into several types, depending on the location and the method of measuring this parameter. First consider the types dependent on the location of the measurement. If the source of the signal and the point of its measurement are located at the same end of the line, one speaks about *near-end crosstalk attenuation* or *near-end crosstalk.* If the source of the signal and the point of its measurement are located at opposite ends of the line, one speaks about *far-end crosstalk attenuation* or *far-end crosstalk* (Fig. 2.6). In the international literature on SCS, near-end crosstalk and far-end crosstalk are traditionally abbreviated as *NEXT* and *FEXT*, respectively. In the Russian literature on cabling systems for intercity, zone, and long-distance communication, these parameters are denoted by A_0 and A_l, respectively.

The higher the values of near-end and far-end crosstalk, the lower the crosstalk level in neighboring pairs and, hence, the higher the quality of the cable. In practice, it is important to know the frequency dependence of the near-end and far-end crosstalk attenuations, as well as the dependence of these parameters on the length of the line.

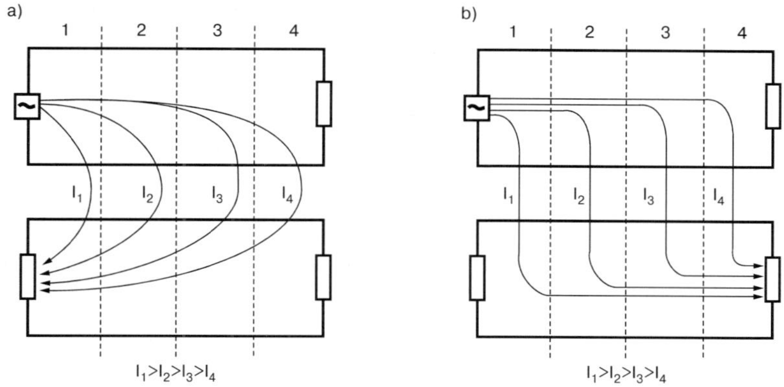

Fig. 2.6. Crosstalk: *a)* near-end crosstalk; *b)* far-end crosstalk; I_i – crosstalk currents

The influencing and influenced pairs lie parallel to each other and are enclosed in a common protective cover. For this reason, the separate conductors of the pair can be regarded as capacitor plates. Hence, the crosstalk attenuation becomes less pronounced as the frequency increases. The standard TIA/EIA-568-A specifies the minimal values for *NEXT* and *FEXT* for cables of length 100 m. The minimum value of the parameter *NEXT* for frequencies exceeding 0.772 is determined by the following approximating equation:

$$NEXT(f) = NEXT(0{,}772) - 15\lg\,(f/0.772), \tag{2.9}$$

where *NEXT*(0.772) is the minimal admissible near-end crosstalk attenuation at a frequency of 0.772 MHz (for cables of categories 3, 4, and 5 this value is specified by 43, 58, and 64 dB, respectively) and f in MHz is the signal frequency. The slope of the function defined by (2.9) is equal to 15 dB/decade in the entire frequency range. To determine the parameter *NEXT* for intermediate frequencies, using linear interpolation on the logarithmic scale is recommended.

Note that the dimension dB/100m used for near-end crosstalk in the standard [28] and some other publications (e.g., [42]) should be interpreted as the crosstalk magnitude measured for a cable of length 100 m. No recalculations for a smaller length are admissible. Thus, if *NEXT*=40 dB for a length of 100 m, then for a length of 50 m, this parameter is also equal to 40 dB.

Additionally, the standard specifies the minimal value allowed for *NEXT* at frequencies lower than 0.772 MHz, which may be necessary for some applications. The corresponding values in this case are presented in a table.

The results of calculations using relation (2.9) are presented in Fig. 2.7. Separate (voltage) components of the near-end crosstalk having the same frequency but different phases are superimposed. For this reason, the real frequency characteristic of *NEXT* has the shape of a noise-like curve with a sharp change of this quantity at close frequencies. The standards specify only the minimal value of the parameter *NEXT*. The cable is considered to meet the standard requirements if the real value of *NEXT* does not drop below the specified value in the entire working frequency range.

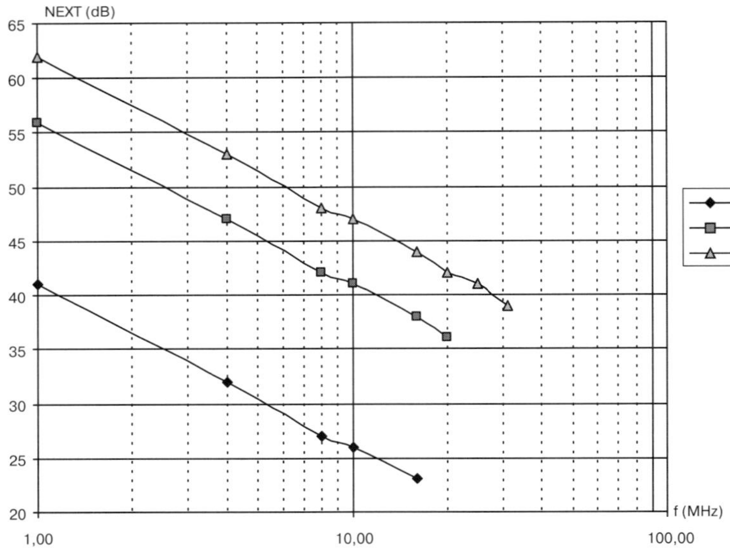

Fig. 2.7. Minimal value allowed for *NEXT* for cables of categories 3, 4, and 5 according TIA/EIA-568-A standart

Figure 2.8 shows a typical dependence of the near-end and far-end crosstalk on the length of the line. As the length of the line increases, the near-end crosstalk first is somehow reduced and then is stabilized. This effect can be qualitatively explained by the fact that starting with a certain length of the line, the crosstalk currents that come to the near end from farther segments (e.g., segments 3 and 4 in Fig. 2.6) are weakened and virtually do not increase the crosstalk between the circuits. Hence, the value of *NEXT* remains constant. One can also come to this conclusion by analytical manipulations. The dependence of *NEXT* from the length of the line can be represented as follows [29]:

$$NEXT = NEXT(l_0) + 10\lg\frac{1-\exp(-4\alpha l)}{1-\exp(-4\alpha l_0)},$$

where α is the attenuation factor, $NEXT(l_0)$ is the crosstalk attenuation for the line length l_0, and l is the distance from the end of the segment (of length l_0).

The aforesaid implies that the values of *NEXT* for two ends of one pair may substantially differ. For that reason, all standards require this parameter to be measured at both ends.

To improve crosstalk attenuation characteristics in balanced cables, one uses different steps of twist in twisted pairs. Apart from the weakening of the electromagnetic coupling between separate pairs, this solution does not allow the pairs to lie close to each other along the entire length, which additionally increases the crosstalk attenuation.

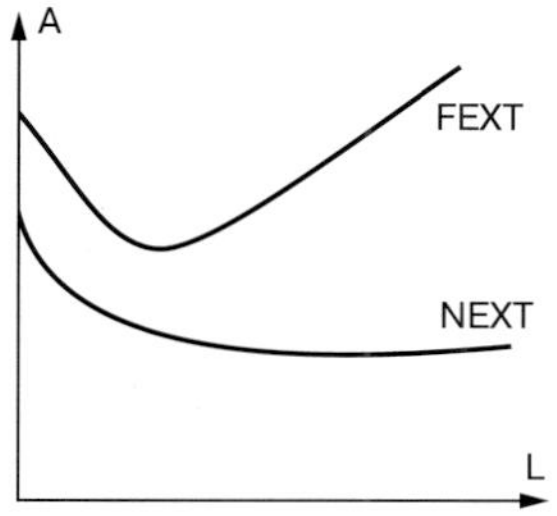

Fig. 2.8. Typical dependence of the near-end and far-end crosstalk on the length of the line

Consider now the kinds of crosstalk attenuation depending on how it is measured. It is well known that network equipment of various intended uses utilizes the balanced cable as the transmission medium in different manners. Therefore, depending on the application and the manner in which the cable is utilized, crosstalk or, equivalently, crosstalk attenuation is standardized in different ways.

By the end of the 1990s, the versions 10Base-T and 100Base-TX of the Ethernet network were the most widespread LANs. In the full-duplex mode, the transmitter and the receiver work simultaneously, each device using its own twisted pair in the same cable. This situation is sketched in Fig. 2.9. The signal, weakened after passing through the twisted pair, is subjected to a powerful crosstalk created by the transmitter working at the same end. For this reason, it suffices to define the parameter *NEXT* as follows:

$$NEXT = P_S - \max P_{XT},$$

where P_S is the signal level and P_{XT} is the level of the crosstalk created by this signal. The value max P_{XT} is taken as the worst case, since it is unknown in advance which two pairs the network equipment will use for the organization of the data exchange.

Fig. 2.9. NEXT definition

By the end of the 1990s, SCS technology had required introducing models that would more adequately fit the potential uses of cabling systems with modern equipment and equipment that is likely to be used in the future. This is accounted for by two factors. First, during the construction of the network equipment, a tendency to simultaneously use several pairs for data exchange in the full-duplex mode had clearly manifested itself. Second, signals of different applications tend to be transmitted in one multipair cable simultaneously without any correlation. In this case, the standardization of only one parameter, *NEXT*, is insufficient, since the receiver, when receiving a useful signal, is affected by crosstalks from several sources simultaneously. To take into account this fact one uses a more complicated design procedure which standardizes the crosstalk attenuation in accordance with the power sum model or the total near-end crosstalk:

$$PS - NEXT = 10\lg \sum_{i=1}^{n-1} 10^{\frac{-NEXT_i}{10}},$$

where $NEXT_i$ is the value of *NEXT* for the *i*-th pair and *n* is the number of pairs in the cable. The design model for a four-pair cable is shown in Fig. 2.10 (each pair is affected by crosstalks from all other pairs). Since the signals transmitted simultaneously through different pairs are not correlated, we sum up the powers of separate crosstalk components to determine the quantity *PS-NEXT*, as well as other "overall" parameters.

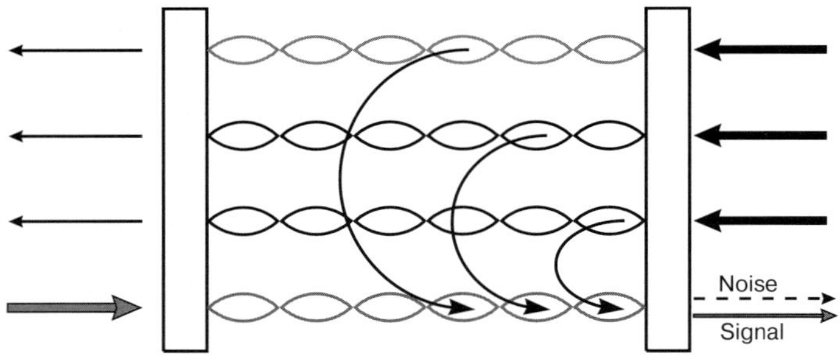

Fig. 2.10. PS-NEXT definition

Unlike the parameter *NEXT*, the value of *PS-NEXT* is not measured directly during testing by means of modern measuring devices, but is calculated on the basis of the values of $NEXT_i$ measured for separate pairs. Because of the difference in the distances between pairs, steps of twist, and other parameters, the difference between *NEXT* and *PS-NEXT* for some specific structures makes up about 2 dB, rather than 4.8 dB. If the quantity *PS-NEXT* is not specified in the list of operating characteristics of the cable, it can be evaluated by the empiric relation

$$PS\text{-}NEXT = NEXT - 3, \text{ dB}.$$

If all four pairs are simultaneously engaged to transmit and receive information, it is necessary to take into account not only the near-end crosstalk but also the far-end crosstalk (Fig. 2.11). To that end, one should normalize the far-end crosstalk as follows:

$$FEXT = P_S - \max P_{XT},$$

where P_{XT} is the level of the far-end crosstalk.

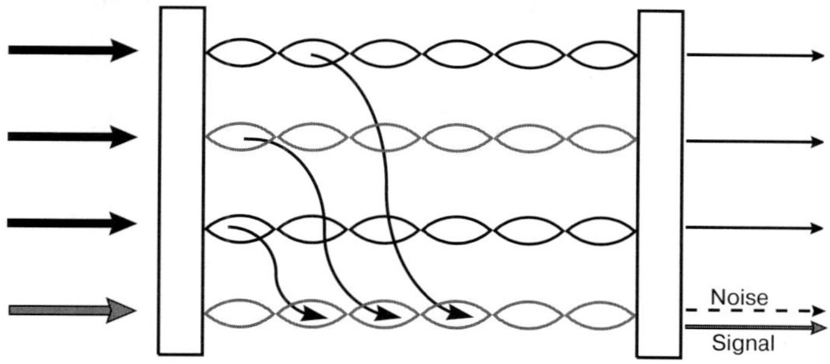

Fig. 2.11. PS-FEXT definition

The working standards do not specify the frequency dependence for the parameter *FEXT*. We choose to indicate only the decrease in the far-end crosstalk as the frequency increases.

When considered as a function of the length of the line, the far-end crosstalk attenuation has an extremum. While the length of the line is relatively small, the increase of the crosstalk power with the increase of the length is the dominant tendency. With a further increase of the length, the increase of the attenuation of the crosstalk components starts dominating, and the parameter *FEXT* monotonically increases.

The far-end crosstalk can be normalized by analogy with the normalization of the near-end crosstalk.

$$PS - FEXT = 10\lg\sum_{i=1}^{n-1} 10^{\frac{-FEXT_i}{10}}$$

The level of the far-end crosstalk is usually lower than that of the near-end crosstalk. However, unlike the near-end crosstalk components, the far-end crosstalk components have zero or small phase difference, which may lead to an additional increase in the total far-end crosstalk power [30].

Apart from the cross-talk components considered above, one sometimes defines and standardizes additional crosstalk characteristics. For example, some manufacturers introduce global crosstalk, GXT, defined as the sum of the crosstalk components induced at both ends of the cable [31].

Horizontal cables are frequently arranged in one cable run and lie very close and parallel to each other for a large length. In this case, it may be necessary to control the alien (intercable) crosstalk attenuation [32]. This characteristic measures the influence of all surrounding cables on one specific cable. By analogy with the intracable crosstalk attenuation, one distinguishes between the near-end alien crosstalk (Alien *NEXT*) and the far-end alien crosstalk (Alien *FEXT*) [33]. With reference to the mechanism of this type of interference, it is clear that the major share of the alien crosstalk power in a specific pair is due to the interaction of this pair with pairs having the same color in other cables, since pairs of the same color have the same step of twist. The specific features of this type of interference do not exclude the case where the level of far-end alien crosstalk substantially exceeds that of the near-end alien crosstalk.

Experiments show that one has to take into account the crosstalk due to coupling of neighboring cables only at frequencies higher than 100 MHz, i.e., only for lines of category 6 that use unshielded cables.

By the middle of 2000, the standard had specified only the values of *NEXT* and *PS-NEXT*, the latter quantity being given only for multipair and combined cables. The values of *PS-FEXT* and other parameters of the far-end crosstalk are specified in the year 2000 version of the standard ISO/IEC 11801.

Immunity

To characterize the quality of the information transmission through wire communication lines, one widely uses a parameter called the immunity to interference (or simply immunity). This parameter is defined as the difference between the level of the useful signal and that of the noise at the point in question [25]. For the design model of Fig. 2.9, the signal level is given by $P_S = P_{tr} - A$, where P_{tr} is the level of the transmitter output signal and A is the attenuation, while the crosstalk level is expressed as $P_{XT} = P_{tr} - NEXT$. The immunity is defined as follows:

$$ACR = NEXT - A \tag{2.10}$$

and, hence, depends only on the magnitudes of the attenuation and the crosstalk attenuation (Fig. 2.12). To be strict, one should speak of the immunity to crosstalk in this case. However, the SCS lines and network transceivers utilizing them are designed so that all other kinds of interference are negligibly small. For this reason, one uses the term *immunity* and the abbreviation *ACR* (attenuation to crosstalk loss ratio) reminds us what kind of immunity is implied.

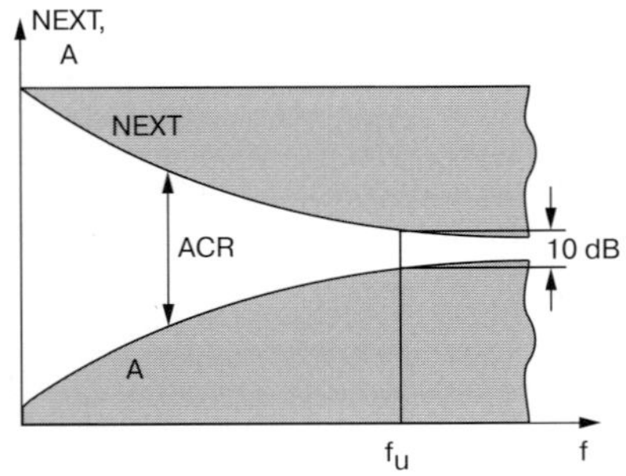

Fig. 2.12. Immunity definition

The parameter *ACR* measures how much the useful signal exceeds the crosstalk. Therefore, it is an integral quantity characterizing the quality of both the twisted-pair cable itself and any data transmission line utilizing such cables. As the parameter *ACR* increases, all other conditions being equal, the signal-to-noise ratio also increases and, hence, communication becomes more reliable. The parameters *NEXT* and *A* on the right-hand side of Equation (2.10) depend on the frequency, the modulus of *NEXT* tending to decrease, and the modulus of A tending to increase as the frequency grows. Hence, the parameter *ACR* also depends on the frequency. The rate of the frequency characteristic of this parameter makes up 20–30 dB/decade for cables of category 5. The specific value of *ACR* is determined by the frequency characteristic of the attenuation. The ISO/IEC 11801 (the 1995 version) specifies the minimal values of ACR allowed for cables of category 5 at frequencies equal to or higher than 20 MHz. The standard TIA/EIA-568-A does not specify the values of *ACR* for different frequencies but one can calculate these values using relation (2.10). The results of calculations for cables of categories 3, 4, and 5 with a length of 100 m are presented in Fig. 2.13. It is apparent from this figure that in the worst case the input signal of the receiver must exceed the crosstalk from a neighboring pair by at least 10 dB. This is equivalent to a signal-to-noise ratio equal to 3.16 when measured with respect to voltage and 10 when measured with respect to power.

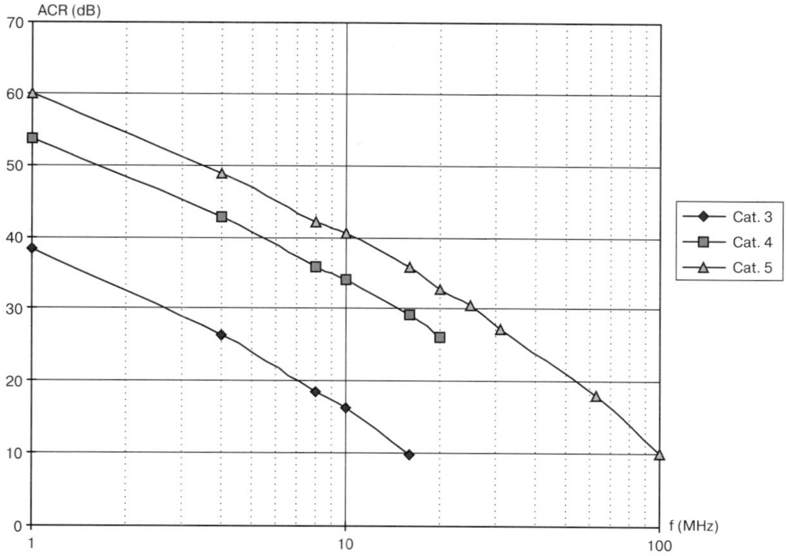

Fig. 2.13. Minimal value allowed for *ACR* for cables of categories 3, 4, and 5 according TIA/EIA-568-A standart

The parameter *ACR* allows one to appropriately define the upper limiting frequency both for any component of the electrical data transmission line and for the entire line. In accordance with the classical definition, the upper limiting frequency is the lowest frequency of a sinusoidal signal at which the transfer constant, being equal to 1 at low frequencies, is equal to 0.707. Recall, however, that the intrinsic noise of the receiver is negligibly small and that the overall noise in

SCS lines is primarily due to crosstalk. (Sometimes, one even says that a receiver works in the crosstalk limitation mode.) Hence, it is reasonable to define the upper limiting frequency as the lowest frequency at which the attenuation to crosstalk loss ratio, *ACR*, is equal to a prescribed value. Thus, the fact that SCS have high immunity to intrinsic noise permits one to substantially expand the bandwidth beyond the limits following from the classical definition by means of conventional methods of radio and communication engineering. These methods include emphasis (decreasing the gain factor at low frequencies) and correction (increasing the gain coefficient at higher frequencies). See also Section 2.3.3.

It is assumed that a cable (as well as a connector, a cord, and the entire line) provides stable full-duplex operation of any application with the upper limiting frequency at a level where the *ACR* makes up at least 10 dB. This assumption is reflected in Fig. 2.12. The exception is for cables of category 4 which have *ACR*=26 dB at a frequency of 20 MHz. One should distinguish between the limiting frequency of the application and the maximal frequency of the cable at which the manufacturer specifies the parameters of the cable. The *ACR* values at the latter frequency are often negative. (This is especially the case for unshielded systems with relatively low values of *NEXT*). The specification of the parameters of cables at these frequencies is necessary to assess the possibility of using these cables for half-duplex or unidirectional (simplex) transmission of some kinds of signals (e.g., TV signals), in which cases the concepts of crosstalk and, hence, that of the *ACR*, make no sense.

For high-rate applications that use all twisted pairs for transmitting data in both directions, the specification of only the parameter *ACR* is insufficient. The noise component due to far-end crosstalk is characterized by the parameter

$$ELFEXT = FEXT - A \tag{2.11}$$

The notation *ELFEXT* is an abbreviation of equivalent level for far-end crosstalk. The parameter *ELFEXT* is an analogue of *ACR*.

To take into account the specific features of the operation of super-high-rate LAN interfaces, one has to use power-sum models and the corresponding parameters

$$PS\text{-}ACR = PS\text{-}NEXT - A,$$

$$PS\text{-}ELFEXT = PS\text{-}FEXT - A.$$

It is convenient to determine the parameters introduced above by means of Fig. 2.14.

This diagram shows two twisted pairs arranged near one another. The generator of the test signal is shown in the upper left-hand corner corresponding to the near end of the influencing pair. To determine a parameter one should subtract the level of the signal at the head of an arrow from the value at the tail of this arrow.

The requirements for the parameters that are supposed to be introduced into new standards for SCS are presented in Table 1.9.

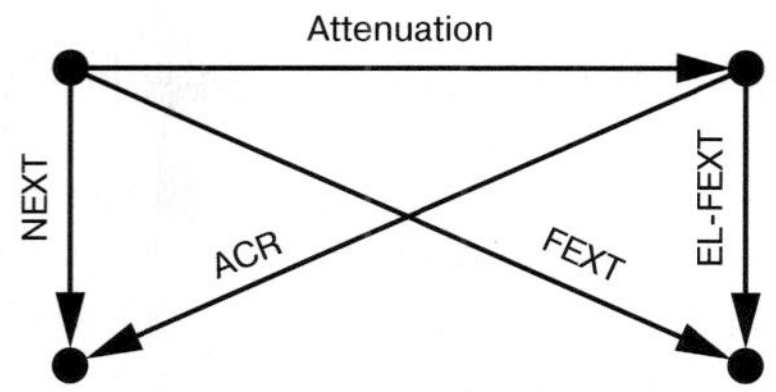

Fig. 2.14 PS-ACR and PS-ELFEXT definition

NVP and Delay

The parameter NVP (nominal velocity of propagation) characterizes the reduction in the velocity of propagation of an electromagnetic wave in a twisted pair. It is defined as the ratio of the actual velocity of propagation of the signal to the velocity of light in a vacuum and is expressed in percent or as a decimal fraction. In the latter case, the index c is sometimes added; for example, 65%, 0.65, and 0.65c are the same value of NVP. Occasionally, the notation VOP (velocity of propagation) is used instead of NVP. The reason this parameter must be taken into account is associated with a delay in propagation of signals. This delay may be significant and critical for some applications, e.g., for Fast Ethernet networks.

Standards of structured cabling systems specify the requirements concerning the parameter NVP only in general terms (Table 2.2). The procedure of measuring this parameter was first defined in detail in the 2000 version of the standard ISO/IEC 11801.

Table 2.2. Parameter NVP for twisted pairs of categories 3, 4, and 5

Frequency, MHz	NVP		
	Cat. 3	Cat. 4	Cat. 5
1.00	0.4	0.6	0.65
10.00	0.6	0.6	0.65
100.00	–	–	0.65

In general, the value of NVP for a twisted pair depends on the diameter of the conductors, the distance between the conductors, and the type of the dielectric. For cables used in SCS, the most important factor is the insulator material. The diameter of the conductors and the thickness of the insulator cover are specified fairly rigidly by standards. To obtain the required level of *NEXT*, the conductors are closely twisted with one another. For example, at frequencies exceeding 10 MHz, the parameter NVP ranges from 0.65 to 0.7 for conductors insulated by polyethylene and from 0.69 to 0.73 for conductors with a teflon insulator, while polyvinyl chloride materials, commonly utilized in cables of category 3, reduce NVP to 0.60–0.64. As is the case for most of the primary and secondary parameters of a twisted pair, NVP depends on the frequency (Table 2.2). For typical values of NVP ranging between 60 and 75%, the signal transit time for a line segment with length 100 m ranges from 370 to 550 ns. The change in NVP due to the change in the state of the cable conductor insulator and environment temperature does not exceed 3% of the nominal value.

The fact that the velocity of propagation of an electromagnetic wave in a twisted pair is finite causes a delay in the arrival of a signal at the receiver after feeding it to the input of the line. For modern high-speed network interfaces, such a delay is notable even for relatively short runs typical of the horizontal subsystem of an SCS. The magnitude of this delay (parameter *Delay*) is critical for a number of applications whose active devices use cable runs as monochannels during the data exchange process. For example, this is the case for all modifications of Ethernet when operating in the half-duplex mode. The value of the delay depends on NVP and the frequency and, in accordance with the 2000 version of the standard ISO/IEC, must not exceed

$$Delay = 534 + 36/\sqrt{f}, \text{ ns/100 m}$$

in the frequency range from 1 MHz to the upper limiting frequency of the cable; the frequency in this relation is measured in MHz.

Unlike this, the American standard TIA/EIA-568-A only requires the parameter delay not to exceed 5.7 ns/m at a frequency of 10 MHz for any pair.

Delay skew

In cables consisting of several twisted pairs, each of the pairs is characterized by its own time of signal transition from the transmitter to the receiver (Fig. 2.15). The spread of signal transition times is primarily accounted for by two factors. These are

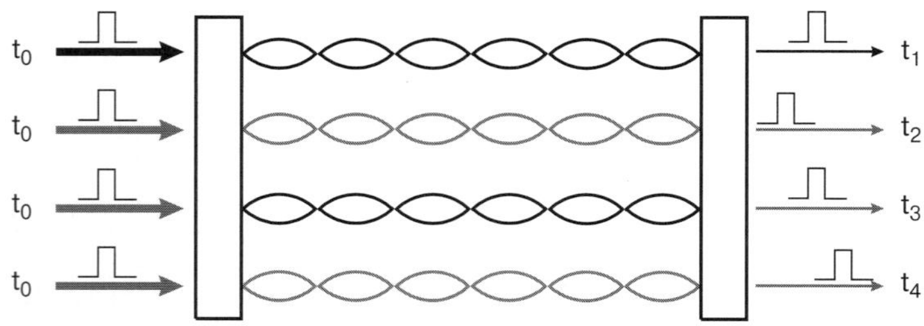

Fig. 2.15 Skew definition

- The spread of velocities of propagation of electromagnetic waves through twisted pairs. This spread is most pronounced when different materials are used for insulator covers (Section 2.1.3).
- The difference of electrical lengths of twisted pairs because of the difference in steps of twist.

The maximal difference of the transition delays among all pairs is called the delay skew and is characterized by the parameter *Skew*. This parameter is defined by

$$Skew = \max_{i,j=1,n} | t_i - t_j |,$$

where n is the number of pairs in the cable to be tested (for the overwhelming majority of cases, $n = 4$),
and t_k is the instant of arrival of pulses at the far end of twisted pairs, provided that all testing signals are fed to all pairs simultaneously.

The standardization of the *Skew* values is useful for several reasons. First, the equipment of a number of high-speed LANs (e.g., 100VG-AnyLAN, 100Base-T4, and 1000Base-T) simultaneously uses all four pairs for signal transmission. The signal is distributed among four pairs on the transmitting side and is assembled into the original form on the receiving side. Normal functioning of such a data transmission scheme is possible only if the signals transmitted through all four pairs reach the receiver simultaneously or with a relatively low spread of time. The violation of this condition may lead to a timing bug preventing the restoration of the original signal on the receiving side. For example, for 100Base-T4 networks, the value of the parameter *Skew* must not exceed 50 ns for any length of the line segment, including the maximal length.

Second, the action of portable cable scanners when measuring the length of cabling segments is based on the measurement of the time delay between the departure of a short testing signal and the arrival of the reflected signal. Having this information and knowing the velocity of propagation of electromagnetic waves along the cable, one can determine its length. If the velocities of signal transmission through different pairs differ, the scanner will give different lengths for different twisted pairs.

Experience shows that the spread of signal transition delays for twisted pairs made of the same material, in a single technological cycle, in accordance with the same technology is rather small. As a rule, it does not exceed 10 ns. In the mid-1990s, some manufacturers started producing cables in which one or two pairs had teflon insulator covers, while the insulator covers for the remaining pairs were made of other materials. (Such a structure improves the fire safety of the cable.) Although the electrical characteristics of such cables formally correspond to category 5, the value of *Skew* may reach 60 ns or more. In this case, it is impossible to give any warranty of normal operation for four-pair interfaces.

The influence of the step of twist on the value of Skew is less significant. As a rule, the spread of the values of *Skew* due to this effect does not exceed 10 ns.

The possibility of changing the parameter NVP by changing the step of twist and choosing the insulator material is utilized in some modifications of horizontal cables to minimize *Skew*. In these cables, the conductors of the pairs that give maximal delay are covered with “fast” teflon [34].

The parameter *Skew* was introduced into official standards in the very late 1990s to take into account the requirements of the equipment for LANs. This is accounted for by a rapid growth of the park of network interfaces that utilize multipair transmission circuits, in which the excessive spread of signal transmission delays may cause a timing bug. The new version of the standard ISO/IEC 11801 specifies *Skew* separately for twisted-pair cables (no more than 45 ns per 100 m) and for communication lines based on twisted pairs (43 ns for a permanent link and 50 ns for the entire channel).

Some companies guarantee lower spreads of delays for their products, as compared with the norms specified by standards. This is the case, for example, for the company Molex. The cabling system Molex Premise Networks (formerly Mod-Tap) utilizes special cables that provide a maximal value of *Skew* not exceeding 35 ns [35].

Return loss and structural return loss

Previously, we considered an idealized homogeneous line in which the signal was not reflected. A real line always has various sorts of inhomogeneities that give rise to reflections. The basic sources of these reflections are manufacturing and operational defects of cables, plug-and-socket connectors, and terminal loads with a resistance different from the impedance of the cabling system. When an electromagnetic wave encounters such inhomogeneities during its propagation along the cable, it is partially reflected and returns to the input. If there are several inhomogeneities, the wave undergoes a series of reflections. As a result, two additional parasite energy fluxes appear in the system: the backward flux formed by the sum of the reflected waves and the forward flux due to double reflections (Fig. 2.16).

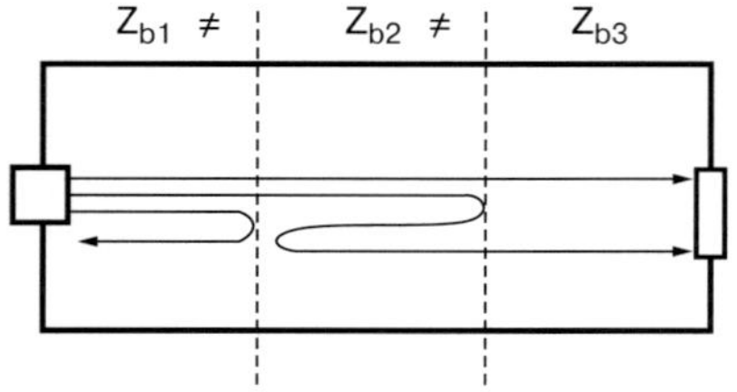

Fig. 2.16. Appearance of the backward and forward fluxes

The backward flux leads to oscillations of the impedance. This complicates matching the cabling system with the terminal equipment, which leads to an increase of working attenuation and causes distortion of the signal. The forward flux distorts the shape of the signal being transmitted and also jams the transmission. The influence of the forward flux is especially significant for the TV image. For normal transmission of TV programs, the power of the forward flux must not exceed 1% of that of the basic flux [1].

The intensity of the backward reflection of electromagnetic waves at places where the impedance is inhomogeneous is characterized by the parameter SRL (structural return loss), defined as the ratio of the power of the basic signal to the power of the backward energy flux. The larger the SRL, the lower the powers of the backward and forward energy fluxes and, hence, the higher the quality of the cable. Since this parameter is small, it is expressed in logarithmic units and specified in technical documents for cables having a length of 100 m.. The standards TIA/EIA-568-A and ISO/IEC 11801 specify virtually the same minimal admissible values of the parameter SRL for all categories of 100 and 120 Ω cables (Table 2.3). All pairs must comply with the norms for SRL.

Table 2.3. SRL for cables of different categories

Frequency, MHz	Category 3		Category 4		Category 5	
	TIA/EIA-568-A	ISO/IEC 11801	TIA/EIA-568-A	ISO/IEC 11801	TIA/EIA-568-A	ISO/IEC 11801
1–10	12	12	21	21	23	23
10–16	12–10lg(f/10)	10	21–10lg(f/10)	19	23	23
16–20	–	–	21–10lg(f/10)	18	23	23
20–100	–	–	–	–	23–10lg(f/20)	23–10lg(f/20)

Sometimes, one uses a more complex specification of the minimal value for SRL. As an example, we present the frequency dependence of the parameter SRL for cables used in the SCS Siemon Cabling System of the company Siemon (Table 2.4).

Table 2.4. Frequency dependence of the minimal structural return loss for SCS Siemon Cabling System [36]

Categories 5e and 6		Category 7	
1 < f < 10, MHz	20+5lgf, dB	100 < f < 300, MHz	21 dB
10 < f < 20, MHz	25 dB	300 < f < 600, MHz	21–10lgf /300, dB
20 < f < 100, MHz	25–7lgf, dB		

The analysis and standardization of the effect under consideration can be extended to the entire electrical channel including a cable, terminal connectors, and patch cords. In this case, one uses the term return loss and the abbreviation *RL*. The determination of *RL* is quite similar to that of SRL. Apart from structural inhomogeneities, an important role in the formation of reflected signals is played by impedance jumps at cable connections to plugs and telecommunications outlets.

The reflection factor at a point where the impedance jumps from Z_1 to Z_2 is determined by the familiar expression [25]

$$RL = -20\lg\left|\frac{Z_1 - Z_2}{Z_1 + Z_2}\right| \tag{2.12}$$

The working standards allow 100, 120, and 150 Ω cables to be utilized in SCS. By substituting pairs of the impedance values with a minimal difference (for example, 100 and 120 Ω) into relation (2.12) we obtain 11.4 dB for the nominal value of the parameter *RL*. This value is substantially less than 16 dB and does not provide parameters required for a signal transmission channel. Because of this, the working standards prohibit the use of pieces with different impedances in one channel.

Transfer impedance

Transfer impedance (Kopplungswiderstand in German) is a measure of the quality of the shield from a shielded cable. The transfer impedance is defined as the ratio of the voltage U_2 induced in the shield by the current I_1 to the value of this current per unit length of the cable (Fig. 2.17):

$$R_k = \frac{U_2}{I_1 l}, \text{ M}\Omega/\text{m}.$$

Originally, this quantity was defined for coaxial cables. In terms of the theory of interferences, it is convenient to treat a shielded balanced twisted pair as a coaxial cable, with the shield and the twisted pair playing the role of an outer conductor and an inner conductor, respectively. In such a consideration, the entire circuit is unbalanced.

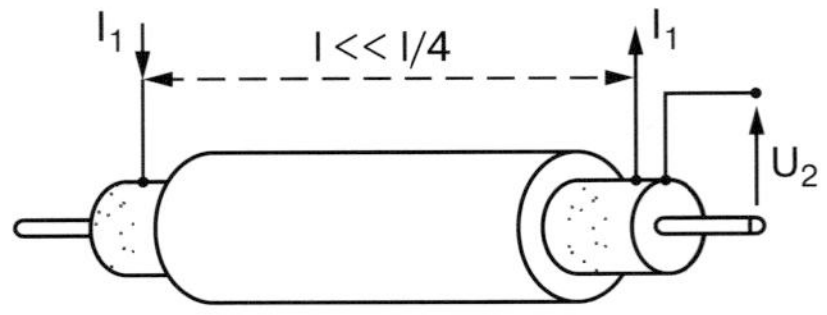

Fig. 2.17. Transfer impedance definition

The working standards specify the transfer impedance for both cables and connecting components (Table 2.5). From the definition of transfer impedance (illustrated by Fig. 2.17) it follows that the lower the transfer impedance, the higher the quality of the shield. The values of the transfer impedance achieved in practice (provided that all instructions for cabling and connection have been complied with) are usually lower than those specified by the standards. For example, for common film shields, the value of this parameter is about 30 MΩ/m, whereas for combined shields we have $R_k < 25$ MΩ/m [2].

Table 2.5. Transfer impedance for both cables and connecting components

Frequency, MHz	Transfer impedance, MΩ/m	
	Cable	Cord
1	50	100
10	100	200

Conversion loss

An ideal twisted pair is designed so that it does not create an electromagnetic field in the long-range zone. This property is due to the twisting of wires, owing to which the current flowing through different portions has the same strength but opposite directions. Separate conductors of a real twisted pair always have an imbalance relative to ground and different DC resistances. In this case the compensation is incomplete, which leads to radiation and the appearance of an electric current induced by an external electromagnetic field in the twisted pair.

An integral quantitative measure of the imbalance is the conversion loss. As is the case for many other parameters, this quantity is determined for the near and far ends of the cable (Fig. 2.18). For the near end, this parameter is denoted by *LCL* (longitudinal conversion loss) and defined as

$$LCL = 20\lg|U_1/(U_3 - U_4)|,\ \text{dB}.$$

In a similar way, this process at the far end is characterized by the parameter *LCTL* (longitudinal conversion transfer loss) defined as

$$LCTL = 20\lg|U_1/(U_3 - U_4)|,\ \text{dB}.$$

The necessity of using different parameters to characterize the conversion loss at the near and far ends of the line is justified by the presence of a significant difference in mechanisms used for the formation of the imbalance in the near-range and far-range zones of the cable. These zones are distinguished by the ratio of the length of the zone (measured along the cable) to the length of the high-frequency electrical signal transmitted through this cable.

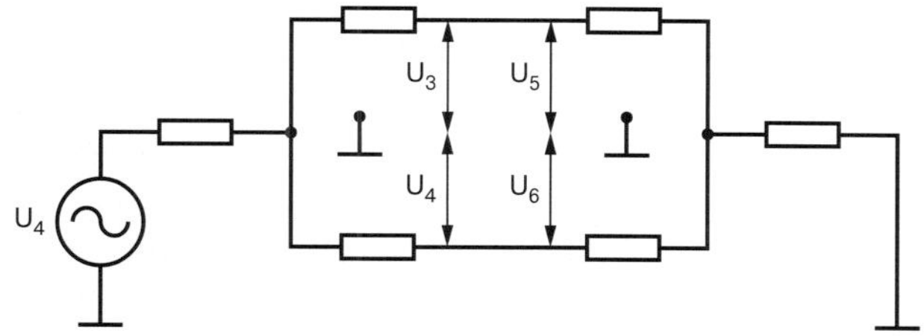

Fig. 2.18. LCL and LCTL definition

In practice, an important property is the dependence of the parameters characterizing the conversion loss on the frequency. Since the processes underlying this phenomenon have a statistical nature, the exact determination of this dependence is impossible. Nevertheless, experience allows us to claim that as the frequency increases, the rates of the decrease in the quantities *LCL* and *LCTL* reach values of 10 dB/decade and 20 dB/decade, respectively.

The standards that were in place by the middle of 2000, do not provide a detailed specification for the parameters characterizing the conversion loss. For example, the standard ISO/IEC 11801 specifies the values of *LCL* and *LCTL* for frequencies not exceeding 10 MHz (Table 2.6).

Table 2.6. LCL and LCTL in accordance with the ISO/IEC 11801 (2000 version) for lines of various classes

Frequency, MHz	Class A	Class B	Class C	Class D
0.1	30	45	45	45
1.0	–	20	30	40
10	–	–	25	30

Interference from external sources of electromagnetic radiation

An imperfect balance between twisted pairs can cause interference induced by external sources of electromagnetic radiation. There are two basic types of such interference.

- Electromagnetic interference (EMI): Low frequencies and high amplitudes are characteristic of interferences of this kind. The main sources of these interferences are electric motors, starters of fluorescent lamps, and power supply cables.
- Radio frequency interference (RFI): This kind of interference is characterized by high frequencies and low amplitudes. The main sources of these interferences are radiotelephones, radio and TV broadcast transmitters, and power sources with high-frequency transformation.

Twisted pairs are fairly immune to interferences of this sort. For this reason, the standards do not specify norms for the level of interference induced by external electromagnetic radiation.

2.2 Transmission of Signals through Fibers

In fiber-optic cabling systems, signals are transferred by a carrier wave of the optical range (usually lower infrared) passing through a fiber (fiber waveguide) made of super-pure quartz glass. Figure 2.19 shows a simplified structural diagram of a fiber-optic communication line. The electrical signal comes to the input of the optical transmitter and modulates the intensity of the output signal of the transmitter. The optical signal propagates in the fiber and comes to the input of the optical receiver. The receiver performs demodulation of the optical signal and restores the original electrical signal. For normal operation, one equips the transmitter and the receiver with sockets for optical connectors. The control circuit of the optical transmitter, the light diode, the light current amplifier, and the matching circuit of the optical receiver are usually integrated in the same housing.

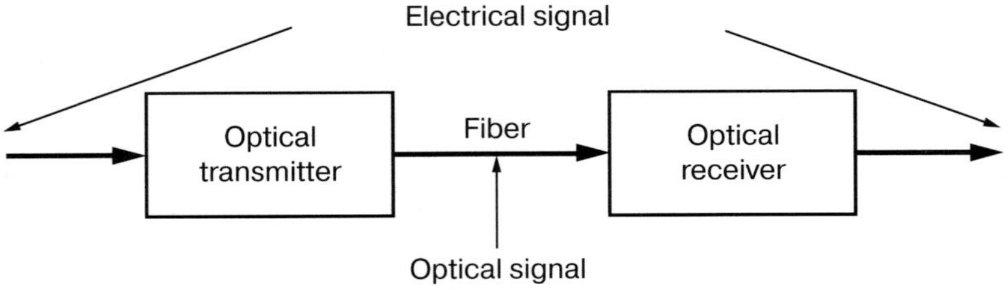

Fig. 2.19. Simplified structural diagram of a fiber-optic communication line

2.2.1 Propagation of Radiation in a Fiber

A fiber is a thin cylindrical fiber glass cable for transmitting electromagnetic waves of the optical range (usually long-wavelength infrared) corresponding to the frequency range 10^{14}–10^{15} Hz.

The action of the fiber waveguide is based on the well-known principles governing the processes of reflection and refraction of optical waves at the interface between two media with different optical properties. The optical properties of a material depend on the relative index of refraction *n*. The medium with a higher refraction index is regarded as having higher optical density.

When the ray is incident on the interface between two media at an arbitrary angle, the refracted and reflected rays appear as shown in Fig. 2.20a (the general case). The angles of incidence (φ_i), reflection (φ_{refl}), and refraction (φ_{refr}) are related by Snell's law

$$\varphi_i = \varphi_{refl};\ n_1 \times \sin\varphi_i = n_2 \times \sin\varphi_{refr}.$$

If the ray goes out of the medium with the higher optical density into the medium with the lower optical density ($n_1 > n_2$), then we have $\varphi_{refr} > \varphi_i$. By increasing the angle of incidence, one can achieve the state in which the refracted ray lies on the interface but does not enter the medium with higher optical density (Fig. 2.20b). The angle of incidence at which this effect is observed is referred to as the *critical angle of total internal reflection* (φ_{cr}). For all angles of incidence exceeding the critical angle, we will have only reflection, and the refracted ray will be absent (Fig. 2.20c). This phenomenon is called total internal reflection; it underlies the transmission of optical signals in a fiber.

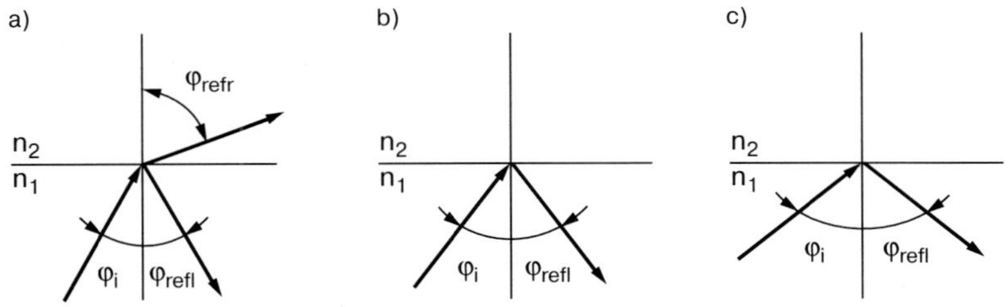

Fig. 2.20 Ray incidence on the interface between two media: *a)* when $\varphi_I<\varphi_{cr}$; *b)* when $\varphi_I=\varphi_{cr}$; *c)* when $\varphi_I>\varphi_{cr}$;

Fibers usually have a circular cross-section and consist of two basic components – a *core* and a *cladding* surrounding it. Both the core and the cladding are made of glass, but the glass of the core has a higher optical density than the glass of the cladding. The diameters of the core and the cladding are conventionally measured in micrometers and are presented in the list of operating characteristics of the fiber in an implicit form with the format diameter of the core/diameter of the cladding. For example, the combination 62.5/125 refers to a fiber having a core with a diameter of 62.5 μm and a cladding with a diameter of 125 μm.

At the interface between the core and the cladding, optical rays propagating along the waveguide undergo reflection. Thus, the core serves for the transfer of the electromagnetic energy, whereas the cladding creates conditions for the reflection on the interface between the two media and prevents the energy from radiating into the environment.

The radiation of an external source, when incident on the input end face of the waveguide, excites in it several types of waves referred to as *modes*. The modes are divided into *launched modes*, *leaking modes*, and *radiating modes*.

The launched modes propagate along the fiber core and provide the signal transmission (rays 1 and 2 in Fig. 2.21). The launched modes are considered the basic type of electromagnetic waves. These are excited by the rays incident on the end face at an angle not exceeding the critical angle Θ_A, called the *aperture angle*. For basic types of modern fibers, the aperture angles range from 11.5° to 17°.

Rays incident on the end face of the fiber at angles exceeding Θ_A reach the core–cladding interface and, losing part of their energy, are subject to significant attenuation (ray 3 in Fig. 2.21). These are the leaking modes.

Finally, part of the rays incident at angles substantially exceeding Θ_A reach the outer surface of the cladding and are radiated into the environment (ray 4 in Fig. 2.21). These are the radiating modes. The radiating modes also appear at irregularities (defects) of fibers. The leaking and radiating modes cause energy loss and distortion of the information being transmitted.

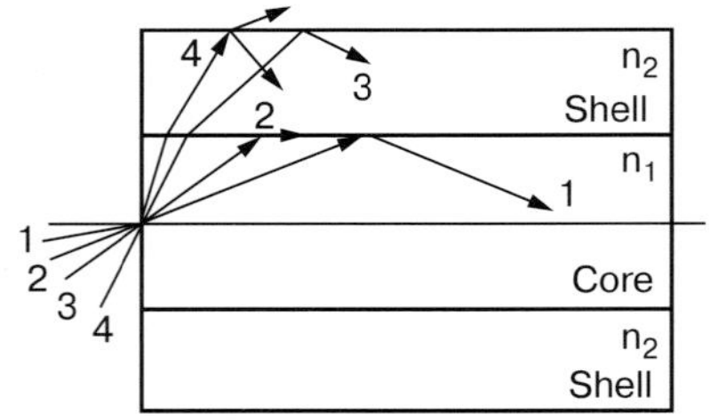

Fig. 2.20. Rays propagation in a fiber

2.2.2 Types of Fibers

The claddings of the most fibers that are in use have a diameter of 125 μm, whereas the diameters of the core may change within a fairly large range. Depending on the diameter of the core, the optical fibers are divided into single-mode and multimode fibers. The multimode fibers can be of step-index or graded-index type. The diameter of the core in multimode fibers is much larger than the length of the optical carrier wave, and the condition of total internal reflection is satisfied for several types of waves (modes), the number of which in standard fibers usually makes up 1000–2000.

The refraction index of the cladding, as a rule, is constant, whereas that of the core can either be constant or change along the radius of the core in accordance with a certain law, referred to as the *profile of the refraction index*.

The simplest type of fiber is the step-index fiber. In fibers of this type, the refraction index of the core is constant along its radius. In multimode step-index fibers, the trajectories of the rays of separate modes have the shape of a zigzag (Fig. 2.22a).

The profile of the refraction index substantially influences the frequency characteristics of multimode fibers. For this reason, profiles different from stepped ones are frequently used in practice. For example, in graded-index fibers, the refraction index gradually decreases when going away from the center in accordance to a law resembling a quadratic parabola (Fig. 2.22b). This profile is created by appropriately distributing alloying elements, predominantly germanium. In such fibers, the trajectories of most of the rays are smooth curves. In the U.S.A., cores with a diameter of 62.5 μm are mostly used, whereas in Europe and in Russia, fibers with cores having a diameter of 59 μm are frequently used.

In single-mode fibers, the diameter of the core (ranging from 7 to 10 μm) coincides in order of magnitude with the wavelength. This provides the existence of only one launched mode in such fibers (Fig. 2.22c).

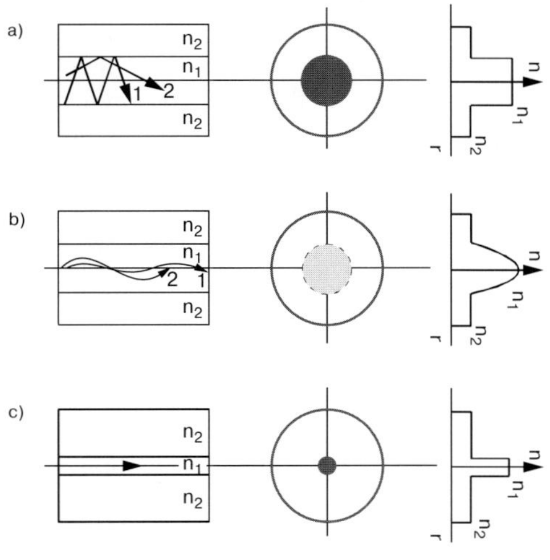

Fig. 2.22. Rays propagation in different fiber types: *a)* step-index fiber; *b)* graded-index fiber; *c)* single-mode fiber

2.2.3 Dispersion of Electromagnetic Radiation

By dispersion one understands an increase in the duration of an optical pulse when passing through a fiber of a certain length (as a rule, 1 km).

Dispersion appears due to the diffusion of spectral and modal components of optical signals in time and determines the fiber bandwidth. This phenomenon increases the duration of the pulse in the process of its propagation through a fiber-optic cable (Fig. 2.23), causes the intersymbol interference (overlap of the fronts of two adjacent pulses), and eventually increases the probability of an error occurring in the signal to be received. One distinguishes between two kinds of dispersion: chromatic and intermodal (Fig. 2.24).

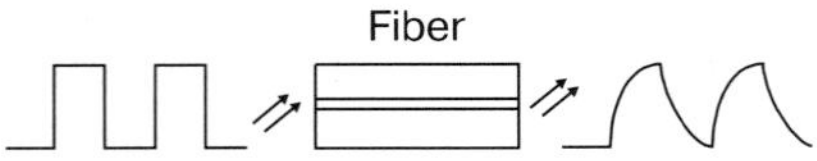

Fig. 2.23. Optical signal propagation through a fiber-optic cable

Chromatic dispersion is accounted for by the dependence of the conditions of light propagation in a fiber on the wavelength. Chromatic dispersion appears because the optical carrier of the transmitter output signal has a finite spectral length. Chromatic dispersion has two components referred to as *material dispersion* and *waveguide dispersion*. Material dispersion is accounted for by the dependence of the refraction index of the core and the cladding on the wavelength. Waveguide dispersion is accounted for by the dependence of the conditions of propagation of a mode on its wavelength. The degree to which chromatic dispersion influences the frequency characteristics of the channel depends primarily on the spectral properties of the source of radiation. For laser sources that have narrow frequency bands of radiating carriers, chromatic dispersion is less pronounced. In noncoherent sources (e.g., light diodes), the carrier bandwidth is substantially broader and, hence, the chromatic dispersion is more pronounced.

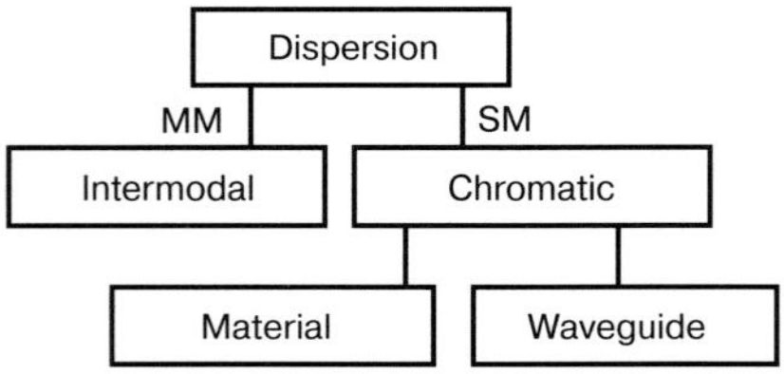

Fig. 2.24. Two kinds of dispersion

Intermodal dispersion appears only in multimode fibers and is accounted for by the fact that different modes have different times of propagation. This difference is due to the scatter in reflection angles and, hence, in the lengths of the channels traveled by separate modes in the core of the fiber. This phenomenon is illustrated in Fig. 2.22a–b. Graded-index fibers tend to keep modes near the center of the core. The farther the rays are away from the center, the longer the path traveled by them. However, the optical density of the material of the core away from the center is lower and, hence, the velocity of propagation of light is higher. Apart from that, the number of modes in the graded-index fiber is less, as compared with the fiber with the stepped profile of the refraction index. The combination of these three factors provides the graded-index fiber with better frequency properties and a considerably larger bandwidth, as compared with the step-index fiber. (The bandwidth of the graded-index fiber exceeds that of the step-index fiber by a factor of more than 10.) At the same time, the graded-index fibers virtually do not differ from the step-index fibers in a number of operating characteristics such as specific loss, radiation input efficiency, and splicing complexity. Because of these advantages, graded-index fibers have virtually displaced step-index fibers in most areas. For this reason, in what follows, with multimode fibers we will imply only fibers with a graded profile. The total pulse-width broadening due to the intermodal, τ_{mod}, material, τ_{mat}, and waveguide, τ_{wg}, dispersions is determined by [37]:

$$\tau = (\tau_{mod}^2 + (\tau_{mat} + \tau_{wg})^2)^{1/2}.$$

With reference to the realistic relationship between the dispersion components, one can take $\tau = \tau_{mod}$, for multimode fibers and $\tau = \tau_{mat}+\tau_{wg}$ for single-mode fibers.

For single-mode fibers, the parameter characterizing dispersion takes into account the dependence of the dispersion on the spectral properties of the radiation source and, therefore, is measured in ps/nm/km. The dispersions of modern fibers range from 3 to 15 or even 18 ps/nm/km. The waveguide, τ_{wg}, and material, τ_{mat}, components of the chromatic dispersion in the wavelength range from 1200 to 1600 nm and, as a rule, have different signs. Since the waveguide dispersion depends on the profile of the refraction index, one can vary this parameter to obtain zero or near-zero dispersion at a prescribed wavelength or spectral band for single-mode fibers. For this reason, for single-mode fibers, one sometimes indicates not the absolute value of the dispersion but the zero dispersion wavelength and the rate of the spectral characteristic of the dispersion in the vicinity of this wavelength. (For standard fibers, these parameters make up 1300–1310 nm and 3.5 ps/nm/km, respectively.)

In multimode fibers, intermodal dispersion dominates, which accounts for a weak dependence of frequency characteristics of such fibers on the spectral properties of the radiation source. Therefore, to evaluate these characteristics, it is convenient to use the characteristic bandwidth. This quantity has the dimension MHz/km and characterizes dispersion in the frequency domain. Typical values of the characteristic bandwidths of modern fibers range from 200 to 500 MHz/km, depending on the wavelength.

The physical meaning of this coefficient is the bandwidth of the optical cable having a length of 1 km. Unlike electrical transmission channels, in optical channels, the proper noise level at the receiver's input turns out to be fairly high. This is due to so-called quantum noise. For this reason, in optical communication engineering, one uses the classical definition identifying the upper limiting frequency for the cable bandwidth with the frequency at which the transfer ratio is equal to 0.707 of the value at low frequencies.

2.2.4 Decay of Signals in Fibers

Units of measurement of optical power

The power of the signal being transmitted through a fiber-optic cable depends exponentially on the distance traveled. For that reason, when calculating parameters of cabling structures, it is convenient to use logarithmic units. These units enable one to reduce basic operations for calculating signal powers to addition and subtraction.

Power of 1 mW is taken as a reference (zero) level in optical communication engineering. (This value is approximately equal to the maximal power of radiation of a typical semiconductor laser or a light diode.) The power level, p, is defined as follows:

$$p = 10\lg(P/1\ \text{mW}),\ \text{dB},$$

where P is the power of the optical signal.

Table 2.7 shows the correspondence between the levels and the powers of optical signals in the range typical of fiber optics.

Table 2.7. Correspondence between the level and power of optical signals

Level, dB	Power	Level, dB	Power
0	1 mW	–30	1 μW
–10	100 μW	–33	500 nW
–13	50 μW	–40	100 nW
–20	10 μW	–50	10 nW
–23	5 μW	–60	1 nW

Mechanisms of loss

An optical signal gradually loses its energy when propagating in the fiber. This effect is called decay. The intensity of the decay influences the maximal communication range between two transceivers. In fiber-optic engineering, decay is measured in decibels.

The signal decay in fibers is due to *scattering loss* and *absorption loss*. Scattering loss is accounted for by fluctuations of the refraction index of the material of the cable core induced by fluctuations of the average density of this material. When optical rays are incident on these irregularities, they change propagation directions and radiate into the environment. This phenomenon is called scattering or Rayleigh scattering. The measure of the scattering loss is defined by the relation [13]

$$\alpha_S = K_S/\lambda^4,\ \text{dB/km},$$

where K_S is the scattering coefficient and λ is the wavelength [μm]. For quartz we have $K_S = 0.8\ \mu\text{m}^4\times\text{dB/km}$.

The absorption loss is due to the infrared absorption; it manifests itself at $\lambda > 1.7$ μm. The quantitative measure of this loss is given by [13]:

$$\alpha_{ir} = C \times e^{-k/\lambda}, \text{ dB/km}, \qquad ()$$

where k and C are constant coefficients. For quartz, we have $k = 0.8 \times 10^{-6}$ m and C = 0.9.)

The losses due to Rayleigh scattering (curve 1 in Fig. 2.25) and infrared absorption (curve 2 in Fig. 2.25) determine the lower limit of loss for a fiber. As is apparent from the figure, the lower theoretical limit of loss in the quartz fiber lies near a wavelength of 1.4 μm. This minimal level of loss is never achieved in practice, because of the following two facts:

- The process of manufacturing optical cables leads to the so-called cable loss, which is caused by deformation of the fiber when applying covers and protective sheaths, as well as twisting of elements of the cable core.
- Due to imperfections in the purification process, the material of the fiber core contains atoms and molecules of foreign substances (impurities). The resonance excitation of these atoms and molecules can cause a sharp increase in the loss at certain wavelengths. In a range of 0.8 to 1.7 μm, the loss is mainly due to water vapor and metals of the first group of the periodic table (e.g., nickel, iron, and chromium). For the loss level to be acceptable, the impurity concentration must not exceed 10^{-9}.

The field of effective utilization of quartz fibers is confined to the visible light and long-wavelength infrared ranges. For medium- and short-wavelength infrared, one should use new materials.

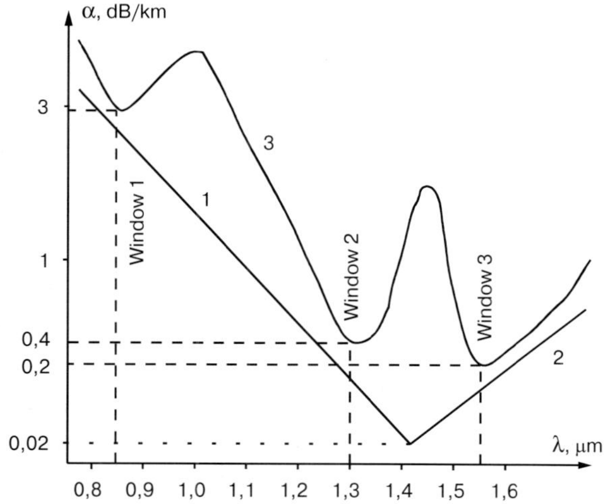

Fig. 2.25. Decay characteristics of the fiber

Curve 3 in Fig. 2.25 shows a typical spectral characteristic of the decay coefficient for real fibers, taking into account both the basic loss and the additional loss induced by impurities. It is apparent from this curve that the operation of fiber-optic cables is efficient not for all wavelengths but only for certain ranges corresponding to minimal loss. The ranges of minimal loss are called *spectral windows*. For quartz fibers, there are three spectral windows that are of interest for practice. It is convenient to define the boundaries of these spectral windows in accordance with the standard ISO/IEC 11801 (Table 2.8). The characteristics of semiconductor radiators and photodetectors are optimized for operation in these windows.

Table 2.8. Typical decay values of optical signals in spectral windows

Spectral window	Wavelength, μm			Typical attenuation, dB/km
	Minimal	Central	Maximal	
1	790	850	910	2–3
2 (MM)	1285	1300	1330	0.7–1.5
2 (SM)	1288	1310	1339	0.4–1.0
3	1525	1550	1575	0.2–0.4

It is apparent from Fig. 2.25 and Table 2.8 that the passage from the first spectral window to the second one provides a considerable gain in the loss, whereas operation in the third window does not give a substantial advantage. On the other hand, the cost of active optic electronic components rapidly increases with an increase in the working wavelength. In fiber-optic communication lines of LANs, serviced mainly by fiber-optic channels of SCS backbone cabling subsystems, the cost of the terminal equipment is relatively high, since the total length of the cable channels is relatively small. For this reason, in SCS engineering, one mainly uses the first and the second spectral windows. The standardization of parameters of single-mode fibers for campus backbone cabling systems is performed with the assumption that single-mode fiber-optic cables designed for city and long-distance communication lines will be used in SCS. The long-distance lines, the cost of which is determined primarily by the length of the regeneration path, operate mainly in the second and the third spectral windows. (These windows, apart from low decay, are also characterized by low dispersion.)

2.3 Transmission of Digital Information along Electrical and Fiber-optic Channels

SCS serve as data transmission carriers for twisted pair network equipment. When choosing the principles and the circuit design of this equipment, the designers always take into account specific features of fiber-optic cables and twisted pairs used as waveguides. This makes it possible to substantially improve the equipment regarding communication range, bandwidth, and immunity to interferences.

2.3.1 Linear Codes of the Network Equipment

Code requirements

Consider an arbitrary digital chain formed by a sequence of separate bits. Each bit is represented in a network equipment by a pulse of duration $C_i/2$, where C_i is the duration of the clock interval. In accordance with the principles of digital representation of signals, the amplitude of this pulse can assume only one of the two fixed values. For example, a positive voltage of the signal can encode the logical zero, and the absence of a signal the logical unity. Such elementary pulses are transmitted along a communication channel consecutively with the clock frequency $f_c = 1/C_i$. The simplest way to deliver this sequence from the transmitter to the receiver involves its direct transmission into the line without any preliminary transformations. This code is referred to as non-return-to-zero code (or NRZ code). The term non-return implies that the signal level does not change during the entire clock interval.

For a number of reasons, which will be discussed later, the utilization of NRZ codes for the transmission of information along SCS channels is not optimal, and a coder is introduced into the transmitter circuit. The coder performs a preliminary transformation of the signal by adding new pulses to the original data sequence and changing polarity and duration of the original pulses. This makes the linear signal substantially better adapted to the transmission along a communication line. We assume that the coder does not change the shape of the pulse and, hence, the pulse remains rectangular. Thus, the code in the line may substantially differ from the code in the original digital sequence. At the receiving terminal, there is a decoder that performs the inverse transformation and reconstructs the original message entered into the receiver's input.

Linear codes are subjected to a number of requirements. The most important requirements can be formulated as follows:

- The content of low-frequency and high-frequency components in the linear signal should be a minimum. In addition, it is desirable that the major portion of the energy spectrum be concentrated in a fairly narrow frequency range; this reduces the signal distortion due to a limited bandwidth of the linear channel.
- Statistical characteristics of the signal must provide a stable operation of receiving apparatuses, as well as the possibility of bit error control in regenerators.
- The code must not constrain the message to be transmitted and must provide the possibility of transmitting any sequence of zeros and units in the original message.
- The structure of linear signal coders and decoders, as well as that of the timing extraction circuit should be as simple as possible.

The relation between the clock frequency of the digital signal and the bandwidth of the communication channel

Most equipment operating through SCS electrical channels uses digital signals for data exchange. It is traditional in cabling engineering that the standards for basic parameters of the transmission channel of SCS electrical subsystems are defined in the frequency domain. For example, the values of the attenuation and crosstalk attenuation are specified for a certain frequency. Fiber-optic communication lines were initially created primarily for transmitting digital signals. For that reason, a large number of parameters of such lines are specified in the time domain. For example, the bandwidth of single-mode systems is characterized by the dispersion parameter (Section 2.2.3). Nevertheless, standardization in the frequency domain is also widely used for fiber-optic systems. For example, the bandwidth of multimode equipment is expressed in terms of the characteristic bandwidth.

To facilitate design calculations and assessment of possibilities for utilizing SCS channels by various network equipment, one needs an effective procedure that would establish a relation between the data exchange rate, the clock frequency of the linear signal, and the required bandwidth of the channel.

Let the communication line transmit a digital signal $f(t)$ formed by a regular sequence of rectangular pulses having a unit amplitude and on–off time ratio $Q = 2$ (Fig. 2.26). This signal is a periodic function of time and, hence, can be expanded into a Fourier series. In the case in question, this series has the form

$$f(t) = 1/2 + \sum_{k=1}^{\infty} a_k \cos 2k\pi f_c t,$$

where $a_k = \sin(\pi k/2)/(\pi k/2)$

From this relation it follows that to restore the signal at the receiving end, it suffices to use the first (fundamental) harmonic of the original signal sequence. This implies an important conclusion that the speed of data transmission along one twisted pair in one direction may exceed the clock frequency by a factor of 2. This observation, not obvious at first, follows from the fact that one half of the clock interval contains two steps (or samples, using the terminology of the Niquist theorem) of the digital signal sufficing the complete restoration of its form. In this case, the other half of the clock interval is used as a protection interval and serves for separating different pulses. Hence, in theory, to provide the restoration of the signal at the receiving end, the upper limit frequency of the communication channel must be no less than half of the modulation rate. This observation is illustrated in Fig. 2.26 by showing the signal waveform for a sequence of rectangular pulses with on–off time ratio Q=2 for various bandwidths of the first order low-frequency filter that models a communication line with a finite bandwidth. It is apparent that the normal operation of the decoder becomes impossible, in principle, if the upper limit frequency of the filter is close to $f_u = f_c/2$. If f_u falls below f_c, one observes, apart from the cut of the leading and tailing fronts of the pulses, a decrease in their amplitudes. This indicates that the time constant of the filter is so large that the output signal cannot reach its maximal value while the input signal is in the high-level state. This fact decreases the signal-to-noise ratio and leads to the deterioration of the immunity to interferences. To prevent this unwanted effect in practice, one broadens the bandwidth of channels of electrical and optical communication systems that use twisted pairs by a factor of up to 2, as compared with the theoretical minimum.

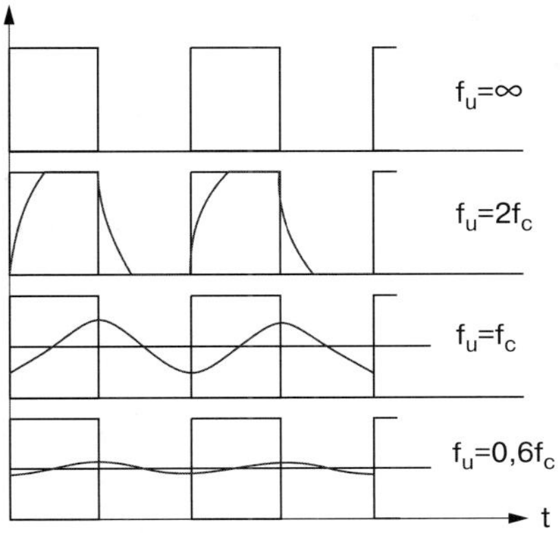

Fig. 2.26. Signal waveform for various bandwidths on the low-frequency filter output

Clock synchronization modes in digital network equipment

One of the problems arising in the transmission of signals through digital communication lines is that of providing synchronous bit-to-bit operation of the transmitter and the receiver (the problem of bit synchronization). Clock synchronization is necessary to enable the receiver to determine properly the time instant for strobing the input signal and decoding the data received. Depending on the realization principle, one distinguishes between the *asynchronous* and *synchronous* clock synchronization. These types of synchronization can be implemented in various manners. The general characteristics of these methods are systematized in Table 2.9.

In asynchronous systems, the clock generators of the receiver and the transmitter have approximately the same frequencies but operate fully independently of one another. The synchronization in this case is provided by additional pulses. Consider one possible algorithm for implementing the asynchronous clock synchronization. (This algorithm is used most frequently in practice.) Immediately before the start of the data transmission, a single starting pulse is fed to the communication channel. Having detected this pulse, the receiver starts the receiving and decoding process under the control of its intrinsic clock generator. Since the frequencies of the master clocks of the receiver and the transmitter are different, this procedure is implemented for a relatively short time interval during which a certain number of digital sequences are processed. The pulse sequence being transmitted to the communication channel always ends with the stop pulse. Having detected this pulse, the receiver stops the decoding process. The next sequence of data can be transmitted only after a pause (protection interval) and also must have the start and stop signals. This algorithm is used in the interface RS-232 operating in the asynchronous mode. Previously, such a method was widely utilized in low-speed telegraphy, where it is known as the start–stop mode.

Table 2.9. General characteristics of clock synchronization modes

Method	Start–stop	Through a separate channel	Predetermined synchrosequences	Self-synchronizing
Type	Asynchronous	Synchronous	Synchronous	Synchronous
Efficiency of the usage of the communication channel bandwidth	Low	Low (a separate channel is needed)	High	Medium
Complexity of circuitry	Low	Low	High	Medium – low[a]
Synchronization rate	High	High	Medium	High – medium[a]
Interference immunity	High	High	Medium	High

[a]Depends on the code used

Synchronous clock synchronization modes comprise

- synchronization along a dedicated channel;
- predetermined syncrosequences;
- self-synchronization.

Synchronization along a dedicated channel involves a single clock generator common to the transmitter and the receiver, the synchrosignals of this generator being transmitted through a separate parallel channel. This mode is widely used in communication lines having a relatively short length, for example, in various modifications of the synchronous interface RS-232. However, for most of the high-speed communication systems operating through SCS channels, the use of adjacent twisted pairs or optical fibers is economically unreasonable. Moreover, for long-distance lines, one can encounter additional technical difficulties associated with the difference in the times needed to travel the cabling channel with the basic and synchronizing signals (delay skew).

Predetermined synchrosequences (or synchrocombinations) are periodically inserted into the signal at the transmitting end during the intervals between transmissions of data sequences. The receiver's clock frequency unit is tuned to the reception of the synchrosequence and utilizes the results of processing this sequence to adjust its own clock generator. This mode is extremely effective as regards the utilization of the communication channel bandwidth but requires rather complex circuits for extracting synchrocombinations from the continuous data flow and processing. In addition, it does not provide the high rate of synchronization that is important for most of the equipment operating in LANs. This mode is widely used in digital telephone systems and in other sorts of equipment for public networks.

The *self-synchronization mode* uses two different clock generators in the transmitter and the receiver but does not use separate signals for their synchronization. The basic idea of this mode is to provide the linear signal with the form that would enable the receiver to continuously tune its clock frequency to the signal frequency, thereby compensating for the joint drift of synchrosignal frequencies. To that end, the linear code is formed so that the linear signal would have rather frequent regular switches between the states with logical 0 and logical 1. These switches are the control action that adjusts the clock frequency of the receiver. The codes based on the outlined principle are referred to as self-synchronizing codes; they are widely utilized in LANs. Unlike the start–stop mode, this synchronization mode does not require a protection interval between separate data sequences, which allows one to increase the efficiency of using the channel bandwidth. As compared with the predetermined synchrosequences mode the self-synchronization mode provides a high rate of synchronization between the transmitter and the receiver.

2.3.2 Codes for Low-rate Electrical Systems

Figure 2.27 gives examples of some types of linear signals corresponding to different codes applied to the digital message that has a length of one bit and the binary representation 11001010.

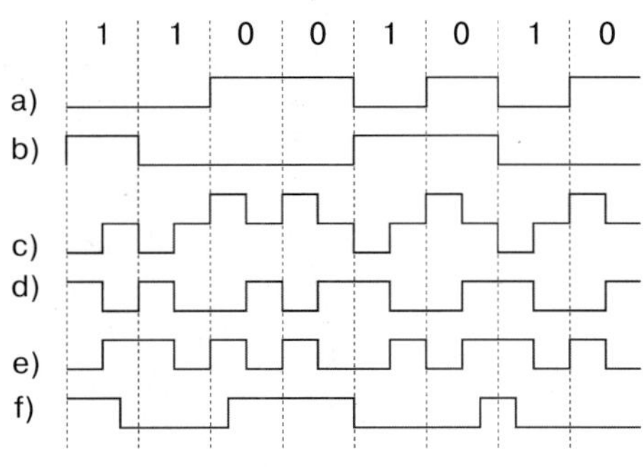

Fig. 2.27. Digital codes: *a)* NRZ code; *b)* NRZ-I code; *c)* RZ code; *d)* Manchester code; *e)* Differential Manchester code; *f)* 4B5B+NRZ-I code

Non-return-to-zero codes

Non-return-to-zero code (Fig. 2.27a). NRZ code provides the simplest implementation for the modulation of signal levels. It is based on the consecutive transmission (without any preliminary transformations) of separate bits of the original message to the communication channel. The basic advantage of this code is that it is simple to implement and that it does not require a large bandwidth of the communication channel. The major disadvantage of the NRZ code is the complexity of the synchronization. This complexity results from the fact that the adjustment of the clock frequency of the receiver is impossible while transmitting long chains of zeroes and units.

NRZ codes are commonly used in digital chips for network equipment. To synchronize these units, one connects the network interface to the common synchronization bus. The NRZ code is commonly used in communication channels in which organizing separate synchronization lines causes no difficulties, for example, when implementing the interface V.24 (RS-232).

Note that the NRZ code has a constant component and, hence, the communication line transmitting this code must provide a gate for zero-frequency signals. To implement the balanced transmission through twisted pairs the network transceivers must have decoupling matching transformers through which the constant component cannot pass. To remove this drawback one utilizes in balanced circuits a modification of the NRZ code called the polar non-return-to-zero (PNRZ) code. In this code, the logical 1 is encoded by a negative voltage and the logical 0 by a positive voltage. However, since the transmission of the constant component is impossible, the PNRZ signal containing a long sequence of zeros and units tends to zero and, hence, will be strongly distorted during the transmission.

NRZ-1 (non-return-to-zero, invert on ones) code. The NRZ-1 code provides the simplest implementation of the variable-strength (differential) coding principle (Fig. 2.27b). The signal does not change during the entire clock interval, the logical 0 and 1 being encoded by the absence or presence of a change in the signal strength at the beginning of the interval, respectively. Differential codes provide certain advantages when organizing the communication channel in the environment subject to interferences. In this case, the determination of the change in the signal strength is more reliable than the comparison of the absolute value of this strength with a threshold value performed by a comparator.

From the above algorithm for forming a linear code sequence it follows that the NRZ code is especially efficient in the case where the probability P of one logical symbol to appear in this sequence is much larger than that of the other symbol, for example, $P(1) >> P(0)$. In this case, a timing bug is possible only for long sequences of zeros. To avoid this drawback, one utilizes scrambling based on interrupting such sequences by an extra logical 1. Both hardware and software implementations of the scrambler are used in practice. The software scrambler implementation is usually entrusted to a higher-level protocol, for example to the coding protocol SDLC/HDLC with the NRZ-1 code. The hardware scrambler is utilized in the networks CDD (TP-PMD) and 100Base-TX. We will consider this type of scrambler later.

Return-to-zero codes

To remove the drawbacks of the NRZ and PNRZ codes, one utilizes more sophisticated coding systems that involve adding new pulses to the original sequence or the inversion of separate elements of this sequence in accordance with an appropriate algorithm. The combination of these modes permits one to match the properties of the signal with those of the communication channel in an optimal way (with respect to a certain criterion), thereby increasing the interference immunity, the efficiency of utilization of the channel bandwidth, and, as a result, the operating speed of the network equipment. The positive effect is achieved primarily due to a redundancy introduced into the linear signal. This redundancy manifests itself by an increase of the clock frequency of the linear signal, f_l, as compared with the data exchange frequency, f_l, i.e., $f_l > f_d$.

Return-to-zero (RZ) code. The RZ code uses pulses of different polarities to represent the logical units and zeros and returns to the medium level at the middle point of the clock interval (Fig. 2.27c). This code guarantees a change in the signal strength for each clock interval. For this reason, it refers to the class of self-synchronizing codes. Another advantage of this code is that it does not contain a constant component and, hence, can be used in systems with balanced transmission. However, the signal changes its level twice during the transmission of one bit, which leads to the doubling of the clock frequency and, hence, to the doubling of channel frequency range, in contrast to the NRZ code. RZ codes require three-level signals to be formed and processed, which somewhat complicates the electrical circuits of the transceivers and makes it practically impossible to use this code in optical communication engineering.

Manchester code. This self-synchronizing code does not contain a constant component, forms only two-level signals, and defines the value of the bit by the direction of the change of the signal strength at the middle point of the clock interval (Fig. 2.27d). The logical 0 is encoded by the change of the strength from the lower level to the higher one, whereas the logical 1 corresponds to the change from the higher level to the lower one. Thus, each clock interval T_c has a pulse of duration $T_c/2$, this pulse belonging to the first half of this interval when transmitting 0 and to the second half when transmitting 1. When switching from 0 to 1, the signal does not change its strength. The absence of the change at the middle point of the clock interval is utilized in some receivers as a trigger signal for the decoder. Thus, this type of code provides a signal jump at the middle point of the clock interval (which substantially simplifies the synchronization circuits of the receivers) and does not have a constant component. However, it also requires doubling of the clock frequency, in contrast to NRZ codes.

The receivers and transmitters of signals using the Manchester coding system have simple circuits and, because of this, are widely used in practice. For example, Manchester codes are utilized by all interfaces for Ethernet networks with a data transmission rate of 100 Mbit/s (10Base-5, 10Base-2, 10Base-T, and FOIRL). These interfaces have a clock frequency of 20 MHz. As has been shown previously, to restore the signal at the receiving end, the upper limit frequency of the communication channel must be no less than half of the clock frequency of the digital linear signal. The category 3 communication channels utilized by the network Ethernet 10Base-T have an upper limit frequency of 16 MHz, which exceeds the theoretical minimum by a factor of 1.6. All components with frequencies exceeding 16 MHz are treated as noise and are suppressed by the low-frequency filter at the receiver input.

Differential Manchester code, as follows from its name, is a modification of the Manchester code. It is illustrated in Fig. 2.27e. In this code, the middle point of the clock interval is used only for synchronization, and the signal strength always changes at this point. The logical 0 and 1 are transmitted by the presence or absence of a change in signal strength at the beginning of the cock interval, respectively. The advantages and drawbacks of this code coincide with those of the Manchester code. The differential Manchester code is utilized, for example, in 4-Mbit/s and 16-Mbit/s Token Ring networks. The 16 Mbit/s modification of this network operates through communication channels based on twisted pairs of category 4. These cables have an upper limit frequency of 20 MHz that exceeds the theoretical minimum only by a factor of 1.25. The small margin requires fairly complex circuits for data processing, which is not the least factor among those determining the high cost of the network equipment for this standard.

2.3.3 Specific Features of Twisted Pairs in High-speed Network Equipment

By high-speed network equipment we understand here interfaces for 100-Mbit/s and 1000-Mbit/s LANs. As compared with low-speed equipment, these interfaces almost entirely utilize the following engineering solutions:

- parallel transmission of the data flow though several twisted pairs simultaneously;
- error correction codes;
- multilevel and block coding.

These principles allow increasing the channel bandwidth by at least an order of magnitude, combined with a moderate (by a factor of 3–6) increase in the clock frequency of the linear signal (Table 2.10). In each specific case, the use of these principles in the interfaces has its own features.

Table 2.10. Coding types in LAN interfaces

Application	Data transmission rate, Mbit/s	Coding type	Clock frequency of the linear signal, MHz
Token Ring	4	Manchester	8
Ethernet	10	Manchester	20
Token Ring	16	Manchester	32
TP-PMD	100	NRZI+4B5B	125
TP-PMD, 100Base-TX	100	MLT-3+4B5B	62,5
100Base-T4	100	8B6T	25
ATM	155	MLT-3+4B5B	96,88
1000Base-T	1000	PAM-5	125

Coding scheme of the network 100Base-T4

The interface 100Base-T4 was developed taking into account the potentials of category 3 cables for transmitting 100-Mbit/s signals in the half-duplex mode. The block diagram of the interaction of two 100Base-T4 interfaces is shown in Fig. 2.28a. In the hardware of this standard, three twisted pairs are simultaneously engaged for signal transmission. Pairs 1 and 2 are used for transmitting signals in one direction, which provides the possibility of collision detection, whereas pairs 3 and 4 can transmit signals in both directions. The direction of signal transmission through these pairs at each time instant is determined by control signals. Using three twisted pairs simultaneously, one can transmit data flux through each of these pairs with a speed of 100/3 =33.33 Mbit/s. To reduce the clock frequency, the linear signal is formed by means of the three-level code 8B/6T (Fig. 2.28b). In accordance with the algorithm implementing this code, the transmitter coder transforms any 8-bit symbol into six three-level symbols to be transmitted pairwise through three twisted pairs. This mode reduces the clock frequency of the linear signal to (100/3)×(6/8) = 25 MHz. The minimal bandwidth of the communication channel to transmit digital signals with such a rate is 12.5 MHz, which is a factor of 1.28 less than the bandwidth of a standard cabling channel of category 3 (16 MHz).

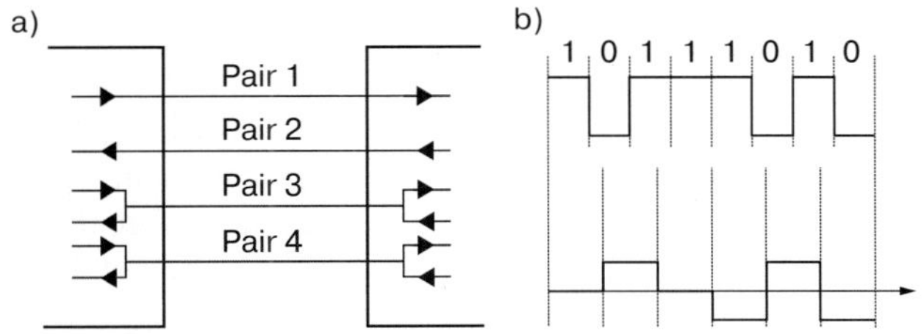

Fig. 2.28. 100Base-T4 network: *a)* block diagram of the interaction; *b)* 8B/6T code

Coding scheme of TP-PMD and 100Base-TX

The modes for coding linear signals of the interfaces TP-PMD and 100Base-TX completely coincide and use the three-level code MLT-3 (multilevel transmission). In this case, only two twisted pairs of category 5 are engaged for data transmission that can be performed in the half-duplex mode (Fig. 2.29).

The coders of these interfaces must provide the compatibility of the network interface TP-PMD (that has been developed first) with the layer PHY of the specification FDDI utilizing the block code 4B5B. (This code will be considered in Section 2.3.4.) Thus, the linear MLT-3 coder is always fed by a signal with a clock frequency of 125 MHz.

The implementation of the MLT-3 code is similar to that of the NRZ-1 code. For this reason, the notation MLT-3 + NRZ-1 is sometimes used for the MLT-3 code. The level of the linear signal changes only if 1 enters the coder input but, unlike the NRZ-1 code, the MLT-3 code is formed so that two adjacent changes always have opposite directions. Formally, MLT-3 is a binary code but when applied for the algorithm outlined above, it permits one to obtain a signal with three levels and a constant component that is close to zero even in the worst case. To emphasize this property, codes similar to MLT-3 are sometimes referred to as a special subclass of quasi-ternary codes.

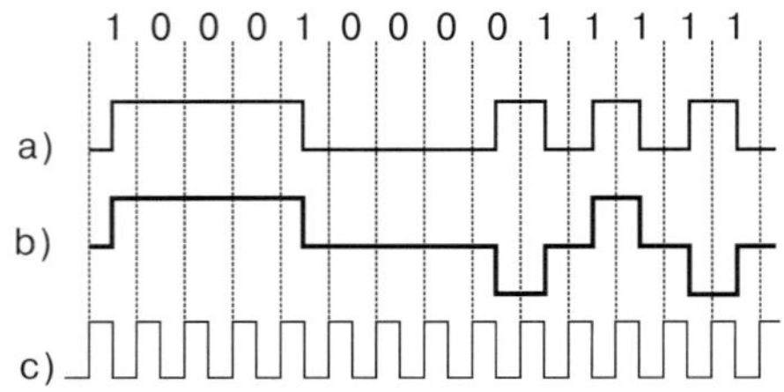

Fig. 2.29. MLT-3 code: *a)* NRZI code signal; *b)* MLT-3 code signal; *c)* clock generator signal

As has been noted previously, the NRZ-1 code does not allow one to achieve a stable synchronization without special measures. In the specific case in question, these measures involve introducing the signals of logical 1 to the input signal of the coder, if long sequences of zeros are detected in this signal. In the interfaces TP-PMD and 100Base-TX, this function is implemented with a hardware scrambler. Technically, this unit is a shift register that generates a quasi-random sequence in accordance with the generating polynomial $x^{11} + x^9$. The input signal of the register is the data sequence itself. The scrambler is placed at the input of the MLT-3 coder. At the output of the receiver's decoder, there is a descrambler restoring the original data signal. The scrambler and the descrambler are synchronized by means of linear state signals. The scrambler permits one to remove excursions from the energy spectrum of the linear signal. This makes it possible to obtain an acceptable level of crosstalk attenuation without using a shielded twisted pair. In addition, the use of the scrambler minimizes the deviation of the constant component of the linear signal from zero.

The three-level code MLT-3 permits one to reduce the clock frequency of the linear signal by a factor of 2, i.e., to 125/2 = 62.5 MHz. This value corresponds with a certain margin to the parameters of the electrical linear channel constructed from the elements of category 5.

A comparison of the network interfaces 100Base-T4 and 100Base-TX shows that for the same data transmission rate (100 Mbit/s), the use of the interface 100Base-TX with a higher category (category 5) channel provides a number of advantages.

- This makes it possible to implement the full-duplex operating mode of the interface 100Base-TX, which in fact is equivalent to an increase in the communication channel bandwidth by a factor of 2.

- This substantially simplifies electronic circuits for data processing, owing to the elimination of such elements as two-directional amplifiers and multiplexing/demultiplexing circuits. According to some estimates, this reduces the circuit complexity by 80–90% [38].
- This does not lead to a loss in the interference immunity caused by nonzero values of the parameter Skew.

These are the reasons why the application area of the interface 100Base-TX is substantially wider than that of the interface 100Base-T4.

Engineering solutions for 1000Base-T

The subcommittee IEEE 802.3ab has developed a standard for the network for transmitting data with a rate of 1 Gbit/s through four unshielded twisted pairs: Gigabit Ethernet 1000Base-T. In accordance with the normative document of this subcommittee, this network is designed for the full-duplex mode operation through a twisted pair communication channel that has an upper limit frequency of 100 MHz and a maximal length of 100 m. In this case, the probability of an error does not exceed 10^{-10}.

To meet the requirements of this standard, the designers of network interfaces invent sophisticated engineering solutions. These solutions are interesting primarily for the fact that they widely and fully use the recent achievements of modern microelectronics to utilize the bandwidth potentials of the data transmission channel as fully as possible.

To implement the possibility of transmitting data flows at a rate of 1000 Mbit/s through SCS electrical channels of categories 5 and 6, network interfaces 1000Base-T use the following engineering solutions:

- All four twisted pairs are simultaneously engaged for data transmission, the data being transmitted in both directions simultaneously through each of the pairs.
- The transceivers are equipped with additional units to minimize certain kinds of noise components.
- A special algorithm is used to synchronize network interfaces.
- The five-level code PAM-5 is applied.

Figure 2.30 shows a simplified circuitry of the transceiver for the first pair of the Gigabit Ethernet network.

High data transmission rates impose very rigid requirements on the synchronization of the 1000Base-T interfaces. To meet these requirements these interfaces always operate in the master–slave mode. Before the start of the operation using the extended autonegotiation mechanism, priorities of the interfaces are defined. The interface with higher priority (as a rule, this is a switch) plays the role of a master unit tuning the far-end transmitter to the cock frequency of its clock generator. If the units to be connected have the same priority, the choice of the master unit is defined by lot.

The input data flow with a rate of 1 Gbit/s is distributed uniformly among all four pairs and, hence, each of the pairs transmits the data with a rate of 250 Mbit/s. To provide the possibility of two-directional data transmission, a decoupling unit (differential system) is introduced into the interface circuit. Since the decoupling provided by this unit is incomplete (imperfect), the interface receives not only the signal from the far-end transmitter but also a signal from the near-end transmitter (an echo signal) that plays the role of noise. To minimize the harmful effect of the echo signal, the signal being transmitted is attenuated by a certain factor, processed by an analogue filter and a digital filter (digital signal processor), and subtracted from the total input signal of the receiver. The attenuation ratio is determined during the tuning of the communication channel before starting its operation.

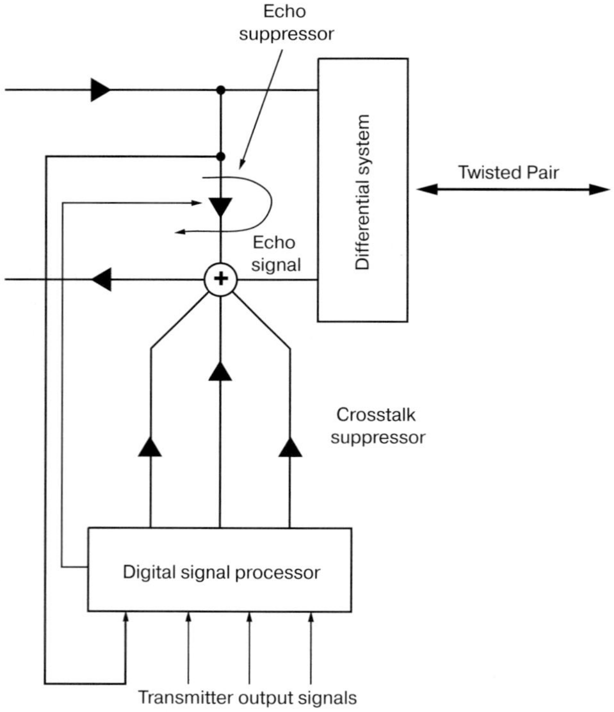

Fig. 2.30. Simplified circuitry of the Gigabit Ethernet network transceiver

This principle is also used to organize the operation of near-end crosstalk suppressors. These suppressors are tuned independently for each pair. This approach permits one to increase NEXT for a combination of separate pairs by approximately 20 dB. Owing to the individual tuning, the100Base-T interface uses $4 \times 3 = 12$ near-end crosstalk suppressors.

The scrambling process in the system 1000Base-T is also organized in a rather original manner. Unlike systems TP-PMD and 100Base-TX, the scramble of each pair in the 1000Base-T interface has its own generating polynomial. This solution turns out to be rather useful for tuning the interface.

The attenuation of a twisted pair depends on its length and increases as the frequency increases. To compensate for the high attenuation of low-frequency components leading to cuts of the fronts of the pulses being transmitted, one uses two approaches. The first approach involves an increase in the amplitudes of high-frequency components in the transmitter (emphasis), while the second approach is based on a decrease of the gain of the receiver at low frequencies (correction). The drawbacks of these approaches are that the first approach increases the radiation and reduces the crosstalk attenuation, while the second one reduces the signal-to-noise ratio. To mitigate these negative effects, the receiver of 1000Base-T utilizes dynamic correction. Dynamic correction implies adjusting the gain at low frequencies depending on the input signal level. This adjustment is performed automatically. To simplify the correction algorithm, 20-m discretization has been chosen to correct the line.

To minimize the crosstalk level, combined interfaces 100/1000Base-T use partial response filters. These units are located at the input of the receiver and add ¾ of the new pulse to ¼ of the previous pulse. The partial response filter permits one to make the shapes of the spectra of the interfaces 100Base-TX and 1000Base-T close to each other and to utilize an output pulse transformer whose characteristics are optimized for the maximization of the crosstalk attenuation.

To reduce the clock frequency to a value allowing one to transmit digital data through twisted pairs of categories 5 and 6, the data in the line are represented either in the Enhanced TX/T2 code or in the PAM-5 code. In these codes, the signal being transmitted has a set of five fixed levels, {–2, –1, 0, +1, +2}; see Fig. 2.31.

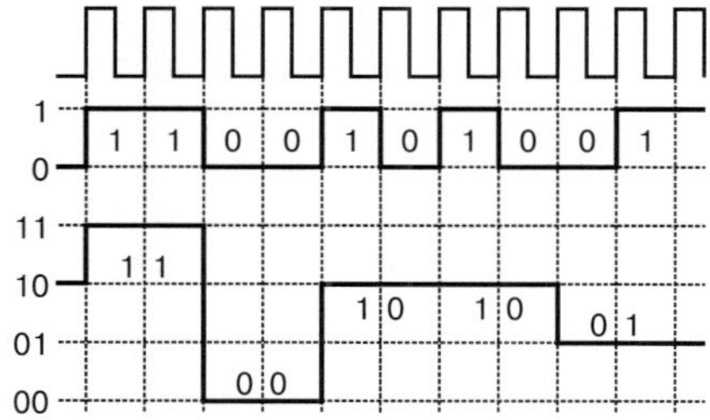

Fig. 2.31. PAM-5 code

Four of these levels are used to code bits of information, and the fifth level is used for error correction. A set of four fixed levels permits one to encode simultaneously two bits of information, since the number of possible combinations of two elements of four ones is equal to four – (00), (11), (01), and (10). Thus, the signal in the Enhanced TX/T2 code has a clock frequency of 125 MHz, which is lower than the clock frequency of separate components of the data sequence by a factor of 2. The quadruples of quinary symbols being transmitted simultaneously through all pairs of the cable form a single unit of information having a length of one byte. For the clock frequency f_c = 125 MHz, the rate of data transmission

For the clock frequency f_c = 125 MHz, the rate of data transmission through all four pairs makes up 1000 Mbit/s (125 MHz × 2 bit/pair × 4 pairs = 1000 Mbit/s).

In accordance with Section 2.3.1, to be able to receive such a sequence without errors, the communication channel must have a bandwidth of no less than 62.5 MHz ($f_1 > f_c/2 = 125/2 = 62.5$ MHz). In Section 12.2.1, we will show that the bandwidth of the channel of category 5 defined according to the criterion of $ACR = 10$ dB makes up about 70 MHz. Thus, signals of the network interface 802.3ab can, in principle, be transmitted through a cabling system of category 5. In this case, the upper limit frequency of the channel exceeds the theoretical minimum only by 12% and, hence, to implement this data exchange one will need interfaces with very complex circuitry and, as a consequence, high cost. For example, the experts of the German company Compu-Shack have estimated that to receive one bit of the Gigabit Ethernet frame the controller of the 1000Base-T network adapter must execute about150 logical operations [39]. The service life of the network equipment is substantially inferior to that of the SCS. For that reason, a common tendency now is to reduce the circuitry cost, thereby reducing the cost of the network hardware, and to compensate for the loss in the immunity characteristics by improving the parameters of the cabling system. This observation was one of the reasons for the development of electrical cabling systems of categories 6 and 7.

At the same time, the intention to preserve the routine practice of production of the components of category 5 gave rise to the development of category 5e. In fact, this standard officially (de jure) sets the technological level that has been achieved de facto for today. This permits one, for example, to increase the channel bandwidth to approximately 85 MHz (defined according to the criterion of ACR = 10 dB). This standard also specifies a number of additional parameters of the ELFEXT type necessary for stable operation of the 100Base-T transceivers through a balanced cabling.

The fifth (redundant) level of the Enhanced TX/T2 code is utilized to construct an error corrector. It is implemented by the Tredpiece coder and the Witterbi decoder. The corrector increases the interference immunity of the receiver by 6 dB.

2.3.4 Specific Features of Linear Codes for Fiber-optic Communication Channels

When choosing codes for transmitting data through fiber-optic communication channels, one should take into account the following specific features of this transmission medium and the hardware components of the fiber-optic transceivers:

- The linear signal can take on only positive or zero values (there is no "negative" light). Hence, a constant component will always be present in the line.
- The intended use of fiber-optic lines is for transmitting high-speed data through large distances. For this reason, the requirement of minimal increase of the bandwidth of the original message is more important for optical systems than for electrical ones.

- The specific features of the hardware components used in fiber-optic communication channels prevent one from widely using multilevel codes. It is, in particular, for that reason that optical radiators (especially semiconductor lasers) demonstrate noticeable time and temperature instability of the output signal power.
- Modern semiconductor radiators cannot generate perfectly monochromatic radiation. Because of this, most of the optic communication lines utilize intensity (power) modulation of the optical signal, since phase or frequency modulation of the radiation is impossible.
- An optical fiber is substantially more expensive than a twisted pair (about 30 cents for fiber 62.5/125 versus 6 cents for a twisted pair of category 5). This makes it economically unreasonable to split the data flow and transmit it through separate subchannels with lower rate. (This mode is widely used in electrical systems.)
- The energy potential of the optical network equipment is substantially lower than that of electrical systems. This is accounted for by quantum noise. For example, the energy potential of a network interface utilizing a twisted pair of category 5 as the transmission medium is no lower than 24 dB, whereas for optical interfaces this parameter makes up 11 dB, which is lower by 13 dB. For that reason, transmission of data in both directions through a single fiber is almost never organized in optical systems. The communication channel in such systems is formed by two fibers, each of which transmits data only in one direction.

Optical communication systems transmitting data at a rate lower than 16 Mbit/s widely use the self-synchronizing Manchester code, since the bandwidth of modern fiber-optic cables suffices for organizing cabling channels with a length less than 3 km, which is specified by the standards. For rates of 100 Mbit/s or higher, the bandwidth of multimode optic cables turns out to be insufficient, and one uses block codes, which are more efficient as regards the frequency range requirements. The possibility of using these codes is provided by fairly high stability of the frequency of the clock generator implemented in modern hardware elements. This allows one to adjust the clock generator frequency less often than for each data bit.

Code 4B5B is an example of self-synchronizing block coding. This code is used in systems FDDI and Fast Ethernet 100-Base-FX. In this coding system, each four input data bits are encoded by five linear bits. The coding rule is specified by the code table. The code table may have different forms depending on the task and required properties of the code. For example, in the FDDI system, the linear signal used for data transmission always has at least two changes in each block. This provides the clock synchronization with stability and minimizes fluctuations of the average optical power. The block coding scheme for the FDDI system is presented in Table 2.11.

Table 2.11. 4B5B coding table

Data block symbol	Binary code	Representation of the signal block	
		Decimal	Binary
0	0000	30	11110
1	0001	09	01001
2	0010	20	10100
3	0011	21	10101
4	0100	10	01010
5	0101	11	01011
6	0110	14	01110
7	0111	15	01111
8	1000	18	10010
9	1001	19	10011
A	1010	22	10110
B	1011	23	10111
C	1100	26	11010
D	1101	27	11011
E	1110	28	11100
F	1111	29	11101

The linear signal in block coding has a redundancy, as compared with the information signal. This redundancy permits one to increase the interference immunity and detect errors, since in this case part of the code combinations is forbidden. When detecting the forbidden combinations, the system generates the command VIOLATION [40].

As compared with RZ and Manchester codes, code 4B5B provides a clock frequency that exceeds the clock frequency of the information signal by a factor of 1.25 rather than 2. This allows a substantial increase in the efficiency in using the communication line bandwidth.

Code 8B10B gives another example of block coding. This code was first developed by the IBM Company for the ESCON equipment. Then it was used in the Fiber Channel equipment from which it was borrowed for the networks Gigabit Ethernet 1000Base-SX (multimode optics) and 1000Base-LX (multimode or single-mode optics) The code 8B10B uses 10 signal bits to represent 8 data bits. Uncoded information consists of eight information bits (A, B, C, D, E, F, G, and H) and a control bit Z. These bits are encoded to ten bits (a, b, c, d, e, i, f, g, h, j) of the transmission character, in accordance with a special table. The control bit takes on the value D* for symbols representing input data and K for special symbols. The code table should be organized so that D symbols do not contain sequences of more than four consecutive zeros or units.

To each input sequence consisting of eight information bits and one control bit there corresponds a name Zxxy, where Z is the control bit, xx is a decimal number composed of the last five bits (D, E, F, G, H), and y is a decimal number composed of the first three bits (A, B, and C). For example, a special (type K) hexadecimal symbol BC has the name K28.5. The receiver decodes the received data into blocks of 10 bits each, after which D* symbols are transformed into one of

256 8-bit combinations and K symbols are used to control the protocol. Symbols that do not belong to type D* or K are identified as protocol errors.

Each D* or K symbol has two, not necessarily different, representations depending on the running disparity (RD) of the symbol. For each new symbol to be transmitted, the transmitter and receiver calculate a new value of RD by comparing zeros and units in the subblocks (the first six bits and the last four bits) of the signal that has just been transmitted. In accordance with the result of this calculation, one determines the representation for the next pulse to be transmitted into the line. Such a mechanism for calculating RD provides the number of changes in the signal level sufficient for synchronization of the transmitter and the receiver.

For more details about the formation of the 8B10B code, see [41].

2.4 Conclusions

The equipment used for constructing LANs and corporate networks as well as public communication network equipment are based on the same data transmission principles. For that reason, the parameters subjected to standardization for SCS mostly coincide with the parameters to be standardized for electrical and fiber-optic networks on the scale of city or higher. Traditionally, most of the characteristics to be standardized in cabling engineering are specified in the frequency domain.

The difference in characteristics controlled by the standards entirely depends on the engineering features specific to SCS. The most important specific features are relatively small lengths of channels combined with the possibility of supporting high-speed data exchange characteristic of modern interfaces.

The basic parameters that are specified by standards and monitored in electrical channels are attenuation and crosstalk attenuation. The frequency range for which these parameters are specified depends on the category of the line.

Most recent applications with a data exchange rate of no lower than 100 Mbit/s widely employ the principle of transmitting the overall information flux through four twisted pairs of the SCS horizontal cable simultaneously. This provides a motivation for increasing the list of parameters to be controlled by standards (e.g., Delay or Skew) and modifying the basic principles and techniques for measuring the parameters that have been specified previously (power sum model).

For fiber-optic lines, one specifies the attenuation ratio and the parameters determining the network performance, such as the dispersion (for single-mode cables) and characteristic bandwidth (for multimode cables). Unlike public networks, these characteristics are specified for the first and second spectral windows. The third spectral window (1550 nm) is of minor use in SCS engineering, since the operation in this window does not provide significant advantages.

The appropriate choice of coding mode for the linear signal provides a more complete utilization of the bandwidth of SCS channels. The major aim of choosing the coding mode is to provide the required interference immunity and synchronization stability. Parallel transmission of data through all four pairs simultaneously and the use of multilevel linear codes permits one to transmit data flows at a rate of 1000 Mbit/s through standard electric SCS channels with an upper limit frequency of 100 MHz.

The fact that optical SCS channels have a relatively small length explains the wide application of multimode hardware for constructing such channels. Single-mode links are utilized in large amount only for constructing campus backbone cabling subsystems with a length of no less than 1500 m.

The prices of the optical hardware characteristic of the late 1990s make the parallel transmission of data in optical SCS channels through several channels with lower speed economically inefficient. Therefore, the basic means for achieving the required quality of data transmission and synchronization stability is the use of block codes.

3 SCS Electrical Components

3.1 Twisted Pair Cables

3.1.1 General Description and Classification

Twisted pair cables with copper conductors are widely applied in SCS for electrical signal transmission. Any cable examined in this chapter includes one or several twisted pairs of wires to be twisted with different steps, and is classified as a balanced cable according to the existing classification. Besides the available twisted pairs, it may also include several auxiliary protective, shielded, and technological elements, which form a cable core. Each connector is insulated by a solid or foamed insulating material. The use of the latter material somewhat diminishes the cable specific mass and considerably improves the frequency performance; however, it increases the cost of the end product. The cable core is insulated by a protective hose to prevent the twisted pairs from environmental exposure and to maintain the cable core during cabling and operating. The presence of the cable core's outer protective jacket is essential to referral of the reviewed product as the cable class. All other electrical products, which are also intended for electrical telecommunications signal transmission, will be further referred to as conductors. Cable products for SCS implemented on twisted pairs may be divided into four basic types depending on the key product area and structure:

- horizontal cable;
- multipair cable;
- patch cable;
- jumper wire.

SCS cables have to meet fire safety requirements. For more details see Chapter 9.

Cables of the reviewed class may be implemented for all three SCS segments; however, strict physical restrictions on the maximum length of the segment hinder their use for D-class high-speed applications of the campus backbone cabling. Regarding that fact the majority of electric cables can be used inside the building. There is a limited number of twisted pair cables for running between buildings. Because of the mentioned limitation on signal transmission channel length for D-class applications, such products have not become widespread for supporting LAN; however, are widely applied for low-speed network equipment signal transmission, for example, PABX.

3.1.2 Horizontal Cable

Modifications of horizontal cables

A horizontal cable[1] is intended for use in horizontal cabling along the segment between switching equipment in floor telecommunications closets to telecommunications outlets of workplaces. The cable type has been so designated because of its horizontal position with minimal vertical segments along the route. Most of the cables being reviewed here have 100-Ω impedance. The 120-Ω cables are quite popular in France. The current standards allow for utilization of cables with Z_b=150 Ω impedance as well. The cable products are widely used in the U.S.A., and share a substantially smaller stake in Europe, as compared with other cable types. It is rare for this cable type to be used in Russia due to the low popularity of the Token Ring equipment.

Products used widely in practice usually consist of four twisted pairs. It is known that part of the network equipment only uses two twisted pairs for data exchange. Accordingly, the standards ISO/IEC 11801 and EN 50173 also allow the application of two-paired cables. This cable type is the most cost-effective solution and possesses superior mass and dimensions parameters, and it is widely applied in several European countries. However, its application considerably restrains the operating resources of the cable system, and specifically, makes it impossible to transmit signals of up-to-date prospective applications of the Gigabit Ethernet type. The two-paired cables are sometimes called the 'oval' ones because of their specific jacket form unlike the four-paired cables, which are called the 'round' ones. Historically, the two-paired variants of horizontal cables are not widely applied in Russia, though they are included as a part of several SCS. Therefore, we assume further that a horizontal cable contains four pairs.

Some companies produce duplex four-paired cables, as 2-port workplaces are widely distributed in SCS. A slang word 'shotgun' is quite often used by SCS professionals to designate a duplex cable because its end part looks like a barrel. It is interesting to note that the term has penetrated even the official catalogues of some cable producers.

The duplex cables come with two main modifications. The first one consists of two hoses to be connected by a string tie. The product is also called a zipcord due to the clearly visible structure of its individual elements (analogy with electrical cord) or the Siamese cable. Occasionally, it can be called the UTP Figure 8 cable (the Cvalim company) or FIG-8 (the Teldor company). The second modification (Type 2x1061C by Lucent Technologies and Uninet 1002 2x4P by Datwyler) comprises two standard horizontal cables integrated as a whole under an overall outer jacket. The external hoses of individual elements are differently colored for their easy identification. Most of the duplex cables constructions include category 5 elements with identical electrical characteristics; however, the cables produced

---

[1] In German technical literature such a cable is sometimes called an installation cable; Russian issues often call it a LAN-cable, although that is not quite a correct designation of the product.

by the company AMP have one element with characteristics corresponding to category 5 and another one to categories 3 or 4. By their electrical and mechanical characteristics the duplex cables do not differ from the common ones; however, their application enable bringing down the general expenses for SCS horizontal system implementation by simultaneous routing the two cable elements to one socket per cycle.

The cables, which include three and more four-paired elements under the overall jacket, referred to as multipaired cables, are reviewed in Section 3.1.3 below.

Conductor materials

Main properties of materials applied during multipurpose communications cable production are given in Table 3.1. Electrical copper is employed in the production of cable conductors. Other materials are only utilized if there are some special requirements for the operating conditions. The insufficient corrosion stability of the copper material is compensated for by using special insulating materials and contacts (see below).

Table 3.1. Main properties of materials to be used in the horizontal cable conductor production

Material	Maximal operating frequency, °C	Conductivity	Soldering ability	Corrosion	Allowable bending radius	Cost
Electrical Copper	+130	Very High	Good	High	Very Small	Very Low
Galvanized Copper	+150	Very High	Very Good	Very High	Small	Low
Silver-plated Copper	+205	Very High	Very Good	Satisfactory	Medium	Very High
Nickel-plated Copper	+250	High	Medium	Very High	Satisfactory	Medium
Nickel	+500	Satisfactory	No	Very High	Medium	High

The horizontal cable conductors are produced from a solid copper wire to diminish attenuation. Individual twisted pairs form a cable core covered by an overall outer protective insulation for all the pairs, approximately 0.5–0.6 mm in thickness. Polymer tapes or straps may be used for pair wrapping to keep the cable core structure free from deformation during production and operating. To facilitate the stripping process of some cables a rip-cord can be placed under the cable jacket. When pulling, the rip-cord makes a slit to allow access to the cable core.

Depending on the type of twisting the horizontal cable wires can be divided into pair and quad types (Fig. 3.1). It is interesting to note that, when using the quad twisting, the wires of one pair are always positioned against each other. Quad twisting provides for smaller outer cable dimensions, higher stability of its architecture, and superior electrical performance; however, a quad-twisted cable is more complicated to manufacture and terminate and, therefore, it is not used widely enough in SCS engineering. In Europe, for example, the quad-twisted cable has become popular only in France. It is prevented from being completely excluded from use by its advantage of a smaller bend radius due to the small outer dimensions and needing less room in the raceway.

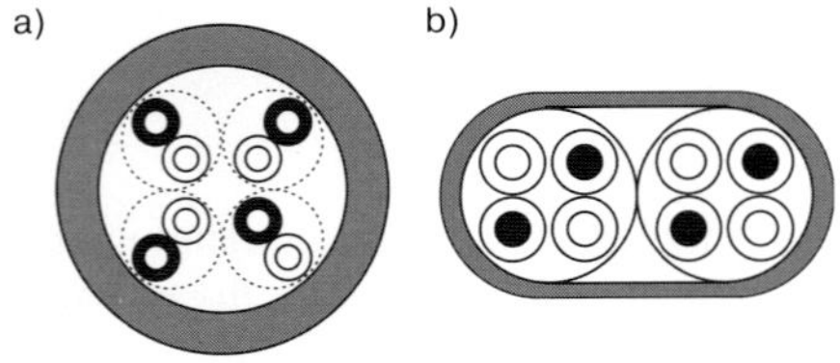

Fig. 3.1. Twisting types of twisted pairs: *a)* pair; *b)* quad

To reduce reciprocal interference of pairs, different and aliquant twisting steps are utilized in pair-twisted cables. As a rule, production equipment (Table 3.2) affects the selection of a specific value of a step of twist, thereby giving a great variety of modifications. The selection implies, that within the effective frequency range of a cable:

- twisted-pair signals have minimum crosstalk (i.e., provide for maximum crosstalk loss);
- the pair itself is free from excessive resonance impacts.

Table 3.2. Steps of twist (in mm) in twisted pairs of category 5 horizontal cable from different producers [42]

Company	Pair 1	Pair 2	Pair 3	Pair 4
General Cable	14	17	12	20
BICC	18	15	20	12
Belden	25	20	16	32
Lucent Technologies	15	13	20	24
Mohawk/CDT	25	17	28	20

The quad-twisted cables have quad pairs to allow them to be more compactly embedded; however, they are electrically discontinued from each other due to the perpendicular orientation of their planes at any point.

Conductor insulation

PVC is regularly used as conductor insulation for category-3 cables. Other high performance materials, such as polyethylene and polypropylene, are widely used in category-5 and higher cables (Table 3.3). One can use both solid and foamed materials. The latter ones enable superior performance because they have much less dielectric loss, but they are expensive and predominantly used in cables with an upper frequency limit above 100 MHz. The insulation thickness is approximately 0.2 mm.

Table 3.3. Basic insulation materials of balanced SCS cable conductors

Material	Latin	Dielectric permittivity ε	Operating frequencies, °C
Polyvinylchloride	PVC	4.0–5.0	–40…+85
Polypropylene	PP	2.4	–10…+100
Polyethylene	PE	2.3	–55…+85
Cellular Polyethylene	–	1.2	–55…+85
Cellular Polyethylene with skin	Foam skin PE	1.5	–55…+85
Teflon	FEP, PTFO, PFA[a]	2.0	–190…+260

[a]Depending on option

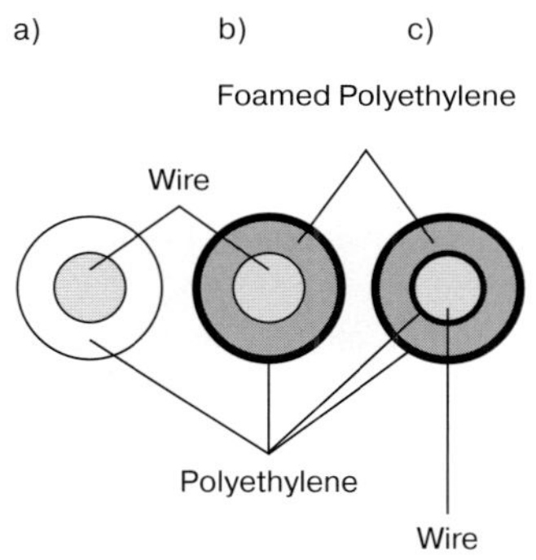

Fig. 3.2. Insulating covers of twisted pair conductors: *a)* solid; *b)* foamed material; *c)* foamed material with an inner layer of solid insulation

A material with a void content (defined as the ratio of the air pockets volume to the total volume of a specimen) of at most 0.5 is preferable for insulation. Otherwise, the insulation cover is not strong enough. This leads to its immediate or gradual deformation by conductors, which results finally in a short circuit when routing a cable with a small bend radius. As a rule, the foamed insulation is covered by a common insulation, the so-called foamed skin (Fig. 3.2b). Along with a reduction of dielectric loss the foam material theoretically allows a lower dielectric constant ε (Table 3.3 presents characteristics of solid and foamed polyethylene), which reduces the twisted pair capacity and improves its frequency performance.

Along with structures shown in Fig. 3.2b, an extra inner solid insulation layer can sometimes be put directly around the connector (Fig. 3.2c) [43]. This solution does not seriously affect the frequency performance of the conductor; moreover, it enables improvement of its bending ability due to the extension of the conductor effective diameter.

Outer jacket

Halogen-free and flame retardant compounds and low-smoke polymers are often used for production of the outer jacket along with the common PVC. Application of flame-retardant materials for the jacket production increases the price of the end product by approximately 20–30%. Therefore, PVC cannot be completely avoided. However, the halogen-free compounds are easily destroyed by the fire. A more detailed analysis of the problem is given in Chapter 9.

The fragility of the outer jacket, which is achieved by mixing a certain amount of chalk in a feedstock, is essential for cable stripping and allows one to cut precisely and accurately at any desired point.

The outer jacket is marked with different gradations of the color grey, but some other markings (dark blue, violet, white, red), specific to each individual producer, are available as well. In most cases an orange marking indicates a flame-retardant jacket, in which case one may use the cable in plenums (see Chapter 9). When designing an SCS the colored jackets may be helpful for identifying cables of corresponding functional sections, outlets, and categories. The color spectrum may vary even within a small production lot, as it is only required to color a feedstock just prior to the jacket extrusion process. To bring down production costs when manufacturing large cable lots, the feedstock of a specific color needs to be ordered in advance.

The products for outer cable routing bear polyethylene jackets, as the material is much more waterproof than PVC and fire-resistant compounds. The outer polyethylene jacket is laid over the common one in order to maintain the original manufacturing method. There are also products by the company Mohawk/CDT with inner core voids filled with a gel (M56871) as an air cushion (M57041) and armored with a corrugated steel cord belt (M57042).

The outer jacket is marked to indicate cable type, diameter and conductor type, producer company and the company's cable labeling, standard type, certification laboratory, and cable length in feet or meters. The two latter parameters differ for the American and European cable companies. The head certification body for American cable producers is the Underwriters Laboratory (UL). The European manufacturers certify products in the Danish organization Delta Electronics Testing. The American cable companies give length markings in feet, but the European producers provide a metric measurement of the parameter.

Horizontal cable shielding

The twisted pair horizontal cables come as shielded and unshielded types depending on whether the individual twisted pairs or core have an auxiliary shielding cover or not. Such classification makes it possible to define four main cable product types, i.e., along with unshielded cables, cables with an overall outer shield, with shields for each pair, and with simultaneous shielding of both the pairs and core. Shielding is basically used to increase parameters NEXT and FEXT, reduce the electromagnetic emission level, and increase the noise immunity. Some types of shields add mechanical strength to cables. The physical architecture of some cable types is given in Fig. 3.3. Figure 3.4 presents their cross-sectional areas.

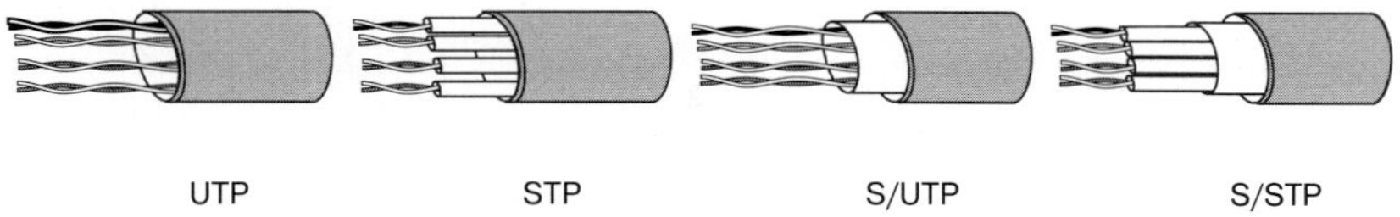

Fig. 3.3. Structure of horizontal cables

Aluminized thin polymer films are widely used for shielding of individual pairs. There are constructions with a metal-plated side both on the inside and outside. The foil edges may be laid over each other by overlapping or joined by a longitudinal seam.

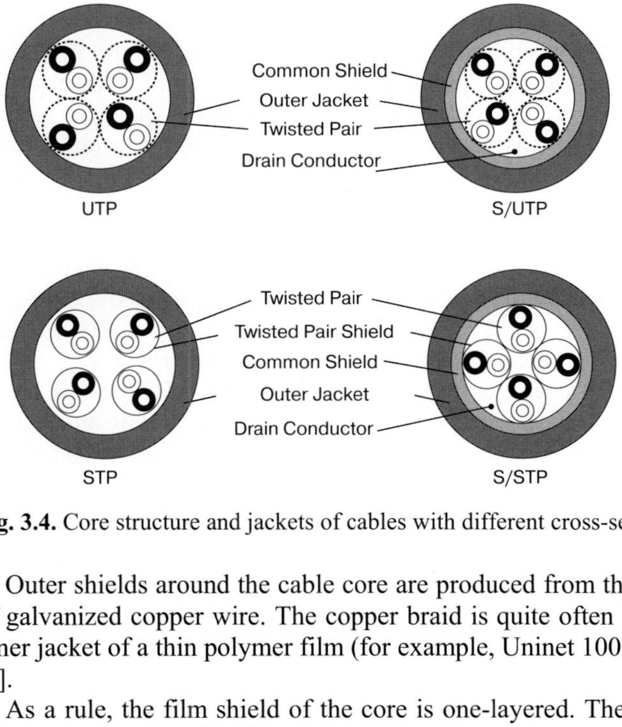

Fig. 3.4. Core structure and jackets of cables with different cross-sectional areas

Outer shields around the cable core are produced from the same film as a braid of galvanized copper wire. The copper braid is quite often laid over the common inner jacket of a thin polymer film (for example, Uninet 1002P cable by Datwyler) [2].

As a rule, the film shield of the core is one-layered. The company Alcatel applied a double film shield in GIGATEK with simultaneous foil winding in different directions in order to reinforce the mechanical strength of the product while retaining its high flexibility. The structure incorporates an additional thin uninsulated tinned copper conductor or a galvanized drain wire with 0.5-mm section. The latter provides for the electrical continuity of the shield when unexpected film fractures occur during cable routing and operating. Some constructions of modular outlets use the drain wire as a key element to ensure the reliable galvanic continuity of the cable and outlet shields. In situations with a two-layered shield configuration the drain wire is laid between separate layers. The braided shield may be of various densities. When manufacturing a low number of cable modifications with shields of different densities, its value is given in the cable technical specifications as a percentage.

The cable with an overall film shield and auxiliary braid over the film is widely applied. This type of configuration is sometimes called the HIGH Screen. The advantage, which contributes to the overall utilization of the two-layered shield, is a much higher mechanical strength of the braid; therefore, when terminating into a connector it ensures a full radial contact with its screening housing. At the same time, film shields protect the cable from radio-frequency interference (RFI) and braided shields from electromagnetic interference (EMI). Thus, the two-layered shield type being reviewed here makes the cable core shielding effective within the total frequency range. It needs to be noted that a drain wire is unnecessary when using a two-layered shield; however, it often remains a part of the configuration in order to maintain uniformity of the entire manufacturing process.

To obtain a much higher cable performance in order to reduce the EMI level and increase the noise immunity, common multilayered shields of film-braid-film or film-braid-film-braid type are utilized. However, the solution leads to a considerable increase in cable costs, its mass, and outer diameter together with its flexibility degradation; thus, cables with such braids are not popular.

Unfortunately, at the time the work for this study was completed, there was no general terminology for constructive modifications of horizontal cables. Producers often use their own terminology in catalogues and technical documentation. Nevertheless, one may create a certain system within the present field. Cables with a film braid are often coded by the letter F (from 'foiled'), a braided shield is marked with S or Sc, abbreviated from 'screen' or BS for 'braided screen'. In addition, quad twisted models (Section 3.1.2) are abbreviated as STQ for 'star quad'. Furthermore, among producers of cable products for SCS, it is widely practiced to provide the shield types, especially the horizontal ones, with both full and abbreviated designations. General information on cable modifications labeling, shield structure, and its functions are given in Table 3.4.

The S/UTP cables application area is a development of a horizontal SCS at an insignificant level of external interfering noises (production departments and other premises with high electromagnetic field sources), or when special requirements to the cable system safety have to be met (unauthorized access protection). S/STP cables, as compared with STP cables, possess a better external noise protection and EMI level performance; however, their key advantage over other solutions is a considerably high NEXT value (up to 10–15 dB and higher when correctly installed). Today, it is assumed that S/STP architecture is the only way to provide a two-way transmission of linear signals with a timing frequency above 250–300 MHz using duplex mode at a distance of 90 m, as required by the standards.

STP and S/STP cables have to be applied in all cases where U/UTP cables are indicated when it is necessary to:

- obtain a cable segment length over 90 m;
- develop data transmission systems, for which the cable electrical performance of category 5 is insufficient;
- meet high requirements to protect against unauthorized access to data being transmitted.

Table 3.4. Horizontal cable types

Symbols		Shield	Shielding results
Main	Alternative		
UTP	–	Default	–
STP		Shielding of each pair	• Decreased EMI factor • Increased protection from external noises
–	PiMF or PMF (pair in metal foil)	Individual film shield of each pair	• Increased Crosstalk
S/UTP	ScTP, FTP	Common shield for all pairs	• Decreased EMI factor • Increased protection from external noises
S/STP	STP, S/STP	Shielding of each pair + outer shield around all pairs	• Decreased EMI factor • Increased protection from external noises • Increased Crosstalk • Increased mechanical reliability

Cable parameters with an individual shield for each twisted pair may considerably exceed category 5 requirements (especially for the NEXT parameter and the ACR). It needs to be remembered, however, that there are still no official international standards either for segment length extension or networks, for which operation the electrical performances of unshielded twisted pairs in category 5, as well as categories 5e and 6 that are in development, are insufficient.

UTP cables, as compared with shielded ones, possess the following advantages:

- Cost efficiency;
- Less work for installation and maintenance;
- No special requirements for building an inner grounding ring;
- Smaller bend radius.

Key advantages of the shielded configurations are their potentially superior protection against external electromagnetic interferences, high mechanical strength when applying braided shields, and better protection from unauthorized access to transmitted data. The high thermal conductivity of metallic elements of some shields provides for effective heat abstraction, which is generated in conductors during data transmission because of the current flow. Based on that, some producers guarantee less attenuation of their shielded products, as compared with unshielded ones.

Moreover, it needs to be noted, that shielded products enable a dramatic reduction of environmental exposure with a high concentration of telecommunications and data equipment [44] that is peculiar to office premises. However, it does not seem to be possible to discuss the issue at the existing level of technological development because of the lack of well-approved and generally recognized medical standards for allowable levels of high-frequency electromagnetic emission on human beings and animals. Comparative data of some mechanical and operating parameters for key modifications of four-paired horizontal cables is given in Table 3.5.

Table 3.5. Standard mechanical characteristics of different horizontal four-pair cables

Cable type	UTP		STP	S/UTP	S/UTP	S/STP
	Cat.5	Cat.6		Film Shield	Combi Shield	
Mass, kg/km	30–33	34–37	42	49	65–85	82–88
Outer diameter, mm	4.9	5.2	5.4	6.2	7.6	8.0
Operating frequencies, °C				–20…+60…70		
Bend radius, mm		30–35		35–40	40–45	

Electrical performance

Key electrical parameters of horizontal cables, which are of practical value and regulated by existing standard editions, are:

- attenuation;
- crosstalk loss or NEXT;
- characteristic impedance;
- DC resistance;
- NVP.

The list will be extended in forthcoming standard versions (Table 1.9).

Standard TIA/EIA-568-A requirements on the peak attenuation of any horizontal cable pair of categories 3, 4, or 5 at a 100-m length at 20 °C are given in Table 3.6. The attenuation is estimated by the formula (2.8), rounding off to one decimal place. Cables are often delivered in fabric packing (using boxes or reels) as 1000-ft (305-m) segments. Table 3.6 also shows peak attenuation for the length segments (according to TIA/EIA TSB-36) in order to easily control the delivery. International standard ISO/IEC 11801 (2000 version) presents almost the same values of allowable attenuation, but with an increased deviation to 0.1–0.2 dB at some frequencies.

When testing is carried out at a temperature above 20 °C, then category 4 and 5 cable values in Table 3.6 have to increase by 0.4% for each exceeding degree, for category 3 cables the values increase by 1.5%.

Table 3.7 presents standard TIE/EIA-568-A and ISO/IEC 11801 (2000 version) requirements for minimum allowable NEXT in any combination of horizontal cable twisted pairs of categories 3, 4, or 5. Since NEXT for long cable segments is only slightly dependent on length (Section 2.1.3), one may apply the same values to verify NEXT of cable segments with lengths up to 305 m.

Table 3.6. Maximal allowable attenuation for horizontal cables of categories 3, 4, and 5 at 20 °C according to TIA/EIA-568-A

Frequency, [MHz]	Attenuation, [dB]					
	Cat.3		Cat.4		Cat.5	
	100 m	305 m	100 m	305 m	100 m	305 m
0.772	2.2	6.8	1.9	5.7	1.8	5.5
1.00	2.6	7.8	2.2	6.5	2.0	6.3
4.00	5.6	17	4.3	13	4.1	13
10.00	9.7	30	6.9	22	6.5	20
16.00	13.1	40	8.9	27	8.2	25
20.00	–	–	10.0	31	9.3	28
31.25	–	–	–	–	11.7	36
62.50	–	–	–	–	17.0	52
100.00	–	–	–	–	22.0	67

Table 3.7. Minimal NEXT value for category 3, 4, and 5 horizontal cables

Frequency, [MHz]	NEXT, [dB]		
	Cat.3	Cat.4	Cat.5
0.772	43	58	64
1.00	41	56	62
4.00	32	47	53
10.00	26	41	47
16.00	23	38	44
20.00	–	36	42
31.25	–	–	40
62.50	–	–	35
100.00	–	–	32

Standard ISO/IEC 11801 is identical to TIE/EIA-568-A in requirements on peak attenuation and NEXT values; however, it gives alternatives for attenuation pairs of values and NEXT of category 5 cables within the frequency range of 20–100 MHz. The ACR parameter value is actually defined within the frequency range. This is done in order to give cable producers a free choice. For example, a cable with a major cross-sectional area of conductor may possess a somewhat higher attenuation performance at some degraded NEXT. If the required ACR value is ensured, such a cable may be certified as a category 5 cable. In the 2000 version of the standard this position was deleted, as the calculated and standard values of ACR coincide.

The ISO/IEC 11801 requirement for the NVP parameter of category 3, 4, and 5 twisted pairs is given in Table 2.2.

Minimum allowable values of SRL according to the standard TIE/EIA-568-A, generated due to reflections from heterogeneities, are given in Table 3.8. The standard ISO/IEC 11801 gives values of the parameter, which are much closer to its American analogue. Frequency *f* in formulas is given in MHz.

Table 3.8. Minimal SRL values for horizontal cables of categories 3, 4, and 5 according to TIE/EIA-568-A and ISO/IEC 11801

Frequency, [MHz]	SRL, [dB]		
	Cat. 3	Cat. 4	Cat. 5
1–10	12/12	21	23
10–16	12-10lg(f/10)/10	21-10lg(f/10)/19	23
16–20	–	21-10lg(f/10)/18	23
20–100	–	–	23-10lg(f/20)

The requirements of ISO/IEC 11801 for other electrical performances for horizontal cables of twisted pairs are given in Table 3.9. They are identical to products of the categories 3, 4, and 5.

Table 3.9. Requirements for other electrical characteristics of twisted pairs

Parameter	ISO/IEC 11801	TIA/EIA-568-A
Wave resistance at frequencies not less than 1 MHz	75–150 Ω	–
Wave resistance at frequencies not more than 1 MHz	100±15 Ω	(100±15%) Ω
Maximal DC resistance of a short-circuited pair of 100 m length at far end at 20°C	19.2 Ω	–
Maximal DC resistance of any twisted pair conductor of 100 m length, Ω	–	9.38
DC resistance skewness of conductors in pair	3%	5%
Maximal capacity of pair skewness to ground (100 m, 1 kHz, 20 °C)	340 pF	330 pF
Minimal DC resistance to insulation conductor-to-conductor or conductor-to-shield	150 MΩ× km	–
Insulation breakdown DC voltage stress of conductor-to-conductor or conductor-to-shield	1 kV, 1 min or 2.5 kV, 2 s	–
Insulation breakdown alternate voltage stress of conductor-to-conductor or conductor-to-shield	0.7 kV, 1 min or 1.7 kV, 2 s	–
Skew parameter, ns	45	–

Mechanical characteristics

Mechanical characteristics for horizontal cables basically depend on the outer jacket and insulation materials as well as the outer cable diameter. The standards specify these parameters (Table 3.10).

Table 3.10. Requirements for mechanical characteristics of horizontal cables

Parameter	ISO/IEC 11801	TIA/EIA-568-A
Diameter of conductors	0.4–0.65 mm	0.5–0.65 mm (24 or 22 AWG[a])
Insulation diameter	≤1.4 mm	≤1.22 mm
Cable outer diameter	≤6.35 mm	≤6.35 mm
Temperatures without degradation of mechanical properties	installation: 0...+50°C operation: –20...+60°C	–
Minimal bend radius • cabling • operation	≤8 outer diameters ≤4 outer diameters	–
Allowable tensile strength during mounting	≤50 H/mm^2	≤400 H/mm^2

[a]AWG values in millimeters, see Section 14.3

The first three parameters define the geometrical size of constructive cable elements. Their violation leads to an incompatibility with modular connectors and IDC (insulation displacement connection) contacts for twisted pair conductors. Current technical specifications recommend the use of conductors not less than 0.5 mm in diameter, as it ensures the compatibility of all parts of the modern connector.

The cable outer diameter is an important value when calculating the capacity of cable raceways and embedded and vertical conduits. Moreover, it directly depends on the minimum allowable bend radius. With other conditions equal thin cables are more preferable for routing and installation.

The minimum bend radius defines cabling requirements. Manufacturers give two parameters for cable specifications: minimum allowable bend radius during and after cable routing. The restriction on the bend radius during cable routing is imposed due to the pulling of the cable by one of the ends. The stressed tension with a short turning radius may cause the cable to be damaged or even break. The minimum bend radius is restricted after routing as dramatic cable bending may cause degradation of its electrical performance. Depending on the cable architecture the minimum bend radius during cable installation may twice as much exceed the minimum bend radius after it, which ensures no damage under dynamic loads. In general, cables with less allowable bend radius are more preferable for bending routing.

The maximal tensile stress establishes requirements on the cable routing conditions. Exceeding the maximal tensile stress, when pulling the cable and violating producer specifications, may cause a jacket blip and distortion in the twisting of pairs. The maximal tensile stress is specified taking into consideration the minimum bend radius during running of the cable. Producers do not always specify the maximal tensile stress for the inner cabling. This means that the cable can undergo all stress at the manual cable routing.

Horizontal cables operate within a temperature range of –20 to +60 °C according to the standard. The standard ISO/IEC 11801 allows for cables with a minimum operating temperature of –30 °C for areas with cold weather conditions. For most cables the temperature should be above 0 °C for routing.

120-Ω cables

Only the international standard ISO/IEC 11801 and its European analogue EN 50173 approve 120-Ω cabling for use as an alternative to 100-Ω cabling, which is preferred in SCS. Cable product modification is in high demand in the market of a number of European countries (for example, France) and, therefore, is approved for cable system design. We mention two reasons that help the 120-Ω cables to be widely utilized for SCS design. The first reason is that having the same absolute spread of Z_b actual impedance value (Equation 2.12), the products provide a lower value for SRL, i.e., it is easier to design cable lines with the upper frequency limit above 100 Mbps. The second reason is that moderate (within a few tens of percent) amplification of characteristic impedance in balanced cable enables significant attenuation without substantial augment of external dimensions (Fig. 3.5) [45].

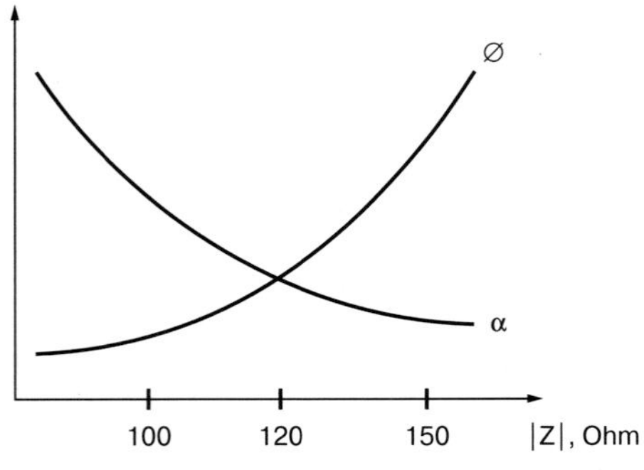

Fig. 3.5. Dependence of outer diameter and attenuation of twisted pair cable on characteristic impedance

This is determined by a proportion, which is known in cable engineering as $\alpha \sim R/(2|Z_b|)$, where R is the active cable linear resistance. At the same time, the characteristic impedance and the cable external diameter within the reviewed parameter range is roughly proportionate to the distance between pair conductors. The fact is specifically described in the standard ISO/IEC 11801, which regulates attenuation differently for 100-Ω and 120-Ω cables; it regulates the latter modification more strictly (Fig. 3.6). The lower attenuation enables improvement of the immunity (ACR parameter) as both cables have an equal crosstalk loss. As a result, the SCS channel capacity increases. Therefore, many SCS of European producers, which are designed over 120-Ω cables (for example, Alcatel, Legrand, Infra+), provide for a significantly broad bandwidth as compared with common channels of the category 5.

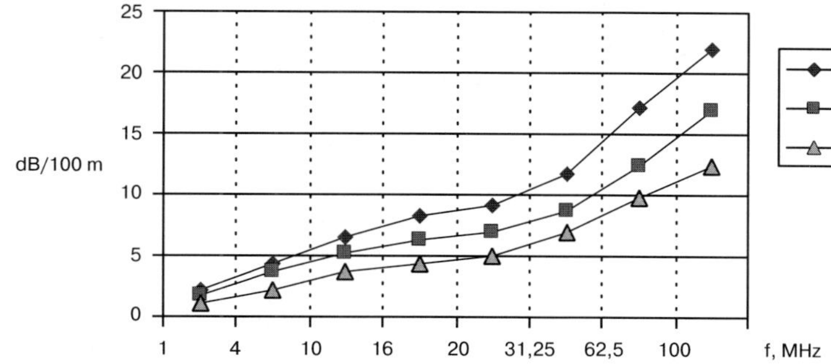

Fig. 3.6. Frequency characteristics of allowable attenuation of cables with different characteristic impedance according to ISO/IEC 118013.1.2.9

Packing of horizontal cables

Horizontal cables are delivered using two different types of packing: cardboard boxes and reels. A cardboard box contains a 305-m (1000-ft) cable to be reeled on an inner cardboard or plywood reel (see a reel-in-box type packing in Fig. 3.7b) or to be formed into a self-supporting winding (see a pull-box or reel-free in Fig. 3.7c). The end of the cable is brought out through a horizontal opening or plastic sleeve (in case of a pull-box). The sleeve positions roll out in the direction of the cable and protect the cable's jacket from rubbing the edges. Marking on the cardboard box provides information on the manufacturing company, cable type, and inner outlet length. Sometimes it also provides for the SCS logotype that the cable belongs to.

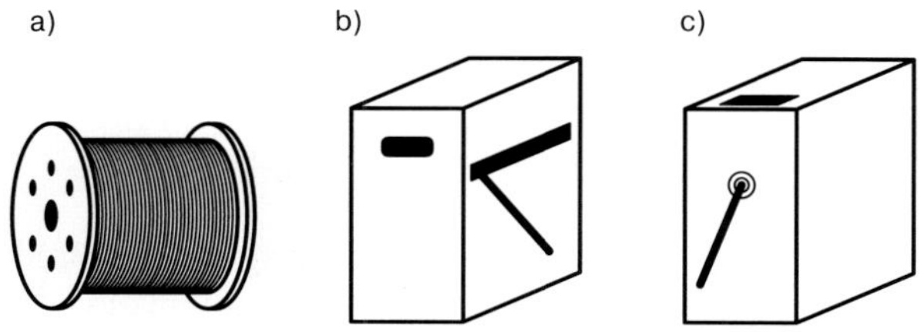

Fig. 3.7. Different packages for horizontal cables; *a)* reel; *b)* reel-in-box; *c)* pull-box

Boxes of another length (for example, BRAND-Rex and RiT Technologies supply 500-m boxes) are rare. Some solutions, which are basically aimed at home networks and SOHO buildings, may require a box with 100–120 m of cable (Molex and Micronet). Boxes are the most popular way to deliver material to SCS installers, as the packing type is convenient for storage and transportation and makes the reeling process easy and avoids auxiliary hardware.

An opening made in the box housing serves as a transportation handle. The box cover has no auxiliary holders such as metallic brackets, adhesive, or paper tape. This provides easy access to the cable drum when needed. Some suppliers provide a tear-off valve on the side of the box to access the other cable end in case of its storage as a reel-free cable.

Some tailor-made solutions for packing of 1000-ft segments are made by the company AESP. For example, boxes can be packed two-by-two into a cardboard package that ensures extra transport protection. The inner cardboard drum can be axially placed between two square pivot plastic frames for reeling the cable without using the cardboard box, if necessary.

There are two solutions to fix the free cable end coming out of the box opening in off mode position when using a self-bearing packing. One can fix the end into the lug or fit the end into the sleeve hole by a hollow plastic cap in the form of a bung.

Besides boxes with a common-length cable, up to 4% of the cable total are supplied by a few cable plants in short-length boxes, i.e., a common cardboard box contains a short-length cable. This is caused by production technological aspects (possible cable breaks, remnants, rejects, etc.). The packing is divided into 100- and 200-m ones, the 100-m packing may contain cable from 100 to 199 m in length and the 200-m one from 200 to 304 m. To encourage the acquisition of such packages and compensate the customer for some inconvenience that may arise from this type of packaging, the price is set as for the 100- and 200-m ones even though the cable may actually be longer.

Reel-in cable has common lengths of 500 and 1000 m (the company ITT NS&S also supplies 2500-ft, i.e., 763-m, reels). Longer cables are possible (the company Lucent Technologies supplies some modifications of the type 1071 reel-in cable for 1200 and 1500 m); however, a 1000-m reel may weigh up to 50 kg and more, which makes the reel awkward on site. The key advantage of the reeled delivery is that it provides for a fewer amount of off-cuts. They are less applicable, as the packing is inconvenient for transport and storage, and complicated when reeling out without using specific hardware. Plastics, wood, and plywood (Fig. 3.7a) are used in the manufacturing of reels.

Horizontal cable production

Four horizontal cables are produced by cable factories employing specific production equipment. This production uses a high technological and automation level; the working personnel's only function is loading machines with materials and intermediate products, takeoff of finished products, performance inspection, and product reset. The technological method splits up into steps in order to make changes between different cable type productions easy, as well as to ensure a selected or total quality control. All stages are successively performed on specific machines or individual conveyor equipment. The list of the main technological steps includes:

- Copper conductor dielectric shielding (the process runs at a speed of about 20–40 m/min).
- Pair twisting of two wires. Each pair is twisted on a pair-twisting machine with a given step of twist, and may be provided with an individual shield of a metallized film (Type PiMF).
- Four twisted pairs form a cable core. If necessary, together with the forming of the core body, it is covered with a film shield including a drain wire. Covering the S/STP cable core with the film shield is performed on a specific wire-wrap gun. As the machine provides for a low output, compared with other ones, the HIGHT Screen class structures are braided over the core film shield.
- Forming the core outer jacket with the help of an extrusion machine. The machine makes the screw feed of a granulated semi-product, which warms up. The process includes marking and control over breakdown voltage strength as well. A water-cooling system, which consists of a number of tubes with gradually lowered temperatures, is used to cool the warmed jacket after extruding. The machine dimensions are the largest ones among the other machinery.
- Cable cutting into 305-m pieces and forming them into a self-bearing winding, or reeling them onto a self-bearing reel, and packing them into a cardboard box.
- Selected or total inspection of products for NEXT and attenuation parameters.

Due to equipment low output capabilities the production of horizontal cable takes 24 h with a break only for maintenance and resetting of the machinery. The production process is almost fully automatically monitored; an operator can only visually control most of working areas.

3.1.3 Multipair Cable

Design specifics

Multipair cables are intended for use in backbone SCS for interconnection of telecommunications closets. The largest part of campus backbone cabling is horizontal while the building backbone cabling is vertical. Cables of such modification have a limited use for horizontal cabling in interconnecting a telecommunications closet with a transition point, consolidation point, or a 12-port outlet.

The reason to refer the cable to the group is that it contains more than four twisted pairs. A general distinctive attribute of the multipair solution is that the cable core has no individual jackets for pairs or quads. Moreover, some producers offer multiunit cables, which may be considered a specific modification of multipair ones. They are distinguished by the cable core, which forms not individual twisted pairs but two- and four-paired elements that are identical to horizontal cables by their construction, and is covered by an individual protective sheath.

The structure of a multipair cable depends on its capacity. If the cable core includes 25 pairs, no other additional elements are usually included. If there are more than 25 pairs, they are grouped in bundles of 25 pairs each, forming the cable core (Fig. 3.8c). Polymer tapes tie conductors of each bundle together. A central glass-fiber retainer can be used to serve as a multielement cable core base in order to increase the strength and tolerance against different mechanical impacts. An overall dielectric jacket protects the core. Along with unshielded multipair cables, S/UTP-structures, which have a shield under the outer dielectric jacket, are produced in limited quantities.

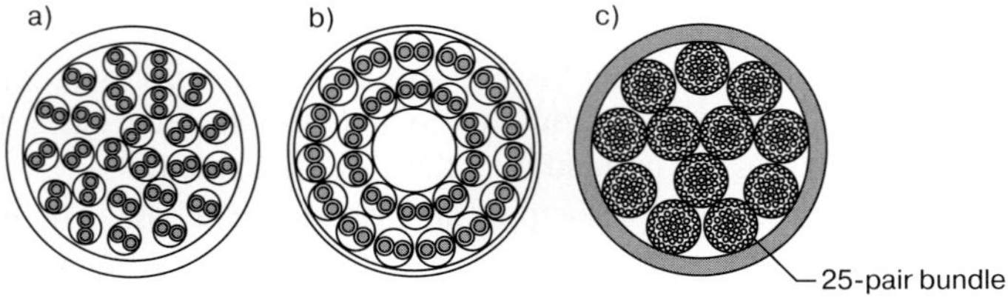

Fig. 3.8. Multipair backbone cabling: *a)* 25-pair category 5 cable (traditional); *b)* 25-pair category 5 cable with central retainer element; *c)* 300-pair category 3 cable

Because of its rather low capacity, the first generation category 5 multipair cable has showed a very irregular inner structure. The only means to organize the cable core in some products was to form five-pair bundles by means of colored tape. The early 1990s marked a rapid popularity of solutions with a regular core structure, which was mainly achieved by using a central power-supporting element (Fig. 3.8b).

Multipair and multielement cable jackets are marked similar to horizontal ones, including indication of type, diameter and number of conductors, testing laboratory, and length in meters and feet.

In order to reduce the attenuation factor conductors are manufactured from a solid copper wire. Similar to horizontal cables they can be grouped into categories 3–5; multipair configurations of category 4 are very rare in practice. Table 3.11 illustrates standard capacity for backbone multipair cables in pairs depending on the category. Sometimes products with a capacity other than that given in Table 3.11 are utilized. For example, the ISCS cable system of the company ITT NS&S does not use 110 panels (see Section 3.3.2); therefore, multipair cables have 24, 48, and 96 pair density. Multielement cables differ from their multipair analogues by having less capacity. For instance, there are solutions, which contain up to 24 two-pair and 16 four-pair elements.

Table 3.11. Backbone cabling standard capacity

Cable category	Number of pairs
3	25, 50, 75, 100, 200, 300, 600, 900, 1800
5	25, 50, 100

The linear mass of category 5 25-pair cable is equal to 180–190 kg/km, the operating range is –20 °C–+60 °C.

The cables of the reviewed group are intended for inner and outer cable routing. The principal distinction of the cable for external routing is using specific facilities and solutions to protect its cable core against moisture. That is achieved by specific methods; for example, polyethylene film is a widely used solution. The inner core voids of some phone cables can be filled in with gel. Since the end of the 1990s the use of this solution for SCS cables has grown. Additional protection of the cable core against moisture and mechanical impacts is accomplished by a protective cover made of aluminum or steel corrugated strip, for example, the ARMM and ARNW series products by Lucent Technologies. However, the internal cables are the only ones produced widely. Multipair cables for external cable routing, which meet the requirements of category 3 and category 5, have not become widely used.

Electrical performance

In general, requirements for the performance of multipair cables correspond to requirements for the performance of horizontal cables. By the way, it is necessary to note that interfering noises from neighboring pairs may sometimes be added to each other. This explains, as compared with horizontal cables, the relative higher weight of the construction. The loss of performance due to crosstalk had been initially classified by the power sum long before the respective standards were established.

Generally speaking, when multipair cable is exploited as a part of horizontal cabling, and used for connection by one cable bundle to several modules of a telecommunications outlet, the requirements for the crosstalk loss it causes become stricter. According to the ISO/IEC 11801 standard the *NEXT* value of such structures has to grow according to:

$$\Delta NEXT = 6 + 10\lg(n+1), \text{ dB} \tag{3.1}$$

where n is the number of neighboring electrical cable elements (pairs, quads) within a bundle.

The difference between the signal power of applications, being connected to a bundle, should not exceed 6 dB, which is corrected by the respective correction index in Equation 3.1.

It follows from the nature of crosstalk and the conditions of Equation 3.1, that the pairs closest to the exposed circuit significantly contribute to the crosstalk level at the near end. Pairs of other bundles do not affect the general crosstalk level as much. Therefore, if multipair cables serve several high-speed applications, then it is recommended to diversify their signals along different bundles. When using multielement units for channel organization such diversity of applications occurs automatically.

The probability of running a multipair cable according to its key assignment (backbone cabling organization) along with power cables is higher than for a horizontal cable. Therefore, the standard TIE/EIA-568-A set up strong requirements for breakdown voltage stress (conductor–conductor or conductor–shield) 3 c at 5 kW DC.

The same conditions exist for electrical performance testing of multipair and horizontal cables. The measuring instruments used are the same as well.

Mechanical characteristics

The standard requirements for the mechanical characteristics of multipair cables are given in Table 3.12. The first two parameters define the geometrical size of conductors. Violating the requirements leads to an incompatibility with IDC connectors in telecommunications closets.

The outer diameter of multipair cables is up to 20 mm and more. The multipair cables display a very high tensile strength. The tensile strength grows with increments of the diameter and stems from the necessity of utilizing additional hardening and supporting elements. It is indicated by the producer in the technical certificate.

With the decrease in the cable bend radius, mechanical loads on its elements increase. It is obvious that short-term loads are less harmful than permanent ones. Nevertheless, as the parameter of dynamic loads forcing the cable during its routing is poorly controlled; a smaller bend radius is permitted during operation as compared with cable routing (Table 3.12).

Table 3.12. Requirements for mechanical characteristics of multipair cables

Parameter	Figure
Diameter of conductors	0.5–0.65 mm (24–22 AWG[a])
Isolation diameter	≤1.22 mm
Temperature range without degradation of mechanical properties	installation: 0...+50°C operation: –20...+60°C
Minimal bend radius	
• cabling	≤8 outer diameters
• operation	≤6 outer diameters

[a]AWG values in millimeters, see Section 14.3

3.1.4 Auxiliary electrical SCS cable products

Patch cable

The patch cable or flexible cable is designated for the production of patching and modular cords. In most cases it contains four twisted pairs[2], and is designed almost identical to a horizontal cable. The key difference between a flexible cable and a horizontal one is that:

[2] One-, two-, and three-paired variants of patch cables are also available on the market in a limited quantity to be predominantly used for telephone applications.

- wires are produced from seven thin stranded conductors with a diameter of about 0.2 mm each (stranded wire)[3]; the Russian national technical literature accepts the term "multiwire structure" to designate such conductors;
- conductor insulation is about 0.25 mm thick. This is thicker than a horizontal cable;
- a high flexible material is needed for outer jacket manufacturing.

It should be noted that a flexible cable with an extra ninth conductor is used in the SMART cable system by the RiT Technologies. Its diameter is 0.4 mm (26 AWG), and it lies on the cable axis and functions as the cable core. It is intended to transmit signals, which are scanned by the cable PatchView system to develop a connection database.

The electrical performances of flexible and horizontal cables are almost identical, except for attenuation. In the first case it is much higher due to the construction of the current conducting core. The standard TIE/EIA-568-A defines that the peak allowable linear attenuation of a flexible cable may exceed the horizontal cable attenuation by 20%. The requirements of ISO/IEC 11801 on attenuation are not so strict. The standard allows increasing the attenuation factor of a flexible cable by 50%, as compared with a horizontal one (Table 3.13).

Table 3.13. Maximal allowable attenuation for patch cables of categories 3, 4, and 5 at 20 °C according to TIA/EIA-568-A and ISO/IEC 11801

Frequency, [MHz]	Attenuation, [dB/100 m]		
	Cat.3	Cat.4	Cat.5
0.772	2.7/3.3	2.3/2.9	2.2/2.7
1.00	3.1/3.9	2.6/3/3	2.4/3.0
4.00	6.7/8.4	5.2/6.5	4.9/6.2
10.00	11.7/14.6	8.3/10.4	7.8/9.8
16.00	15.7/19.7	10.7/13.4	9.9/12.3
20.00	–	12.0/15.0	11.1/14.0
31.25	–	–	14.1/17.8
62.50	–	–	20.4/25.5
100.00	–	–	26.4/33.0

In this regard, some products and, particularly, the standard ISO/IEC 11801 operates with such concepts as electrical and mechanical lengths, which form the prime relation as follows:

$$\textit{Electrical length} = (1+\Delta\alpha) \times \textit{Mechanical length} \qquad (3.2)$$

where $\Delta\alpha$ is the value of the maximal relative increment of linear attenuation (Table 3.14).

[3] In some foreign products this is emphasized by the fact that conductor diameter is indicated according to the following scheme: AWG24/7 or AWG24/7x0.2.

Table 3.14. Interaction between electrical and mechanical lengths

Mechanic length	Electric length	
	TIA/EIA-568-A	ISO/IEC 11801
1	1.2	1.5

In practice, the patch cord almost never exceeds 10 m in length; however, the attenuation performance is given as if for a 100-m cord so as to follow the same standard as for horizontal cables.

Along with horizontal cables the patch cords are also divided into categories according to their performance. According to the statistics from the second half of 2000, the cable industry manufactures products in the 5, 5e, and 6 categories. The products have displayed no other, except for those listed above, specific constructive distinctions from horizontal cables. We may only notice that in order to improve the performance of category 6 cables, a central separator is incorporated into the product (Section 3.1.4.), which leads to a degradation in flexibility, and therefore less demand for such a solution.

Mechanical performances of flexible and horizontal cables are almost identical, except for the fact that the minimum allowable bend radius, taking into consideration standard operating conditions, shrinks to 20–25 mm. For the optimal mounting of an 8-contact modular plug the diameter of the patch cord conductor's insulation has to be between 0.8 and 1 mm.

The patch cord experiences considerably higher and more frequent bending and tensile loads, as compared with a common horizontal cable. Because of this the patch cord has an extended working range (and thus an extended storage temperature range). Thus, for example, the patch cord R322900 of RiT Technologies operates within the temperatures of –20 and +60 °C (the storage temperature between –30 and +70 °C), while the operating temperature of the horizontal cable R3729070 by the same company is between –5 and +50 °C (the storage temperature between –20 and +60 °C).

The patch cord is produced in shielded (S/UTP and S/STP) and unshielded modifications. The color code of the conductors (Section 3.1.5) has to match with the color code of the horizontal cable, although the American manufacturers (for example, Quabbin Wire&Cable) may use other codes. The outer jacket bears almost the same marking and identification labels, as well the length label.

It needs to be noted that the patch cords being manufactured from the patch cable are used in telecommunications closets and user workplaces, which do not belong to a plenum class; therefore, fire-retardant and low-smoke jackets were not initially provided for them. At the end of the 1990s, with an increase in the number of ports for cable systems, and as a result, a considerable growth of the volume of cable products in equipment rooms, most manufacturers started offering jacketed patch cords that fully meet fire safety requirements.

Jumper wire

Jumper wire comprises one unshielded twisted pair of category 3 without a protective outer jacket. Conductors are produced of solid copper wire with a diameter of 0.51 mm and PVC insulated. One jumper is colored white according to the standard TIA/EIA-568-A; the other is usually dark blue or red. The jumper is basically used for the interconnection between patch panels and 66 type IDC contacts. Sometimes it may serve for wiring on type 110 boards. The standard packing used for a two-pair jumper wire 305 m (1000 feet) or 201 m (660 feet) long is a reel with a 15-cm diameter.

Along with the one-pair jumper wire, there are also other modifications: two-pair, three-pair, and four-pair ones. Its core is formed by twisting of twisted pairs with each other. Some manufacturers have recently started producing a jumper wire with 2, 3, and 4 pairs under an overall protective jacket, which is in general a traditional horizontal cable by its design. According to the designer's point of view, such a jacket helps to maintain the twisted architecture of a pair between connecting panel contacts and, therefore, a jumper made of such wire may possess the same performance of cables from category 4 and even 5.

The color code of multipair cable conductors completely corresponds to those used for labeling horizontal cable conductors, and is being revised (Section 3.1.5).

Under-carpet cable

The majority of horizontal cables are run to telecommunications outlets in cable ducts under raised floors and suspended ceilings by the use of specific hardware (inside decor columns, in floor and under floor ducts, etc.; Section 5.2.3). Sometimes it is not advisable or possible to apply some of the above solutions in practice. In which case, the only solution is to run cable over the floor. Conduits lying under raised floors, floor ducts, and special cables may be alternative solutions to this problem. Under-floor conduit solutions are designed beforehand in the project stage (see Section 10.2). Floor ducts are discussed in Section 5.2. We will concentrate our attention here on special cables for floor mounting without using additional protective hardware.

A special feature of floor or under-carpet cabling without using additional protective hardware is that a ribbon-structured cable is laid with a maximum pressure distribution along its surface. This can be achieved by using side wings with a tapered section to be attached to the outer jacket at their wide base (Fig. 3.9).

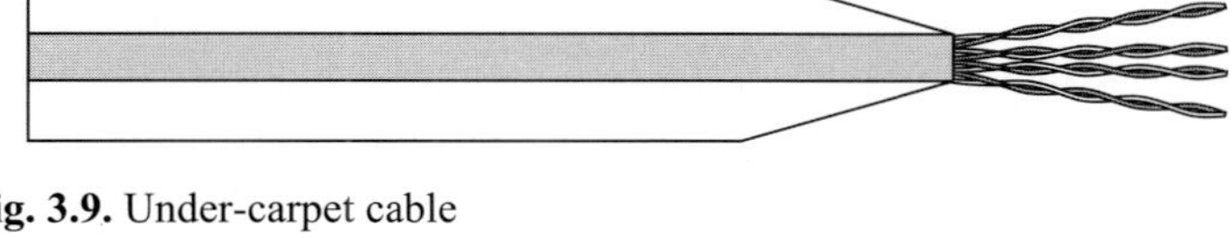

Fig. 3.9. Under-carpet cable

The maximum thickness of the wings is chosen to be equal to the outer diameter of the cable. The presence of the element makes it easier to mount the cable on the floor using brackets or adhesive tape and prevents the formation of kinks during routing.

General rules for designation of the cable type correspond to rules for common horizontal routing. Some restrictions are set out in the standard TIA/EIA-568-A. According to this document, cables for under-carpet cabling are prohibited for use in premises with high moisture, exposure to corrosive liquids, and routing along power lines with a distance less than 152 mm (6 in.). Telecommunications cables have to be laid where crossing a power lead. The under carpet cables are divided into categories 3 and 5. A brief list of some types is given in Table 3.15.

The common packing is quite small and usually contains cables such as ones 200–250 ft (61–76 m) in length because of the short length of cable needed under carpet runs.

Table 3.15. SCS cables for under-carpet routing

Producer	Article	Category	Package, [m]
Hubbell	DCTP04200	3	61
AMP	557874-1	5	76.2
	555318-1	3	76.2
PCnet	65504-98	5	61
	65304-98	3	61

Horizontal cables with frequency limit above 100 MHz

Today, the market for SCS components offers a number of fabric-made horizontal cables, whose performance considerably exceeds the category 5 standard requirements. General features for unshielded products of this group are:

- All the products have an ACR parameter of approx. 10 dB within a frequency range of 150–200 MHz and even higher, i.e., they correspond to the cable performance of advanced category 6 and often with a considerable margin.
- The ACR parameter is increased mainly due to the improvement of the NEXT parameter (by 10 dB and higher); though, a certain part may be achieved by degrading linear attenuation. Cables with attenuation equal to category 5 cables were widespread before. Cables with an essentially higher attenuation performance are in wide demand.
- Cable performance is certified up to frequencies of 350–550 MHz in order to transmit signals of one-way applications, which usually imply, in practice, multichannel on-air and cable TV.

At the same time, one can definitively divide the reviewed cables into two subclasses with frequency limits of 350 MHz and 550–600 MHz. The low subclass models frequently differ from the standard category 5 ones because of their higher NEXT and PS-NEXT parameters, while high frequency products show a high crosstalk loss together with a considerably degraded linear attenuation. At present, cables with the 300–350 MHz frequency limit are used to build category 5e channels, while cables with the upper frequency limit of 550–600 MHz are used as a part of category 6 systems.

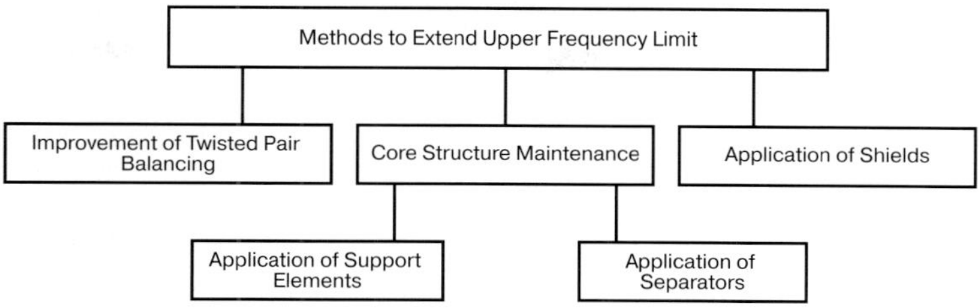

Fig. 3.10. Solutions for increasing the upper frequency limit for horizontal cables

A brief description of high-performance UTP cables is given in Table 3.16.

Table 3.16. UTP horizontal cables with frequency limit above 100 MHz

Producer	Type	Frequency limit f_l at ACR=10 dB, [MHz][a]	Upper standardized frequency f_u, [MHz]	Attenuation to f_u, [dB/100 m]	SRL to f_u, [dB]	Source
		Subclass 350 MHz				
Alcatel	LANmark	210	350	37.6	13.1	[46]
AMP	FUTURE LAN 350	140	350	44.9	16	[47]
Belden	1700A	160	350	37.7	30	[48]
Mohawk/ CDT	Mega LAN 400	160	400	48.5	12	[49]
Pirelli	DX1100-3	175	200	37	–	[50]
BICC Brand-Rex	Turbo-LAN 350	150	350	41.2	–	[51]
NK Cables	UC 400 24 4P	240	400	45	–	[52]
Elgadphon	Giga-Star	200	250	33	17.3	[53]
PCnet	Category 5 plus	100	350	44.3	–	[54]
		Subclass 550–600 MHz				
Lucent Technologies	1071, 2071, and 3071	180	550	55.3	15.9	[55]
Mohawk/ CDT	Advance Net 500	180	550	53.7	17.8	[49]
RIT Technologies	Giga UTP	185	600	55	–	[57]
Belden	1872A	165	550	55.3	16	[58]
BICC Brand-Rex	Turbo-LAN 1000	390	550	52	13	[51]

[a] f_l values of all the cables given in the table have been calculated from attenuation and crosstalk characteristics presented in materials from the companies

To decrease attenuation one increases the diameter of the copper conductor of a wire up to 0.55–0.6 mm (compared to the typical for category 5 twisted pairs of 0.51–0.53 mm) and employs insulated coatings, particularly made of foamed materials, with decreased dielectric loss. By the way, this leads to the somewhat larger linear mass and outer diameter as compared with the category 5 cables (Table 3.5).

The *NEXT* parameter may be increased in two ways (Fig. 3.10). The first way is to maintain the cable core structure during cable routing and operation by incorporating an auxiliary load-carrying element into the cable core. A central plastic retainer (type 1711AA cable by Belden) or a plastic separator of the C^3 type (central crosstalk cancellation) with a cross-sectional area in the form of a four-rayed star (Fig. 3.11) can be used as the element. The first cables of this structure appeared on the market in 1998. However, since 2000 their popularity has grown rapidly (LANMark cables by Alcatel, 7812A by Belden, 81-type by Lucent Technologies, R3729XXX type by RiT Technologies). Configuration with the central separator provides additional diversity for each pair by its positioning in an individual slot, which leads to a considerably high NEXT, and thus ACR (it is clearly shown in Table 3.16 for the LANMark cable performance).

The second way is to support a high balancing accuracy of twisted pairs, i.e., a step of twist. The following approaches can be realized here. The companies Lucent Technologies and NEK/CDT accurately control a step of twist during cable manufacturing with a slight decrease in the step of twist, as compared with category 5 products. The company NEK/CDT uses an additional twist of pairs in the core, which also provides for an increase in the NEXT parameter. The company Belden promotes the second approach by bonding conductors in the pair. This provides a very high flexible cable for different kinds of minor radius bending; however, it makes its connection into outlets and panels more complicated. In order to maintain the core structure during cable running, the cable, if supplied without a central load-carrying element, is often delivered on reels. A specific PS-NEXT gain is ensured by the diversity of individual twisted pairs. To achieve this PS-NEXT gain the outer jacket with an oval-shaped sectional area, the so-called flat cable, is used instead of a traditional round one (Fig. 3.12). MediaTwist by Belden and 9C6R4 by Siemon [26] serve as examples of such a cable.

A restriction on the margins of the conductor diameter, as well as the conductor eccentricity and insulation (up to ±1% compared with the standard, ±3% for category 5 cables), will minimize the structural return loss.

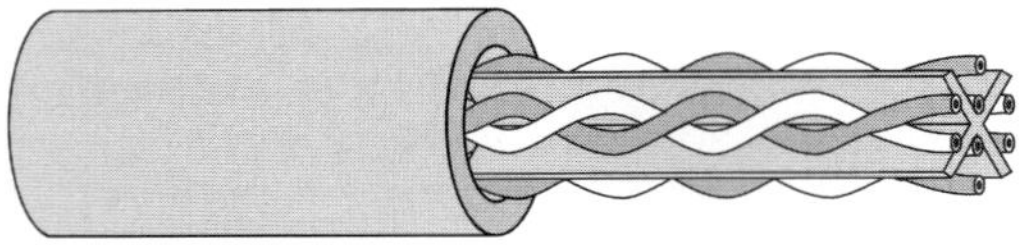

Fig. 3.11. Horizontal cable with separator

Many technicians admit that the current technology enables a further increase in the standard SCS horizontal cable operating frequencies, without individual parameter interpolation for pairs, with the possibility of using them in data transmission systems (i.e., with the criterion of ACR=10dB), but only for shielded cables.

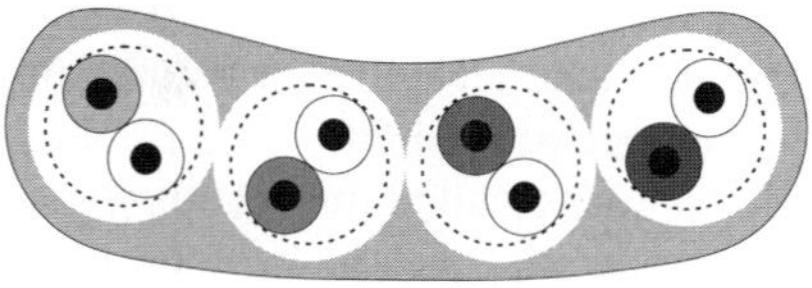

Fig. 3.12. Flat cable (MediaTwist type) by Belden

Utilization of S/STP shielded cables enables raising the upper frequency limit significantly. Thus, the major cable manufacturers commercially produce this cable type, and its pattern is certified for frequencies up to 1.0 and even 1.2 GHz. Its application areas are SOHO systems, class F applications signal transmission, and the development of category 8 systems, which at present are being discussed within the academic community [58]. We have to note that the upper limit frequency of the cables for the parameters above 900 MHz enables the cable to be widely applied, for example, during the development of multichannel cable TV systems for an end user, with an upper frequency of 862 MHz in Europe, 855 MHz in USA, and 765 MHz in Japan [59].

Two factors hamper the mass utilization of this cable type:

- A lack of official standards for norm parameters above 100 MHz in communication lines.
- A lack of high-quality widespread high-frequency connectors, which would make it impossible to use potential cable performance as a signal transmission platform in the full range (for details, see Section 3.2.6).

Essentially, higher requirements applied to the manufacturing technique used in mass production have significantly increased the product cost (approximately by 30% for UTP cables with a limit frequency of 350 MHz) as compared with a category 5 cable; by the middle of 2000 the cost of the cable exceeded the category 5 product roughly by 10–15%.

A considerable increase in production and utilization of minor subcategory cables was noticed at the end of 1999 after the major leading SCS manufacturers stopped the promotion of category 5 systems in order to transfer to the category 5e ones.

Bundled cables for SCS horizontal cabling

Cables that include a few types of cable elements, i.e., optical fibers (including single-mode and multimode), twisted pairs of different categories, coaxial tube, and other cable elements, are called bundled, composite, or hybrid cables [60]. The bundled cables, which include optic fibers and a twisted pair, are utilized in SCS construction. The reason for using those is the fact that a telecommunications outlet in a certain work area is supported by one optical and one electrical port in some cable systems. The key advantage is that the cables are smaller in their dimensions and their routing is cost effective as compared with the routing of several individual cables.

Apparently, American companies have become pioneers in the mass production of bundled cables for specific use in SCS engineering. Thus, by the middle of the 1990s there were cables, which included a four-paired element of UTP, one element of type 1 IBM, and a 2- or 4-fiber optic cable (so called dual- and tri-media-cables). Thus, the elements could be placed under the jacket, or each element could be individually covered and joined together by string tapes (Fig. 3.13). More specifically, they were mixed components formally integrated into a whole. By the end of the 1990s bundled cables were not wide spread and used only for specific communications needs.

The need for special products has emerged from the above-described increase in solutions based on optical line routing directly to a work area. As a result, the standard set of equipment of some SCS manufacturers contains bundled cables. The new products differ from the previous ones by a distinctly high packing density due to the product-specific design. For example, the MNC880FO cable included as a part of a SCS from the company INFRA+; it also contains two 62.5/125 fibers along with two four-pair elements of category 6 (quad twist).

The second option stemming from a bundled architecture to build up SCS horizontal cabling is the blotwist cable of the Blolite system. It will be discussed in detail in Section 4.7. We have to note the highly increased popularity of hybrid cables with a coaxial tube and a four- paired element of category 3 for use in the cable configuration in a residential area.

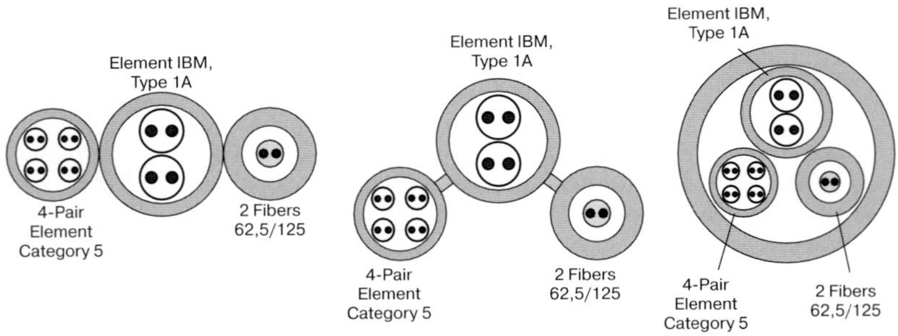

Fig. 3.13. Varied constructive solutions for bundled cables

A bundled cable may be used as one of the main SCS elements of a particular manufacturer. For example, the solution has been chosen by specialists of the companies Ackermann and Huber+Suhner. The AMCS (all media cabling system) is based on a bundled cable with two twisted pairs of category 5 and four optic fibers. The solution has the most advantages when structuring centralized optic architectures with optic fibers for the transmission of LAN signals and twisted pairs for connecting telephones.

3.1.5 Color Code of SCS Electrical Cables

When installing SCS cables it is important to create an effective identification method for twisted pairs and their conductors in electrical cables and for optic fibers in optic cables. Because the dimensions of the elements are so small it is inefficient to employ marking inscriptions; therefore, a color code is used to mark insulation and other elements of the cable core in different colors. Alphanumeric labeling, abbreviations, symbols, and various icons are used solely for outer jackets.

The standards IEC 708-1 and TIA/EIA-598 are the main regulatory documents for defining norms for color coding and marking. They follow the same approach and specify twelve basic marking colors (Table 3.17). Electrical balanced cables apply the first five colors to designate a pair in a group, the last seven for the color of the group. Thus, the first conductor insulation of a pair is colored like the group, but the second conductor insulation is colored like the pair.

Therefore, insulation of the first insulated conductor of a cable pair in four-pair cables is colored white (cells with color coding of some widely used cables are marked dark). See Table 3.18. The compact twist with the main conductor defines, which specific pair the white-colored conductor belongs to in UTP and S/UTP cables of categories 5 and 6; its identification is not complicated.

Table 3.17. Color codes of conductor insulation and fiber cladding of SCS cables according to TIA/EIA-568-A and ISO/IEC 11801

No.	1	2	3	4	5	6
Color	Dark blue	Orange	Green	Brown	Grey	White
No.	7	8	9	10	11	12
Color	Red	Black	Yellow	Violet	Rose	Blue

Collar markings, which are arranged on the cover every 10–15 mm, are used according to the standard to identify a second conductor of a pair in cables with step of twist of twisted pairs above 38 mm, i.e., actually category 3 cables, and the color code has to match with the color of a cover of another pair's conductor. This implies a "two-colored" method to mark insulating conductors; the term "two-colored" is frequently used in specifications and among professionals. For example, a W-BL conductor or "white of blue" (or white–blue) means a white conductor with blue markings (or, a white conductor twisted with a blue one). Several cables of category 5 (by companies Belden, Plastic Insulated Cable, and others) employ the same method, probably due to an unwillingness of manufacturers to change the common manufacturing method. In practice, the cover of a white conductor of a pair may be also marked by a color line arranged along its axis. This is especially convenient for modular cords, as it facilitates identifying conductors where they terminate into plugs.

The IEC 708-1 specifies norms for the color code of backbone cabling, which are given in Table 3.18. Within each of the 25-pair bundles the twisted pairs are divided into 5 groups of 5 pairs each. One of the connectors of each pair within the group has a specific color of insulation, a separate color for each group. The color numeration of groups begins with color 6 (white) according to Table 3.17. The second conductor of the pair has its own color depending on the number of the pair within the group. Each cable bundle containing more than 25 pairs is additionally wrapped by two-color polyethylene tapes and straps to color code the cable with the same method as is used for pair numeration marking.

Along with those given in Table 3.17 some other color codes are possible, which are defined by national regulations and companies' standards. The study in [1] presents a full range of color coding guidelines from different manufacturers.

It seems to be an easy process to color the insulation of conductors in twisted-pair cables, however, it is a rather complicated procedure in practice. One needs to keep in mind two factors when coloring. One of the factors is that the polyethylene used in most cases as a material for conductor insulation is quite color-proof. The other factor is that colors contain metallic particles, which may cause a degradation in electrical performance in both the twisted pair and the cable. For instance, a brown color causes a considerable growth of capacitance between conductors and essentially increases the attenuation in twisted pairs.

Table 3.18. Color code of pairs in multipair cable bundle

Group	1st wire color of pairs in group	2nd wire color				
		Pair 1	Pair 2	Pair 3	Pair 4	Pair 5
1	White	Dark blue	Orange	Green	Brown	Grey
2	Red	Dark blue	Orange	Green	Brown	Grey
3	Black	Dark blue	Orange	Green	Brown	Grey
4	Yellow	Dark blue	Orange	Green	Brown	Grey
5	Violet	Dark blue	Orange	Green	Brown	Grey

3.2 Electrical Cable Connectors

Twisted-pair connectors provide SCS cable connections with switching equipment in telecommunications closets and telecommunications outlets in workplaces, as well as for network equipment. These elements have to meet following technical requirements:

- minimum attenuation;
- high crosstalk;
- minimum SRL;
- small DC resistance;
- timing and temperature stability of characteristics;
- easy cable mounting;
- easy connection;
- good figures for dimensions.

A number of solutions that are discussed below were created in order to meet the requirements.

3.2.1 Mechanical and Electrical Parameters of Connectors

Termination to connectors

The method of terminating the cable conductors to contacts of a connector plays an important role for electrical, frequency, and maintenance performance of an electrical signal transmission channel. It is known that termination is carried out by different methods. For instance, the standard IEC-352 specifies winding, crimping, pressing, and screwing terminations. Soldering and screw-joint termination also have to be included here. The IDC method is used for termination within the SCS. Such a connection is further referred to as an IDC contact[4]. This technology, as compared with other technology, is simple and possesses high temperature and temporal stability as well as vibration strength. In some cases, multiple termination and operation in environments highly polluted by aggressive industrial contaminants are available.

The IDC method is always based on the application of a double spring cutting-edged contact with a conductor to be terminated into its opening. The cutting edges fit through a slit in the insulating cover to provide an electrical contact to the conductor (Fig. 3.14).

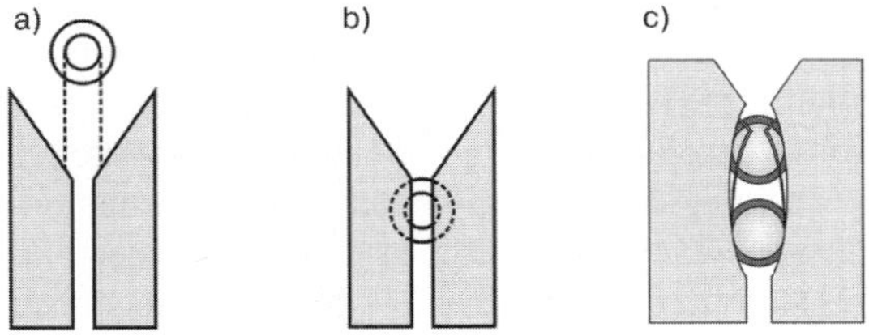

Fig. 3.14. Termination to connector using the IDC method: *a)* before termination; *b)* after termination; *c)* termination of two conductors at a type KATT contact

Cutting into the copper of the conductor provides for a not-so-high transfer impedance value. The effective area of the connection is extended with time due to diffusion, and that results in some improvement of the electrical performance. A fine blade and no mechanical tension provide an airtight connection. Based on this, a further revised KATT contact by Molex is sometimes called in some com-

[4] The IDC method has not received an exact interpretation in Russian technical literature. Various journals being translated or written by domestic authors offer such terms as “in-cut method”, “method of in-cut contact though insulation”, “insulation displacement method”, “through insulation contact method”, and others. None of the terms, from the point of view of the authors, is perfect, and therefore the term IDC-contact will continue to be used.

panies' specifications the airtight IDC. There is no ingress of oxygen, and, therefore, no oxidation and galvanic corrosion effects appear. Several principal modifications of IDC contacts, which differ from each other by shape and cutting-edge disposition, (Fig. 3.15) have come into wide practice:

- type 110;
- type 66;
- type Krone or LSA-Plus;
- type KATT.

The use of IDC-contact tubular modifications is quite limited. The general performance of principal contact variants is shown in Table 3.19; the termination method and contact disposition are shown in Figs. 3.14 and 3.15, respectively.

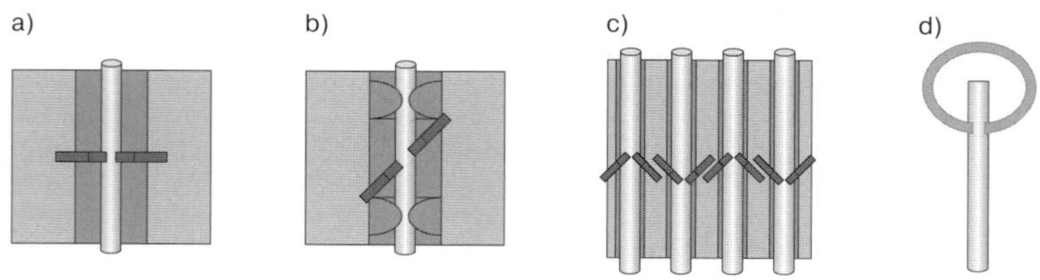

Fig. 3.15. Disposition of contacts in different IDC connectors: *a)* 110 and 66 types; *b)* Krone; *c)* KATT *d)* Tubular contact

All IDC contacts discussed below are designed to ensure the electrical and mechanical performance at a minimum of 100 on–off cycles. Some IDC contacts for a single application were popular at the end of the 1990s. The designers ensured a high performance of connections and other switching elements to operate on frequencies far above 100 MHz. The single-use solution enables a deformation-free termination into the IDC contacts, which is impossible at multiple connections. At the same time, the cutting edges survive multiple terminations unlike its housing.

Table 3.19. General characteristics of different IDC contacts

Contact type	Cutting-edge disposition	Possibility of connecting more than one conductor	Category
110	Right	Impossible	3–5
66	Right	Possible	3[a]
Krone	Angular	Possible	3–5
KATT	Angular	Possible	3–5

[a]Category 5 in variant 66M1

The cutting edges of 110 and 66 contacts are right disposed unlike Krone and KATT contacts with angled cutting edges. The 110 contacts are not intended for the connection of more than one conductor, the rest of the contacts enable (in cases with certain limitations) a parallel connection of several conductors, which is extremely important for telephony. The possibility to connect several conductors to a contact is realized by an extension of the cutting element (less cutting-edge diversity) or use of a cutting edge to be specially shaped. A sample of a standard product, representing the first approach, is the 66 contact. As using the approach leads to inevitable deterioration of the contact electrical performance due to high self reactance, it is no longer used in up-to-date solutions for high-speed data transmission channels. The best known example of the second solution is the KATT contact with crescent-shaped cutting edges of the blades optimized for the termination of two conductors (Fig. 3.15c).

The cutting edges of the 66 type IDC contacts are arranged in parallel and have a sufficiently large length (about 25 mm). The type 66 IDC contacts are primarily intended for the connection of jumpers and widely used in SCS to serve low-speed applications. For example, they enable the creation of parallel telephone lines for PABXs, etc.

The 110 type contacts differ from the 66 type ones by fewer geometric dimensions and an initial orientation for use in high-speed circuits of category 5. When they were developed it was not intended to create parallel connections; therefore, they are intended for use at a connection of only one conductor. Similar to the 66 contact, blades of the 110 contact are in a parallel disposition and are perpendicular to the conductor axis in the on-position.

The contacts with an angled position of blades differ by the fact that the working surface of the blades cuts through the conductor copper with its angular cutting edge, not with its plane as in the 66 and 110 contacts. Because of this, no wedge-shaped slits, which are vulnerable areas for corrosion, appear at the point where the blade connects with the conductor [61]. These contacts are known in two basic modifications.

Krone elements have blades located in parallel and rotated at a 45° angle to the axis of the wire[5]. Due to the fact that cutting-in points of the active edge of the blade are axially located, the cross-sectional area of the terminated conductor diminishes very slightly. Permanently acting torsion forces resulting from blade flexibility and distributed along the conductor enable achievement of a higher stability against possible mechanical impacts. Silver plating its blade working surfaces also increases the maintenance reliability of the contact.

The principal drawback of the Krone contact is that after pressing the conductor into the contact, an irregularity in the given orientation of the conductor occurs due to the angular position of the blades. This caused the designers to extend the length of the working element itself by incorporating a long plastic guide stub, which negatively affects the mass and dimensions of the product. This drawback is eliminated with the KATT[6] contact from the company Mod Tap by use of angularly positioned blade-edged working elements; however, in contrast to Krone contacts, the contact surfaces make a close to 45° angle. A supplementary distinc-

[5] Even the term LSA-Plus 45° to designate such a contact feature is used in some publications.

[6] The contact was designed in 1992 and initially was aimed at operating at frequencies up to 350 MHz

tion from analogues is that the KATT module has the angle of its neighboring contacts turned out in opposite directions (Fig. 3.15c). This provides for a very high stability against axially pulling mechanical impacts.

The idea of introducing supplementary clamping elements to improve service reliability is present in the 110 contact by Reichle & De Massari. Like the Krone contact, the contact uses a plastic guide-stub in a long groove; however, it differs by making several lugs available that directly contact the conductor jacket at several points. As a result, the present modification of the contact shows more vibration stability.

The main advantage of a tubular contact is its potentially small dimensions. However, the tubular contact has not become that popular because of the strict cutting allowances of the conductor and the impossibility of applying group instruments for the installation of several conductors of the same type. Some improvement in the field performance of the tubular contact was made by the company Raychem. Its product has a cutting-edged tube top that cuts out the excessive conductor and makes the mounting procedure substantially more comfortable.

Electrical characteristics of twisted-pair connectors

The NEXT values specified by current versions of basic SCS standards (TIA/EIA-568-A, ISO/IEC 11801, and EN 50173) are given in Table 3.20. Figure 3.16 presents a bar diagram with figures of NEXT attenuation for cables and connectors of different standard and potential categories. Their comparison shows that standards specify stricter requirements for NEXT than for cables. This can be explained, obviously, by the necessity of creating a certain crosstalk margin, as its value is strictly affected by the installation quality, which cannot, naturally, be controlled by connector manufacturers.

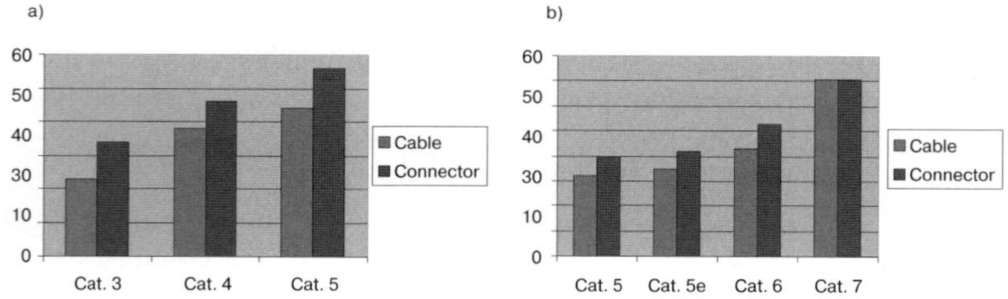

Fig. 3.16. Requirements of current and prospective standards on the NEXT parameter of different categories cable and connectors: *a)* at a frequency of 16 MHz; *b)* at a frequency of 100 MHz

Due to the linear disposition of contacts, NEXT values for different combinations of pairs are different (see the example in Table 3.21). The standards specify only the worst crosstalk value, as do the majority of manufacturers. A hexagonal diagram makes it easy to estimate the actual NEXT value (Fig. 3.17).

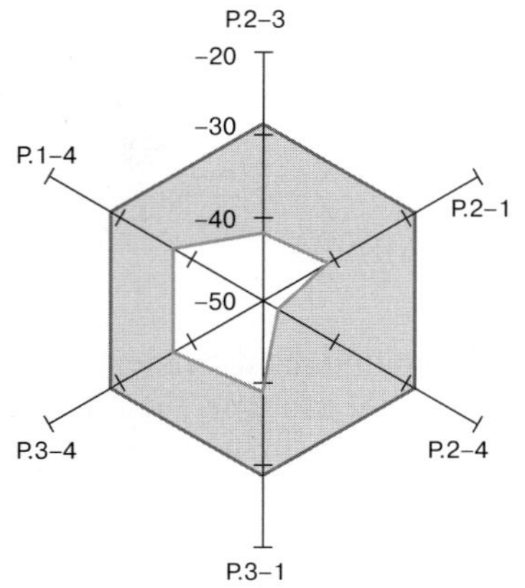

Fig. 3.17. Hexagonal diagram for NEXT

Each line of the diagram corresponds to one of the pair combinations, and shows a given and actual NEXT value at any fixed frequency (mostly at the upper frequency limit). The indicated points are joined by straight lines, the space between them (marked by a color) indicates the actual value of the guaranteed crosstalk margin. The hexagonal diagram demonstrates the quality of a specific connector; if it is necessary to get an extended noise-immunity margin for any application, it enables the selection of the respective pairs immediately.

Table 3.20. Maximal allowable attenuation and NEXT for connectors, dB

Frequency, [MHz]	Category 3		Category 4		Category 5	
	Attenuation	NEXT	Attenuation	NEXT	Attenuation	NEXT
1.00	0.40	58	0.1	>65	0.1	>65
4.00	0.40	46	0.1	58	0.1	>65
10.00	0.40	38	0.1	50	0.1	60
16.00	0.40	34	0.2	46	0.2	56
20.00	–		0.2	44	0.2	54
31.25	–		–		0.2	50
62.50	–		–		0.3	44
100.00	–		–		0.4	40

Table 3.21. NEXT value at a frequency of 100 MHz for different pair combinations of SMART 24 series panel modular connectors by the company RiT Technologies

Contacts	1–2	3–6	4–5	7–8
1–2		42.4	54.0	54.0
3–6	42.4		42.1	42.0
4–5	54.0	42.1		46.7
7–8	54.0	42.0	46.7	

Besides table and graphical presentation of the frequency-dependent crosstalk parameters, it is sometimes useful to analytically define them by approximation formulas, similar to those used for cable description. Thus, for example, the frequency performance of NEXT for a high-frequency connector GG-45 may be calculated as:

$NEXT > 80$ dB at $f < 31.25$ MHz;
$NEXT = 80 - 20\lg(f / 31.25)$ dB at $31.25 < f < 600$ MHz,
where f is the signal frequency (given in MHz).

Designers of network equipment take into consideration the dependence of the crosstalk value on numbers affecting each other. Thus, for example, the ATM system provides for signal transmission along pairs at their maximum diversity from each other in a connector (it corresponds to 1 and 4 pairs for 568 series arrangements).

Attenuation values of different category connectors are specified by the standards within the operating frequency band, which is given in Table 3.20. For category 3 connectors the standard ISO/IEC 11801 additionally specifies values of both maximum and average attenuation within the operating frequency band (up to 16 MHz) after the installation procedure, which is equal to 0.4 and 0.2 dB, respectively, irrespective of frequency.

The current versions of technical documentation specify values of SRL only for connectors of categories 4 and 5 and, according to the ISO/IEC 11801 2000 version, are shown in Table 3.22. At the same time, frequency values are, as usual, given in MHz.

Table 3.22. Minimal allowable return loss level in a connector for twisted pairs according to ISO/IEC 11801, dB

Frequency, [MHz]	Category 4	Category 5
1–20	23	26
20–100	–	26 – 20lg(f/20)

Input to output DC resistance of SCS twisted-pair connectors should not exceed more than 0.3 Ω by the standard ISO/IEC 11801. An excess of the recommended value indicates poor termination and degradation of other electrical parameters.

The coupling resistance value, which is important for screened structures, is 10 mom at a frequency of 1 MHz and 200 mom at a frequency of 10 MHz according to the standard ISO/IEC 11801, 2000 version. It is supposed that the given values and the parameter value itself may be more accurately defined at higher frequencies.

Some companies guarantee superior values of electrical parameters for their products, as compared with the given values. Thus, for example, the company Telegartner specifies for its connectors MJ-45LFS of category 5 a maximum attenuation of 0.14 dB at a frequency of 100 MHz (given value of 0.4 dB). The value of the secured NEXT of production connectors of major manufacturers is usually given as much as 1–3 (rarely 5) dB as higher than the recommended requirements in order to create a crosstalk margin.

It is planned that new versions of the standards will extend the list of the specified parameters for electrical connectors. As an example, requirements for NEXT and FEXT values for different categories of connectors are presented in Table 3.23 [62].

Table 3.23. Requirements of prospective standard versions of crosstalk attenuation value at a frequency of 100 MHz of electrical connectors in different categories

Category	NEXT	FEXT
5	40	Not specified
5E	43	35
6	54	43

Mechanical characteristics of twisted-pair connectors

The current versions of the main technical documents specify a long list of mechanical parameters for twisted-pair connectors and other functionally related elements without specifying a particular type of connector. The nominal diameter of a wire may be selected to be 0.5–0.65 mm; it may be a solid or stranded wire. The number of conductors terminated to a plug module of a telecommunications outlet at a workplace is equal to 8. In other places connectors are intended for an integer number of pairs. The contact element of a connector is not to have less than 750 on–off cycles without a degradation of its electrical parameters. The standard TIA/EIA-568-A also specifies connector operating temperatures between –10 °C and +60 °C.

3.2.2 Modular Connectors

General description

The modular connector was initially designed for use in telephone systems. The process of its connection and disconnection is very easy (one linear action without using special hardware). This made the connector very popular in the data transmission sphere, and in 1987 it was certified for ISDN systems. At present, this type is often used for SCS construction and is applied by all three SCS systems both for switching of telecommunications equipment and for the connection of network devices for different purposes (Fig. 3.18).

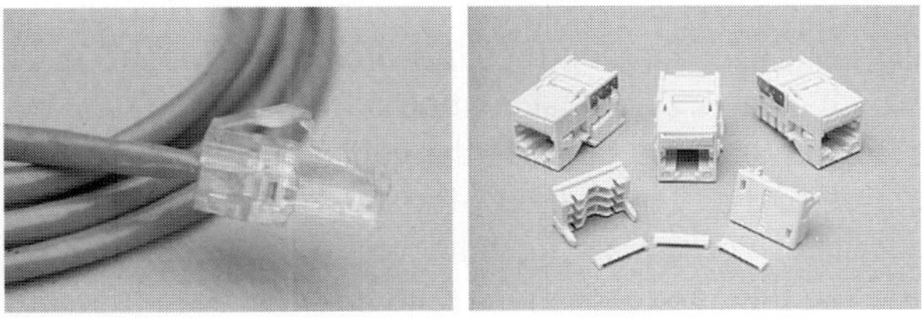

Fig. 3.18. Plug and jack of an 8-position modular connector

The connector consists of two parts: plug and jack, and is structured under the "contact bar" principle. According to the principle, plug contacts when connecting slide along jack contacts, and provide for a reliable galvanic continuity due to their flat, long construction, and at the same time push back dirty particles to make the contact ready for the next connection. Unfortunately, the lower disposition of jack contacts promotes their easy pollution during operation, and, therefore, the jack is additionally protected by different methods that are discussed below.

The housings of the plug and jack are made of thermoplastic. The plug and jack contacts, interacting with each other when the connector is assembled, are usually made of beryllium alloyed copper (beryllium copper). The working surface of the contacts may be finely gold plated. The method provides a guaranteed high-quality termination. A number of companies present several modifications of modular connectors, which are gold plated with different thicknesses. Thus, their cost and durability will differ as well. The contact parts of the plug and jack of modular connectors made by Lucent Technologies and Mod-Tap have a two-layered plating of the working surfaces: a gold-plated layer of 1.27 mcm over a nickel layer of 2.54 mcm (50×10^{-6} and 100×10^{-6} in.). Besides the nickel undercoat, brass can be used in some cases as the undercoat material in two-layered coatings (the DeltaCom company).

When designing contacts, along with the selection of materials, special attention is also paid to their form, as only these parameters provide the required elasticity and durability. A high-quality contact survives several hundreds and more on–off cycles without degradation of electrical characteristics. The majority of the leading manufacturers specify a value of 2500 for the parameter, which complies with the current versions of the standards. The company Panduit guarantees 1000 on–off cycles for its MBX products.

It should be noted that modular connectors were initially defined by the international standard IEC 603-7 [63] and specified only for the frequency of 3 MHz. Their design was so good that after a number of design improvements it become possible to increase the upper frequency limit by almost twice as much. The best models could demonstrate up to 200–250 MHz.

The architecture of a modular connector enables the production of a shielded option for assembling on shielded twisted pairs. To do so the plug housing is metal coated by metal sputtering or embedding to provide electrical contact with the cable shield after crimping. The reliability of the cable and plug shield's galvanic continuity is achieved by applying special techniques, which enable the housing of the mounted plug to be fully crimped around the cable screen. The jack contact as well as the IDC contacts to be terminated are shielded by metallic housings. When connecting the plug into the jack their metallic elements are in close contact with each other, which creates electrical continuity of the shield.

Eight-contact modular connectors, the same as the cables, provide for characteristics of the categories 3, 4, and 5. The market offers a wide range of modular connectors, whose characteristics correspond to standards of the 5e and 6 categories. When terminating into the plug and jack special attention is paid to maintain the original twist of the twisted pairs, which cannot be damaged more than 13 mm for category 5 elements and 25 mm for categories 3 and 4. Breaking this rule leads to a dramatic decrease in all types of crosstalk.

Plugs of modular connectors

The plugs of modular connectors are intended first of all to be mounted on flexible cable, though they may also be mounted on common horizontal cable. Geometrical parameters of the cable are shown in Table 3.24.

Table 3.24. Geometric parameters of cables to mount an 8-position modular plug

Diameter of conductors in pair, [mm]	0.5–0.65
Outer diameter of conductor insulation	0.8–1.0
Outer diameter of outer jacket	4–6

The conductors of cable twisted pairs are connected to the plug contacts by the IDC method. The contact plate is equipped with sharp edges, which cut through the insulation when crimping, and provide electrical contact with the copper conductor. The modular connectors have the plugs with contact-blades architecture, which differ for cables with solid and twisted wires.

The contact blades of a solid wire have a half-round form with two notches at the ends. It cuts through insulation when crimping, and tightly surrounds the wire from above and the sides, pressing it to the plug body. The contact blade of a stranded wire has a sharp notch in the middle; when crimping the plug the blade presses the conductor to the body, and the lug fits between its wires. Mounting the plug on a cable with an incompatible type of wire leads to poor contact during operation. There are also plugs with multipurpose blades (Fig. 3.19).

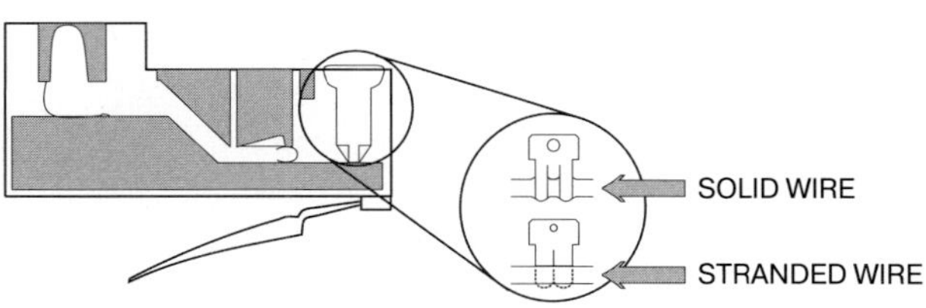

Fig. 3.19. Form of blade edges of plug contacts for modular connector

The housing of the majority of plugs is made of a colorless transparent plastic, which enables visual control of the position of conductors and the wiring quality. The use of colored transparent plastic is extremely rare. The material can be used to meet special requirements for color coding patch cords. Because of low production, the plug's cost increases by approximately 20–30%.

A special manual crimping tool is used for the field termination of a plug on a cable. The companies ITT Datakom, Panduit, AMP, Lucent Technologies, Reichle & De Massari, and some others have designed plugs whose construction enables termination without using any tools (tool-free termination solution). A removable plug cap performs the functions of a pressing element, which provides a galvanic continuity of conductor and contact. The plug IDC contacts may sit in one or two rows. The majority of such products have electrical characteristics of category 3, some companies guaranty characteristics of category 5 and even higher.

The use of guide channels at contact blades provides for a correct sliding entry of conductors into the plug. This can be accomplished by two different techniques. The first technique, which is popular in practice, enables the use of thin parallel lugs 5 mm in length, which are formed immediately in the mounting slot of the plug housing. The other technique enables lacing conductors through a special mandrel with the same inner lugs as beforehand, and then fitting it in the housing.

The standard TIA/EIA-568-A specifies schemes and procedures for cable stripping to crimp the modular plug. The stripping scheme is shown in Fig. 3.20. The cable jacket is cut about 20 mm from the outer end. The cable end is made flat. The pairs are untwisted prior to the jacket edge and grouped in the correct order according to the selected wiring scheme. Conductor 6 is made to cross over conductors 4 and 5. The distance from the jacket where conductor 6 crosses over conductors 4 and 5 should not exceed 4 mm. The trimmed length of the conductors should be about 14 mm, measured from the jacket. Prepared for termination cable is inserted in a plug so that its jacket penetrates the plug by at least 6 mm. The plastic strain relief ensures a reliable holding of the cable jacket.

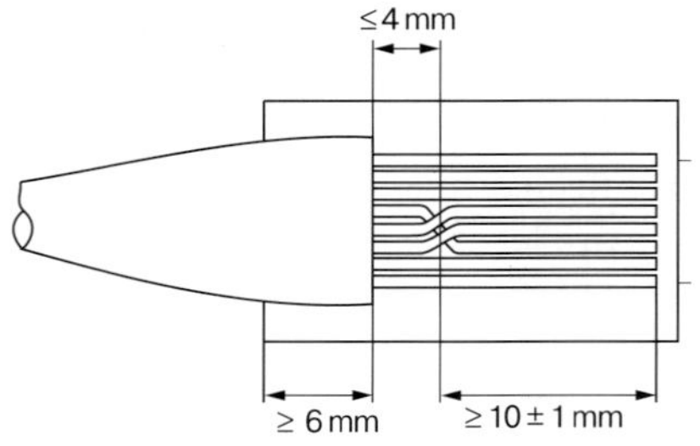

Fig. 3.20. Cable stripping scheme to terminate into modular connector

The procedure requires a high cutting and trimming accuracy of conductors prior to their entry into the plug. The WPT-8 is a tool made by the company Panduit, which comprises a template to set in individual connectors for exact trimming with common cutters or clippers. The trimmed cable terminates into the plug with a slot, shown in Fig. 3.21g.

In a number of up-to-date products within the field, where blades of the contact plates make an in-cut, the dual-in-line disposition of connectors in two different planes with its displacement half the distance between conductors, is applied in order to improve the electrical performance of the connector by maximally maintaining the fabric-made twisting of wires in twisted pairs. To insert insulated conductors correctly, the opening for a cable entry is specially designed (Fig. 3.21g). Such a solution was accomplished, for example, by Lucent Technologies and Panduit. Lucent Technologies has made the following steps to improve crosstalk attenuation performance on plugs for D8GS cords:

- twisting length of a cable pair is decreased by up to 6 mm;
- conductors are positioned at a 90° angle in two IDC contacts groups, located at the rear plug housing[7] (analogue of a two-part terminator assembly of jack, revised below in Section 3.2.2.3);
- ribbon conductors, connecting IDC contacts with plug contacts, have a biplanar location with increased space diversity.

All these innovations required a new design for the tail part of the plug particular, namely, an increase in length and height, and made it impossible to install the plug in field conditions.

It is interesting to note a solution of the company Steward Connector, which may be considered an evolutionary development in the multilevel disposition principle of conductors. Unlike the previously described solution the displacement of conductors to the second plane is used for the maximum diversity of pairs 3-6 and 4-5, which are the most critical pairs for crosstalks.

The mechanical reliability of a plug and patch cable connection can be achieved by two approaches. The first approach is intended for long body plugs using a primary strain relief, which is an integral part of the plug. The strain relief is fixed by a crimping or punch-machine tool. The long body plugs are designed so that the jacket protrudes beyond the back of the plug by at least 6 mm. Some companies produce several modifications of the long body plugs with different forms of the contact entry. The geometry of these long body plugs is selected to match the cable type. Thus, for example, there are plugs for round, oval, and flat cables (Fig. 3.21).

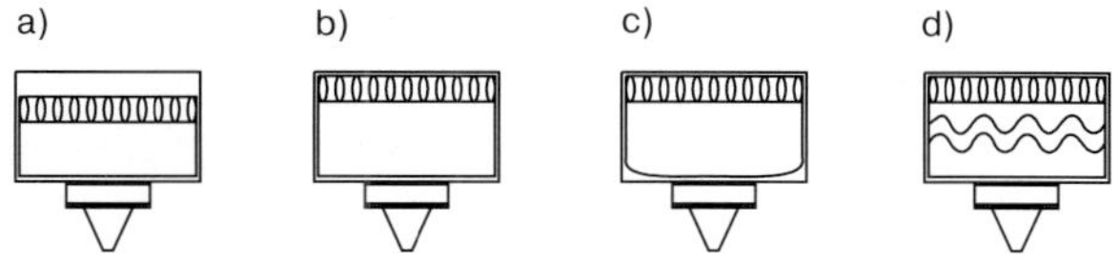

Fig. 3.21. Entry geometry of modular plugs to fit: *a)* flat cable; *b)* round cable; *c)* oval cable; *d)* category 5e and 6 cables

The second approach is realized on short body plugs, when strain relief is made of a soft plastic that is poured over the point of termination with further pressing, and is used in fabric-made cords (Fig. 3.22).

The plug snaps into the jack by a latch, which is integrated into the plug and sits on the opposite side of the contacts. A rubber or polymer boot is put on the plug, and covers 10–15 mm of cable with its rear end. Boots with rear openings of different diameters are manufactured, ensuring a flexible mounting environment. The boot predominantly serves to maintain a certain bend radius of the cable entering the plug body, which ensures high electrical performance. There is a latch guard on the front part of the boot. The latch guard does not hamper the movement

---

[7] The cover is made to go over all the wires. In Outlet Connection Module by the company Reichle & De Massari two covers are used, each covers two pairs of wires.

of the latch and simplifies the disconnection of the plug. To provide more flexibility, the rear end of the boot is corrugated or has notches. Boots for the short body plugs are formed during mounting of the cable. The latch lever of a traditionally styled plug (Fig. 3.22a) protrudes up the body too high and causes some inconvenience during SCS operation as it snags on conductors and other objects. To eliminate this disadvantage a number of methods were developed that can be grouped into two main groups. The first group is aimed at the application of protective specially shaped boots. There are the following solutions:

- Boots with a molded latch guard on the front, which covers the lever end of the latch (snag-free cap, Fig. 3.22b), are often used. The latch guard protects the plug latch from snagging when pulling cords through the pathway or cable managers. Sometimes, the latch guard is replaced by a semi-ring. This probably makes the solution cost effective and easy –to make, but the solution loses functionality, as compared with traditional products.
- The so-called port identifier of series OR-20300082 by the company Ortronics has a boot not covered by a cap but a bow, which overlaps the latch (Fig. 3.22c). Because the latch lever is hard enough, it has to be made longer; otherwise, it is not convenient to plug off.

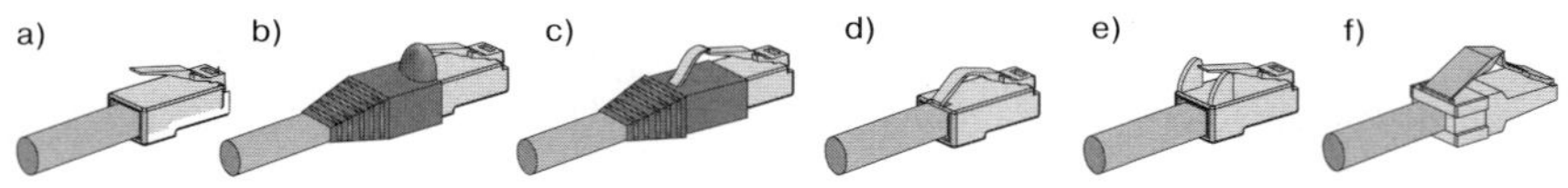

Fig. 3.22. Profiles of modular plugs; *a)* traditional; *b)* with boot with molded guard-latch; *c)* with boot with bow; *d)* with U-level; *e)* with protective legs; *f)* with protective tab on housing

It should be noted that the rear end of the boot when separately delivered has a round slot with a diameter that can be customized when making the order and selected to match the cable. A flat slot is possible when terminating a low-frequency cable.

The second group of solutions is based on modification of the rear of the plug body in order to keep the compatibility principle of traditional and improved plugs to make it possible to plug into the standard modular jack. The following solutions are used:

- The PAN-PLUG by the company Panduit has the latch with an U-lever, which is made of elastic plastic and thinner than that of traditional solutions (Fig. 3.22).
- The series Tangle Free also by Panduit uses cords with plugs, which have two small triangular legs (Fig. 3.22d) on the rear housing, enabling them to keep the form of the lever.
- The cord D8GS by the company Lucent Technologies has designed the plug with a flexible straightforward slat on the rear hosing to be fixed at a 30° angle to the long axis of the housing, and its end overlapped by the latch lever (Fig. 3.22e).

Modular outlets

Cable connection to outlets. The outlet with connecting elements is produced in the form of an outlet connection module, which consists of two main functional parts: a jack and so-called contact terminal, whose main function is the termination of conductors (Fig. 3.23). The contact terminal in the majority of solutions is realized as a set of IDC contacts. The conductors may be connected to the outlet connection module by following methods:

- the IDC 110 contacts (often these contacts are grouped in two groups with four contacts each in order to decrease the general length of the module; the other term is 2-component terminal);
- individual IDC contacts for a single position impact tool or termination cap;
- clamps K10 turned at ¼ for single or a pair of connectors (solution of the company Legrand);
- Krone contacts (outlets RJ-K by the company Krone);
- single tubular IDC contacts;
- screw clamps for a stripped connector.

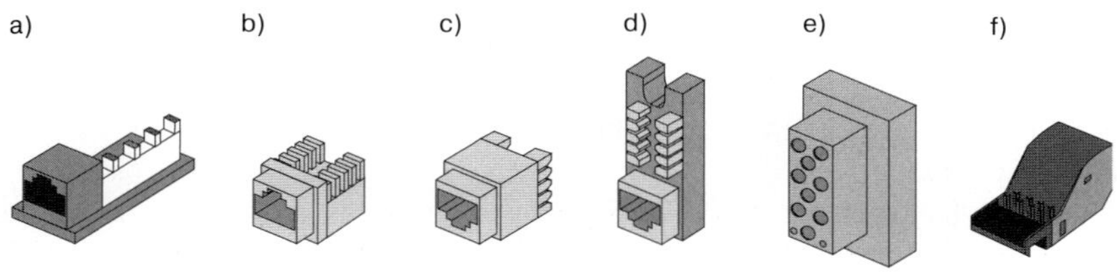

Fig. 3.23. Outlet connection modules with: *a)* IDC 110 contact; *b)* 2-component terminal; *c)* 2-component terminal rear-positioned; *d)* perpendicular jack; *e)* rotated clamps; *f)* individual IDC-contacts

The last method was only used in data transmission systems with earlier produced outlets of category 3. At present, it is out-of-date: it does not ensure high reliability because of a corrosion hazard and it is not convenient because of the high fragility and relatively low flexibility of a solid wire without insulation. It is used by an extremely limited number of companies and only in telephone outlets as it is unable to obtain a high-frequency performance.

To connect conductors:

- Terminals based on common or 2-component contact bar use a single-wire impact tool in most cases.
- The rotated clamp by Legrand with an inserted single or a pair of conductors is activated by a screwdriver.
- Pusher insertion integrated into the protective cover is used for individual IDC contacts. Once conductors are prewired into the contact guides, the next step of their installation may be finished with a special pressing tool used as pliers. The tools PJT-X and CJT-X are used in Pan-Jack and, respectively, Mini-Jack by Panduit, which represent a small lever to be fitted by its tip in a special opening

on the housing. The same company has suggested using a CGST tool with the lever principle when assembling its Mini-Jack outlets, which is a functional analogue to the crimping tool, although its dimensions are larger.

The majority of outlet connection modules have the contact terminal located in upper or lower parts of the housing. Thus, it is fairly simple for twisted pairs to enter from the rear. The entry from other sides often happens during mounting in constrained conditions, for example, in conduits and wall outlets. It inevitably results in a dramatic bending of twisted pairs at a 90° angle and often considerable degradation of connector electrical performance, predominantly by the NEXT parameter. Two solutions can eliminate this drawback by turning the jack or contact terminal at a 90° angle relative to a traditional position. The first solution presented by Krone positions the terminal in the rear housing with the plane oriented perpendicular to its axis. The second solution was suggested by the company BICC in the slime-line-module of type CAT5C-WDO-SL450-001, and installs the jack perpendicular to the contact plane of the terminal. Finally, the dramatic cable bending in traditional solutions can be partially eliminated by using a small half-round groove on the rear terminal between the contact bars.

The necessity of decreasing the mounting height of the outlet connection module is required during SCS under-carpet cabling. To solve the problem the company AMP produces modules 558053 of category 5, whose contact terminal is turned at a 90° angle; however, unlike the solutions discussed here the turn is made sideways.

The cable jacket needs to be stripped off as close as possible to the contacts during wiring, i.e., the length of uninsulated twisted pairs should be minimal. Neglecting this rule is possible only for some SCS manufacturers, and only in cases when the covered cable is not free to bend with the minimum allowable radius in a narrow conduit or certain types of outlets.

The outlet connection modules of different categories produced by one company are quite often similar in form and dimensions. To solve the problem of their visual identification the company Lucent Technologies marked the front face of series M100 modules as CAT5. The company AMP uses modular insertion units NetConnect with modular jack in its SCS The dimensions of the decorative cover of the insertion are large enough to allow for textual marking. The insert unit indicates its type, category, and number of equipped contacts. A similar solution was used by the company Siemon. Some modifications of the CT series insert units are traditionally made as a digit placed inside the rhomb.

One of the key parameters of a modular connector is the NEXT parameter. To provide a required crosstalk level it is necessary to maintain the original shape of all current-carrying elements of the jack and cable, to minimize the untwisting length of pairs, and not to allow a dramatic bending of conductors. This is achieved by a number of measures, which may sometimes be bundled. To maintain the structure of the current-carrying elements one may use the following:

- maximum shrinkage in dimensions of all details, including current-carrying elements, in order to decrease their own reactivity; for example, in jack 808 by the company ITT NS&S the length of pair untwisting is just 9–10 mm instead of the usual 13 mm;

- placement of IDC contacts in two rows, the contacts are often of different height and may be separated by dielectric bars to balance and compensate crosstalk;
- placement of IDC contacts of the pair not close to each other, but with a small displacement, in order to decrease untwisting during wiring (high-frequency connector TERA by Siemon);
- tool-free termination into IDC contacts. The termination is performed by the use of a termination cap, which is set in place by one's fingers or a tool in the form of a lever, pusher, or pliers;
- use of solutions designed only for a single termination. The above-mentioned modules of type 808 by the company ITT NS&S and S100 by Siemens serve as examples;
- protruding the cable jacket inside the jack housing to maintain the core structure;
- use of termination caps for single-position conductors;
- cable entry to the jack parallel to its axis for cases where the terminal is located in the rear or perpendicularly to its axis for traditional solutions;
- in two-module jacks intended for mounting in raceways, the cable entry is possible at a 45° angle using a clamp (solution of the company FMT, Fig. 3.24);
- use of a clamp rotated ¼ for two wires in one pair to minimize untwisting (Legrand solution).

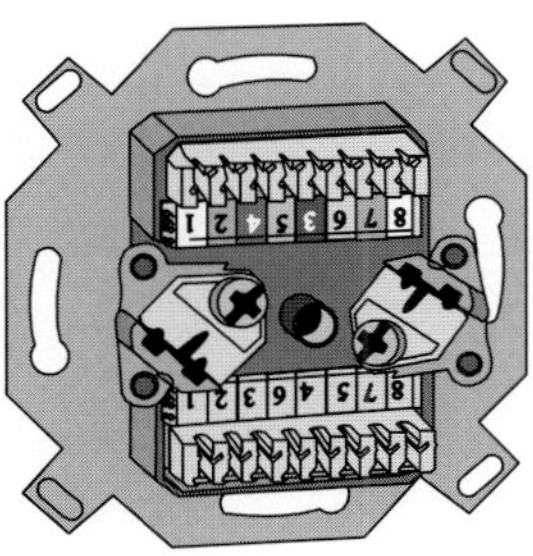

Fig. 3.24. FMT two-module jack

Constructive features of jack housings. According to the current standards all 8 conductors of a horizontal cable or cord should be wired in an 8-contact modular connector.

The wiring accuracy of twisted-pair conductors to the jack is confirmed by the use of color tapes or digital labels of contact elements. Some types of outlets are equipped with a double color marking of jacks, indicated with letters A and B or abbreviations 568A and 568B in order to wire according to 568A and 568B wiring schemes. The markings are placed next to the terminal slots and may be set both between IDC contacts arrays and on external housing.

The connection of IDC contacts of the terminal with jack contacts is done by three different schemes. The first two options are widely applied in traditional solutions, where a direct connection is used or a connection by means of a plated circuit card. At the direct connection the contacts are made of one copper-alloy sheet by punching. To improve crosstalk performance of the connector the middle part of such contacts may be laid in different planes separated by dielectric plates. At connection through the plated circuit card the contact physically consists of two parts, which are connected via a current-carrying path made of a copper foil.

The third solution is realized by means of a modular insert unit, which inevitably leads to insertion of an auxiliary plug-and-jack connector into the channel. This results in some degradation of its parameters. This is the reason for the limited production by SCS product manufacturers.

There are two main solutions for connectors with a modular insert unit (Fig. 3.25). The first one, promoted by the company AMP, comprises a modular insert with plated circuit card and plated conductors, which insert into the connector (Edge Connector). At the beginning of 2000 this solution has two modifications with very similar schemes: Edge Connector operates in frequencies up to 300 MHz, while the Edge Connector Mark II operates up to 1 GHz. The licensed copy of the Edge Connector is also used for modular insertions of the ACO series used in the cable system VDU by Legrand.

The company Datwyler has jack plates equipped with pin contacts. Because pin contact is less reactive, it is possible to extend the upper frequency limit of the connector. A large mounting plate in the modular insert unit of the company Datwyler enables them to realize the principle of cable sharing (see Section 1.5).

a) b)

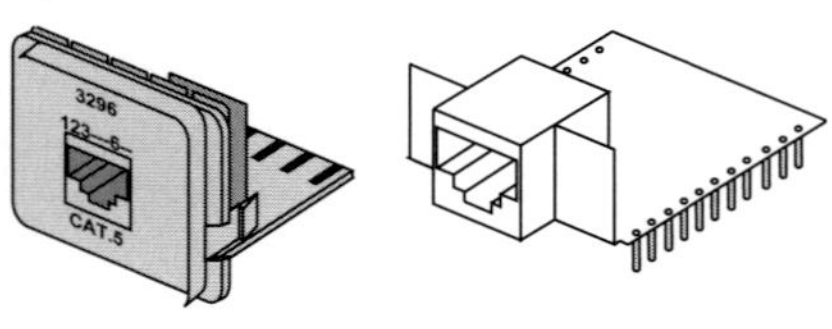

Fig. 3.25. *a)* AMP and *b)* Datwyler connectors with a modular insert unit

Some modifications of modular insert units with two jacks have the contacts connected to pins via a block. The cable specific wiring type is given by a switching element inserted into the block.

An increase in operation reliability and stability of jack contacts against possible cable pull-outs is achieved by using strain-relief covers to put on the IDC contacts after wiring, which additionally fix the conductors to their insulation.

There are solutions using the strain-relief covers for one, two, and four pairs of conductors. The cover, as a rule, is designed as a single component. In jacks from the FTJ series by the company Hubbell the adjustable on the side axis plate functions as the cap. The company Reichle & De Massari has jacks where the axis of the adjustable cover is placed in the front part of the outlet connection module and is perpendicular to its long axis. The covers for 8-conductors are quite often provided with side latches to provide for a reliable fixation. Outlets M100 series by Lucent Technologies with a 2-component terminal have U-shaped covers, which fix two pairs of conductors on different parts of the terminal and also functions as the key holder of the outlet connection module in case it is installed on the face-plate of a decor conduit.

The cover of the HD58 series jacks by the above-mentioned company Hubbell also covers two pairs of conductors, but on one side of the terminal.

The solutions described above ensure good protection against galvanic continuity loss between the cable conductors and IDC-contact blades at different pulling and bending stresses; however, they do not ensure the maintenance of a twisting structure at pull-out stresses. To eliminate this drawback, engineering applies the following methods (Fig. 3.26):

- additional fixing of the cable by a plastic piece, which is fixed on a T-shaped tab or jack opening;
- additional fixation of cable with a wedge, inserted in a special slot of the housing of the screened jack; the wedge keeps in its on-position by the force of friction (the solution by Datwyler); a similar idea is used in jacks 7700 by the company INFRA+; however, the wedge is equipped with a ratchet holder;
- utilization of small, sharp and long, cutting-edged lugs, which cut through or press into the cable jacket and protect it against side-to-side movement (the solution of the company Siemens, applied in outlet connection modules S100 and U100, respectively; Fig. 3.26b);
- in ALPHA 824 boards of the company Elgadphon the holder is designed in the form of a split tubular spring clamp, which tightly wraps the cable in the on-position (Fig. 3.26c);
- use of metallic clamps or removable U-shaped holders. One of such components is a required accessory of any screened product and functions as a grounding element.

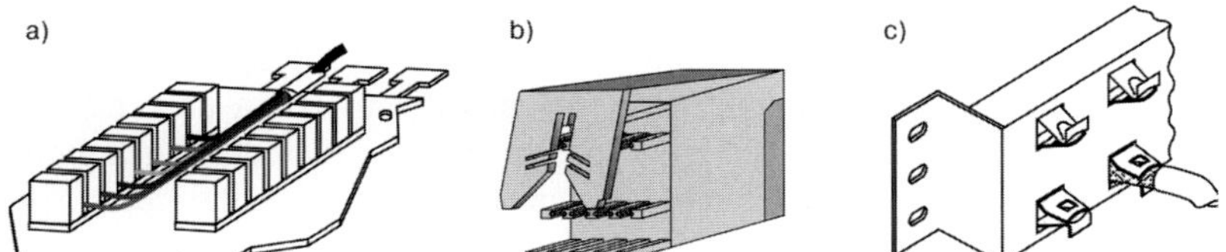

Fig. 3.26. Technical solutions for additional cable held in a modular jack: *a)* plastic tie and T-shaped lug; *b)* cutting-in elements of outlet S110 by Siemens; *c)* clamp by Elgadphon

The jacks may accept different wiring schemes with a special grouping assignment of individual contacts. To improve impact parameters in up-to-date products a direct connection to contacts without the untwisting of cable pairs is often used. The internal wires are crossed according to the required wiring layout. These dielectric plates, which separate the wires, are made of specially selected materials and have a specific thickness. This solution achieves both continuity and a given capacitance between the conductors, and balances internal circuits in the module. Before, such an effect was achieved through an installation of discrete capacitors in the form of add-on components; however, this solution was unable to achieve a high puncture strength of the construction and is not permitted by current versions of the standards. There are also balanced capacitors in the form of a planar element to be formed at the expense of selecting specially designed current-carrying paths.

Jacks for the connection to a screened cable have the same design as plugs for the UTP cable. The presence of a screening housing and other elements providing galvanic connectivity between plug and jack screens makes them differ from the latter ones. The screening housing is most often designed in the form of a cover, which covers the terminal part once the wiring is completed. The company Panduit uses the MUSJC588 series, a screening housing made of a thin sheet metal put around the jack after wiring the cable, for its jacks. In modules of type S100 by Siemens the screen is designed as a compound of two elements: the first one provides a contact with the cable screen, the second functions as the carrying base, which fully covers all other elements of the jack. Depending on the amount of jack elements covered by the screen after its mounting, jacks by the corporation Alcatel consist of screened modules (screening only the jack for the plug connection) and modules of the EMC[8] type (full screen of the jack and terminal).

The element that provides shield galvanic connectivity is designed as a metallic insert to be fitted into the jack. To decrease the transfer impedance value the insert is equipped with spring self-cleaning contacts in the form of taps. The majority of products are equipped with two such contacts, in the modules S100 mentioned above three taps are applied. The company AMP uses four contacts in outlets of the EMT system. The latter innovation is necessary because allowances for the geometric size of the connector elements, specified by the current version of the standard IEC 603-7, are too large. The increase in the amount of contacts enables elimination of possible gaps when mounting along all three axes and provides a reliable radial contact of screens during the entire period of operation.

The design of the jack housing strongly depends on its application and may be finished in two main modifications. The first modification is a universal outlet connection module, which is placed in its specific position by locks. Such modules can be of three styles: angular, flat, and keystone (Fig. 3.27).

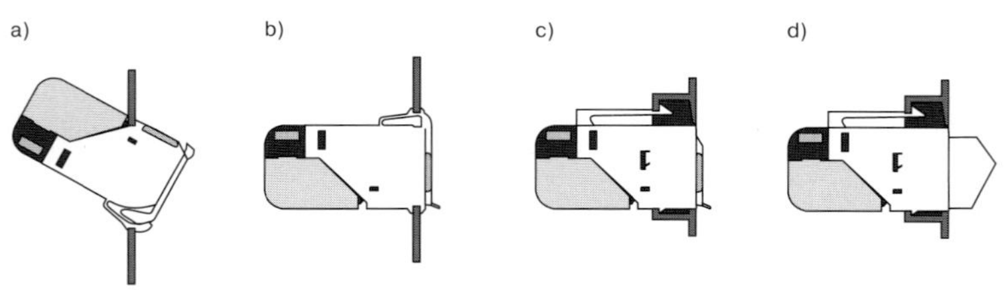

Fig. 3.27. Installation of outlet connection modules in conduit or outlet: *a)* with angled profile; *b)* flat; *c)* keystone style; *d)* angled keystone

[8] EMC is abbreviated from electromagnetic control

Mounting into a raceway requires a front adapter. Installation in the housing of a telecommunications outlet is made directly without using any auxiliary elements. The analysis of Fig. 3.27b, c shows that realized along these schemes the outlet connection modules differ from each other only by the way they are fastened. Therefore, it is possible to realize the modules in key-stone style with outer angular mounting of the jack, which results in a certain improvement in performance (Fig. 3.27g, see Section 3.3.3). In practice the modification is used, for example, by the company HCI [64].

The outlet connection module with a faceplate serves as an example of the specialized construction. It is aimed at mounting in raceways, so-called connection panels (see Section 3.3.3), and multimedia outlets. Due to the extended use of 2-port workplaces in SCS (the standard requirement), modules of that type are finished as 2-ported, and they consist of three main components: jack, contact terminal, and mounting circuit plate card. The terminal has a 110 type bar. These bars are in most cases arranged parallel to each other. In Europe a wide majority have 2-ported modules with a plastic or metallic shaped component, which functions as a support element. In this case the decorative faceplate is fixed by screws or latches.

Some modifications of modules with a faceplate support a folding or adjustable spring shutter over the jack to protect its contacts from dust in the off-plug mode. Sometimes to mount the protective shutters on a universal module, an adapter may be attached to its faceplate. Some constructions of the adapters may be of a special design, which enables encoding them for connection to a certain jack, if their plug is equipped with a matching element. Note also the solution of type zero-footprint shutter by the company Molex. It comprises a protective shutter, which is fixed inside the jack, and when sliding into the plug it moves back without an angular turn [65]. Its main priority is better protection against dust, as compared with traditional solutions.

Outlet modifications. The numbering of connector contacts goes sequentially from left to right from 1 to 8, if one looks on the jack, as is shown in Fig. 3.28. The figure shows some modifications of connectors.

The traditional 8-contact connector (Fig. 3.28a), which is called Western Plug, 8PMJ (8-position modular jack), 8P8C (8 position 8 conductor), or somewhat incorrectly RJ-45[9], is used widely in SCS practice. It is installed in telecommunications outlets of workplaces and in switching equipment of telecommunications closets. In addition, the interface of network equipment for its connection to SCS is realized on a modular connector.

The keyed 8-position connector (Fig. 3.28b) was designed by the company DEC in order to eliminate the wrong termination of its plugs to a traditional 8-position jack. These connectors as well as the 6-position modified ones (Fig.

---

[9] The abbreviation for registered jack, RJ defines a particular wiring scheme of individual wires into outlet contacts. For example, a 6-position outlet may be wired to RJ-11C scheme (one pair), RJ-14C (two pairs), or RJ-25C (three pairs). An 8-position jack is wired to RJ-61C and RJ-48C, for an 8-position keyed jack there are schemes RJ-45S, RJ-46S, RJ-48C, etc. In a technical catalogue or documentation the RJ-45 means a traditional 8-contact modular connector.

3.28g), also called MMJ connectors (modified modular jack) have not become widespread, and they are not recommended for SCS.

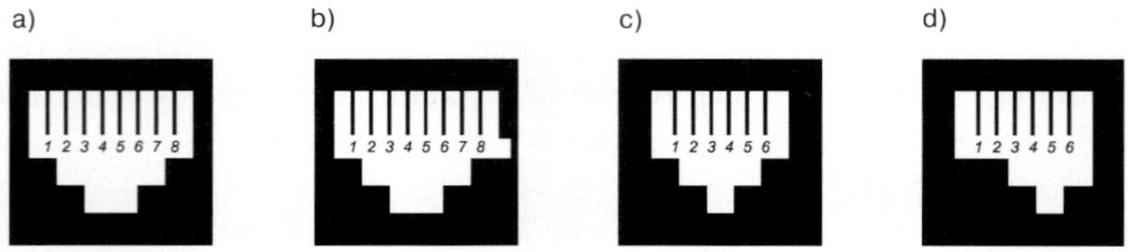

Fig. 3.28. Modifications of connectors: *a)* 8-position; *b)* keyed 8-position; *c)* 6-position; *d)* modified 6-position

The 6-position connectors (Fig. 3.28c) as well as 8-position ones, which are sometimes not quite reasonably called RJ-11 or RJ-12, are widely used in non-structural cable systems to connect telephones on workplaces. In order to achieve conformity in the SCS, the telecommunications outlets have to be equipped only with 8-position modular connectors. The modular connectors are designed in such a way as to make it possible for a 6-position plug of a telephone cable to be connected to an 8-position telecommunications outlet leaving its 1 and 8 contacts unterminated.

In the SMART system of the company RiT Technologies a special 10-contact modular connector is used. Two auxiliary conductors (at present only one of them is used) enable a so-called scanner to control the interconnection of two ports by a patch cord on the panel, and provide two-way transmission of control and diagnostic information along the unbalanced circuit. The conformity of the cable system SMART and its compatibility with other equipment is reached by a specific design of the jack: it may fit a standard 8-position modular connector. It is obvious that control data exchange is impossible when cords equipped with such plugs are applied. For more details see Section 13.1.3.

Wiring schemes of modular connectors

The balanced cable conductors may be wired in the jack in different ways. Because of the flat design of a plug choosing the best wiring layout is complicated. Any selection is a trade-off solution, which results in the improvement of one or another parameter.

Three main wiring schemes of twisted pairs (Fig. 3.29): T568A, T568B, and USOC have become the most widespread in practice. The SCS uses the two first ones; however, applications using the USOC scheme may also occur. Therefore, some companies manufacture patch cords and equipment cords with modular connectors to be wired according to the USOC.

Information on the wiring of modular plug and jack wires according to their color code in accordance with the T568A, T568B, and USOC schemes is given in Table 3.25 and is also shown in Fig. 3.29.

The advantage of the USOC is the connection of pairs 1, 2, and 3 when terminating a 6-position plug into an 8-position jack. The main disadvantage is that it has the worst, as compared with the T568A and T568B, balance of pairs 3 and 4 due to the significant diversity of the contacts.

Table 3.25. Wiring of twisted pairs in a modular connector by different wiring schemes

Contact	T568A		T568B		USOC	
	Pair	Signal[a]	Pair	Signal[a]	Pair	Signal[a]
1	3	T3	2	T2	4	T4
2	3	R3	2	R2	3	T3
3	2	T2	3	T3	2	T2
4	1	R1	1	R1	1	R1
5	1	T1	1	T1	1	T1
6	2	R2	3	R3	2	R2
7	4	T4	4	T4	3	R3
8	4	R4	4	R4	4	R4

[a]T – signal transmission, R – signal reception

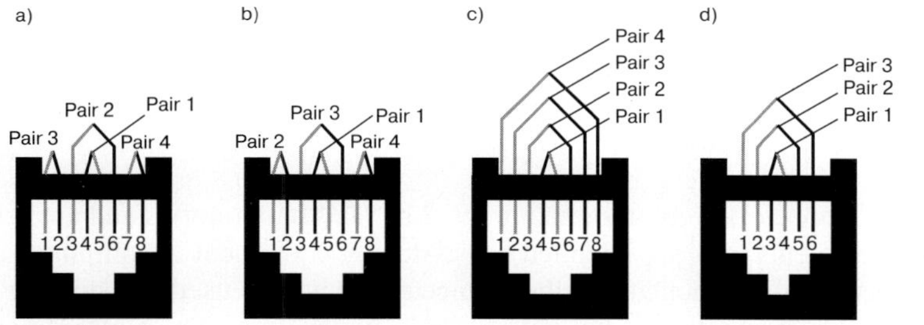

Fig. 3.29. Wiring scheme of twisted pairs in a modular connector: *a)* T568A; *b)* T568B; *c)* USOC 8-position; *d)* USOC 6-position

The wiring schemes T568A and T568B are nearly identical except that pairs 2 and 3 are reversed and colored as orange and green, respectively. An advantage of the T568A is the compatibility of pairs 1 and 2 with the USOC. However, the T568B (AT&T 258) is widely used in Russia, which is, obviously, explained by its utilization in the Systimax SCS by the company Lucent Technologies, which had a leading position in the segment of the national market in the first stage of its development. Therefore, the use of the 568B wiring scheme is recommended for SCS construction. The SCS standards do not recommend using only one and the same wiring scheme. These are customized wiring schemes. The only restriction is the standard TIA/EIA-568A, which prohibits the parallel use of two different wiring schemes within one SCS.

Besides the schemes T568A and T568B, there are several other termination schemes specified by different documents. Some of them are compatible with T568A and T568B (for example, Token Ring, 802.5), some are even compatible with the same indication of pairs (10Base-T, 802.3; TP-PMD, X379.5). Table 3.26 gives examples for how to use the T568B in several up-to-date data transmission systems.

Table 3.26. Application of T568B wiring scheme for some data transmission systems

Application	Data transmission rate, [Mbps]	Active contacts	Pairs used (568B)
IEEE 802.3 (Ethernet)	10	1236	P2P3
IEEE 802.3u (Fast Ethernet)	100	1236	P2P3
IEEE 802.3ab (Giga Ethernet)	1000	12345678	P1P2P3P4
IEEE 802.5 (Token Ring)	4/16	3456	P1P3
IEEE 802.12 (100VGAnyLAN)	100	12345678	P1P2P3P4
3270	2.36	45	P1
S/3X и AS/400	1	45	P1
Telephone sets	<0,1	3456(78)	P1P3(P4)
V24 – V.28 (RS 232 C)	0.0192	3456(1278)	P1P3(P2P4)
Apple Talk – Local Talk	0.23	3456	P1P3
Arcnet	1/20	45	P1
Telex, Fax	<0.1	45	P1
ATM (TP)	155	1278	P2P4
TP-PMD	100	1278	P2P4
ISDN	2.048	12345678	P1P2P3P4

3.2.3 Type 110 Connectors

The type 110 connectors were designed in 1972 by Bell Laboratories and are standard "panel" elements, i.e., are intended for switching equipment in equipment areas. Some individual components of the connector are widely used in a number of telecommunications outlets for horizontal cable connection, consolidated-point connectors, and other similar elements. It is almost never used as the interface element of the active equipment, unlike the modular outlet. That is, obviously, explained by its large linear size in three- and four-pair modifications.

Prior to the appearance of the switching elements of the type 110 the connection was performed by jumpers, which did not allow a compact twisting of the wires. This increased the attenuation and degraded NEXT, so that the system performance was not higher than for category 3. The type 110 connectors enabled using patch cords with plugs to switch signal transmission channels (Fig. 3.30). The cord possesses a steady compact twisting of wires in pairs and potentially superior electrical characteristics due to the available outer jacket. In addition, patch cords are more convenient for multiple connections than jumpers.

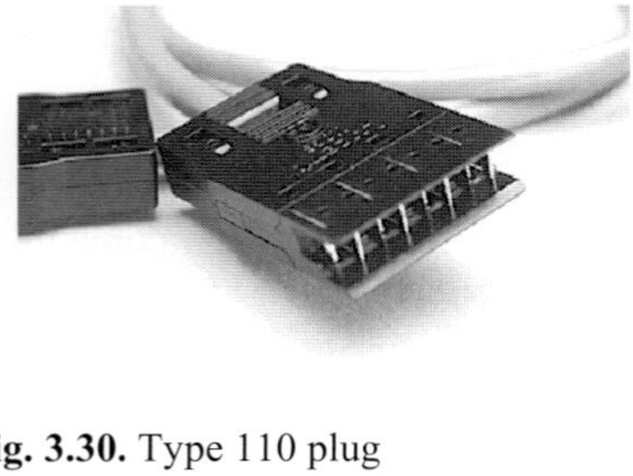

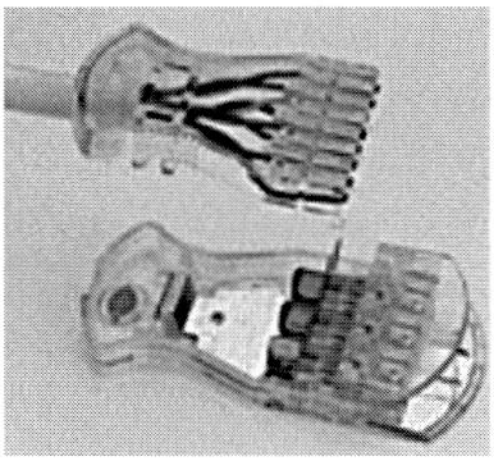

Fig. 3.30. Type 110 plug

The type 110 connector consists of two parts: a plug (Fig. 3.30) and a contact bar. The bar functions as a jack after a so-called 110C block is mounted. The plug contacts display metallic plates, which are perpendicular to the housing plane and fitted into the opening between the contacts on the bar. When designing the connector along with frequency characteristics of category 5, special attention was paid to the protection of current-carrying plug and jack elements by plastic housing components against external mechanical impacts in order to increase service reliability. The plug's interactive contacts and the 110C block may be gold plated some micrometers thick to prevent working surfaces from oxidizing and improve the reliability of the contacts between them.

The plug housing has protective lugs to avoid connecting it incorrectly to the bar. An arrow on the upper plug indicates the right installation position. There is a solution for the 110 connector that is also designed by the company Siemon. The housing of the product has an asymmetrical form and is connected to the bar with its visor on top. It should be noted that a four-pair plug by the same company has special corrugated indentations on the housing for easy handling and connecting.

The type 110 plug is produced for 1, 2, 3, and 4 pairs, which enables the connection of a patch cord to signal transmission channels of different capacity. For example, a one-pair plug is enough for the connection to an analogue telephone channel. Using four-pair plugs for the connection to LAN data transmission channels is recommended.

The housing of the type 110 plug is assembled from two components connected by plastic latches. The patch cord conductors are terminated to the IDC 110 contacts, which are installed in one part of the housing by means of a common single-wire impact tool.

The contacts of the plug are numbered from left to right in the on-position for connection to the block. There is a strong correspondence between the contact numbers and pair conductors of the patch cable, i.e., the first pair of conductors is connected to the first pair of contacts, the second pair to the second one, etc. This feature makes the cable mounting of the plug easy and enables minimization of the crosstalk at the expense of an almost complete elimination of the untwisting in pairs.

The majority of manufacturers make plug connector housings for industrial mounting on cords of an opaque plastic. Plugs, which are intended for cable mounting in field conditions, are often made of a transparent plastic. This makes it easier to visually control the accuracy of the conductors' positions. Both parts (products of series SD110P by Siemon and ICNP110XPR by ICC) of the housing as well as only upper part (P110PC-4 by Panduit) can be transparent. In the latter case, it provides supplementary visual control of the proper position of a plug when mounting it on the 110 block.

The big plastic housing of the type 110 plug (especially in the four-pair option) in principle causes no problems marking different self-adhesive labels. The housing may be provided with an extra window. Oval plastic icons of 10 different colors may be snapped in for individual identification of the type 110 plugs made by Siemon.

Connecting blocks of type 110C are mounted on the contact bar after wiring pairs of horizontal or backbone cables to it. The block construction may include metallic X-shaped elements with IDC contacts on one side. Their quantity corresponds to the number of conductors served. Blades on the side facing the contact bar cut through the insulation of the conductor with a four-pair impact tool. The block is tightly fixed on the contact bar by means of eight tight clips, and in this way presses the connectors to the contact bar. To perform a new connection, one only needs to remove the connecting block from the bar with pliers. The second half of the block includes 110 connectors to connect patching cords and jumpers. The availability of blocks with different capacity enable flexible administration of the SCS.

The 110 connectors cannot provide shield connectivity for shielded cables. To improve the electromagnetic performance of the products the connector made by Siemon has an internal shield that separates individual twisted pairs from each other right up to the point where conductors are connected to the IDC contacts. It also provides a galvanic coupling with the screen of cables S/UTP and S/STP when assembled.

3.2.4 Other Types of Connectors for Signal Transmission

Other than the main types of connectors reviewed, there are some other connectors for twisted pairs that are rarely used in practice (mostly in SCS of European manufacturers).

4-position connectors of type MIC (number 8310574 IBM) have been designed for use in IBM cable systems, and are intended for a cable connection with characteristic impedance of 150 Ω. They differ from other connectors by their androgyne design, i.e., the connector uses only plugs and the components are not divided into a plug and jack. The plug has an asymmetric design and is terminated directly to the other plug. The plug has not become popular due to the complicated installation, large dimensions (it has half the packing density in comparison with traditional modular connectors), and no possibility of a connection to four-pair cables.

Series DB connectors are intended first to support LAN applications, and are aimed at operation with 100-Ω cables. They are produced for 9, 16, and 25 positions. The 9-position connectors (D-sub type) are the most popular among them. Their jacks have almost the same dimensions as modular ones; however, the serviceability of the connector itself is essentially reduced. This is explained by the fact that assembling and disassembling procedures of the plug consist of screwing and unscrewing two captive screws. The design of the connector provides effective shield connectivity for spliced cables, which enables extending the effective frequency range (up to 300 MHz) for some types.

The DB connectors series may be installed in switching equipment in equipment areas, as well as in telecommunications outlets. As an example, we should note shelves 12443 XX series (capacity for 16 and 24 ports) and the one- and two-port jacks 27689 XX series by Ackermann. To simplify the connection of conductors to jack contacts, one can use a crimping solution as well as wiring of IDC-contacts of a Krone type.

Connectors G4 (4-pole connector) and CAG 8 (8-position connector) are also a part of the equipment manufactured by the company Ackermann. The products are intended for cabling with a characteristic impedance of 150 Ω. Some modifications of the jacks are made for these connectors to equip them with break contacts. This enables the formation of ring structures without corresponding connections in telecommunications closets and supports the operation of ISDN systems.

3.2.5 Type 110 Connectors of Alternative Circuits

The use of switching equipment based on modular connectors has become widespread for electrical channel construction for SCS available on the Russian market. The lower popularity of the 110 connector is explained by drawbacks such as poor aesthetic characteristics, large dimensions of the patching plugs, as well as stricter requirements on the system administrator regarding the knowledge of rules and principles of SCS construction. An important fact is that the solution based on panels with 110 connectors is rather expensive due to the high price of patching cords in the four-pair option (approximately 2.5 times as much), so even the decreased cost of the panel does not cover the expenses. These factors lead to the fact that there is no such type of switching equipment within the numerous SCS groups available at all (for example, ICCS of the company Corning and FreeNet of the company Reichle & De Massari), but these panels have been included in the list of the standard components for SCS of some manufacturers much later than the equipment with modular connectors (the latest example is the company RiT).

The factor that hinders the 110 panels from being completely excluded from SCS construction is that a higher functional flexibility of the switching equipment modification compensates for all its drawbacks. It is possible to administrate pairs without using external adapters and other attached components. The fact that there is no overlap of conductors in pairs 36/45, which is used in a modular connector and primarily popular in wiring schemes of the 568 series, makes it possible to get a potentially superior crosstalk performance. The untwisting of wires to a shorter length and their wide diversity lead to an additional improvement of the parameter. We have to mention a fact that is usually missed: in spite of the considerable width of a four-pair plug, the lower height of the array and the possibility of their dual-in-line disposition without a loss in maintenance operability enable, for standard 19-in. panels, obtaining a port density equal to the port density on the panel with modular connectors (i.e., 24 per one unit of height).

It follows from the facts stated above that switching equipment with 110 connectors possesses considerable technical and economic benefits, which, however, may be extracted in full only if the cable system will support a considerable, at least proportionate to the number of LAN workstations, number of telephone units and other middle- and low-speed network devices. In other words, it is effective when the cable system serves a large number of devices, which do not use the four-pair data transmission circuit in general or, at least, in a majority of cases.

The change to high-capacity equipment by no means affects the list of applications supported by the cable system. Taking into consideration the facts stated above, and the beginning of a mass introduction of SCS with electrical parameters that exceeded requirements for category 5, it was expected that several new modifications of switching equipment with the 110 connectors and improved performance would be launched into the market in 1989–1999.

All solutions analyzed below differ from traditional ones by at least one of the following constructive features:

- they are fully or partly based on a 110 connector;
- their characteristics according to the full scope of electrical parameters exceed requirements for category 5;
- all of them maintain the functional flexibility of the prototype, including the possibility of administrating a pair;
- they are almost free from the prototype drawbacks because of the use of some technical solutions.

VisiPatch system connectors

The system VisiPatch was suggested by the company Lucent Technologies in 1999. The main task for the designers of the connector was to considerably improve installation, aesthetics, and maintenance parameters of the switching equipment together with the maintenance of electrical characteristics of the system GigaSpeed. The given task was achieved because of substantial changes in design and construction of the majority of the functional elements of the system equipment: plugs, bar, installation frame, and cable managers.

The most radical changes were made in plug design. These products have the contacts laid on the turnover plane relative to the traditional placement, but because of that the contacts are laid above the housing level and oriented to its working side towards the cord. As a result, at the connected plug the patch cord fits close to the panel (Fig. 3.31) without any stand-off. It is called the reverse patch cord solution in some publications. The product has an unusual shape and its rear part is approximately 17 mm thicker, which makes it visually different from previous products.

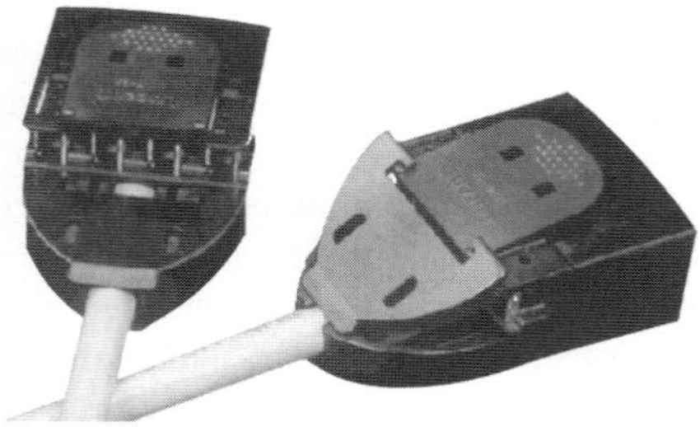

Fig. 3.31. VisiPatch plug

The 110 connectors are equipped with a more convenient spring lock, as compared with the prototype, which keeps the plug away from an unexpected disconnection from the connecting block.

The side housing of the plug has a significant surface area and is free because of the constructive circuit selected. This makes it efficient to individually mark the plug with self-adhesive labels. An auxiliary cover, which is immediately adjacent to the plug housing in front of the contacts in the on-position, makes connections easy. Fixed in an upward position by a footprint lock of the connector, the cover covers the contacts making a sufficient gentle slope. The solution protects against snagging the plug by other conductors when removing the plug from the patch panel. The slot bar of the system VisiPatch is an integral part of the 110UW-28 wiring block, which is installed on its front part and does not have any substantial differences from a traditional 110 bar.

Type S210 connectors

The type S210 connectors were designed by the company Siemon specially for high-speed applications. The product is one of the main functional elements of System 6 SCS equipment by the company Siemon, and it exceeds all characteristics of the existing category 6 and class E projects existing in the middle of 1999. For example, the NEXT margin is 25 dB, as compared with category 5, and 11 dB, as compared with the category 6 project (the figures are given for a frequency of 100 MHz). Such an essential advantage was achieved through the use of a new disposition of contacts within a traditional connector circuit to be grouped two-by-two and along all their length, and additionally, partly or fully covered in the connecting block by an overall screen for each pair.

Like a traditional 110 connector a S210 connector consists of two main elements: plug and connecting block, which is installed on the bar during mounting. Unlike its prototype the wire pair is not turned around the guide color-coded lug when mounting on the bar, but is wired into two slots on the left of the lug by their color wire to the right slot (the lug is in an upwards position above the bar plane). The connecting block housing of the S210 connector is provided with decorative guide lugs, which guide the block and make the installation procedure by means of an impact tool convenient.

Inside the housing there is a screening metallic plate, which separates the pair contacts from each other and is efficient when mounting plug on the cable S/UTP and S/STP. Thus, the twisted pair wires at the termination point into IDC contacts, i.e., along the whole length of the untwisting, are screened, which ensures very high connector crosstalk parameters. The cable entry of the housing has a somewhat smaller width and its grip is adapted for the thumb and index fingers, which makes the connection procedure easier. Its upper part has a space to install a snap-in color-coding icon of 12 colors, which is traditional for equipment by the company Siemon.

The plug and the connecting block of the connector have a width that exceeds similar parameters of a traditional 110 connector by approximately 1.5 times (Table 3.27) because of the use of an additional diversity of wired pairs and the mounting of circular shields. Therefore, the standard bar accepts only four connectors instead of six as is traditional for the 110 type. Simultaneously, the connecting block is somewhat thicker. As for the rest, i.e., the mounting and marking elements, as well as the installation technique, are completely the same as for the solutions for the 110 type.

Table 3.27. Geometrical parameters of connecting blocks for connectors

Block type	P110C4W	GPCB4	S210C-4
System	110 and VisiPatch	Giga-PUNCH	S210
Width, mm	30	22	45.5
Depth, mm	22.6	22.6	24.5
Thickness, mm	6.2	8.8	8

Giga-PUNCH system connectors

The key idea of the Giga-PUNCH system from the company Panduit is the creation of equipment infrastructure based on the 110 panel with electrical characteristics that comply with the category 6 project. Thus, it was created to fulfill the demand to make the plug dimensions comparable with those of a plug of a modular connector, while maintaining the key advantage of the prototype, i.e., the possibility of administrating pairs. The problem was solved by moving away from the traditional linear constructive circuit of the connector, and using a dual-in-line grouping of contacts with their approximately half-a-step reciprocal displacement (a checker-wise disposition). The latter solution is necessary despite its somewhat lower technological effectiveness, as this is the only way to ensure a linear wiring in guide slots without bending.

Among the additional technical features of the connector we have to note the rounding of the angular parts connecting block working surfaces and the bar lugs that ensure easy mounting and further maintenance of the panel.

An interesting technical feature of the system Giga-PUNCH regarding the connector itself, is that it includes two modifications of the panels. The so-called high-density panel has a density of 144 and 432 pairs diversified on four and twelve bars, which is traditional for the 110 type equipment. The standard panel density for the same quantity of bars is 96 and 288 pairs, i.e., it practically matches standard panels of the 110 type. The standard density panels are intended for flat cables and differ visually from their more spacious analogue by an opening between groups of guide slots of the bar, which is stipulated by a smaller connector width.

The connecting blocks of the connectors for different systems are shown in Fig. 3.32 and their performance is given in Table 3.28.

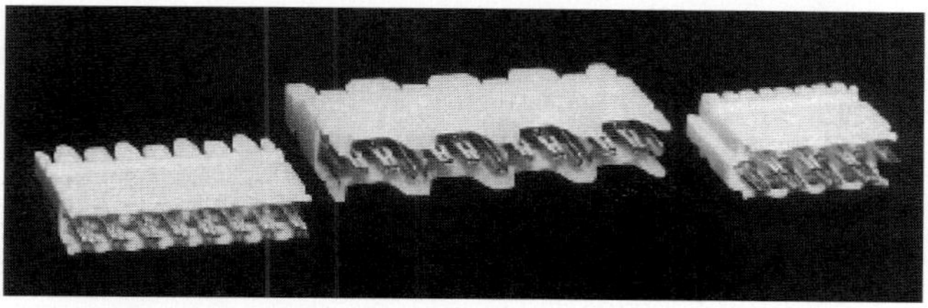

Fig. 3.32. Connecting blocks: left – systems 110 and VisiPatch, center – system S210; right – system Giga-PUNCH

Table 3.28. Electrical characteristics of connectors at a frequency of 250 MHz

Parameter	Category 6 project	Giga-PUNCH	S210	VisiPatch[a]
NEXT	>45	52	58	49.6
PS-NEXT	>42	49	55.2	44.9
FEXT	>35	47	–	no data
PS-FEXT	>32	44	–	no data
ELFEXT	–	–	61.5	no data
PS-ELFEXT	–	–	59.2	no data
Attenuation	<0.32	0.03	no data	no data
Return loss	>15	24	24.3	20.7

[a]For the system VisiPatch all figures have been calculated using the standard model of a channel with three connectors

3.2.6 High-Frequency Connectors for Category 7 Project Solutions

Applied solutions and schemes

A standard 8-position modular connector is distributed mostly for the construction of different SCS. This is encouraged by features such as standard interface, well-designed solution, wide range of manufacturers, and extensive application in active equipment. The key drawback of the connector is its parallel and closely placed contacts and their shape, which is not suitable for operation in frequencies of several hundred megahertz. The electrical performance of existing versions does not meet the requirements of category 7 and class F applications for most pair combinations. Thus, for example, solutions of the companies AMP and Telegartner enable the transmission of 600 MHz signals only along contact pairs 1-2 and 7-8 with maximum diversity, which provide a minimization of crosstalk. Thus, required NEXT characteristics may often be achieved when conductors of the suspended pair are connected to mass, which breaks the principle of cable system uniformity.

Some solutions for high-frequency connectors commercially produced, and more or less applied in engineering, are presented below. We have to distinguish between contact schemes with one or two rows, shown in Fig. 3.33. It is interesting to note that the dual-in-line scheme, which provides a greater diversity of contact pairs while maintaining the traditional connector design, is very often used.

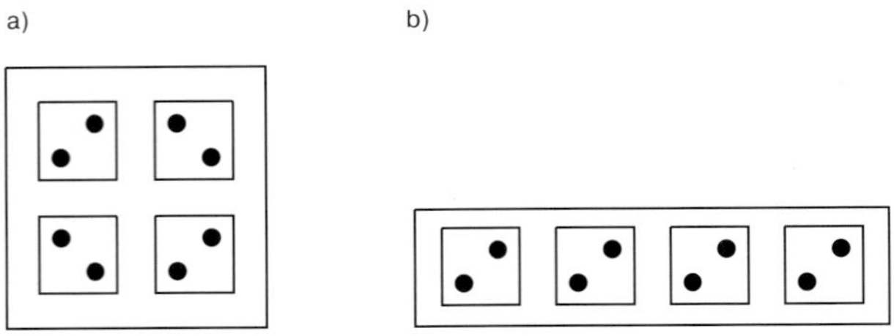

Fig. 3.33. Disposition of pair contacts in high-frequency connectors; *a)* two rows; *b)* one row

Modular solutions

The main feature of modular solutions is to provide straight and reverse compatibility with traditional connectors. Two main solutions were known in this field by the middle of 1999.

The first is the GG-45/GP45 type connector suggested by the company Alcatel[10]. The plug GP-45 (GigaPatch) and the jack GG-45 (GigaGate) housings are of the same design as their prototype solution (Fig. 3.34).

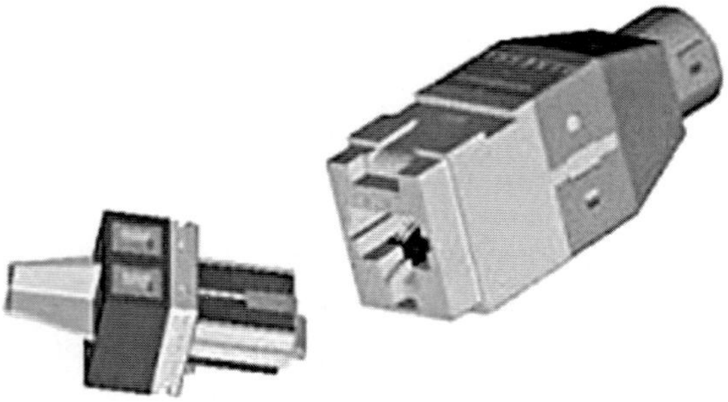

Fig. 3.34. Type GG-45 connector outlet by Alcatel

The main discrepancies are two auxiliary pairs of pin contacts, arranged in the lower side angles of the housing and a slide-switch with a side working element. The key idea developed for high-frequency operation is to diversify contact pairs at a maximum possible distance by using a dual-in-line scheme. Following the backward compatibility principle required incorporation of a mechanical switch,

[10] On 29 July 1999 the WG3 of the ISO/IEC committees had a meeting where it was decided to assign the GG-45/GP-45 connector as the base connector for category 7. It has to be used when compatibility with standard modular connectors of categories 5 and 6 is required.

which has become the main drawback of the solution [66]. The pair separation inside the plug and jack housings of the connector is achieved by means of a cross-head separator.

The second solution was designed by the company AMP, and at present is known as Category 7 modular plug and jack connector system. The solution is based on the application of internal printed conductors. The printed circuit plate is made to provide an efficient diversity between conductors by forming grounded conductive screens; each screen is covered by a grounded cover.

The cover serves as the screen, but the connector itself is finished in the form of a coaxial mini-cable. For efficient plug and jack shield connectivity, two boundary grounds (contacts 0 and 9) have been added to the eight telecommunications conductors. The solution differs visually from the traditional modular ones because of a greater housing length (Fig. 3.35), which enables installation of a printed circuit with connectors.

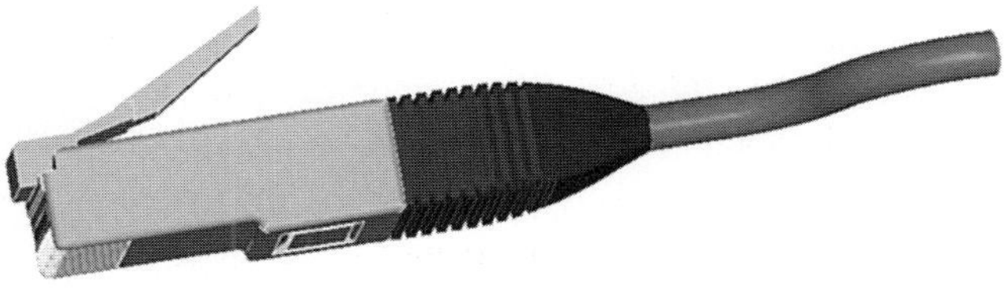

Fig. 3.35. Plug of the category 7 modular plug and jack connector system by AMP

The valuable advantage of the solution developed by AMP is wide range of Y-adapters. The adapters allow transmitting signals of several applications across one cable. The drawback is the complicated manufacturing due to engineering factors.

Alternative solutions

The solutions presented here, in contrast to previous ones, are not compatible with the traditional modular connectors. The connector plugs and jacks are generally configured according to the dual-in-line scheme (the only known solution accomplished along the one-line scheme is type CF8 connector made by Ackermann) and specified for S/STP cables. They have to provide a full screen around a pair along its whole length, which meets a specific requirement for such connectors. Thus, a cross-head metallic separator fits into the jack to separate cable pairs prior to their entry into the contacts.

In solutions that do have to meet high requirements for the operating frequency, the round screen may be replaced by a common metallic separating screen. The example is ALL-LAN of the company Montrose/CDT, which provides, in particular, an ACR value not less than 24 dB at a frequency of 300 MHz.

The third solution for the high-frequency connectors operating on frequencies of about 1 GHz is based on a screen balanced scheme. This implies that the plug and jack include the same open components that form a closed chamber when assembled. The connector MiniC by the company IBM is configured according to this scheme.

The plug and jack may both have flat (for example, connectors TERA by the company Siemon and 7777F by INFRA+) and pin (connector CMG by Telesafe) contacts; the latter ones often possess more superior performance characteristics than are recommended by the category 7 standard requirements and are certified up to frequencies of 850–1000 MHz.

Another feature of the given subclass of connectors is a wide distribution of their two-pair versions along with the four-pair ones configured according to the scheme "two-halves". Because of this, if required, two two-pair plugs may be plugged into one jack close to each other. The lines that meet the requirements of the category 7 project are often used for cable sharing.

The procedure of the plug and jack mounting on cable S/STP is a very complicated process, as it is necessary, first of all, to create a minimum transfer impedance between screening components over the entire operating frequency range. To simplify the mounting procedure the company Siemon designed a universal tool, which eliminates additional crimping and is applicable for both plug and jack.

3.3 Switching Equipment

3.3.1 Patch Cords

The patch cord is used for the manual connection of different SCS cable segments with each other. The patch cord is made up of a multipair cable segment with two connectors mounted on the ends (Fig. 3.36). Depending on the type of switching equipment the patch cord may be reinforced by two plugs from modular connectors, two plugs of 110 connectors, and combined by plugs of modular and 110 connectors on different ends. Table 3.29 presents different types of patch cords and their applications. Besides the products specified in the table, the bundled patch cords with a modular plug on one end and an electrical MIC connector jack of the system Token Ring on the other, which are used in SCS to transmit signals of the same network equipment, have a limited distribution.

Table 3.29. Types of SCS patch cords

Connector type		Number of pairs	Application field
First end	Second end		
110	110	1–4	Connection by one or several pairs of 110 switching equipment
110	Modular	2 or 4	Connection by two or four pairs of 110 switching equipment to modular or outlet panels
Modular	Modular	4	Connection by four pairs of modular or outlet panels

Some manufacturers of SCS equipment also include in their standard catalogues patch cords equipped with modular plugs, or a modular plug on one end and one-pair plug of a type 110 connector on the other end. These cords are used to transmit signals of high-speed applications, whose equipment functions along the two-pair scheme.

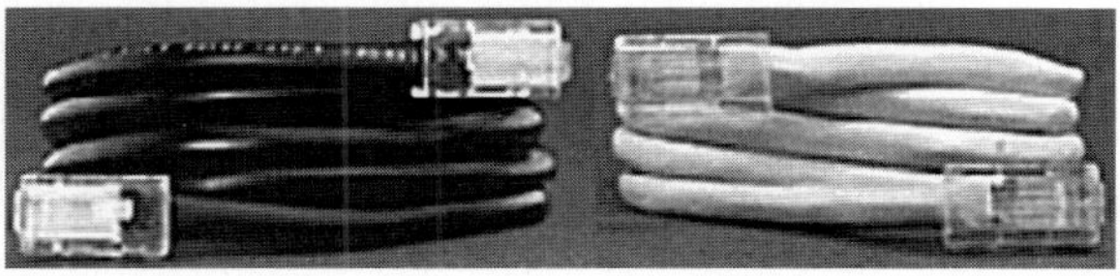

Fig. 3.36. Patch cord

The patch cords are classified as categories 3 to 5. The summary of category 5 patch cord types produced by different companies is given in Table 3.30.

The minimum allowable bend radius is an essential parameter for patch cords, which are often used in the limited space of a 19-in. mounting rack. The parameter value for most of them is within 20–25 mm.

Table 3.30. Unshielded patch cords of category 5

Producer	Cord Type[a]	Connector Type[b]		Length, [m]	
		First end	Second end	Min.	Max.
Homaco	MA5-84T-XX-S	Modular	Modular	0.31	10.0
Lucent	D8AU2-2	Modular	Modular	0.61	3.1
Technologies	110P8CAT52X	110	110	0.61	4.6
	119P8CAT5-2X	110	Modular	0.61	4.6
Ortronics	OR-827DTP80XXDE	Modular	Modular	0.31	5.49
Panduit	UTPCX	Modular	Modular	0.61	6.1
Siemon	MC5-8T-(XX)-B(XX)	Modular	Modular	0.91	7.62
	S110P1-P1-(XX)	110	110	0.91	7.62
	S110P4A4-(XX)-(XX)	Modular	Modular	0.91	7.62
Hubbell	C504P24JXXDESN	Modular	Modular	0.61	7.62
ITT NS&S	195500-XXXX	Modular	Modular	0.3	10.0
ICC	ICPC9XXBL	Modular	Modular	0.91	7.62
	ICPC5XXBL	110	Modular	0.91	3.05
	ICPCSBXXBL	110	110	0.91	3.05

[a]X indicates floating index number.
[b]Cords with modular connectors specified in the table are wired according to the scheme 568B

The patch cord is indicated with an alphanumeric index (this is sometimes very complicated: see Table 3.30), which contains encoding of the plug type on ends, cord length, color of encoding elements, and sometimes the name of a company (for example, Ortronics and ICC, see Table 3.30). We have to note that the color spectrum of elements may be given as a numeric (for example, the company Siemon) and 2-letter alphabetical index (for example, BL-blue, BK-black, etc.).

Using patch cords with modular connectors wired according to the scheme T568B in the SCS is recommended. The number of pairs in cords with two 110 connectors is defined by a specific application to be served by the SCS. A screened option is possible only in patch cords with two modular connectors.

To simplify the administration of a cable system, producers offer patch cords of different length. A number of discrete values with some foot or decimetric step, depending on the producer, are usually applied. It should be noted that the standard TIA/EIA-568-A does not recommend using patch cords longer than 6.1 m (20 ft).

After connecting switched ports the extra patch cord length is laid in a standard or auxiliary cable manager of the patch panel. One of the problems that may occur during the connection procedure in practice, is in trying to make the procedure easy, patch cords longer than required are used. This makes it difficult to lay cable in the cable manager and results in numerous loops. To eliminate this drawback the company Perfect Patch has designed a device with the same name (Fig. 3.37) in the form of a soft lock, which fully complies with the requirements of the standard TIA/EIA-568-A. Once plugs are connected, the cable excess is looped several times and locked by the Perfect Patch. Thus, the device functions as an efficient soft controller of the patch cord length.

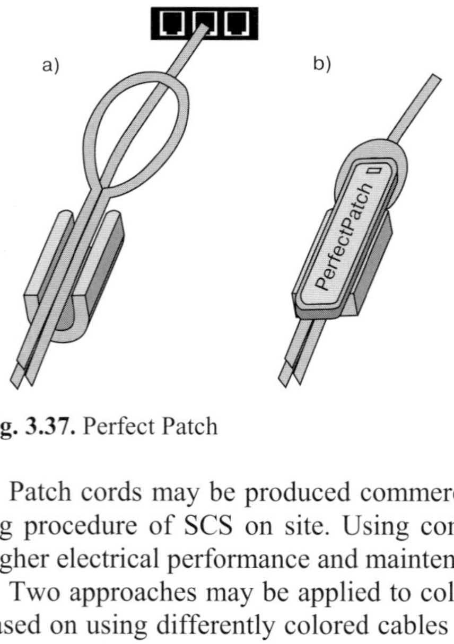

Fig. 3.37. Perfect Patch

Patch cords may be produced commercially or immediately during the mounting procedure of SCS on site. Using commercial patch cords with a potentially higher electrical performance and maintenance reliability is recommended.

Two approaches may be applied to color code the patch cords. The first one is based on using differently colored cables or plug terminators during cord production. An additional marking of plugs by means of small oval icons is not excluded, though it is very rare in practice (for example, in some cords of MC5 and MC6 series by Siemon). The second approach is promoted by the company ITT NS&S, which applies a color collar in the form of a lock to be fixed on the rear housing of a modular plug during manufacturing or immediately on site. At present, six types of collars are standard, which are delivered in packages of 50 pieces each. A similar solution is used by the company IBM; however, to improve reliability of the collar fastening an auxiliary fixing strap is added. Marking the reviewed solution by replaceable inscriptions is also allowed. Thus, there is an additional space for labels on collars of OR-827DTP800(X)DE series cords by the company Ortronics, which is covered by a transparent cover.

3.3.2 Patch Panels

Patch panels are mounted in telecommunications closets and equipment rooms, which are used for different SCS cable termination and the manual connection of cable system segments with each other by patch cords or jumpers. The patch panels of any type of SCS element have to comply with basic requirements:

- possessing a maximum density of ports, i.e., a maximum number of modular connector plugs, 110 bars, etc. per height unit;
- providing an easy connection procedure with patch cords or jumpers;
- making it possible to use an effective symbol and color marking of ports and the whole patch panel;
- providing for cable entry complying with existing bending standards, tension strength, etc.;
- having fasteners in a 19-in. rack directly or by means of a standard adapter;
- providing easy mounting of both a single port and the whole patch panel.

The patch panel for connecting screened cables has to meet additional requirements of providing a full round cable screen coverage and low resistance.

The cable patching to the panel is done by means of IDC contacts. To connect patch cords type 110 or modular connectors can be used. Jumpers are applicable for type 110 connectors or type 66 panels.

The solutions used in telephony were widely applied along with constructive solutions adapted by computer network engineering when designing panels. It resulted in a great number of various products. All the products may be divided into three main groups:

- 110 panel;
- 66 panel;
- modular panel.

We would like to review connecting panels with Krone connectors, which are mainly used for connecting panels and the distribution blocks of telephone systems in nonstructured cable networks, and have not yet become widespread in SCS. Other types of connecting panels, in particular, the above-mentioned (Section 3.2.5.2) type S210 connectors, will be made available once new versions of standards are adopted and the category 6 is officially adopted.

There are also some patch panel models being produced by only one or a limited group of manufacturers, and which have not become widely distributed, at least, in Russia. As an example, we mention a connecting panel BIX of the company NORDX/CDT described in the study [1], as well as panels with DB series plugs, which are popular in several European countries.

110 panels

The 110 panel (Fig. 3.38) is formed by identical connectors reviewed in Section 3.2.3. The key advantage of the product as a connecting element is its possibility of reconnecting each pair, which provides a very flexible SCS. The drawbacks of traditional connecting panels are a less aesthetic surface finish and that administrator of SCS should possess a strong knowledge of their organization principles. Most of the 110 panels satisfy characteristics of category 5. As an example of a product with a higher performance we shall note PAN-PUNCH panels by Panduit (category 5e) and 110 Patch Panels by Lucent Technologies.

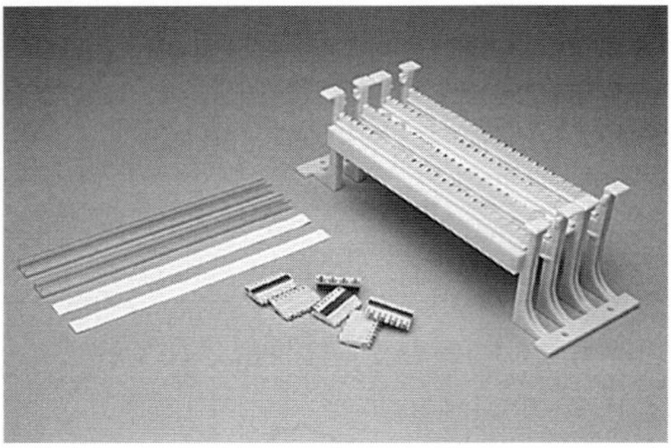

Fig. 3.38. 110 panel for wall mounting

The key elements of 110 panels are:

- 110 wiring block;
- connecting block;
- coding tape;
- cable managers;
- fixation elements.

The 110 wiring block is a base constructive element of a 110 connecting panel. It represents a plastic support with protruding contact plates. Each contact plate has 50 slots, which serve for the arrangement of conductors and IDC contacts of 110 connecting blocks. The capacity of the plate is calculated with respect to the cabling of one 25-pair bundle of a backbone cable or six horizontal cables. The width of the slot, which separates two adjacent contact plates, and height of the contact plates are defined taking into consideration this factor. A multipair cable is usually inserted in the separation slot between the plates via several elongated openings, almost oval-shaped, located to the left and to the right of the slot. Some modifications of the 110 panels have auxiliary openings in the center slot, which make the entry and connection of horizontal cables easy. Most products recommend stripping the jacket to a length that enables only individual twisted pairs to be inserted in the slot. Well-known exceptions from this rule are panels from Panduit and ICC, which have four-pair cable jackets to be stripped only to the length to be wired on the contact plate.

To make the mounting procedure easy, such blocks of some equipment producers additionally group contact pairs of the plate into groups of 3, 4, and 5. The faceplate of the raised part, which separates the slots of one pair, is marked by the color of twisted-pair second wire. 100-pair wiring blocks with four contact plates are the most distributed ones. There are also half-height 50-pair solutions with two contact plates.

The mandatory elements of the block are plastic lugs with a square space on the end, which are located on each side of the contact plates. Space is left for coding. It may be a manufacturing coding in the form of digits from 1 to 12, or it can be coded directly on site during the SCS mounting procedure by means of different labels or a marker. The features of wiring blocks of alternative constructive schemes appeared at the end of the 1990s; these are reviewed in Section 3.2.5.

Connecting blocks are reviewed in detail in Section 3.2.3. We have to add that they are produced for two (110C-2), three (110-3), four (110C-4), and five (110C-5) pairs.

One of the requirements for 110 connecting panels is, that even if the 25-contact plate is partly filled, all its slots have to be covered by connecting blocks, i.e., if, for example, five blocks 110C-4 are installed to the left on the 25-pair contact plate, the rightmost block will then be 110C-5.

Coding tapes are required for replaceable markings (Section 13.1.4) and take the form of paper strips to be fixed between the contact plates of a connecting block by means of a transparent cover, and serve for the identification of a signal transmission channel.

Cable managers serve for the management of extra-long patch cords and equipment cords, which enable one to avoid haywire and loops, and provide good visible access to coding tapes. The cable managers additionally protect the patch cord against deflection under its own weight that may result in electrical performance degradation of contacts in the connector. The use of cable managers is mandatory for category 5 cable line construction. These elements are of a different capacity and orientation in the on-position. The high-capacity cable managers for 110 panels have a vertical on-position, and have a special position in the equipment list of the existing SCS. A detailed view of the components is given in Section 5.1.4. 100-pair wiring blocks of telecommunications towers are separated by low-capacity horizontal cable managers, which are footprint elements of the construction. The capacity of a 110 panel footprint cable manager is calculated for the management of all patch cords that are patched in adjacent wiring blocks (i.e., four plates). To improve the reliability of cord fixation and to give the panel a good aesthetic form, some types of the elements are provided with decorative removable plastic housings.

Fixation elements serve for the mounting of one or several wiring blocks in a wall-mounted panel or 19-in. tower.

To fix the 19-inch component in place metallic brackets of different heights with openings for a 19-in. pattern are usually applied. Individual 100-pair wiring blocks and cable managers are installed on the brackets in two vertical columns by means of rivets or screws.

Universal installation elements are metallic support frames made of an anodized black-colored steel with screw holes, knock-outs for cable ties to fix individual cables or their bundles, and other auxiliary elements. A plastic bracket, which is assembled from two components, is used in the system VisiPatch by Lucent Technologies instead of the metallic one. The assembly is more convenient for transportation. The required durability level is achieved by the increased thickness of the product as compared with steel. For the convenience of installers and customers the plastic is dark "metal colored". The present solution ensures not only galvanic continuity, but a build-up of the support because new sections are easy to attach.

The construction made of installed wiring blocks with separating cable managers on such a support frame is often called a tower. The support frame may be wall mounted; assembling into the 19-in. mounting tower is done by means of brackets. As a rule, one pair of the brackets enables mounting two telecommunications towers next to each other. The towers on the support frame may accommodate up to 900 pairs (standard values are 300 and 900 pairs), see Table 3.31.

Table 3.31. 110 Towers with four-pair management

Company	Panel type	Pair quantity
Lucent Technologies	110PB2-300FT	300
	110PB2-900FT	900
Ortronics	OR-806003522	300
	OR-806003525	900
Panduit	P110KT3004	300
	P110KT9004	900
Siemon	S110MB2-300FT	300
	S110MB2-400FT	400
	S110MB2-500FT	500

The wall-mounted panels differ from other panels because they have mounting plastic legs with screw holes and a standard density for 100 and 300 pairs, though sometimes other values of the parameter occur (Table 3.32). The mounting legs may be incorporated into the panel construction, and, if required, supplied as a detachable component for mounting on a wiring block.

We have to note the following two solutions as options for wall mounting. The company Molex produces a type 36.C0010 frame for seven blocks, which is a functional analogue to the metallic plate for wall mounting, and differs from a metallic plate only by mounting wiring blocks and cable managers in one column. The company Lucent Technologies has produced a mounting element type 110A1 hanger bracket made in the form of a bar 110.5 cm in length with screw holes and fixing tabs. Once two such plates are vertically wall mounted at a given distance from each other, up to five towers of type 1210 with 300- and 900-pair options can be attached.

The majority of 110 panels are designed for horizontal mounting. There are some models of panels with vertical contact plates, principally designed for wall mounting. At the same time, a wall-frame type 66 (Section 3.3.2) is used, but the panel density is limited to 50 pairs.

The majority of 110 panels for a cross-connect tower variant are supplied with blocks of the same type. The cross-connect towers, which are assembled from blocks designed for a different number of pairs, most often for four- and three-pair ones, have a limited distribution. The panel option is convenient to use in networks, where functional sectors of cross-connect equipment are strictly divided into computer and telephone ones beforehand.

Table 3.32. 110 panels with four-pair management for wall mounting

Company	Article	Pair quantity
Panduit	P110KB1004	100
	P110KB3004	300
Ortronics	OR-30200006	100
	OR-30200007	300
Lucent Technologies	110AB2-100FT	100
	110AB2-300FT	300
AMP	588841-1	50
	588842-1	100
	588843-1	300
Siemon	S110AW1-50	50
	S110AW1-100	100
	S110AW1-200	200
	S110AW1-300	300
BICC	CAT5-S110-AW1-50	50
	CAT5-S110-AW1-100	100
	CAT5-S110-AW1-300	300
ICC	IC110WF050	50
	IC110WF100	100
	IC110WF300	300

If protection against unauthorized access, fire, water, sunlight, and other aggressive factors is also required, the 110 switching equipment may be placed in a metallic protective cabinet. Telecommunications equipment producers, including Russian companies, produce a great variety of such products. The 3A cable terminal section by the company Lucent Technologies is an example.

Traditional 110 panels are designed for reconnection by means of cords with 110 plugs. This is somewhat inconvenient for small networks, as it inevitably leads to a diversification of the cords. To eliminate this drawback combined panels have been designed, which include contact bars parallel to the respective contacts of the modular jack, which enables only the use of modular cords. The solution lowers the density of the ports; however, the density does not play an essential role in low capacity networks with fewer than 100 workplaces. The modular plugs in such combined panels may be located parallel to the 110 (S100DB1-24RJPA by Siemon) bars, as well as grouped as an individual block located at the side of the bars (MJWC5-8-39TB by Homaco). In this case the 110 bars are covered with a decorative protective cover.

There is a limited assortment of the 110 panels with bar contacts of two connectors printed along internal current-carrying lines on the printed circuit board. The line also has a break contact. The contact may be broken by means of a special adapter, which is sometimes necessary when testing equipment is connected. Such an interface is especially important when the panel is used for telephone signals transmission. The panel is very convenient for the organization of transition points. The panel type S110T(X)1 by the company Siemon serves as a good example of such equipment.

As it was noted before, one of the drawbacks of the 110 panel is that it has bad aesthetic characteristics. These parameters may be improved if the standard horizontal cable managers are covered by removable covers (a solution by the company Siemon). The panel of the VisiPatch series by Lucent Technologies serves as the best example. It has improved aesthetic parameters due to the use of plastic wiring blocks covered when assembled, as well as specially designed plugs with return orientation of the contact plates.

Type 66 panel

The 66 panels have been in use for several decades. They differ from similar products reviewed above by the fact that they do not need patch cords. Channel switching is done with jumpers; IDC contacts S66 (Fig. 3.39) are used for the termination of cables and jumpers. Because of the lack of a plug-and-jack connector the 66 panels are aimed, first of all, at the operation of such applications that do not need frequent connections. Therefore, their key application area is telephone systems.

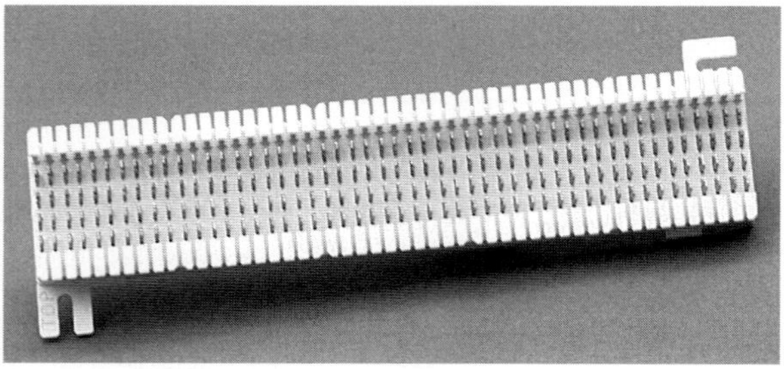

Fig. 3.39. Type 66 panel

The standard 66 panel includes the following components:

- connecting block;
- coding tape;
- cable managers;
- fixation elements.

The connection block is the base constructive element of the 66 panel. It is formed by a plastic frame, which has a varying height, and incorporated 66 IDC contacts of varying capacity. The latter ones are oriented, first of all, on the wiring of a 25-pair backbone cable. It is also possible to wire other density cables with solid wires. Using multipair cables is not recommended. There are also connecting blocks with fewer contacts on the line. The contacts may be one-, two-, four-, and eight-sectional, and sometimes are supplied with a tab for bonding or winding of a wire for internal interconnection (Fig. 3.40). To make the wiring easier some types of contacts have a guide slot prior to the cutting edges. The upper part of one of the working elements is supplied with a hook, which prevents the wire from dropping out during insertion.

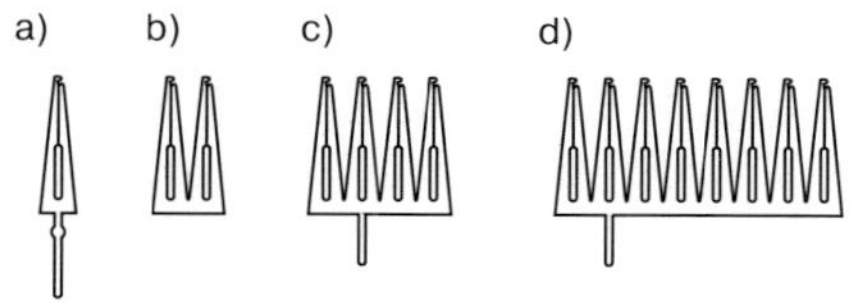

Fig. 3.40. 66 Panel contacts; *a)* one-sectional; *b)* two-sectional; *c)* four-sectional; *d)* eight-sectional

Some solutions for connecting blocks are completed with a standard protective cover installed once the wiring procedure of the cables and jumpers is completed. The cover can be put direct on the block housing or hinged. The hinges are installed into slots of the jumper cable manager. It is possible to color covers in different colors in order to code functional sections. Thus, for example, the company Siemon offers nine differently colored covers for its panels.

The functions of coding elements are performed by a paper tape, which sits under the transparent part of the protective cover, and serves for signal transmission channel identification.

Cable managers are responsible for accurate jumper management during the termination of contacts on the connecting block. The majority of solutions look like fanning strips along the long side of the connecting area. The fanning strips provide management for the jumper wire without any fixing or tension-relief effect.

Panel fixation elements are intended for wall mounting or for use in 19-in. equipment.

To wall mount a single block, a plastic bracket with fixation latches is used, which is the standard footprint element of the panel types reviewed. The right on-position is marked as TOP, sometimes supplied with an arrow. The bracket also functions as support in mounting several panels close to each other on a big metallic wall frame. Special metallic cabinets may be used on sites with unauthorized access. There are also plastic cabinets with a transparent cover for the organization of consolidated points. Mounting in the 19-in. racks is very rare in practice.

The key advantages of the 66 panels are:

- a high density of contacts and good dimensions for jumpers as cross-connect elements;

- the termination of more than one pair of cables and more than one jumper to each contact, which provides a very flexible SCS configuration;
- an aesthetic profile because the cable bundles are hidden.

The drawbacks may be related to the facts that the administrator has to possess a strong knowledge of SCS structure and that the cross-connecting procedure by means of jumpers is not as convenient as other types of cross-connect equipment. In most cases, the great length of a pair twisting prior to its termination to a 66 contact, as well as the large dimensions of the contacts, do not enable achieving electrical characteristics of the 66 panel higher than category 3. Therefore, they are widely distributed in telephone systems. There is also a limited number of the 66 panels with a substantially improved frequency performance, by the NEXT parameter, due to downsized contact elements S66M1-50 and an increased distance between them. Thus, for example, the height of the contact element is diminished from 25.7 mm (1.01 in.) to 19.7 mm (0.775 in.). This enables their use in category 4, 5, and even 5e systems [67].

Modular jack panels

Basic components. The principal differentiating characteristic of the reviewed patch panels is 8-position modular jacks, located on the faceplate, and intended for patch plugs of corresponding patch cords (Fig. 3.41).

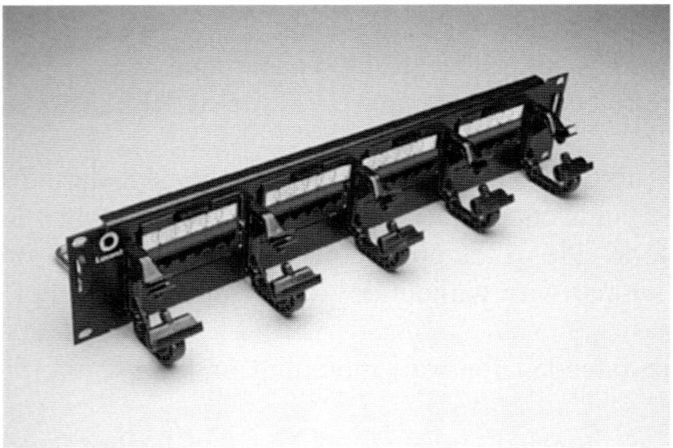

Fig. 3.41. Modular jack panel

The modular panel consists of the following components:

- modular jack block;
- designation elements;
- cable manager;
- fixation elements.

The modular jack blocks are base constructive elements of a patch panel with modular jacks. It is configured as a plate with premounted modular jacks of the respective category. The base material of the support is anodized aluminium. The support plate also functions as a faceplate; therefore, it has a specially designed aesthetic profile. The majority of SCS equipment manufacturers place their company logo on the left side of the panel, with the designation of the product and its category on the right.

The majority of patch panels are configured for modular jacks horizontally arranged in one (up to 24 ports) or several (a maximum of five in existing solutions) lines. At the same time, the minimum recommended port density is 16 per 1U of height and 48 per 3U of height [68]. There are a sufficient limited number of panel types, which have line assemblies of jacks vertically arranged. Such assemblies are usually colored black. It is possible to use a color code for the easy identification of panels of different functional sections. Thus, for example, the company ITT NS&S has produced category 3 panels in black, red, and green.

The modular jack blocks are divided into demountable, permanent, and intermediate (Fig. 3.42).

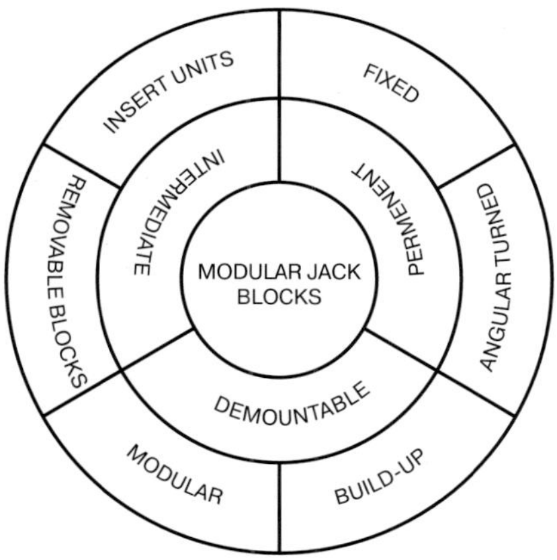

Fig. 3.42. Modular jack blocks and their options

It is interesting to note that in contrast to telecommunications outlets (Section 3.3.3) the vast majority of the reviewed panel types have a straight-through arrangement of jack modules. Only a few panels with angularly positioned jacks are known, for example, products of the 4224 series by the company Telesafe. Convenient connections and the easy reading of enlarged markings support the solution.

The panels reviewed in this section have the most developed variety of designation elements among their functional analogues. Besides numeration, individual jacks of patch panels may be marked with write-on coding in squared-shaped fields in the left lower corner. The marking fields may be provided for each individual jack as well as for a group of jacks. The write-on one is made by a soft pencil or a special soft-tip marker. Solutions using replaceable labels, when the faceplate of a modular jack is equipped with a labeling window and replaceable transparent latched cover, have not become popular. The panels of the ALPHA series by the company Elgadphon have a port digital identification and a marking field combined due to a splayed marking window in the left upper corner. It is possible to use a double-marking of ports when there is also a place to snap in a color coding icon in addition to the marking field. This solution is used in some types of panels produced by the company Ortronics.

Cable managers serve for accurate horizontal and backbone cable management as well as patch cords. In a vast majority of cases the cable manager is designed as a single element, reviewed in detail in Section 5.1.4 below. There are some options for the class panels with integral elements, basically, in the form of rings mounted on the faceplate under modular jacks. In a vast majority of cases the cable managers to be wired on the faceplate are designed as detachable brackets in order to downsize packaging during storage and transportation. The brackets are mounted on the plate housing by fitting them into special openings or by screws.

The panel fixation elements are used for wall mounting or in the 19-in. racks. The second option is more popular. Some manufacturers include an adapter with the standard panel kit to provide for mounting to 23-in. racks.

When wall mounting, the panel, as a rule, is fixed horizontally. Solutions for a small number of vertically mounted ports (as a rule 12) have become widespread (for example, 558260 by AMP and HD5-89D-12 by Siemon). The products are sometimes called compact patch panels. They are installed using a plastic bracket identical as for the panels.

The panel usually has only differently configured IDC contacts on its rear side. In some panels of the SMART series by the company RiT Technologies and by the company Elgadphon the faceplate is equipped with modular jacks. These are mounted instead of traditional jacks where IDC contacts of the bottom row are replaced by modular connector pins, i.e., so-called in-line adapters (Section 3.4.2). The engineering solution enables using common modular cords as a functional analogue to the electrical pigtail (Section 3.5.2). Connection of the plug from the back panel releases the faceplate from a superfluous bulk of patch cables, which improves the aesthetic characteristics and makes marking reading easier. In so-called intellectual panels of the company RiT Technologies there is an additional connector on the back side to connect the scanner of the PatchView system by a ribbon cable (details in Section 13.1.3).

Modular jack blocks and their options. Permanent blocks (Fig. 3.42) have preinstalled modular jacks. This solution provides a very high density of jacks, and respectively, the possibility of multiple cable wiring. Signals from IDC contacts are transmitted to telecommunications outlets by means of printed conductors. Modular jack and contact terminals assemblies, which are located on one printed circuit, may possess a different density. The most widespread of the assemblies in practice are the 6- and 8-port options.

Due to the used printed circuit method the reviewed panel type made by some companies is wired only by a single-wire impact tool. The use of a four-pair tool is prohibited due to the risk of damaging the circuit by a hard impact during the positioning of a wire into a IDC contact. The majority of permanent blocks are installed according to a given scheme, which causes some inconvenience during mounting. Swivel modules are used to eliminate the drawback.

Demountable blocks enable the mounting of modular jacks direct on site. In cases where the installation is done on site, stackable panels are used. The panels are configured for one specific module type. Their key advantage is a high density of ports, which is not less than in the permanent blocks (i.e., 24 per 1U of height).

Modular jack panels have their blocks installed via adapters; the blocks are sometimes compatible with telecommunications outlet blocks on workplaces. This increases the functional flexibility of the device because it is then possible to fit any desired modules with any quantity. Designers note one more advantage of such panels: they are applicable for both electrical and optical solutions, as well as during system maintenance and its development.

Both options of the panel have free openings to be covered by detachable blind covers to improve the aesthetic appearance, which may be used to augment the system if required. There are two options for the delivery of demountable blocks. Panels and jacks are most often delivered separately and installed on site. Sometimes the panel has jacks to be fabric premounted in openings.

The vast majority of panels with demountable blocks are intended for the traditional mounting of modular jacks, i.e., with the contact on the upside and the groove fitting the latch lever downside. The solution with a side-mounted jack is very rare, and is only used in situations when a vertical installation is impossible due to design features on the rear side of the jack. Some panel types of the company Quante serve as examples of this approach.

Due to design features the commercial use of patch panels with demountable blocks in major networks is not practical, as the permanent ones have better cost effectiveness and port density (an inevitable consequence of universality and a high functional flexibility). Thus, for example, 96-port panels HD5-96 (permanent option) and CT-PLN-96-ID (demountable option) by the company Siemon have a height of 4U and 7U, respectively. It is assumed that the advantages of the demountable option are revealed when constructing compact cable systems or when some regular changes in the SCS cabling are required.

Panels that combine the main features of both permanent and demountable jack mounting schemes with a more or less side-to-side displacement depending on the solution are called intermediate panels. According to this, there are two basic options for their design. The first one, promoted by the company AMP (NetConnect panels), has cable to be terminated to the panel contacts, but the module accepts an insert with printed conductors. The application of the solution called edge connector provides for the flexible interface (8- and 6-position modular pins, MIC electrical jacks of the system Token Ring, etc.) depending on a specific situation, while its product cost is highly increased, as for any modular solution. Thus, the solution tends toward the permanent scheme.

The demountable blocks may be also designed as multiple assemblies (for example, CPP panels by Panduit). They are mounted in the housing, if required. To improve the aesthetic appearance of the panel, free openings are covered by blind covers. In panels of the ALPHA series by Elgadphon images of modular jacks are depicted on 8-module blind covers.

Application and design options. It is efficient to use modular jack panels in part of an SCS for the maintenance of LAN applications. It is sufficient to administrate it with full four-pair data transmission channels in this case. The pre-mounted panels, in the authors' opinion, possess superior aesthetic characteristics among similar products, and have an easier and simpler patching process. At the same time, if administrating each individual pair is required, they make the SCS cost ineffective, especially for backbone cabling. Therefore, the key application area for patch panels is horizontal cabling.

The typical density of the patch panel is for 12, 16, 24, 32, 48, 64, 96, and 120 ports. Depending on the specific producer some values can be skipped. The use of panels with a large number of ports is not advisable, as it hampers the management of patch cords. The panel with capacity per 1U of height includes a maximum of 48 ports (the company Krone) in a majority of solutions. Multiport panels are produced with increased height.

A majority of producers manufacture the panel type in two options: with 568A and 568B wiring. Using the T568B scheme for the termination of cables to modular jacks is recommended when installing the panel.

General information on panels of some SCS equipment producers is given in Table 3.33. SMART series panels by the company RiT have jacks where each slot is equipped with a red indication LED for the convenience of the cable system maintenance. For the transmission of controlling signals special 9-conductor connecting cords are used. The panels are reviewed in detail in Section 13.1.

Table 3.33. Patch panels with modular jacks of category 5 and wiring scheme 568B

Company	Panel type	Number of ports	Height
Siemon	HD5-16A4	16	1U
	HD5-24A4	24	1U
	HD5-32A4	32	2U
	HD5-48A4	48	2U
Lucent Technologies	1100AA1-24	24	1U
	1100AA11-48	48	2U
Alcatel	ACS-503.124	48	3U
MOD-TAP	IDC-KATT1U-24xRJ45 cat.5	24	1U
	IDC-KATT1U-48xRJ45 cat.5	48	2U
Homaco	MJPC5-8-37TB	24	1U
	MJPC5-8-35TB	48	2U
Hubbell	MCC5802110A19	16	1U
	MCC5806110A19	48	2U
	MCC5812110A19	96	4U
	MCC5815110A19	120	5U
ICC	ICMPPOXX5B	24	1U

Cable termination to modular panels. The modular jack IDC contacts of the panels are designed in 110 and Krone options. At the same time the American producers tend to apply 110 contacts, but contacts LSA+ by Krone are more popular in Europe. Sometimes some companies suggest two panel options with different contact types. The contacts, as a rule, are placed on the rear panel in one or two rows, which causes some inconvenience during wiring. To eliminate this drawback the following solutions have been designed:

a) The application of 19-in. rack solutions, which enable folding back the mounted panel at an angle closer to 180 degrees by means of different swivels and hinges. This provides a convenient access to IDC contacts on its rear side (Fig. 3.43a).
b) The company RiT Technologies has designed mounting brackets to hook on rails of the 19-in. rack in order to mount the panel with a downwards turn of approximately 135° relative to the standard working position. Once all cables are preinstalled, the panel is removed from the brackets, the brackets are removed, and the panel is installed in their position (Fig. 3.42b).
c) The modular panel PATCHMAX by the company Lucent Technologies includes four or eight (depending on the option) 6-port, so-called module adapters DM2150. The modules are installed in the panel housing by latches, and if it is necessary to install some cables they may be turned horizontally up to almost 180°. This provides free access to the IDC contacts 110 on the rear printed circuit of the module.
d) There are solutions with IDC-contact lines to be turned by 180° and placed on the front panel, positioned close to the modular jacks. After wiring, the contacts are covered by a hinged or snap-on decorative cover. The convenience of gained in mounting the panels makes up for the slight decrease in port density, as the IDC-contact lines take 0.5U of height per modular line. In panels of the company Reichle & De Massari there is no decrease in the density of ports, as compared with traditional solutions, due to the fact that the IDC contacts are covered by hinged covers with designation fields.
e) Front access modular for 110 panels of the company Ortronics, which can be considered as an option of the previous solution, have modular jacks to be installed on a protruded base of the panel faceplate.
f) Due to that fact, cables may be routed to guide pins 110 without using the rear side. This solution is not highly aesthetic; however, it is very efficient when mounting additional panels into a highly loaded rack.
g) A retractable type panel with a front inclination up to 30° is provided by the companies Alcatel, Ackermann, and some others (Fig. 3.43c).

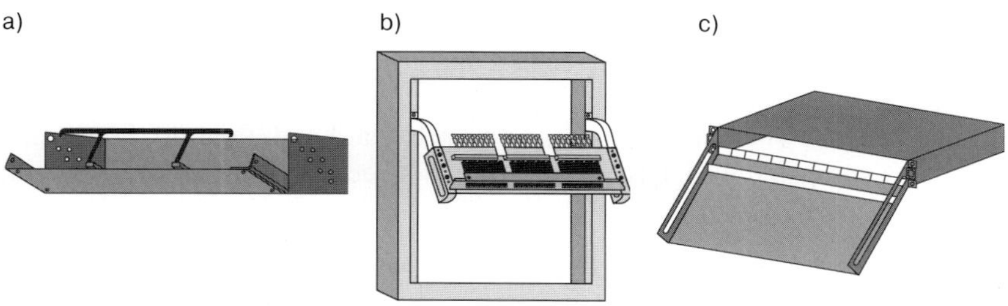

Fig. 3.43. Solutions to simplify the installation of patch panels; *a)* fold-back panel; *b)* mounting brackets; *c)* retractable panel

Some solutions use Telco connectors (Table 3.34) to connect multipair cables to modular panels; the panel is manufactured with two options with a Telco plug and jack. The reviewed options are usually designated in technical literature by the terms patch panel 110 and patch panel Telco.

Table 3.34. Patch panels with Telco connectors and modular jacks

Company	Panel type	Number of ports	Category
Lucent Technologies	2500 Cat5 Modular jack panel	24, 48	5
Mod-Tap	27.6E.241.B004G	24	–
Ortronics	OR-808004926 OR-808004927	24	3
Siemon	CT-PNL-XX-XX-01	16–96	3
Telesafe	4145-0000	24	5
	4143-0000	24	3

Besides the traditional installation into mounting racks in equipment rooms, some SCS producers recommend Telco panels as a base element for a consolidated point.

Features of modular panels for screened cables. The panels for screened cable S-STP differ from those of unscreened cables by using screened outlet modules and cable screen grounding. Cable S-UTP panels only have common screening housing to completely cover the outlet modules. The housing may be both screw- and latch-mounted. One prefers the latter mounting option. Most constructions with a given configuration are grounded by means of a detachable and integral U-shaped bracket with clamps (grounding bracket), which are mounted on the rear of the panel. Sometimes clamps may be installed on the back fold shelf of the panel housing. The bracket also functions as a cable manager. Low transient resistance with a grounding bar is sometimes only achieved due to the direct contact of the screening housing of an outlet module with a slot on the panel in structures with demountable blocks.

The low resistance between the connecting panel housing and a protective or telecommunications grounding ring is provided when mounting the panel into a 19-in. rack and tightening with screws. Leftovers from the color and varnish may sometimes distort the parameter value; therefore, the screened panel kit includes a grounding cable.

Due to the necessity of using screening and grounding facilities the port density in a screened panel is usually low, as compared with its unscreened analogue of the same density. There are a few serial models that break the rule. For example, the Russian company ITco. has a panel with 24 screened ports per 1U of height.

Preterminated and cordless panels

The installation of connecting panel is a rather complicated and long procedure, and is performed by highly qualified personnel by using special tools and expensive testing equipment. The desire to protect the end customer from having to problem-solve has resulted in a number of technical solutions, which may be divided into two groups. The first group contains products with a fabric-made wiring, the second one does not need wiring at all.

The first approach is realized by so-called preterminated or prewired panels, which serve to ease the termination process of telephone stations to an SCS. The product mostly includes one or several 110 or 66 blocks with a prewired number of multipair cables. The solutions with modular jacks are less widespread. The far end of the cable is equipped with a plug or Telco jack. The optimal capacity of the preterminated 110 panels may be up to 900 pairs. Some options of 66 panels are wired by means of a cross-connect wire. In this case, a modular jack may also be used along with Telco jacks; both options for the jacks are mounted directly on the rear or side of the panel housing, and are used on one panel individually or together. Patch panels with modular connectors, whose outputs are connected to TELCO contacts, may be symbolically referred to as the preterminated panels.

The vast majority of preterminated panels are intended for installation in telecommunications closets and used for the connection of multiple-access network equipment. There are few products of the type to be used as multiset MUTO outlets in open work areas, for example, the panels OR-60950092 and OR-854044831 by Ortronics.

The de-wiring principle is used in so-called cordless panels, which contain a number of modular pairs with short internal interconnections along current-carrying printed circuit lines. The jack may be wired along different schemes with mandatory matching with the cord type connected. The product Cordless patch by Siemon is an example of this.

The development of the cordless wiring principle may be realized based on a plug-type connection of two wires to be inserted into one slot, which is widely applied in telephone systems and also used in panels of the SMART series by the company RiT Technologies. Some models of the series, sometimes called a switching patch panel are equipped with modular connectors to be connected in pairs along current-carrying printed circuit lines. The same solution is used in panels of the company Elgadphon. The products differ from their analogues by the presence of an additional slide switch, which enables breaking the connection (Fig. 3.44).

Designers used here the prohibition of SCS standards for the parallel connection of any piece of equipment to a signal transmission channel. This makes unnecessary for the element, which closes the pairs to have its electrical contacts being accessible from the faceplate of a panel. This allows instead of a common outlet fitting for parallel connection to use a switch as a more convenient element.

At the design stage one may manage without using patch cords at all by use of a technical solution and careful planning of cable wiring (at least during the initial maintenance period). The panels with incorporated switches, besides their main application, enable an effective connection to LANs of different testing equipment, due to the available parallel connection of similar wires of two outlets.

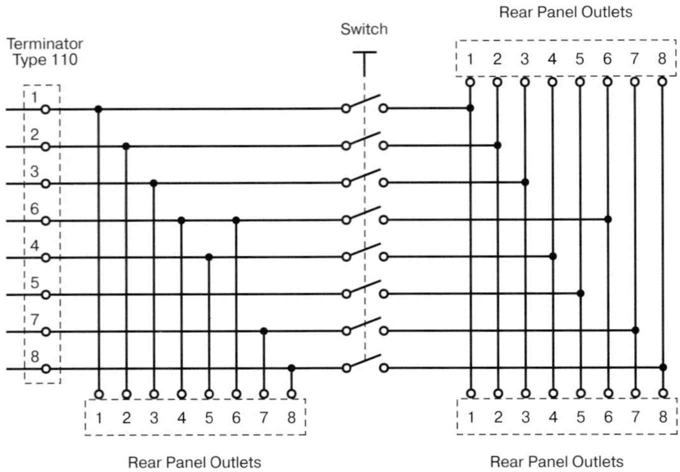

Fig. 3.44. Panel module wiring scheme with incorporated switch

Other options for connecting panels

Besides the main connecting panels that have been already reviewed, there are a small number of other products with similar functions. Panels with baluns (for definition, see Section 3.4.2.3) are intended for use in telecommunications closets when serving high-capacity network equipment with coaxial and triaxial interfaces. They are usually produced as a 16-port option; the 32-port option is less popular. Depending on the profile the faceplate may be equipped either with a modular jack or a coaxial or triaxial tube. Such equipment is produced, for example, by Hubbell and AMP.

Panels with elements of active equipment do not refer to the standard SCS products, and had a limited distribution at the end of 1990s. More detailed information about the products is given in Chapter 7.

Distributors

The distributors formed a specific class at the very end of the 1990s in connection with a rapid growth of SCS to serve the needs of small work areas and home networks (SOHO sector). The element functionally comprises a switching device, which was initially intended for mounting without using any support components and out of equipment rooms. The distributor is designed as a plastic housing containing 4–6 modular jacks. Its profile looks like a multiport telecommunications outlet (Section 3.3.3) with the distinction that modular jacks are arranged only on one narrow wall of the housing.

The distributors are configured for easy mounting on a wall, in raceways, and even on a user's desk. If required, several distributors may be installed vertically by means of a fastening clamp, or next to each other. It is designed that the cable entry is available from the rear, side, and bottom of the housing. The housing is made of thermoplastic with a low content of halogens; metallic housings are also available. The decorative cover of the housing has a special place for designation labels and icons.

Legs with an undercover of anti-sliding material and magnets for mounting on steel surfaces are available as options. An example of a distributor being a commercial part of an SCS are the MPD series products by Telegartner [69].

3.3.3 Telecommunications Outlets

Traditional products

Telecommunications outlets are installed in work areas and intended for horizontal cable connection. The electrical outlet module is an integral part of horizontal cabling. The jack includes a housing and one or several (maximum 12) 8-position modular jacks. The standard ISO/IEC 11801 recommends that one telecommunications outlet has to serve approximately 10 m^2, and to provide the following minimal configuration of jack modules:

- one module of category 3 or higher;
- one module of category 5 or an optical connector.

When using two-pair cables their jacks have to be designated. The telecommunications outlet housing is usually made of plastic and, depending on the fastening method, may have a different configuration. The most popular are housings for wall mounting and on a raceway routed along the wall as well as in floor conduits. The general classification of telecommunications outlet housings is given in Fig. 3.45. By analogy to common electrical outlets, we shall further classify such housings as outer (Fig. 3.46a) and inner (Fig. 3.46b–d) ones.

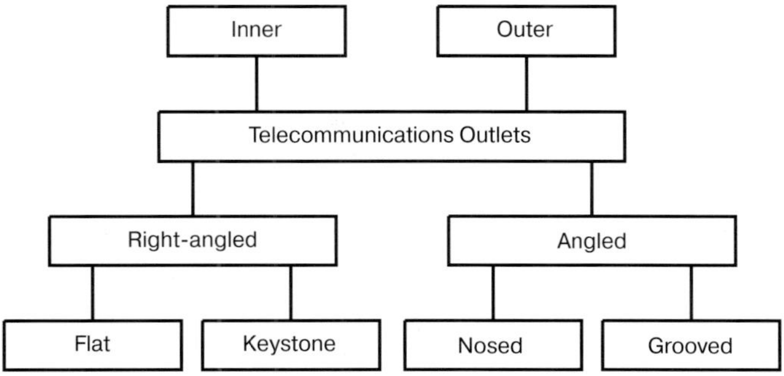

Fig. 3.45. Telecommunications outlets solutions

Inner and outer housings have an almost-similar configuration, and the only distinction is that the outer housing covers the jack module all around while the inner housing is designed in the form of a decorative faceplate or another mounting support leaving the rest of the module open. The profile and fastening elements of the faceplate are designed taking into consideration national and regional utilization. Thus, for example, there are Italian-, French-, German- and Scandinavian-styled panels. The panel leaves the rest of the jack module open. The open parts are protected from mechanical exposure and dust from other constructive elements.

The housing has to meet the following standard requirements:

- aesthetic appearance;
- designation space;
- resistance to mechanical exposures in case of unexpected snagging by equipment, etc. (for long life all freely accessed open parts have to be highly durable);
- it is desired that it is not possible to dismount jacks without a screwdriver or similar tool, or at least, that it should be complicated (curiosity-proof).
- the housing of an outer outlet is mounted by: anchor bolts or similar elements, two-sided adhesive tape, and magnetic fasteners. The company Molex delivers housing for telecommunications outlets equipped with magnets to fix on steel partitions and desks. A similar solution is used by Lucent Technologies for housings with not less than a four-module capacity. In contrast to tape, the magnet is detachable here, one magnet is used to fix housings for four modules, for six and twelve modules four magnets are used.

Inner plug housings are placed in their standard position, as a rule, by means of latches. Screw fastening is more rare, and mostly in shielded heavy solutions.

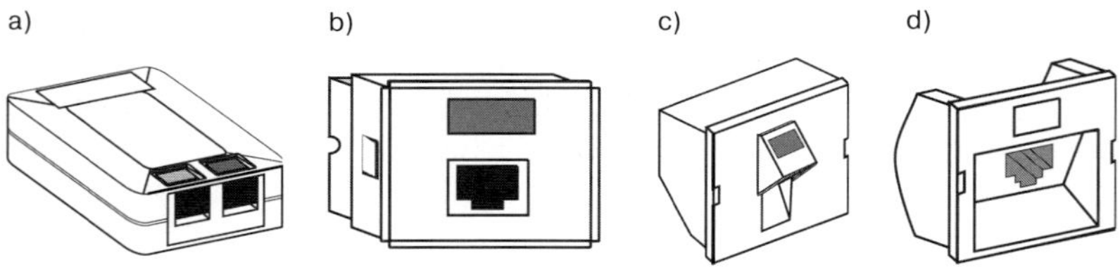

Fig. 3.46. Telecommunications outlets: *a)* outer right-angled; *b)* inner right-angled; *c)* inner with the angular nose; *d)* inner with the angular groove

To protect the jack against dust and other airborne contaminants while in the disconnected position, hinged or folded-down spring-loaded shutters are often used, the latter ones may open up or down. The hinged shutter is part of a plug housing, while the folded-down door is often part of a modular connector jack, and sometimes is designed as a removable adapter. For easy opening of the shutter they are often provided with several tabs protruding from the housing in the fold-down option or lugs in the hinged option. The vast majority of jack modules have individual protective covers. Sometimes there are protective doors to cover simultaneously two jacks of the double module (e.g., some types of outlets of the ICCS cable system by Corning).

The outlet connection modules are designed with different fastenings for telecommunications outlet housings and horizontal cables. The modular jack is usually fixed with plastic latches, sometimes there are also screw-mounted fastenings through the metallic faceplate or simply by a faceplate decorative plastic pressed into the cover. In all these cases, the jack has to be located perpendicular to the housing (Fig. 3.46a, b) or at an angle position (Fig. 3.46c, d). One may use housings with both an angular nose or groove for the angular-positioned jack. In a majority of cases the jack is mounted at an angle of 45°, sometimes other options are possible (15°, 75°, etc.). The groove-equipped system (Fig. 3.46g) has smaller dimensions. The option with a nose (Fig. 3.46c) is normally easy for designation reading. The main technical benefits of the angular option are the following:

- It provides a large bend radius for a modular cord at the point of connection to an outlet (the cord will be in an almost vertical position). This reduces the probability that it is damaged by the user and improves its electrical performance. The same aspects are applicable to horizontal cables when cabling in conduits.
- In case of unexpected patch cord pull-outs, the outlet connection module is less exposed to mechanical loads due to its smaller lever.
- It provides a more efficient protection of a modular outlet jack from dust and other particles when in the disconnected position.

If telecommunications outlets with an angular-positioned nose need to be installed, then two approaches are used in practice. The first one is based on the application of a special module for an angular installation (Fig. 3.25a), which is mounted on a flat decorative faceplate. The second approach is aimed at using a module for a flat mounting, but the nose is finished on the faceplate. In addition, we have to note that a telecommunications outlet or a functionally similar element with an angularly positioned jack module is the only option for some under-carpet conduits.

Most of the telecommunications outlets have one or two jack modules. According to information available in the middle of 1999 the configuration with three modules is a very rare option. Thus, these modules may be both triangular-shaped (4671 series outlets by Telesafe) and placed in one line (MINI-COM series products by the company Panduit) in right and angular nose options.

Solutions to enable the adaptation to specific installation conditions are rather rare and exist only as two options. The first option, promoted to the market by a number of German companies, comprises outlets for conduit mounting with a vertically movable (within 5.4 mm) upper part. The second option, designed by the company Lucent Technologies, allows the fitting of modular connector outlets, with a double latch depending on a specific situation, into the faceplate at 90 or 45° angles.

There are constructions of the telecommunications outlets with elements for the protection of network equipment against high voltage and impulse noises. Gas dischargers and different diode circuits as well as the grounding of individual wires through RC circuits are used for this protection. The normal operation of some types of network equipment requires the use of load resistors, which ground individual conductors or connect them to each other. Such resistors are installed as the standard kit elements of a telecommunications outlet, for example, in some types of modular inserts of the NETConnect systems by the company AMP.

Horizontal cable may be wired on the 110 connector line in the outlet connection module or on individual IDC contacts with various modifications. The IDC contacts are often covered by a plastic cover, which serves as a mounting element for conductors in some solutions. To increase the reliability of operation the cable is often additionally fixed. The cable is then installed in the module by a plastic hoop, which comes in a delivery kit and is sometimes already preinstalled by lacing it through the respective slots.

A cable may enter the housing of an external telecommunications outlet from the rear, side, or bottom into respective openings. To improve the aesthetic parameters the opening is covered by a thin plastic separating piece, which has to be snapped out when lacing the cable.

Outlet connection modules for shielded cables basically differ from unshielded ones by a shielding housing and a conducting insert into the jack that ensures the electrical continuity of the shield. The cable shield was often fixed to a special housing tab by its winding in earlier solutions. As the solution failed to provide a full round contact for outlet and cable shields, the fastening elements for the most part are today made in the form of a clamp or a hoop to provide the full round contact. There are also two-part assemblies with one part fixing a cable jacket and the other covering the peeled part of the screen; Along with an additional fixation this also provides electrical continuity [70]. The housing construction depends on the cable type, which is managed in the outlet connection module. When using S/UTP cables an overall shielded housing is often installed for all outlet connection modules. In outlets mainly designed for S/STP cables, it is mandatory that each module is covered by a shielded housing. Procedures to ensure a reliable electrical contact with a metal-plated surface of the plug housing when the patch cord is connected are reviewed in Section 3.2.2.

Outlet connection modules of telecommunications outlets are divided on the basis of their electrical performance characteristics into categories 3, 4, 5, 5e, 6, and 7. The scope of normed parameters corresponds to the scope of the performance for electrical characteristics certified by the standards. Producers of some types of outlets with two-outlet connection modules also specify crosstalk values between individual pairs of neighboring modules in their technical specifications. The outlets may be designated by:

- color strap-in icons (with pictures of handset and monitor) or the inscriptions DATA, PHONE, LAN, VOICE, etc. Such straps may be inserted in jacks of an outlet housing or an outlet connection module as well as on a module protective cover;
- protective adapters and detachable hinged colorful covers;
- outlet connection modules with differently colored faceplates (standard colors: white, grey, black, dark-blue, red, and yellow);
- windows on the housing for designation with signs and symbols by means of adhesive labels or removable inscriptions;

- digits to be made during outlet manufacturing in multiport (6 and 12) housings;
- special adhesive markers and correspondent inscriptions or pictures to stick on the housing.

Some companies offer insert units for their telecommunications outlets, which serve as a 6-position modular jack after their installation. This insert unit makes the connection of a telephone cord convenient and also may serve as a designation mean of the outlet. The producers usually offer the telecommunications outlets housing in several colors, though the housings with the highest distribution are white with different tints.

Outlets of telephone devices

Special-purpose outlets for the connection of telephone sets are very rarely applied in practice, and mostly on sites where the connection of other network equipment seems to be impossible (for example, security posts). Due to national specifics for switching products and their incompatibility with standard modular connectors (Fig. 3.47) the majority of SCS producers avoid including the present equipment as a part of their systems and avoid certifying solutions based on that. Using telephone adapters for the connection of telephone sets is recommended (Section 3.4.2).

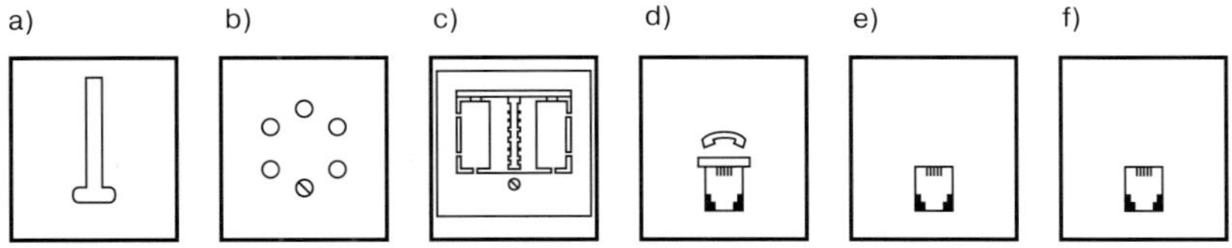

Fig. 3.47. Outlets profile for the connection of telephone sets with national standards [71]: *a)* french; *b)* belgian; *c)* german; *d)* portugese; *e)* spanish; *f)* international

It is possible to use specific solutions only if those incorporate a modular jack. Thus, in particular, there are special outlets for telephone set wall mounting, which are carried out with a right-angled installation of one outlet connection module (flat and key-stone mounting schemes), and differ in that they have two round plastic head-lugs with standard fastening holes to hook up a telephone set on the faceplate. Usually the assembly of these outlets is based on a faceplate adapter for inserting a common outlet connection module. Along with 8-position universal modules, 6-position modules may also be used in this case. An example of such a product is the steel adapter MX-WP-SS and MX-WP- (XX)-SS for modules of the MAX series by the company Siemon.

3.3.4 Open Office Solutions

Outlets for under-carpet conduit mounting

Under-carpet conduits are widely applied to connect workplaces that are located in spacious premises and far from the walls to an SCS (for more details see Section 5.2.4). Housings of the conduits are designed to mount standard internal telecommunications outlets. In some cases a producer of the conduits manufactures only housings to be adapted to a specific use depending on local conditions and with the use of its engineering facilities. Under-carpet modules (in German, bodentank) function as the engineering facilities, as applied to the telecommunications outlets. The under-carpet modules include an adapter housing matched with the box of a specific producer with preinstalled external outlets [72].

MUTO outlets and consolidated points

Telecommunications outlets with 6- and 12-outlet connection modules most often function as MUTO outlets in practice. The construction of the outlet housing is also used for the entry of 25-pair backbone cables along with standard horizontal ones. The majority of the products are usually designed for installation in work premises and, therefore, has special aesthetic characteristics. There are also options with more primitive designs and, respectively, a lower cost for mounting under raised floors.

There are special-purpose products for use as MUTO outlets. For example, the company AMP produces a 6-port outlet modular distribution box. Its specific features are that there are options for 568A and 568B wiring schemes. They can be used for a horizontal cable Telco connector connection, and there are spaces for plastic icons for port designation. Some switching products of this type have a common constructive profile. The type of exit connector indicates their specific use. Thus, for example, the company Lucent Technologies produces so-called zone wiring boxes aimed at the entry of six four-pair or one 25-pair cables to be grouped on 110 contacts. Six modular connector outlets (MUTO outlet) or type 110 (consolidated point) are used as the exit connectors depending on the option. There is a standard kit of fastening elements for wall mounting the product. In the AHB1 housing by the same company a 25-pair cable is connected by a Telco connector. Six modular jacks serve as ports.

Besides electrical outlets the market has also started offering optical (Section 4.3.6) and combined outlets of consolidated points. In the latter case a modular construction principle is applied, which enables adapting the configuration of the product for the specific needs of the office or its part. The majority of consolidated points realize individual connection schemes in practice, i.e., each telecommunications outlet is connected to its outlet connection modules individually and by an individual cable. The company Molex has presented a 6-port group device RICI (rapid inter-connect interface) (see Fig. 3.48).

The device includes two symmetrical housings with DataGate outlets to be placed on one of them. The other one is equipped with WE8W plugs, which provide electrical characteristics of category 5e. The housings are made of strong plastic Lexan, and easily assembled by means of holders. The RICI device is equipped with a universal mounting unit, which allows an easy mounting of the consolidated points on a wall or behind a suspended ceiling. The easy reconnection of the pin field of the device and the possibility of serving six ports at once enable, in the designers' opinion, saving up to 30% of the personnel time in automation department during SCS re-configuration in an open office.

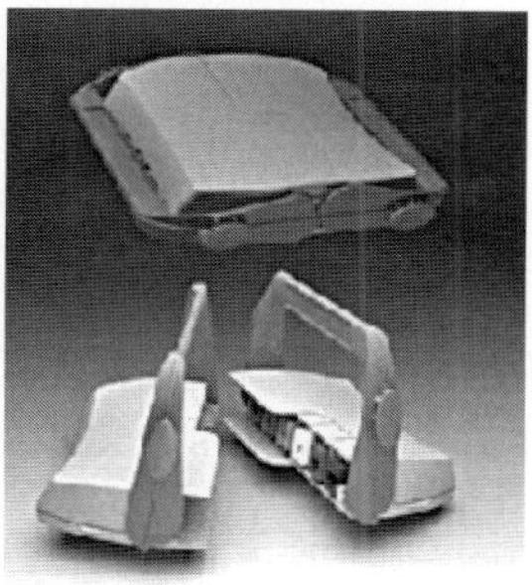

Fig. 3.48. RICI device

3.4 Modular Cords, Adapters and Line Extensions

The elements reviewed in this section are used for the connection of network equipment to an SCS, and are officially beyond the SCS standards. Thus, they are not specified in valid normative documents. The documents only provide recommendations for their configuration and application. However, a more detailed review of the constructive features and functions of these elements are needed for two reasons. The first reason is that modular cords, adapters, and line extensions are often a part of the auxiliary equipment included by an SCS manufacturer in its catalogue and also supplied by a system integrator company, which realizes SCS. The second reason is that the right choice of these elements leads to many more applications, which may use the SCS as a platform for their signal transmission, i.e., which directly define the technical and economic efficiency of the cable system created.

3.4.1 Equipment Cords

Specific constructive features

These cables are intended for the connection of different network equipment to an SCS in work areas and telecommunications closets. The device is not specified in detail by the valid versions of the SCS standards, which specify only the type of their connector and determine some common restrictions in length. In a majority of cases, equipment cords and modular cords have a similar construction, and if they have the same length they are interchangeable. A piece of patch cable with installed 8-position modular plugs on both ends is used to produce such an element. Such cords are classified according to the standards as categories 3–5. There are also cords of categories 5e, 6, and 7 on the market. The cords may be unshielded and shielded. The latter ones are made of a shielded patch cable (STP, S/UTP, or S/STP) and plugs for shielded twisted pairs described above.

In some cases the shielded systems are construed along a one-way grounding scheme according to the fact that the shield of the elements forming a horizontal subsystem is connected with the ring of a telecommunications grounding ring in a telecommunications closet. To realize the scheme, shielded patch cords, with one plug that electrically discontinues the shield from the cable shield, are applied. To provide the correct connection of such a patch cord its plug has a special designation labeling. The drawback of the approach is the probability of an incorrect connection during maintenance, which would result in a dramatic drop in shielding efficiency. To eliminate this, the company Telesafe designed a SAFEGROUND® type patch cord, whose shield discontinuity is accomplished in the middle of the cable. Two parts of the shield are connected by a capacitor of 150 nF with breakdown voltage stress not less than 230 V. The solution guarantees the alternate current continuity of the shield together with the galvanic discontinuity.

The equipment cords may be factory produced with special automatic or semiautomatic equipment, as well as directly on site during SCS mounting. Two approaches may be applied for field manufacturing. In a majority of cases one may use a hand crimping tool. The second approach is realized, for example, by the company Lucent Technologies, which has produced a 700A8 type plug. This element is similar to a common household electric plug in design. Its housing consists of two parts and is fixed by a cutting screw when assembled. The housing contains eight IDC contacts to be fitted in one row inside one of its parts. The other part functions as a protective cover and also as a mounting working element. The simplicity of the construction and assembling is achieved at the expense of a considerable degradation of electrical characteristics. According to specifications the maximum rate of the signal transmitted across the plug has not exceeded 10 Mbps.

Many SCS producers prohibit home-made cords to be used in certified systems. This is explained by their lower reliability and difficulties in meeting parameters of category 5 and higher (especially for the NEXT value). The producers offer the simplest equipment cords of different lengths. A number of discrete values with a certain step are usually used depending on the manufacturer. The American companies mostly tend to integer foot values of the parameter. If a considerable volume is delivered to the European market, or if the production is located in Europe, then a decimetric discrete value is usually used.

As has been stated above, equipment modular cords do not structurally differ from patching cords, and often are a part of SCS specifications as an element for termination of workstations and other similar network computer equipment. Cords with four-position plugs for a telephone set connection are included in its delivery kit and, therefore, are not included in the specifications.

Equipment cords in telecommunications closets and equipment rooms connect ports of network equipment and connecting panels. Most constructions connect the panel on the faceplate (Fig. 3.49a). Figure 3.49c presents a solution, which may be practiced for some panels of the SMART series by the company RiT Technologies. A similar constructive scheme is also applied to some panels of the ALPHA series by the company Elgadphon. Its key benefit is that the front part of the connecting field is released from the excessive cables of equipment cords, which results in an improvement of aesthetic characteristics of the panel and easy designation reading. The use of so-called electrical pigtails (Fig. 3.49b) provides these benefits as well.

Similar to patching cords, the PerfectPatch may be used to adjust the length of equipment cords in telecommunications closets and equipment rooms.

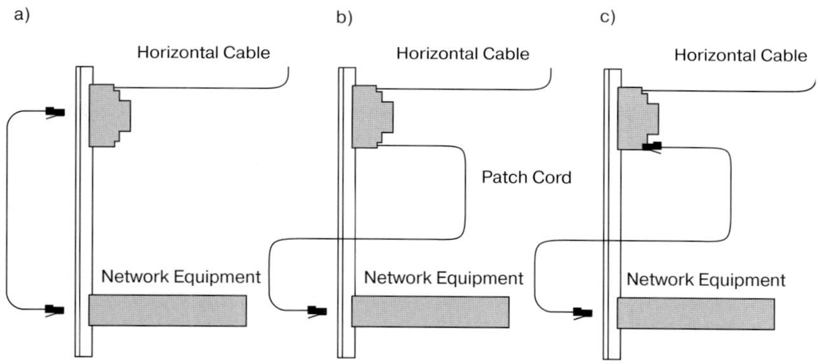

Fig. 3.49. Options for the connection of network equipment to an SCS: *a)* by equipment cord; *b)* by electrical pigtail to patch panel; *c)* by equipment cord and SMART panel by the company RiT Technologies

Options for four-pair equipment modular cords

The modular cords are divided into straight-through, reversed, and special ones. The straight-through modular cord has cable wires connected to the plug pins in order to connect the plug pins with the same numbers (Fig. 3.50a). If the colors appear in the same order on both plugs, the cord is wired "straight-through". The reversed cords may be required for the connection of certain types of network equipment to an SCS. They have pins in the second plug that are connected in reverse order from those of the first plug, i.e., the 1 pin of the first plug is connected to the 3 pin of the second plug, the 2 pin of the first plug is connected to the 6 contact of the second plug, etc. (Fig. 3.50b). The reversed modular cords are rarely used; they are used mostly for the direct connection of two workstations and other devices of a peer-to-peer local network to be configured without using a concentrator.

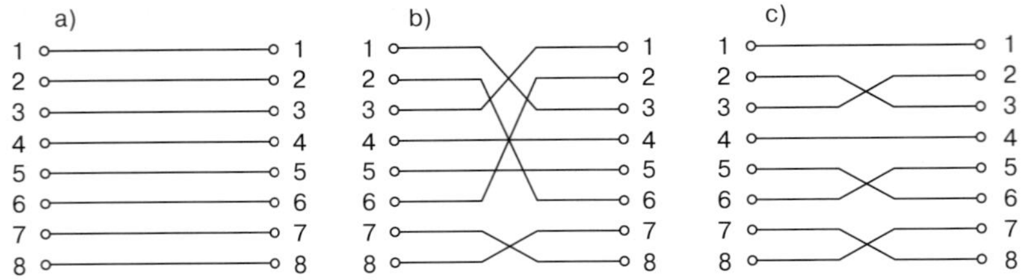

Fig. 3.50. Contact scheme of modular connector plugs by different types of cords: *a)* straight-through cord; *b)* reversed cord; *c)* zero-modem cord

Using special cords is usually required when forming an S/STP horizontal cable wiring with an operating frequency 600 MHz or higher. It is known that to get the required NEXT characteristics in communications lines such cables use boundary contact pairs of a modular plug, but wires of the rest of the pairs are used to increase noise immunity and are connected to the grounded, i.e., it is impossible to use them for data transmission. To connect Ethernet, Fast Ethernet, and Token Ring to such lines one may use adapters, which are reviewed below. Somewhat more convenient for maintenance solution allows getting special cords, which due to internal crossing of individual conductors connect the working pairs of network equipment interfaces to 1 and 4 pairs of an outlet (Table 3.26). For example, Fig. 3.50c depicts the wiring of a so-called zero-modem cord used for the direct interconnection of two computers via a serial port. Cable jackets along with designation labels are differently colored for a more accurate identification of the cord types.

In fiber optic SCS subsystems the so-called bundled or hybrid cords supplied on both ends with optical connector plugs of different types are widely applied. Their use enables easy coordination of a port type of network equipment and an outlet within a switching device. As there is a deficiency in the variety of electrical connector types, as compared with optic ones, the bundled cords are much less distributed in electrical SCS subsystems. Thus, telecommunications closets and equipment rooms use cords with type 110 and modular connector plugs (Section 3.3.1), which are sometimes used as termination plugs.

The group of special cords also includes so-called modular patch cords produced, for example, by Telegartner. These products realize the cable sharing principle. The cords differ from the standard ones by the crossing of their pairs and, in different options, are intended for simultaneous 10Base-T (yellow jacket) and 100Base-T (green jacket) signal transmission.

Different testing and measuring devices have to be connected to the SCS along with the network equipment. The cords that are often used for the procedure relate to the special ones. One end of the cord is equipped with a plug of a modular or type 110 connector, the other end has standard screw clamps or "alligator" type clips.

Cords for the connection of telephone sets to an SCS may sometimes be ordered separately. They are user friendly, and often designed as a soft spiral with coils approximately 1.5 cm in diameter. This type of cord fails to enable electrical characteristics required for high-speed data transmission systems; however, it is very convenient to use, as its coils shrink in a tensile-free mode, and the cord takes little space on a desk.

Electrical pigtail and one-end terminated cables

The majority of electrical pigtails, in contrast to the optic ones, are switching devices of a special type. The element comprises a piece of patch cable with a modular connector plug, which is mounted to only one of its ends. Twisted pairs of another unterminated end are grouped on contacts of the type 110 or modular connector of the connecting panel. In the latter case, as the wiring runs from the back side of a panel it enables halving the amount of cables on the front part of the connection field. This improves both the aesthetic characteristics of the panel and makes it easy to read the designation of ports (Fig. 3.49b). One more application field for electrical pigtails, which is rapidly growing, is the interconnection between connecting panels of consolidated points in an open office and telecommunications outlets.

There are some types of electrical pigtails that are manufactured unshielded. Their electrical characteristics enable their use as a part of category 5 lines. The electrical pigtails based on a shielded cable are much less widespread (for example, we note products of the 9KE020XX series by the company Kerpen); however, they may operate on frequencies considerably exceeding 100 MHz because of the shielding.

Electrical pigtails cost somewhat less, as compared with patch cords; however, their use strictly configures the equipment to be mounted in telecommunications closets. Replacement of one electrical pigtail by another with a different length requires access to the rear of the panel with modular connectors, which is not always convenient or even possible. The mentioned facts lead to the fact that electrical pigtails are manufactured by a limited but sufficient number of producers of SCS equipment (for example, Lucent Technologies and RiT Technologies); the solution based on their use has not become widespread. Besides the electrical pigtails with modular plugs the company Lucent Technologies has also suggested an electrical pigtail with an MIC plug for the Token Ring system. Electrical pigtails may serve as material for the production of bundled patch cords of desired form. This is required if a connector that differs from the modular one serves as a base for a certain SCS.

The utilization of an electrical pigtail directly for the termination of horizontal cables in electrical subsystems is associated with an additional connection in the signal transmission channel, which is prohibited by current versions of the standards. Besides that, the installation of a connector on an electrical cable is technically an easier procedure as compared with optic cables. Therefore, electrical pigtails in their traditional option are used very rarely, and basically only for special needs. Thus, for example, the company Telesafe produces a so-called WTP pigtail. It comprises a short splitting cord to be mounted inside a wall outlet and intended for realizing the cable sharing principle. It operates at frequencies up to

862 MHz and has an incorporated balun to change to a coaxial cable. Connection to an S/STP horizontal cable is carried out by means of a type CMG connector.

The so-called preterminated assemblies are widely used in optic communications systems (Section 4.4.2). They comprise a cable piece of a certain length with connector plugs to be mounted on its ends. Functional analogues of the assembly for the construction of SCS electrical subsystems are so-called connectorized cables. The product often comprises a 25-pair cable of category 5 with TELCO connectors to be mounted on its ends. As an example of the product we indicate products of the CC525PP series by the company Lucent Technologies. The connectorized four-pair cables are very rare, and are used only in cases where a patch panel of a modular type serves as a base for a specific realization of SCS. In this situation an SCS producer offers its users a range of cable lengths up to 90 m. The cable system ICCS by the company Corning serves as an example of such an approach.

Bundled and multipair equipment cords

During the maintenance of an SCS bundled and multipair cables are widely used. The bundled cord's second end, depending on its application, is mounted with either a 15-position connector DB-15, a 25-position DB-25, or several pin connectors. The second end of the equipment cord is wired according to schemes that depend on a specific application. The most frequently used multipair cable has a standard length of up to 3.05 m (10 ft) and Telco connectors (RJ-71). Switching parts of the connector are based on two plates of 25 contacts each. The plug and the jack of the connector have almost-similar constructions and dimensions, and are easy to use for mounting on multipair cables. Before these connectors were not higher than category 3. Recently, a number of manufacturers, for example, Lucent Technologies and AMP, have launched a production of Telco connectors to meet requirements of category 5. The Telco connector is installed both on one or two cable ends. A cable with one connector has its second end wired on IDC contacts of switching equipment in telecommunications closets. The equipment cord with a Telco connector plug or a jack on both ends is often used for the connection of PABX to SCS.

An interesting feature of the cords reviewed here is that at present they are the only electrical switching products in commercial use. It is mandatory for the contact elements of the connector in the off-mode to be covered by protective covers (usually dark-blue for the outlet and red for the plug).

The splitting cord or hydra type cord (see Fig. 3.51) is used if the connection of multiport equipment to modular panels is required. It comprises a Telco 50-position plug or jack, which accepts six four-pair cables with modular plugs on the other end. There are also cords of this type with 8, 12, and even 24 6-position modular plugs, which enable making the necessary connections for each individual pair of a multipair cable. Samples of the product type industrially manufactured have electrical characteristics of category 3. They can be made of UTP and S/UTP cables. The majority of the splitting cords are made of unshielded cable.

The cable system Mod-Tap includes splitting cables 23A-212-816 and 23A-208-811, which are intended for connection to a shielded 25-pair cable.

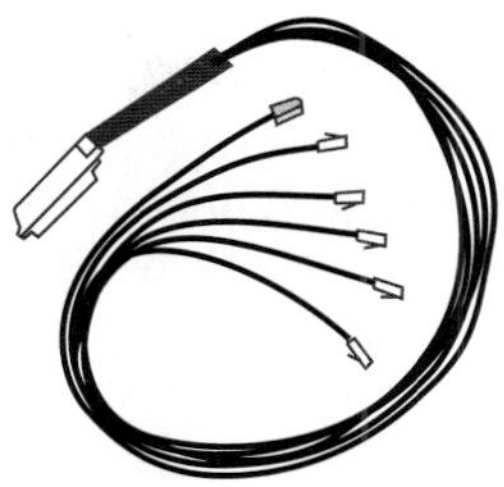

Fig. 3.51. Hydra type cord

3.4.2 Adapters

The adapters as well as the equipment cords are not a part of the SCS, and are used for the connection of network equipment. According to the standard TIA/EIA-568A the group includes those elements, which realize at least one of the following functions:

- termination to other connectors of incompatible sizes and types (adapter);
- a change in wiring scheme;
- splitting one multipair cable to several cables with a fewer number of pairs (splitter);
- connecting cables to each other

Device adapters

T-adapters are used to change a connecting scheme to a modular connector, for example, from T568A to USOC. The adapter usually consists of a short piece of cable with a mounted modular 8-position plug and a housing with one modular jack. There are also known "strict" options in the form of a plug-and-jack type insert unit. The housing has plug and jack contacts to be mutually connected inside, as required. T-adapters provide a system performance not higher than 10 Mbps. As an example of the device types we note products of the 950 series by the company Hubbell.

Adapters from modular connectors to interfaces RS-232 and RS/6000 (sometimes called modular adapters) are used when it is necessity to connect devices with connectors DB09, DB15, or DB25 of interface RS-232 (V.24) to SCS telecommunications outlets. They comprise a housing with a mounted plug or jack of one of the above-mentioned DB series connectors and a telecommunications outlet. There are shielded and unshielded options of the device.

Very often, a user may configure an adapter of this type to meet the specific conditions needed directly on site. The procedure runs by connecting DB connector pin contacts to contacts of a modular jack by short conductors, which enables the contacts from one jack of the DB connector to jump over another. The contacts are numbered by a color code of connecting conductor insulation. Use of a special pushing tool makes mounting and connection procedures easier.

Adapters from modular connectors to interface Token Ring comprise a housing with an MIC connector of the system Token Ring to be mounted on one side and a modular jack on the other side. There are shielded and unshielded options of the adapter. Their use enables changing a standard interface of network equipment Ethernet, Fast Ethernet, CDDI, and ATM with a modular jack to the interface of the system Token Ring, and transmitting their signals across 150-Ω STP cables to be used in IBM cable systems. Since SCS based on 150-Ω cables are not popular, such adapters are almost out of use in engineering practice in Russia.

Splitters

Y-adapters, twins, splitters, or duplex couplers divide in two (rarely three or four) telecommunications outlets those pairs of a cable, which are connected to contacts of one modular plug. The majority of the Y-adapters include an 8-position modular plug and a housing with two modular jacks. The plug may be tightly fixed on the housing (adapters 400B2 and 400K by the company Lucent Technologies) or connected to it by a short piece of cable (modular adapter by the company Siemon) (see Fig. 3.52). To increase the fastening reliability some models of the Y-adapters by the company Quante are supplied with an additional knurling head screw to be set up on the front surface between the jacks. It is obvious that such an adapter may be used only together with special outlets.

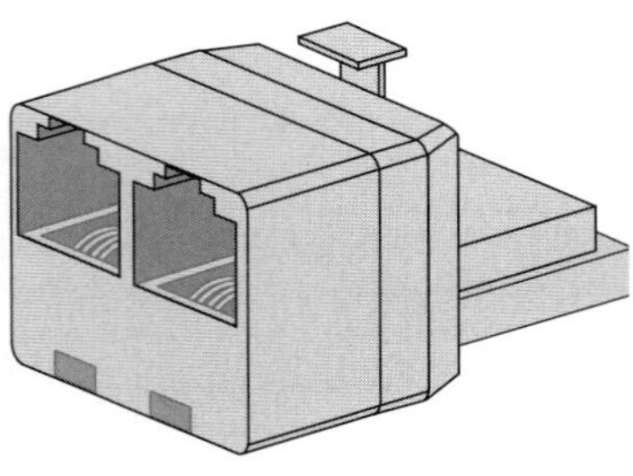

Fig. 3.52. Y-adapter

Modular jacks may be placed both close together and one above the other. The key benefit of the second option is that when a double outlet is used, the housing of the adapter does not overlap the second modular jack in the more popular horizontally oriented configurations. The so-called easy flex adapter by the company Reichle & De Massari has a hinged modular plug with a possible turn within 60°. This enables easy connection of the adapter to outlets with an angularly installed jack. Once the Y-adapter is installed in the outlet, its housing is fixed in the on-position only by a fastener of the plug. In some cases the construction is additionally fixed by screws. To perform the procedure an appropriate outlet construction is necessary. An example of the product is the adapter of the 75XX series by the company INFRA+.

As has been mentioned above, there are also individual examples of this class of product with a large number of modular jacks. For example, the so-called female duplex coupler by the company ICC has three outlets. The distinction between the two options is that the first one is plugged in the outlet and used for operation with a horizontal cable and the second one is aimed at splitting patch cords. For a Y-adapter with four outlets we indicate the so-called modular adapter (an adapter for one pair modularity), which provides connection of four 6-positioned modular plugs to the jack of a high-frequency connector, produced by the company AMP (Section 3.2.6) and differs from the other because of an available integrated cord with a plug.

The adapter has the plug and jack contacts mutually connected inside its housing, as required. The Y-adapter is most often used for the splitting of a four-pair horizontal cable into two groups of two pairs each. This allows connecting two telephone sets or one telephone set and a computer with a network adapter Ethernet with one horizontal cable.

The Y-adapters support data transmission rates not higher than 10 Mbps; therefore, the majority of the devices are manufactured as unshielded. Shielded options of this type are rare in practice [73].

In some cases "terminal" outlets may serve as Y-adapters. Thus, for example, in a dual-port modular jack of BRWP by the company Hubbell there are two 6-position modular jacks on the front panel, as the rear panel positions eight IDC contacts for the wiring of horizontal cable conductors. The option breaks the principle of wiring universality and is not recommended for wide use. The most successful use is the modular insert units of the system NetConnect by the company AMP, as they are detachable and, if necessary, the original wiring required by the standards may be easily recovered.

The harmonica or multiline-adapter may be considered as a Y-adapter for a multipair cable. The element is called harmonica because it looks like a mouth harmonica. It is aimed at splitting a 25-pair cable into several groups of wires for 2, 3, and 4 pairs.

There are two main options for the constructive finish of harmonicas. The first option, which is the most popular, has 8-contact modular jacks on a plug or a 25-pair Telco outlet; the second option has six four-pair lines of the 110 type. The number of modular outlets depends on the chosen wiring scheme. In case of a two-pair wiring the number of modular outlets amounts to 12. For more convenient maintenance options, harmonicas with a Telco connector to be mounted parallel and perpendicular to the mounting plane of modular connectors may sometimes be offered by one producer. An example of the solution is the adapter of the 258 series by Lucent Technologies.

Harmonicas may be made as a separate device (Fig. 3.53) as a module to be installed in different conduits, cabinets, and other similar elements, as well as on panels (so-called harmonica patch panel) for mounting in a 19-in. rack.

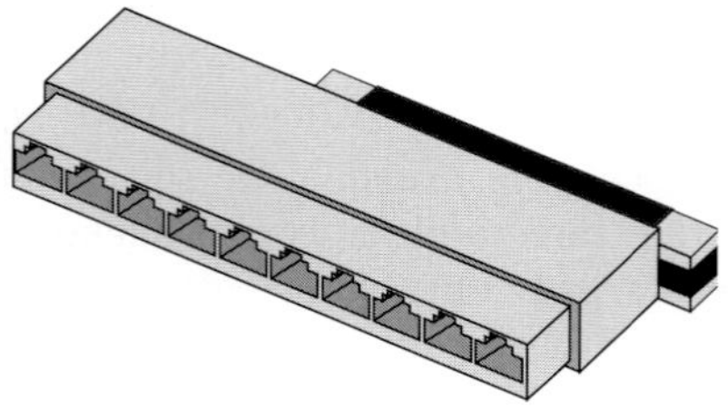

Fig. 3.53. Harmonica

Baluns

Balun stands for BALance–UNbalance and comprises a device for the connection of a twisted pair and coaxial or twin-axial cables. Besides the physical connection it also serves for the transition from an unbalanced transmission scheme to a balanced one, and for the coordination of the characteristic impedance of different signal transmission mediums.

There two main options for baluns: coaxial and twin-axial. The coaxial baluns are mostly used when SCS cabling is used for television signal transmission, and in this case they provide a high level of transmission characteristics with a wide frequency range. For example, the balun of the company Telesafe operates within frequencies of 47 to 862 MHz has a loss not higher than 2 dB at a return loss not lower than –10 dB. The most popular application area of the twin-axial baluns is the connection of terminal equipment with a host computer, whose interface is realized by means of the twin-axial connector, via the SCS.

The coaxial and twin-axial baluns often have a similar constructive finish for the housing. To identify the devices a color code is applied to the housings. Thus, for example, baluns for the connection of 3270 type terminals have red housing (red balun) and the housing of twin-axial components to serve the system AS/400 is colored green (green balun).

Baluns for twin-axial cables of 365A by Lucent Technologies consist of two elements. The 365 Balun adapter is intended for its own transition from a twin-axial cable to a balanced one, and the 365A terminating adapter minimizes the return loss ratio at a disconnected workstation. The device of this group may be finished as a housing with integrated components of modular and coaxial connectors. (so called mini-balun) In the second traditional option one or both of these elements are mounted on pieces of a corresponding cable of a certain length (for example, in the 384A type balun by Lucent Technologies the length of a balanced cable is 3.6 m).

The majority of baluns are finished as a separate add-on device. The company AMP produces a series of baluns for the system NetConnect, which look like modular insert units for mounting in a universal outlet or panel. The key advantage of the solution is a dramatically decreased possibility of mechanical damage during maintenance. The company offers a multipair cord 555540 of hydra type with 16 plugs or outlets BNC and incorporated matched transformers.

The standard horizontal cable includes four twisted pairs with good transmission characteristics. Many designers take advantage of the main benefit of baluns in connecting different television devices to the SCS. Thus, for example, there are baluns with two and three coaxial connectors. The first option is used for a common TV and voice signal transmissions, the second one is for signals of main colors.

Baluns for TV signal transmission across a baseband are sometimes called baseband video adapters, while devices that serve multiprogram TV signals are called broadband video adapters. The balun outlet to a twisted pair may be both in the form of an outlet on the housing and a plug. In the second option, in order to provide easy maintenance, in a majority of cases the plug is terminated to a piece of standard horizontal cable of an appropriate length. The company Hubbell produces 16- and 320-port panels with premounted baluns. Depending on the option the faceplate may be equipped with either an 8-position modular outlet or a BNC type coaxial outlet.

A balun has a modular connector with shielded or unshielded elements. In the shielded option the shield has electrical connectivity with the outer conductor of the BNC connector. The outgoing conductor of the matched transformer may be terminated to different contacts, most often to pairs 1-2 and 4-5, in a modular outlet.

The appearance and general scheme for one of the possible solutions for a balun is depicted in Fig. 3.54.

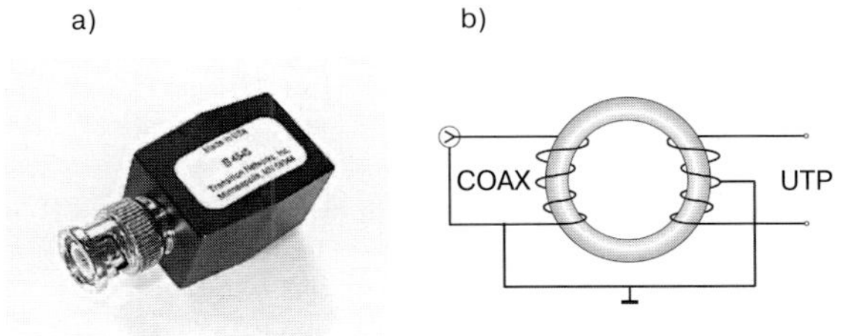

Fig. 3.54. Balun: *a)* appearance; *b)* generic scheme

Alternative adapters

Modular Adapter. A modular adapter provides a user-friendly interface when a connection to a 66 and 110 panels is required. It comprises a housing with one of the sides with a modular jack, the other with elements for direct connection to the panel. Connection to the 110 panel is an alternative to a bundled cord. Sometimes one and the same company offers two options of an adapter for three or four pairs to a 110 connector (products 1110A1-3 and 1110A1-4 by Lucent Technologies). The key application area is the connection of different measurement equipment to base lines. The majority of the adapters are designed for a maximum four-pair connection. The company Siemon offers a TAP-50 device for simultaneous connection to 25 pairs of the 66 block. The second end may be terminated to a plug or Telco outlet.

Measurement Adapter. A measurement adapter has a similar physical configuration, constructive design, and functions as a modular adapter. It differs from a modular adapter by its mounting position. It is mounted not to 110 and 66 contacts, but to breaking jumpers of special panels, and may be supplied with matching load resistors. At the breaking of a controlled channel these two options differ by testing directions (A and B). A parallel connection of the measurement equipment to the channel is also possible (Fig. 3.55c). In this case the adapter to be used is called a monitor adapter. The most preferred application area of the equipment is a consolidated point [26].

I-adapter. This element is also widely known as an in-line adapter, and is used to increase the length of patching and equipment cords. Its application inevitably leads to a dramatic growth of crosstalk and the return signal (a decrease in NEXT and SRL) and, therefore, is possible only in an SCS with a sufficiently large allowance for the parameters.

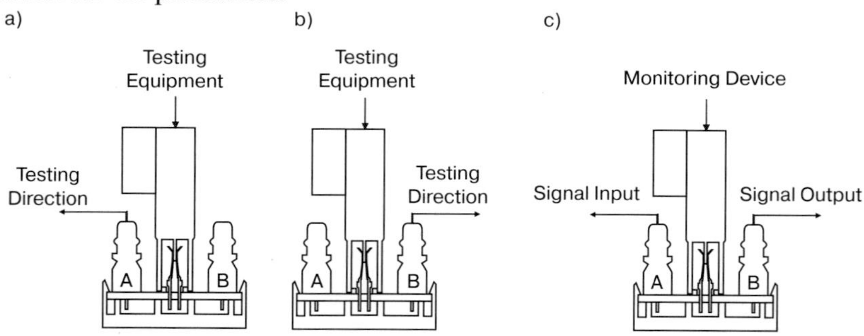

Fig. 3.55. Measurement adapters and options; *a)* option A, with breaking; *b)* option B, with breaking; *c)* measurement monitor adapter

The housing of the adapter is made of plastic. A color code is often used for the easy maintenance of links. A metallic housing to provide shield continuity of shielded cords is also available. Modular jacks may be placed both on the side of the housing (direct adapter), and on the adjacent side (angular adapter). The angular adapter simplifies bending. A short list of adapters is given in Table 3.35. The majority of the devices are designed for stand-alone use. The Canadian company UNICOM produces PATU5-816D-RA panels to be mounted into a 19-in. rack, which have 16 slots to fit in-line adapters. The company Hubbell produces flanged housings with a special clip for easy mounting of its adapters on plates, panels, and other similar constructions.

Table 3.35. I-adapters

Producer	Article	Type	Max. transmission rate, [Mbps]
Lucent Technologies	451A-50	Direct	10
	451A-60	Direct	10
AMP	555052-1	Direct	16
	555051-1	Angular	16
UNICOM	ILC-U508A-XX	Direct	100
Hubbel	BRIA4P	Direct	16
	BRIAJF8	Flanged	16
ICC	ICMMA3508DR	Direct	–
Panduit	MMC	Direct	–

Bridging adapter. A bridging adapter is used when several work stations or other devices are arranged according to the point-to-point principle. This device includes several modular jacks with a parallel connection to matching contacts. The adapter is called a Y-bridge adapter when there are two jacks. As an example of a traditional bridging adapter we point out the 367A device by Lucent Technologies, which is oriented to support seven workstations for the host computer IBM System 36/38 or AS/400 via balun 356A (Fig. 3.56). Due to specific use the modification of the adapter is not produced as a stand-alone device; although, it is not recommended by current standards. In practice, it is most often produced as an insert unit to be mounted in a plug or raceway with appropriate color coding. In cases where the adapter is designed for external use, its entry port for connection to SCS cables may be designed as a plug or jack.

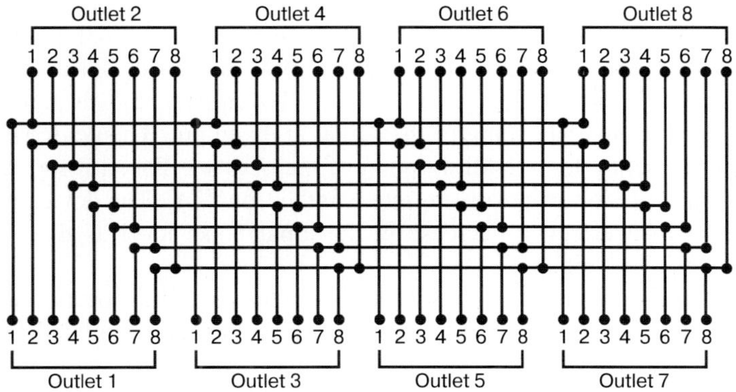

Fig. 3.56. Scheme with internal connections in a bridging adapter of type 367A

The ADS plate by the company CHZ Neubauer functions as a bridging adapter. The product comprises a 19-in. panel with two separated groups of nine modular jacks each. One of the connectors serves as an input–output port, the rest are ring-structured. The feature of modular jacks in these products is that, when the plug is disconnected, the internal conductors are closed to maintain the ring structure.

ISDN adapters. This device is used as its name suggests for the connection of termination ISDN devices to a cable system. Its plug is aimed at the connection to a standard jack of an 8-position modular connector; the outlet is configured for a 6- or 8-position plug. Inside the housing there is a load resistor to provide normal functioning of the ISDN bar. Sometimes this adapter is arranged by the bridging scheme, and differs from the standard ISDN adapter by the availability of several internal load resistors. In this case one housing has several outlets and serves several ISDN devices.

Telephone adapters. Some countries produce telephone sets with patch cords to be equipped with a plug differing from those of a 6-position modular connector by its design. The connection of the devices to an SCS happens via telephone adapters. This product is designed as a housing with a jack to be matched for the connection of a telephone cord plug. Connection to a telecommunications outlet of an SCS is performed using a plug, which is rigidly mounted on the adapter housing or on a short piece of cable. A wide variety of these adapters is manufactured by the company ITT NS&S.

Media filter. The media filter is a filter of lower frequencies with a threshold frequency value of 25 MHz, and a gain slope of 60 dB per decade. Use of the device enables an essential increase in the NEXT value because of an efficient suppression of a high-frequency signal out of the network equipment operating band. It is recommended for the transmission of Token Ring networks signals across SCS. The medium filter is not an adapter, but has a very similar construction, connection principles, and functions. It provides, in particular, an efficient interface between a 100-Ω UTP cable and a 150-Ω connector of the system Token Ring. Initial specifications for Token Ring equipment recommended shielded twisted pairs of types 1 and 2 with rather efficient protection against external parasitic EMIs and high internal noise suppression. The medium filter enables a dramatic decrease in parasitic emissions when transmitting the Token Ring signals across a UTP cable due to additional balancing of the signal at its output. The device provides the highest efficiency for a 4-Mbit option of the network Token Ring. At a 16-Mbps rate of transmission its positive effect is rather low [74]. The device provides one more positive effect: a decrease in the input noise level as the filter winding functions as a locking coil.

The device comprises a plastic housing (with outer metallic housing in a shielded option). One of the housing side surfaces incorporates a mounted DB9 outlet; the opposite side may place an 8-position (sometimes 6-position) modular jack or entry for a cord up to 8.6 m in length also with a 6- or 8-position modular plug, for example, products 558280 and 558281 by the company AMP.

3.4.3 Line Extensions

A line extension functions in an SCS as a common household extender in the power supply system. It is used in cases when there are no equipment cords of an appropriate length available, but network equipment being connected to the cable system does not need to comply with high requirements for the bandwidth of the signal transmission channel.

Similar to the household analogues, the SCS line extensions may be of two different options. The first option has a patch cord with a modular plug to be incorporated in the device. A plug with a housing terminates the second end of the cord. The equipment is manufactured by the company Deutsche Electraplan (products of type 687 521 860 with an 8-position outlet and 687 602 with a 6-position outlet).

The second option is sometimes called a distributor, and uses a standard equipment cord for the connection to an SCS outlet. The device construction looks like a housing with five modular connectors, one of which is used for the connection with an SCS outlet. Thus, the distributor, in contrast with a traditional household extender, presents itself as a multiport device. The distributor may be designed to serve the same or different devices. An example of the first product is a distributor of the Swiss company Reichle & De Massari, which is intended for the connection of four analog telephone sets. On the other hand, an MCS device by the Swiss company Telena, depending on its constructive finish, enables the connection of both analog telephone sets and workstations as well as other analog devices of the network Ethernet.

We have to note that a typical equipment cord with a mounted I-adapter may also function as a one-port extension (Section 3.4.2).

3.5 Auxiliary Equipment

Some modifications of equipment, which have not become widespread in the construction of SCS and are manufactured by a limited number of companies, will be reviewed below.

The majority of these products are formally beyond the standards for some reasons; however, their use enables within certain limits (sometimes very substantial) the extension of the functionality of the cable system.

3.5.1 Installation Kits for a Cable System

For a rapid deployment of compact cable systems the companies (Molex, Lucent Technologies, Alcatel, etc.) produce special kits consisting of components and installation materials (Table 3.36). The kit includes certain elements sufficient for the construction of a small SCS that, first of all, supports a LAN. The kit's required elements are subscriber outlets, fabric-made patching and equipment cords, and connecting panels. These elements may be supplemented with cables, mounting racks, testing equipment, a tool kit, etc. The user is provided with a detailed installation manual and even videotape (Molex). If increasing a number of ports is required one may use several kits or a special extension kit (Molex solution). The cable system created quite often provides characteristics not higher than those of category 3.

Table 3.36. Cable system installation kits

Company	Article	Cable	Patch panel	Telecommunications outlet	Patch cords	Rack	Category
Alcatel	OffiSys	UTP or FTP	2 × 10	10 double	20	12U	5
Lucent Technologies	Systipack	UTP, 305 m	12-port outlet	12 single	24	-	3
Ackermann	Net-bundle-12/-24/-48	no data	1-4 × 24 ports	12, 24, or 48 double	24 – 96	6 or 12 U	5
Mod Tap	28.D0030	UTP, 120 m	8-port box	8 double	8	–	3

The kit by the company Ackermann includes a wall cabinet and concentrator Ethernet for more functionality. These kits, which are called bundles by the company, are manufactured for 12, 24, and 48 ports, and primarily aimed at creating infrastructure for the connection of individual working groups to a server, as well as equipping the workplaces with telephone communications. The latter is possible because the kit includes 2-port subscriber outlets [75].

Producers of lower-level network equipment also offer cable system mounting kits for SOHO assembled from the OEM products by third-party companies along with SCS producers. We note the Wiring System by the company Micronet, as an example. They differ from products of the SCS producers by a wide range of network adapters, modems, mini-concentrators, and other similar products available as auxiliary devices.

3.5.2 Splice Modules

Splice modules are intended for splicing horizontal cables. Their use enables extending a four-pair cable or splicing a damaged cable without attaching a new segment, i.e., the module serves as a plug-and-jack connector. The device may be used for cable repair to quickly restore communications and as an element to organize a transition point. It has a housing with fixations to fasten spliced cables and two kits made up of a cut-in contact, which when connected to each other via current-carrying printed circuit lines are used for cable wiring. To protect the device against external EMI its body may have an inner shielded housing.

The use of splice modules is conditioned by possibility to incorporate an auxiliary permanent connector of a transition point or its analogue into horizontal cabling. Based on that, the use of a splice module in a signal channel does not lead to a degradation of its attributes by category. However, the current SCS standards prohibit the use of any elements for splicing of horizontal cables.

Therefore, the products are produced in small amounts and by a small number of companies; they have not become widespread. A module VM 8-8 by the company Telegartner, which enables splicing cables with an outer diameter of up to 10 mm serves as an example. The module is based on IDC contacts LSA-Plus by the company Krone and certified to operate in frequencies up to 600 MHz. A 4732-100 splice module by the company Telesafe has similar characteristics.

3.5.3 Automatic Distributors

Automatic distributors comprise a device with several inputs and outputs to be interconnected by means of a stimulus sent to the driving port. The distributor allows the performance of [20]:

- circuit breaking;
- through circuits activation;
- auxiliary equipment connection;
- measurement of line parameters;
- change in network configuration.

All switching is done physically, i.e., it is independent from applications and, in particular, it is possible to switch signals of analog telephone stations, which have a high voltage of ringing currents. The capacity of distributors may achieve 49X98 ports. The switching time is 4.5 ms. Electrical characteristics match those of category 3. Independence from the power supply is achieved by incorporating bistable electromechanical relays. Automatic distributors have a limited use when constructing SCS with a large quantity of telephone lines. The company Krone has produced a great variety of these products with the FSA-Plus series.

3.5.4 Demo Equipment

Demo equipment is used for presentations, exhibitions, and the training of certified specialists. The equipment includes the most typical and interesting SCS components from the producer's point of view. A part of the elements may be produced in transparent housings or as enlarged models to demonstrate the interaction of individual details during maintenance.

Demo equipment is listed in catalogues as an element of SCS and produced with two main options. Demo racks are made in a 19-in. format, and relate to stationary equipment. The rack has electrical and optical panels, cable managers, distribution blocks, and other elements mainly for use in equipment rooms. In contrast to that, sample cases basically include outlets, connectors, and similar small components to be applied to workplaces or specified for a certain technology. Instructors use the sample cases for distance training, seminars, and other arrangements. In some cases salespeople use them during marketing campaigns. It should be noted that some SCS producers require the purchase of demo equipment when signing a partnership agreement with a system integrator.

3.6 Conclusions

A wide range of serial electrical and switching products as well as matching accessories are designed for different applications. They are capable by means of standard facilities and solutions of solving any situation occurring during SCS construction with a capacity of up to a few tens of thousands of ports.

The main scope of cable and switching SCS products has 100-Ω impedance; therefore, they are recommended for the construction of cable systems. The 120-Ω cables have high transmission parameters; however, they are less preferable because of their low availability on the Russian market.

The characteristics of unshielded SCS cables comply with standards that require parameters with considerable allowances. Due to mounting complexity and high cost, shielded constructions are suited for SCS construction only at a considerable noise level or where special requirements for the protection against unauthorized access to transmitted information are present.

The horizontal four-pair cables constitute the majority of SCS cable products. Constructive options and modifications of the product line enable the improvement of technical and economic parameters and the increase of the maintenance flexibility of cable systems within a great variety of architecture and the designs of up-to-date office buildings.

The most perfect serial models of shielded cables have parameters certified up to 1.2 GHz. The standardization of the parameters for high-frequency unshielded cables is set up to 600 MHz. This opens prospects for designing high-performance cable systems of category 3 and higher.

The main type of connectors in an SCS (up to a category 6 project) is a modular one. Connectors of the 110 type are efficiently used when administration by individual pairs is needed. Connectors of the prospective category 7 may be realized by the same scheme as for the modular connectors or they may have an incompatible interface.

Switching equipment of an SCS comprises modular, assembled, or fixed schemes, and have a standard capacity from 100 to 900 pairs per device. The standard kit technical facilities enable easy mounting of connecting panels both in a 19-in. rack and on a wall. This enables the creation of all kinds of cross-connect joints both in equipment rooms and in open offices.

The development of an infrastructure of jumpers and adapters makes it possible to use the SCS infrastructure to support the normal operation of network equipment with coaxial and twin-axial interfaces, as well as to perform different kinds of measurements. In some cases the functional flexibility and convenience of SCS maintenance may increase using auxiliary passive equipment that is beyond the SCS standards.

4 Optical Fiber SCS Components

Optical fiber links are mainly used in structured cabling system (SCS) backbone subsystems. The block diagram of the link used to design a campus backbone subsystem is shown in Fig. 4.1. The basic elements of this link are shown and are considered in this chapter in detail.

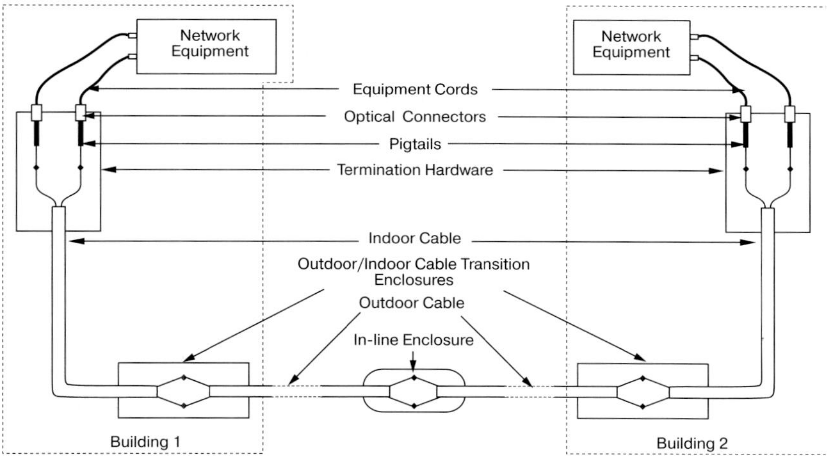

Fig. 4.1. Fiber-optic campus backbone subsystem layout

4.1 Optical Cables

4.1.1 Field of Application and Classification

The fiber-optic cables used in SCS are intended for the transmission of optical signals within and between buildings. They enable the implementation of all three SCS subsystems, though the application of fiber optics for LAN operation support is still limited in horizontal subsystems. In building backbone cabling, optical cables are used as frequently as twisted pair cables, and in campus backbone cabling they play a dominant role. Depending on their main utilization, optical fiber cables can be subdivided into three main types: outdoor cables, indoor cables, and cables for cords.

Outdoor cables are used in campus backbone subsystems to link separate buildings. Indoor cables are mainly used for building backbone subsystems. Cables for cords are intended for patch-cord manufacturing and for horizontal cabling when implementing "fiber-to-the-desk" and "fiber-to-the-room" designs. The general classification of optical SCS cables can be represented as a diagram (Fig. 4.2).

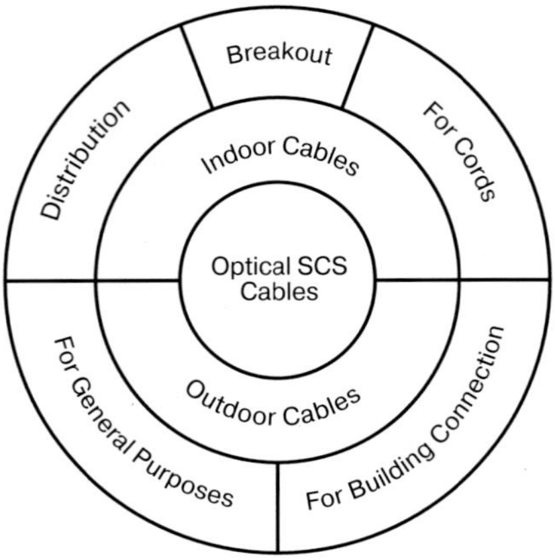

Fig. 4.2. SCS optical cable classification

4.1.2 Optical Characteristics of Fiber-Optic Cables

Fibers made of quartz glass are the basic components of optical cables. Quartz glass has rather low mechanical strength and poor resistance to external atmospheric effects, therefore all other parts of the optical cables are intended to protect the fibers from external mechanical effects and moisture under the operating conditions for the cables.

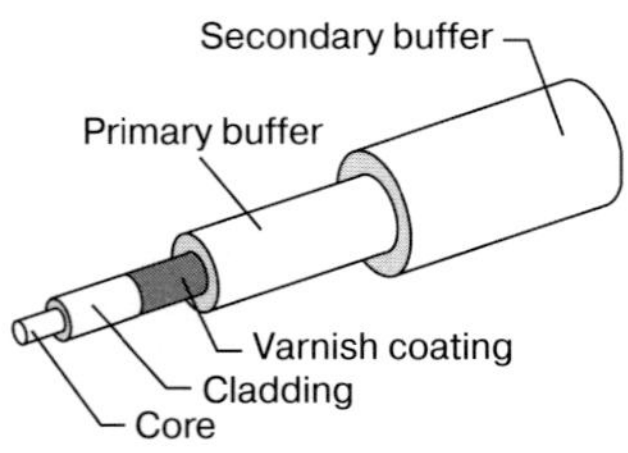

Fig. 4.3. Optical fiber structure

The standard structure of a fiber is shown in Fig. 4.3. The light conductor has a cylindrical shape and consists of a series of concentric layers. The core and the cladding are the basic layers of this cable. The core of single-mode fibers, according to the IEC-793 standard and to guideline G.652 of the International Telecommunications Union (ITU), has a diameter of 9.3±0.5 mm; some fiber manufacturers use other diameters in the range 7–10 mm with the same tolerances. The diameter of the core of multimode fibers is, in accordance with the IEC-793 and G.651, 50.0 or 62.5 mm with an admissible spread of ±3 mm.

According to the working standards, it is possible to use all three types of fibers for optical SCS cables. It is necessary to note that a 50/125 (core/cladding diameter) multimode fiber is considered an alternative type by the international standard ISO/IEC 11801, whereas the American standard TIA/EIA-568-A in the present edition does not allow utilization of this fiber type. In comparison with 62.5/125 fibers, the 50/125 fibers have much better frequency parameters, which is very important for signal transmission by means of the Fiber Channel and Gigabit Ethernet equipment. Therefore, the popularity of multimode optical cables with fibers of this type increased considerably in the late 1990s all over the world, including the USA.

The outer diameter of the cladding of multimode and single-mode fibers is equal to 125±2 mm. This size was chosen for as a standard. The cladding is covered with a 2- to 3-mm layer of a polymer coating. The polymer layer thickness is a part of the nominal width of the cladding. The main purpose of this coating is to protect the quartz glass against effect of atmospheric moisture and corrosion. Fiber flexibility is provided by the primary protective coating of epoxy acrylate with an external diameter of 245±15 mm. In 1999, a number of fiber manufacturers of telecommunications cables strengthened the tolerances for this parameter up to ±5–10 mm. The fiber in such a coating is considered insufficiently protected from mechanical effects; therefore, it is necessarily supplied with additional strengthening elements, which are described in Section 4.1.3. In addition, the core noncircularity (no more than 5%), the cladding noncircularity (no more than 2%), and the core–cladding eccentricity (no more than 3 μm) are also standardized. Fiber manufacturers usually use different colored protective jackets from both ends of the cable for easy identification of fibers (see Table 3.17). This color does not influence the optical characteristics of the fibers.

The optical cable may contain only multimode fibers, only single-mode fibers, or fibers of both types. Information on combined (composite) cables is seldom provided in manufacturers' catalogs, but custom cables are available. Information about the number of single-mode and multimode fibers in standard composite cables is given in Table 4.1.

Table 4.1. Number of fibers in composite cables [76]

Number of fibers		
Total	Single-mode	Multimode
18	6	12
24	6	18
24	12	12
30	6	24
36	12	24
60	12	48

In some cases, multimode optical cables are used in horizontal SCS subsystems, usually in the process of implementation of "fiber-to-the-desk" projects. Multimode cables are often used as a basis for building backbone subsystems, however, single-mode cables can also be used. Depending on the required distance and the frequency range, cables with multimode or single-mode fibers are utilized in building backbone subsystems.

Table 4.2 lists the maximum permissible values of attenuation and bandwidth coefficients of multimode optical cables used in SCS. The requirements of the international standard for SCS optical cables are stricter than those of the American standard. Note that the standard ISO/IEC 11801 (2000) permits utilization of standard TIA/EIA-568-A multimode cables with a bandwidth coefficient of 160 MHz/km for a wavelength of 850 nm, but this requirement limits the maximum length of a link to 1.6 km.

Table 4.2. Maximum permissible attenuation and bandwidth coefficients of SCS multimode optical cables

Wavelength, [nm]	850		1300	
Standard	TIA/EIA-568-A	ISO/IEC 11801	TIA/EIA-568-A	ISO/IEC 11801
Attenuation coefficient, dB/km	3.75	3.5	1.05	1.0
Bandwidth coefficient, MHz km	160	200	400	500

In addition to the SCS standards, the requirements for optical cables are included in other widely used normative documents. In particular, the requirements for multimode fibers, which are defined in the standard IEC-0793 G.651, are listed in Table 4.3.

Table 4.3. Requirements of standard IEC-793 for multimode fibers

Wavelength, [nm]	850		1300	
Fiber type	62.5/125	50/125	62.5/125	50/125
Attenuation, dB/km	3.2	2.8	0.9	0.8
Bandwidth coefficient, MHz km	200	400	500	1000

For reasons of economic expedience, the same type of cable production is used in SCS and PSTN. If we compare Tables 4.2 and 4.3, we can see that the values of the attenuation and bandwidth coefficients have appreciably the best values in comparison with the requirements of the SCS standards. This situation is also reflected in the specifications issued by manufacturers of this type of cable. For example, multimode cables of the BICC Corp. have a guaranteed bandwidth coefficient of 900 MHz/km instead of 500 MHz/km required by SCS standards, and cables by the Mohawk Co. provide attenuation of 0.8 dB/km instead of 1 dB/km (all values are given for a wavelength of 1300 nm and 62.5/125 fibers).

It is necessary to note that network equipment used for local and corporate telecommunications networks and based on multimode optical cables use transparency windows of both 850 and 1300 nm equally. Therefore the cables used in SCS have attenuation and passband performance optimized for these wavelengths.

Characteristics of single-mode optic cables are more completely specified by the standard TIA/EIA-568-A. The requirements are based on the specification TIA/EIA-492BAAA, which specifies non-dispersion-shifted single-mode optical fibers. The attenuation coefficients should not exceed values given in Table 4.4, which presents requirements for attenuation coefficients and dispersion given in standard IEC-793 (G.652). The analysis of the data shows that, like the case for multimode cables, the achieved attenuation values are considerably lower, since the designers of standard TIA/EIA-568-A used standard values for maximum permissible dispersion that were unchanged from the basic specifications.

Standard TIA/EIA-568-A also specifies:

- the wavelength of zero chromatic dispersion of single-mode optical cables, which should be 1300–1324 nm at a dispersion slope of not more than 0.093 ps/km×nm^2;
- mode-field diameter[1] for the 1310-nm wavelength should be 8.7–10 μm;
- the cutoff wavelength[2] must not exceed 1270 nm (the international standard defines the value of 1280 nm for this parameter).

Table 4.4. Maximum permissible attenuation and dispersion for single-mode optical cables according to TIA/EIA-568-A and IEC-793

Wavelength, [nm]	Attenuation, [dB/km]	Dispersion, [ps/nm km]
1310	≤1.0/0.5	≤3.5
1550	≤1.0/0.3	≤18

4.1.3 Tight Buffers of Optical Fibers

The three basic kinds of fibers are shipped to optical cable manufacturing plants with a primary buffer with external diameter 0.25 mm. This buffer layer is insufficient for protection from external mechanical stresses, which may occur during cable laying and operation. Therefore, it is necessary to supply the fibers with additional tubular protective elements, which are generally called secondary protective buffers. Today, three basic types of secondary buffers are used in the design of optical cabling.

---

[1] In single-mode fibers a considerable amount of the optical energy is transmitted in the cladding layer. Therefore, it is advisable to indicate the so-called diameter of the mode field rather than the core diameter. This parameter is equal to twice the distance from the fiber axis to the point where the density of the optical capacity decreases by a factor of 2.72 as compared with maximum value.

[2] Actually, it presents itself a wavelength, which cable (correctly fiber) uses for operation in the single-mode

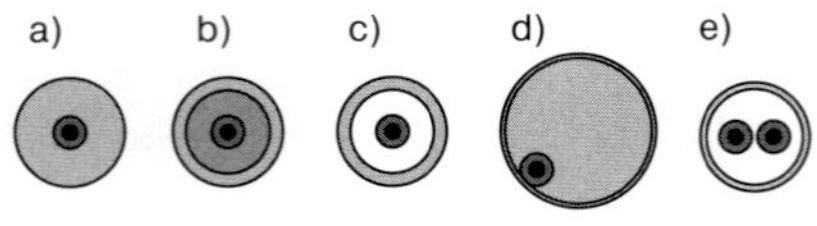

Fig. 4.4. Structure of fiber buffers: *a)* secondary buffer of 0.9 mm; *b)* two-layer buffer with 0.9-mm external diameter; *c)* micromodular structure; *d)* modular structure; *e)* mini-breakout structure

The outdoor backbone cabling widely uses modules (Fig. 4.4d). The module is a plastic tube of different rigidity with a diameter of 2–3 mm, in which one optical fiber or a bundle can be routed (the maximum number of fibers is 12 in available production samples). In this solution there is no direct mechanical connection between the fiber and the buffer. This provides very small attenuation sensitivity to temperature fluctuations and stretching forces. The module pipes have a free space filled with a waterproof gel. This structure is intended for the protection of fibers against moisture in case of damage to the external jacket. The gel is made of materials containing mineral oil products that essentially reduce the fire resistance of the cable products. The large external dimensions of the tubes require special sealing in terminal and in-line enclosures. Large external dimensions of the tubes also hamper the installation of optical plugs.

The 0.9-mm external diameter tight buffer, which is stowed without a backlash on a primary buffer having diameter of 0.25 mm (Fig. 4.4a), is widely applied for indoor cables. Small external diameter, high flexibility, and lack of internal waterproof filler enable easy installation of optical plugs. The main drawback of this type of buffer is poor protection of the fiber against moisture, poor cable dimensions and weight values, and an increase in attenuation due to losses on microbends caused by mechanical deformation of the fiber by the external buffer.

Semitight buffers are the result of the aggregation of all principal advantages of the reviewed solutions. At present, there are two main types of semitight buffers. The first type, called the micromodular solution, represents a tube with external diameter of 0.9 mm with one loose-fit internal optical fiber (Fig. 4.4c). The free space is filled with a waterproof gel. In the second case, a two-layered secondary buffer is used. The second internal layer usually has a diameter of 400–500 μm and is made of soft silicone (Fig. 4.4b). Both versions provide quite efficient mechanical insulation of the external layer and fiber surface that considerably reduces the losses caused by microbends. The structure provides for a certain improvement of moisture resistance. Therefore cables consisting of such optical fibers may be used for outdoor installations of up to several hundreds meters.

The general drawback of the 0.9-mm diameter secondary buffer is its significant external dimensions, which lead to a considerable increase in the cable external diameter. To eliminate this drawback, a mini-breakout solution has been suggested. The solution may be considered as the penetration of the principle of modularity in the field of indoor cabling. The solution represents a thin tube with external diameter of 0.9 mm with two primary buffered optical fibers laced through without gel filling (Fig. 4.4e). It is advisable to use this option when a high-density installation of some types of connectors is required (see Section 4.2.6). In particular, this solution enables manufacturers to improve the dimensions and weight parameters of cables for cords [76].

4.1.4 High-quality Multimode Fiber

Standard optical cables intended for SCS have optical fibers with parameters specified in existing normative and technical documents. The analysis of the characteristics specified by these documents shows that they actually correspond to the engineering level of the late 1980s and the early 1990s, when the optical fiber channel of any LAN was considered as a medium that did not technically restrict the channel performance. This situation dramatically changed in 1998 as a result of the commercial development of Gigabit Ethernet network interfaces in the construction of telecommunications and computing systems. At the same time, it became apparent that optical cables whose parameters fully comply with the working editions of the SCS standards (transparency window of 850 nm) sometimes did not ensure communications for distances greater than 220–275 m. These values are obviously insufficient to comply with the technical requirements for an optical pathway that may appear during structured cabling systems construction. In particular, such severe restrictions essentially limit the possibilities for the design of building backbone cabling and for the construction of systems with centralized optical architecture, according to TSB-72.

Table 4.5. Parameters of fibers and VCSEL for network interfaces

Oscillator	Wavelength, [nm]	Spectral width, [nm]	Modulation current, [mA]	Output, [dB/m]	Maximum data transmission rate, [Mbps]
LED	850	50	60	–15	125
Laser with VCSEL	850	0.5	10	0	2000

As far as SCS optical pathways are concerned, the Gigabit Ethernet system has several principal features that radically distinguish it from its low-speed analogues of the same standard. First, a semiconductor laser is used as an oscillator, as modern LED do not react quickly enough for the specified application area. In addition, specification 802.3z set parameters for optical interfaces in first and second transparency windows, however, for cost-efficiency it is preferable to use the first transparency window (850 nm or an SX range).

We comment on these specified positions in detail below. The majority of the transmitting blocks of optical interface 1000Base-X use vertical cavity surface emitting lasers (VCSEL), which operate at 850 nm. The comparison of this type of oscillator with an LED (see Table 4.5) clearly shows that the laser is more advanced by all the key parameters. In addition, this of type oscillator less expensive than lasers of the LX range (1300 nm). According to Alcatel, the LX range of lasers was 5 times more expensive than VCSEL in mid-1998. The benefit from this cost reduction is illustrated in information from the company ITco.(Russia): at the end of 1999 the average scatter in recommended cost by nine leading network equipment manufacturers for LX- and SX-range transceivers exceeded $1000 per interface. The present study gives almost the same values [78].

Thus, the practical need to adapt optical signal transmission channel parameters to the technical features of the oscillator type has resulted in the design of new types of fibers. The following primary issues were solved during their development:

- optimization of performance to operate at 850 nm, where dispersion distortions are mostly observed;
- suppression of the differential mode delay effect that occurs in fibers, as specified by the TIA/EIA-568-A and ISO/IEC-11801, when optical transmitters of laser oscillators are used.

Fiber performance can be optimized for data transmission. The spectral performance of the bandwidth coefficient ΔF of multimode fiber tends to a maximum. Thus, the boundary coordinates depend on the refractive index profile and the concentration of alloying additives. At 100–155 Mbps it is economically feasible to use light-emitting sources, which provide superior channel characteristics at 1300 nm, where the bandwidth of standard fibers is maximized due to a minimum of chromatic dispersion. During the manufacture of new fiber types, the coordinate for this characteristic maximum can be shifted to shorter wavelengths by means of certain engineering techniques (Fig. 4.5). This results in the improvement of the frequency properties of an optical fiber in a window of transparency at 850 nm and in the alignment of option values ΔF in the SX and LX ranges (the last property is important for universality of a cable pathway).

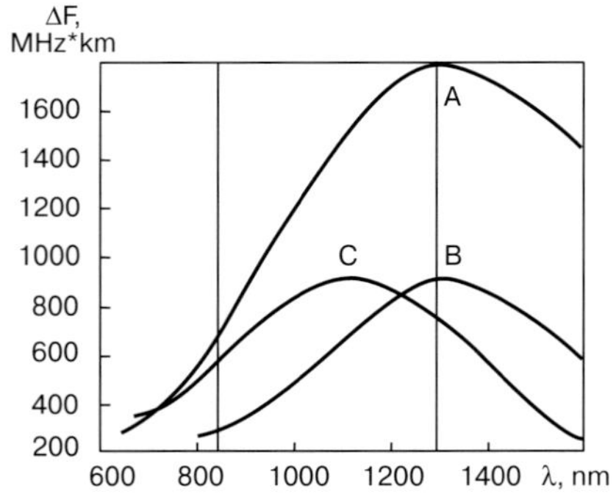

Fig. 4.5. Spectral performance of multimode fiber bandwidth coefficient: *a)* standard 50/125 fiber; *b)* standard 62.5/125 fiber; *c)* high-quality 62.5/125 fiber

The phenomenon of differential mode delay occurs from the actuation of a relatively small number of modes in a multimode fiber by a laser radiation source, whose beams propagate near the axis of the fiber core (the so-called low-order modes). If the axial part of the fiber has a small defect, such as poor response of the refraction caused by manufacturing technique, part of the radiation is transformed into high-order modes, which propagate near the cladding and carry the majority of the optical signal capacity. Because of the various propagation speeds of these modes, impulse fragmentation occurs at a certain length of the fiber, which leads to input error and an equivalent reduction of bandwidth. From the stated mechanism of differential mode delay, methods to eliminate the undesirable phenomenon can be clarified: the creation of special conditions for radiation input in the core, and the elimination of the modal conversion origin by improvement of the fiber manufacturing method. The first method implies a small shift of the radiation input point from the fiber axis. This is carried out with the help of a special MCP-cord (mode-condition patch cord). This is the most preferable method when working with older types of fibers. The curve, which describes the structure of refraction parameter, is made smooth by means of special techniques and without a downward excursion in the center of the fibers, especially those optimized for high-speed applications (Fig. 4.6).

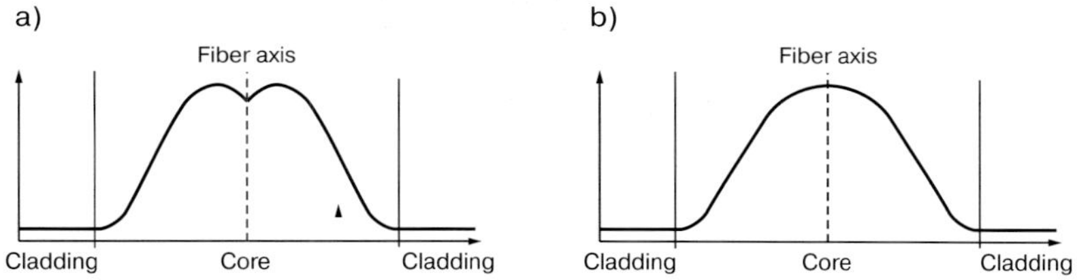

Fig. 4.6. Refractive index profile in multimode fibers: a) standard fiber; b) high-quality fiber

The bandwidth coefficient of multimode optical fibers is basically determined by intermode dispersion, whose intensity at a first approximation is proportional to the number of directed modes in the given fiber type. Because of the very small number of directed modes in 50/125 fibers, it is much easier to achieve higher values of the coefficient $*F$ than for 62.5/125 fiber. Taking into consideration this situation, the majority of SCS manufacturers located in the USA deliver equipment with cables that include this fiber type, although such cables are not permitted by the working edition of the American standard TIA/EIA-568-A (Table. 4.6). We note, for example, such known manufacturers of SCS as AMP and Siemon. A further positive argument for the application of this type of production is its low cost (approximately by 5% lower).

The microwave optical interfaces of the Gigabit Ethernet, Fiber Channel, and ATM types, which use optical lines for their signal transmission, use much less power than does equipment working at speeds below 100–155 Mbps (approx. 7~11 dB). In view of this fact, somewhat lower attenuation coefficient values are ensured for new fiber types in comparison with standard ones, which enables users to obtain certain noise immunity margins in the channel.

According to standards TIA/EIA-568-A and ISO/IEC-11801 the manufacturing techniques for fibers must comply with the requirements for broadband capability. To achieve the parameters required for the functioning of Gigabit Ethernet systems and similar systems, it is necessary to elaborate new production techniques.

In particular, Corning applies outside vapor deposition (OVD) technology to the manufacture of intermediate products used to make Infinicore fibers. Alcatel uses advanced plasma and vapor deposition (APVD) technology to manufacture GigaLite fibers.

Table 4.6. Multimode optical fiber with improved frequency characteristics

Company	Name	Fiber type	Attenuation, dB/km, at wavelength 850/1300 [nm]	Bandwidth coefficient, [MHz km]	Ensured operation distance, [m], for the Gigabit Ethernet system at wavelength 850/1300 [nm]
Alcatel	GigaLite	62.5/125	3.2/0.9	500/500	550/–
Alcatel	GigaLite	50/125	2.4/0.7	700/1200	750/–
Corning	Infinicore 1000	62.5/125	3.0/0.7	–	500/1000
Corning	Infinicore 2000	50/125	2.5/0.8	–	600/2000
Lucent Technologies	LazrSPEED	50/125	–	–	600/600

Designers of new fiber types emphasize their full compliance with the fiber standards for optical and mechanical performance. Sometimes, one may designate such fibers as fibers for laser network interfaces in order to emphasize the novelty of the solutions and to take into consideration their specific area of application. The term "fibers for LED optical transmitters" is used for traditional fibers.

The application area also imposes the list of parameters given by a company's specifications, as there are no standards or approved recommendations thus far within this relatively new area. In particular, Corning guarantees that its Infinicore fibers can achieve a certain communications distance without specifying the bandwidth coefficient via the standard Gigabit Ethernet interface. This approach is justified by the strong influence of the measurement conditions on this parameter. It is interesting to note that manufacturers of this series of fiber cables set a bandwidth value in their specifications apparently for uniformity reasons and to follow tradition.

The majority of fibers examined in this section are used in applications with the maximum data transmission speed of 1 Gbps. According to the specifications, LazrSPEED fibers and Blolite systems produced by BICC Brand-Rex support data transmission at the rates of 10 Gbps if the principle of spectral condensation for four wavelengths is applied (the platform for the prospective 10-Gigabit Ethernet technology). Simple calculation shows that their bandwidth coefficient will not be less than 2.2 GHz/km.

4.1.5 SCS Optical Cables

Outdoor cables

Design features of various kinds of cables. External or outdoor cables are used in SCS campus backbone cabling subsystems. The basic design requirement for the cables is their high mechanical strength (resistance to stretching and pressing forces), along with small attenuation and high quality, moisture resistance, and a wide range of working temperatures. Small dimensions and low weight also are important factors. To obtain the level of parameters required for efficient operation, one might use various approaches, each of which is based on the cable structure.

At present, there is a great variety of design solutions for outdoor optical cables, which can be conditionally divided into four groups (Fig. 4.7).

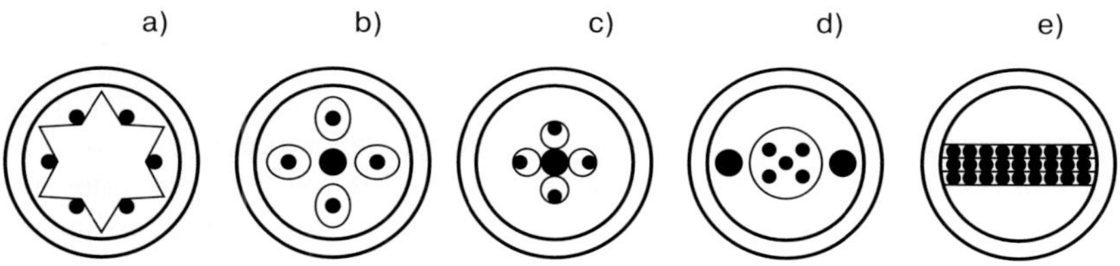

Fig. 4.7. Standard core s of optic cables: *a), b)* with profiled core; *c)* modular; *d)* with central tube; *e)* ribbon

A figured element serves as a platform for a cable with profiled core (Fig. 4.7a,b). Its internal plenums or openings fit the optical fibers. This solution was widely used in the 1980s. At present, it is rarely applied due to its limited capacity (usually no more than 16 fibers).

Cables of the so-called modular design (Fig. 4.7c) have a traditional lay twisting. Each lay is composed of modules with diameters of about 2 mm (see Section 4.1.3). The number of fibers that can be placed in the module varies from 1 up to 12. Fibers are loosely laced through a tube of the module during the cable manufacture. A manufactured cable has fibers that are slightly twisted in spiral and lay inside along the tube. Thus, the cable can be stretched and bent somewhat during its routing without any deterioration of its optical characteristics. The majority of the cables under reviewed that are at present offered in the market have a single-layed design. Six-modular solutions are widely used, but eight-modular variants are rarely applied. To ensure higher capacity, modules are placed in two lays, or a central enforcement element of greater diameter is used to fit around a large number of modules. Some companies designate this design solution as a multitube cable.

A tube of greater diameter located on the cable axis (Fig. 4.7c) may also serve as a core. This cable type is very convenient for stripping. Superior protection from crushing is ensured due to the maximal distance of fibers from the external jacket, however, the tube possesses somewhat lower operating temperatures and stretching stability, as compared with a traditional multimodular design. NK Cables (formerly Nokia) has applied a spirally shaped channel in its Spiral Space tubular elements to improve fiber protection.

In practice, the majority of modular cable solutions fit no more than 144 fibers. Today, they have taken a leading position in the total production volume of outdoor cables. This is explained by the ensured protection of fibers against mechanical impact and the environment as well as the simplicity of their stripping and installation.

Those cable products that possess both the primary features of cables with profiled and modular cores are also available in the market, but in limited quantities. The profiled element, whose external openings fit tubes of modules with fibers, is used as a core. The profiled core provides high stability to crushing (for example, the specification for GNSLWLV-series cables produced by Ericsson gives the parameter as 6 kN instead of 1.5 kN specified for a traditional modular design [79]). The modular tubes ensure operational convenience and improved sealing along the tubes.

Ribbon cables (Fig. 4.7e) are used mainly for backbones of major telecommunications networks of city due to their compact fiber arrangement. This arrangement dominates over other solutions when large (some hundreds and more) numbers of fibers are used. Application of these cables for SCS construction is not advisable at the present time since the application area does not require such high capacity. Moreover, because of their extraordinary design, installation of plug&jack connectors and the manufacture of permanent connectors require expensive engineering hardware and highly qualified personnel.

The types of outdoor cables (Fig. 4.8) can be divided as follows: cables containing metal strengthening elements and/or electric conductors, and completely dielectric cables. In comparison with completely dielectric cables, those cables with metal strengthening elements possess greater mechanical strength to crushing and stretching forces, their fibers are not damaged by rodents, and they have smaller external diameter at equal breaking strengths. The main drawback of the cables with metal strengthening elements is that they do not provide a full galvanic discontinuity of connected points.

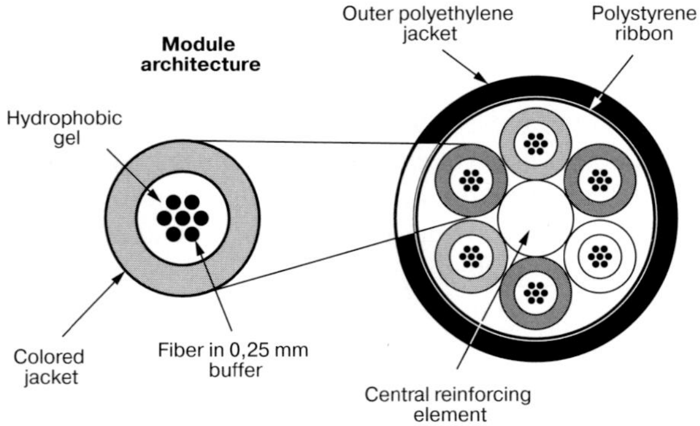

Fig. 4.8. Outdoor cable structure

The standard temperature range of widely applied outdoor optical cables is from -40°C up to +70°C. There are special frost-resistant solutions that function normally at temperatures up to -60°C. The minimum temperature is determined by the rigidity of polythene (this material becomes fragile at -70 °C), which is used in external jackets, and by the temperature at which the waterproof gel becomes hard. The maximum operating temperature is basically determined by the polythene performance. The melting point of this material is near 120°C. Keeping this in mind, some manufacturers guarantee the unchanged operating characteristics and parameters for cables on rapid heating to 90°C, if there are no mechanical loads. Special solutions designed for oil storage, petroleum engineering, and other similar objects can sustain higher temperatures. Because of their use of heat-resistant materials, they tolerate heating to several hundred degrees over several hours [7].

Strengthening covers and elements. Optical cables face stricter requirements for mechanical strength than do the balanced electrical cables. This is due to the low plasticity of glass in comparison with copper and aluminum. Glass can extend by stretching forces up to 2–3%, as compared with a copper wire, which can stretch up to 5–6%. Application of strengthening glass-fiber-reinforced plastic and/or metal elements, which stand deforming forces during cabling and maintenance, provides the cable with necessary mechanical strength.

Strengthening elements can be integrated in the cable core or may penetrate into the external jacket. Most frequently, a central reinforcement element (glass-fiber-reinforced plastic filler, a steel lead in a polymeric loose tube, or an ordinary wire) is used in the cable core. In modular cables some modules can be replaced by strengthening fillers. The force elements, which are external to the core, are usually Kevlar braids, wires, or glass-fiber-reinforced plastic fillers that penetrate the material of the external loose tube (an example is schematically shown in Fig. 4.7d), and armored covers from a steel wire of various diameters.

When steel wire is used, typical maximum stretching loads are 10,000 N (in the solutions with a two-layer armor these maxima reach 20,000 N or higher); if other armor types are used, the maximum stretching load can reach 2500–3500 N. This can be explained by the fact that, unlike ribbon and braid, the steel wire successfully stands the stretching.

Armored covers give the cable an additional breaking durability and protect it from crushing. Sparse or dense metal braids, ribbons of corrugated steel, or round zinced steel wires of various diameters serve as the armor.

There are two kinds of armor made from steel ribbon some tenths of a millimeter thick. Most frequently its seam runs parallel to the axis of the cable. In this case, the ribbon necessarily has small corrugations that allow high flexibility of the cable. Armor made by the winding of a steel ribbon (in this case the axes of the ribbon and the cable core are located at a certain angle) is rarely used. In the second case, a surface of the ribbon is made smooth, which somewhat reduces the cable external diameter. The ribbon surface in modern designs is frequently covered with a polymeric corrosion-resistant material.

Armor made from steel wires of various diameters is used in poor conditions and significant tensile loads may occur. If is necessary, two layers of wire may be applied on the cable with the axes of the wires positioned at low angles to the cable axis, but with different winding layers. Wire braid armor with wire diameters of some tenths of a millimeter substantially differs from the steel corrugated ribbon by its smaller overall diameter and greater flexibility, however, because of the low efficiency of wire-wrapping machines, it is rarely used.

The primary armored covers are frequently supplemented with braids of glass-fiber-reinforced plastic threads; sometimes the ribbon armor is additionally strengthened by two or four steel wires intruded into the material of the external jacket (this design is used, for example, by Lucent Technologies and Siemens). The latter ensures cable-stretching stability close to that of cables with common wire armor, with only slight increases in the weight and external diameter of the cable. We note that according to the information from some cable manufacturers (in particular, Siemens) only layers of round steel wire are considered as the armored covers of cables. Other strengthening components refer to elements to protect against rodents.

Further protection against crushing is mainly carried out by various covers and jackets, which provide protection against forces of 1000 N or higher that affects a cable with 1-cm length [80]. A marked increase in crushing stability is provided by use of a profiled element as the core. This a solution of Ericsson, which is applied to GNSLLDV cables and is based on the so-called amortization layer made of thermal transfer ribbon. The ribbon is usually placed between two external jackets and serves as crushing damper.

Typical mechanical characteristics of modern serial cables for outdoor cabling are listed in Table 4.7.

Table 4.7. Typical mechanical characteristics of modern serial cables for outdoor cabling

Parameter	Value
Number of fibers	4–144
Cable outer diameter, mm	10–20
Operating temperatures, °C	
• Installation	–10...+50
• Operation[a]	–40...+60
Minimum bend radius	
• Installation	20 × outer diameter
• Operation	15 × outer diameter
Maximum allowable tension during installation, N	2500–10,000
Maximum allowable compression, N/cm	2000–4000

[a]Special frost-resistant cables with lower working temperatures of –60°C are available.

Elements providing moisture resistance. Moisture resistance of a cable is determined by the presence of elements providing longitudinal and cross sealing in its structure. A waterproof gel, which fills the plenums of the cable core and the spaces inside the modules, is the primary aid for longitudinal sealing. The material of a waterproof gel is selected so that it:

- does not lose water-repellent properties within operating temperatures;

- does not harden at low temperatures;
- has a viscosity that does not hamper fiber movement inside a tube of the module and does not flow out during stripping and maintenance procedures;
- is chemically neutral and nontoxic.

Today, it has becoming popular to route cables inside utility pipes, including water pipes in western Europe. In cables for such applications, the waterproof gel should comply with an additional requirement of absolute neutrality to potable water.

High-molecular weight hydrocarbon compounds are most often used as waterproof gels. As a rule, a homogenous gel fills the free space inside the module and other plenums in the cable core. There are also solutions where gels of mixed composition are applied for this purpose.

External jackets define are lateral cable sealing. Various modifications of polythene are most often used as materials for external jackets. To provide high resistance to UV-radiation, the jackets are black in color. In some modern designs, an additional layer of cellulose paper is applied; this layer swells when moistened and seals small holes in the jacket.

It is well known that the polymeric materials used in cable engineering have a finite degree of moisture resistance. Given time, moisture leaks develop in various plastic jackets. A high degree of protection from diffusing water molecules is provided by a waterproof gel. It is assumed then, that utilization of cables in typical moisture and wet environments does not require special measures of protection against moisture. When cables are used in extreme humidity environments, that is, in conditions when the cables are permanently surrounded by water, the most effective protection against moisture is a thin metal jacket, which is absolutely impenetrable for moisture. The jacket is made of aluminum or copper.

The cable lateral sealing is also increased by some types of armored covers. The most effective type is the armor from a corrugated steel ribbon with additional gluing of the longitudinal seam by a water-resistant glue or by welding.

Auxiliary structure elements of outdoor cables. In addition to the basic elements considered above, cables used for outdoor cabling include in their structure a number of additional components that improve some of their properties and their maintenance.

To simplify the stripping of cables, a ripcord is laid under each external jacket. When pulled, the ripcord makes a slit on the jacket to allow access to elements of the cable core. In some cables a ripcord is also used to open a glue seam in an armored cover made from steel corrugated ribbon.

One cable maintenance problem is protection against rodents. Several methods have been developed to solve this problem, for example, the impregnation of the external jacket material with poisonous and repellent components. The most effective and low-cost solution of this problem is the use of armored cables. If it is necessary to provide galvanic discontinuity, nonmetallic elements are used. The solutions are based on the use of braids of glass-fiber-reinforced plastic wires and solid tubes. These elements are placed directly under the external loose tube.

Indoor cables

Optical fiber cables for indoor cabling are used for the construction of horizontal subsystem and SCS building backbone cabling subsystems. They differ from cables used for outdoor cabling by two key parameters:

- they have a smaller external diameter and weight in combination with higher flexibility because they lack a waterproof filler[3] and because of the application of strengthening covers without armors;
- superior fire safety characteristics

Like twisted pair cables, the indoor optical cables used in SCS must meet fire safety requirements. The fire-safety properties of a cable are determined by the dielectric materials used in its structure (in general, the materials of the external jacket). In catalogs published by cable manufacturers the indoor cables are usually divided into plenum- and riser-type. The aspects of SCS fire safety are discussed in detail in Chapter 8.

Cable fibers of this group are necessarily supplied with a secondary polymeric buffer with diameter 900 μm, which overlaps tightly the primary buffer with diameter 250 μm. Fibers covered with such buffers enable direct installation of optical plugs without the use of any auxiliary hardware. The convenience of plug&jack connector mounting procedures is achieved at the expense of some increase in attenuation in comparison with the outdoor cables. However, this is not important, as the length of a building backbone cable shall not exceed 500 m, according to the standards.

To protect the cable core from mechanical exposures, indoor cables have a layer of Kevlar threads, which is located directly under loose tube. Unlike outdoor cables, a free stacking of these threads without plaiting is used in this case.

There are two basic constructions of indoor cables. Products of the first group are known as distribution cables. They contain optical fibers in a 0.9-mm buffer. These fibers fit in a common buffer together with Kevlar strengthening threads and are terminated in termination hardware (see Section 4.3). In breakout cables each fiber is placed in an additional protective loose tube with external diameter of 2–3 mm. Thus, the described product may be considered as the analogue of a multielement electric multipair cable. Such solutions possess large external diameter and mechanical strength that is determined by two factors: the presence of a central force element and an additional layer of Kevlar threads under each individual protective loose tube. These solutions are focused basically on the manufacture of preterminated cables (see. Section 4.4.2) and the implementation of individual fiber breakouts without the use of splitting enclosures. The breakout implies a branching point of a polycore cable. The manufacturing of polyfiber patch cords can be also be considered as a possible choice, though it is rarely used in practice.

An example of indoor cables structure is shown in Fig. 4.9. The Table 4.8 illustrates typical mechanical characteristics of modern serial products of this type.

---

[3] Some manufacturers refer to these cables as so-called dry structure cables.

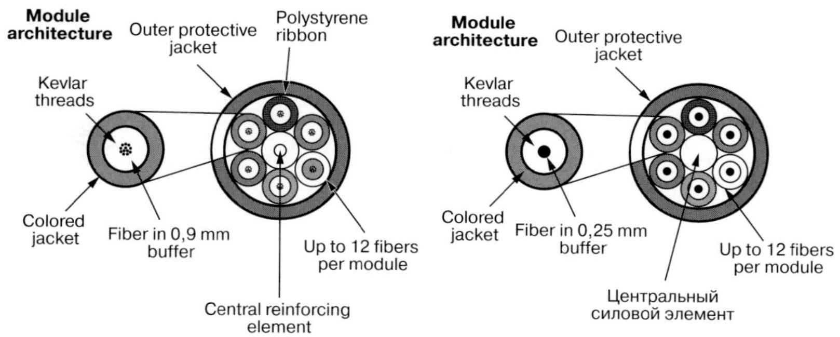

Fig. 4.9. Mohawk indoor cables: left, distribution: right, breakout

Table 4.8. Typical mechanical characteristics of modern cables for indoor cabling

Parameter	Value
Number of fibers	2–36
Outer cable diameter, mm	5–15
Operating temperatures, °C	
• Cabling	0...+30
• Operation	–20...+70
Minimum bend radius	
• Cabling	15 × outer diameter
• Operation	10 × outer diameter
Maximum allowable tension during installation, N	400–3000
Maximum allowable compression, N/cm	1500–2000

The standard maximum value of indoor cable capacity does not exceed 12 fibers. As a rule, such cables have no central reinforcing element, and their mechanical strength is provided only by a layer of Kevlar threads. A few companies, Ericsson, for example, use a profiled core as the basis for distribution cables (cable type GNSLBDV). This element functions as the load-resistant reinforcing.

If increased capacity is required, a solution similar to that for outdoor modular cables is applied: the central element, which serves as the load-resistant carrier, fits around several common cables (often 6, rarely 12). Then, an overall buffer covers the cable core. This solution increases the capacity to 144 fibers. If it is necessary to decrease the capacity of the cable, some of these modules are replaced with strengthening bars and/or fillers. Cables of a similar design are usually custom-manufactured.

The presence of additional fiber cladding in combination with low density also leads to significantly increased core size of an outdoor cable. This is shown more clearly in breakout solutions. On the whole, the external diameters and the weight of the these cables are much lower than those of the outdoor cables of the same capacity due to the lack of armor and the use of facilitated strengthening covers. To reduce the external dimensions of this type of cables, one can use a ribbon solution. An example of this solution with a four-fiber ribbon is shown in Fig. 4.10. The operating temperature of indoor cabling is from -20°C up to +70 °C.

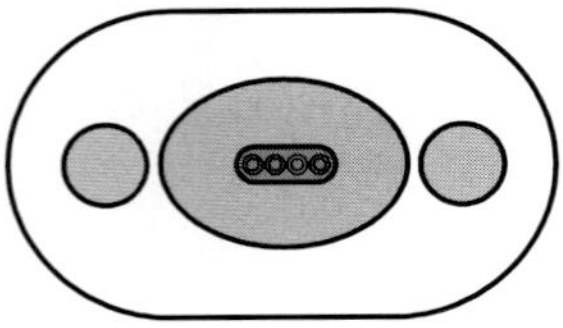

Fig. 4.10. Ericsson GAXLBD indoor cable

Cables used for building connections

Cables used for building connections have features between those of indoor and outdoor cables. Cables of this type were widely sold at the end of the 1990s as a result of the rapid growth of local area networks and an increase in outdoor subsystems during SCS build-up. The indoor cable serves as a platform for such solutions, however, after some special improvement measures, their stability to environmental exposures was greatly increased. Therefore they can be used to connect separate buildings at a common length of the cable run of up to several hundreds of meters. However, these cables are more expensive than outdoor cables.

The main features of cables used for building connections is that their structures include materials that enable operation from -30–-40 up to +70–+80°C (that is, they possess an extended temperature range), and that increase their moisture resistance. The solutions based on a two-layer external jacket are the most widely used in these cases. The external cover of these cables is made of low-smoke halogen-free material that provides the fire-retardant level required by the standards for indoor cabling. A second internal layer provides the necessary moisture resistance. Some improvement of mechanical properties is achieved in this solution from the placement of a second layer with Kevlar threads between the jackets. Such cables sometimes are called reinforced jacket cables or simply reinforced cables. The second solution is implemented in some Ericsson cables, where a waterproof strip is laid under the external jacket. This strip inflates when moistened and seals small punctures in the external jacket.

It should be additionally noted that, according to recommendations of some SCS manufacturers, when indoor cable normally functions at temperatures from -40°C to +80°C, it can be applied to campus backbone systems with relatively short pathway lengths. The only limitation is compliance with the requirement to ensure protection against moisture, which is usually achieved by laying the cables inside a tube.

Cables for cords

Cables for cords, which are called minicables, are intended for the manufacture of patch cords and equipment cords. They can be used in “fiber-to-the-desk” and “fiber-to-the-room" solutions for horizontal cabling. Sometimes this type of cable is applied for local cabling in equipment rooms and telecommunications closets.

This type of cable product is actually an indoor cable with one or two buffered fibers. The buffer is 0.9 mm in diameter. Because of their wide distribution, cables for cords are segregated in a separate group.

We note that fiber with a 0.25-mm buffer is typically not used for cables for cords and indoor cabling. By the middle of 1996, there were cords made of OKG cable manufactured in the Moscow Electroprovod plant (this design was implemented in this factory) in the Russian market, however, the production of these cables has stopped due to the complicated installation of the optical plugs.

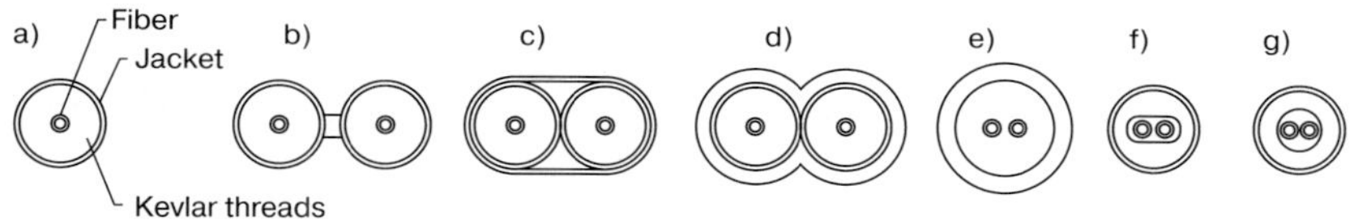

Fig. 4.11. Cables for cords: *a)* simplex; *b)* zip-cord; *c)* dual-subunit duplex/heavy-duty duplex; *d)* Duplex+ by Brugg; *e)* round duplex; *f)* with duplex ribbon; *g)* mini-breakout

The structures of cables for cords are shown in Fig. 4.11. First generation cables were either simplex cables (Fig. 4.11a) or duplex cables. Duplex cables have no additional overall jacket (Fig. 4.11b), or have an overall jacket of various forms and thicknesses (Fig. 4.11c–f). Common duplex cables without jackets are frequently called zip-cord or zip-cord duplex, and cables with a common external jacket are called heavy-duty duplex. Kerpen designates duplex cables without a jacket applied in FLine cabling systems as duplex figure 8, and jacketed cables - as duplex figure 0. Alcatel applies the term "dual fiber oval cable" to the latter solution, and Teldor uses the term flat duplex.

There are two kinds of heavy-duty duplex group cables. The first kind, which is widely applied, has the following features: its jacket is thin and covers protective loose tubes of separate fibers (cables M9X080 and M9X081 by Mohawk and type 1861 by Lucent Technologies, Fig. 4.11d). The second kind includes cables with thick jackets and partially fits in the opening between the loose tubes (the cable Duplex+ designed by Brugg, Fig. 4.11e). The common distinctive feature of this type of cables is their structure: they consist of two simplex cables with loose tubes without any special elements for fixing or fastening (they are fixed only by their overall jacket). It is assumed that cables in overall jackets provide superior protection against mechanical exposures and are more convenient, however, such structures as zip-cords are used more often in SCS implementation due to their cost-efficiency.

Like indoor cables, cables for cords have a polymeric buffer with diameter of 900 μm for fiber protection. A layer of Kevlar threads, which is located under the external jacket and surrounds the fiber in the 0.9-mm buffered cover, provides such cables with necessary mechanical strength.

There are some products in the market that are generally application-specific. An example is the DX-type Ultra-Fox duplex cables by Optical Cable Corp. This product has a flexible polyvinyl chloride jacket. The external dimensions of this cable are optimized for installation of optical MIC plugs.

Figure 4.11f,g shows two structures whose popularity has increased rapidly in SCS engineering. Their distinctive feature is the presence of an overall jacket for a standard simplex cable with a diameter of 2.5–3 mm. The buffer covers two fibers, which may form a ribbon (Fig. 4.11f) or lay inside a thin-walled tube with a diameter of 0.9 mm. In the latter, the optical fibers have only a primary tight buffer with an external diameter of 0.25 mm. Such cables are generally used for the production of patch cords with high-density connectors (Section 4.2.6). Typical mechanical characteristics of modern cables for cords are listed in Table 4.9.

Table 4.9. Typical mechanical characteristics of cables for cords

Parameter	Value
Number of fibers	1–2
Protective polymer fiber buffer diameter, mm	900
Each fiber outer protective buffer, mm	2.5–3.0
Working temperatures, °C	
• Cabling	0...+30
• Operation	–20...+70
Minimum bend radius	
• Cabling	15 × outer diameter
• Operation	10 × outer diameter
Maximum allowable tension during cabling, N	For simplex: 350 For duplex: 700
Maximum allowable compression during cabling, N/cm	200

The important feature of fiber-optic cables for cords, which essentially differ from electrical balanced cables used for similar purposes, is that despite their high flexibility, their basic transmission parameters (attenuation and bandwidth coefficient) are equal to those of backbone cables. These enables designers to vary the lengths of connecting and patch cords within much wider limits and to increase their lengths by more than 30 m from the appropriate reduction of backbone cable lengths.

4.1.6 Color Markings of Optical Cables

Color marking of optical cables is based, in general, on the same principles as that of electric cable markings. The external fiber buffers, module tubes, and bundled fiber elements (strips and tapes) are painted in basic identification colors. The following special conditions must be taken into consideration:

- marking colors are not divided into groups of colors for marking separate fibers and their bundles;
- elements that facilitate a pair grouping of fibers are usually not applied. There are only a few examples of cables where two optical fibers have the same ex-

ternal buffer color, with the second fiber of the pair marked with a hoop marking; the distance between these labels is 20–30 mm. Formally, this marking means that the fiber belongs to the next group, but practice shows that if the marking is present, fibers of the same color terminate a pair of outlets;

- in optical cables produced by foreign manufacturers, color marking is used much more often than in electrical cables. The color marking used by the foreign manufacturers is distinct from that listed in Table 3.17 (see examples in Table 4.10).

Table 4.10. The color-coding of fibers and modules of the European optic cables manufacturers

Manufacturer	**1**	**2**	**3**	**4**	**5**	**6**
Brugg, Switzerland[a]	Red	Green	Yellow	Dark blue	White	Violet
ABB, Norway	Black	Brown	Red	Orange	Yellow	Green
Siemens, Germany[b]	Red	Green	Dark blue	Yellow	Colorless	–
Ericsson, Fabryka Kabli	Red	Dark blue	White	Green	Yellow	Grey
Ericsson, Oszarow, Poland	Red	Green	Dark blue	White	Violet	Orange
Ericsson, Helkama, Finland	Dark blue	White	Yellow	Green	Grey	Red
Manufacturer	**7**	**8**	**9**	**10**	**11**	**12**
Brugg, Switzerland[a]	Orange	Black	Grey	Brown	Pink	Turquoise
ABB, Norway	Dark blue	Violet	Grey	White	–	–
Siemens, Germany[b]	–	–	–	–	–	–
Ericsson, Fabryka Kabli	Brown	Black	Orange	Violet	Pink	Turquoise
Ericsson, Oszarow, Poland	Grey	Yellow	Brown	Pink	Black	Turquoise
Ericsson, Helkama, Finland	–	–	–	–	–	–

[a]The same color-coding principle is approved by the Swiss standard PTT CH 840.05.02.
[b]The DIN 47002 and IEC-304 standards approve the same color-coding principle.

Besides standard color coding of pipes of the outdoor cables modules, the Russian Federation and some European countries employ the principle of color coding, which can be considered as the key coding layout. The principle of such coding implies that there are only two painted modules of different colors in every lay, which can be placed not necessary next to each other. The first number is assigned to a module of one color (red, for example), and other modules are numbered in order from the first colored module to the second one (the reference module).

In the German-speaking countries the cable type is defined according to DIN VDE 0888, and therefore cables of similar designs produced in different factories have the same markings. To specify the manufacturer of the cable, a color identification tape is laid under the outdoor cable jacket. Thus, for example, in cables produced by Siemens, two white, one red, and one green thread are used, whereas in cables produced by Siecor (Siemens and Corning), two red, one green, and one black thread are used [81].

Application of color marks for external jackets of indoor cables and cables for cords is not specified by the working editions of SCS standards. In practice this are many marking layouts that are defined by the manufacturing company's standards. We note that the use of orange jackets made of inflammable materials is wide-spread (similar to electrical cables), but yellow is used practically everywhere to mark jackets of single-mode cables for cords.

The cable manufacturers use an individual system of optical cable identification marking based on the IEC-794-1 recommendation. Usually the mark implies an alphanumeric index to fit the basic data of cable design, designation (function), and optical characteristics in a more or less detailed form. The foot or meter labels of length are necessarily added to the index on the external jacket. To mark the external optical cable, special paint or a thermal-transfer marking method is often used. Thermal transfer marking is very durable and, for example, possesses superior durability after cabling in a conduit run. Among additional marking elements that are sometimes put on the external jacket, we would like to highlight the symbols of a wave (optical cable) or a telephone handset (communications channel, Fig. 4.12). Russian cable manufacturers use thermal-transfer marking without filling the marking signs with color, which makes the cable inconvenient, especially for storage warehouses.

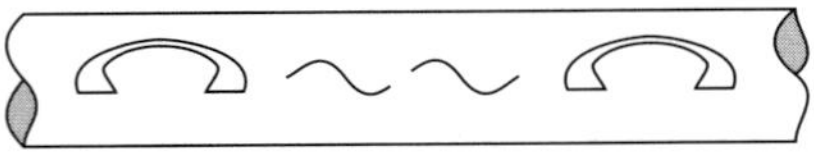

Fig. 4.12. Identification of optical outdoor cabling by DIN 0888

4.2 Optical Connectors

One of the basic problems that must be solved in the design of any optical communications line is the necessity of splicing the optical fibers. Generally, this problem may be resolved in two essentially different ways: with the help of plug&jack connectors or with permanent optical connectors (Fig. 4.13). The permanent connectors, which sometimes are called splices, are widely applied during implementation of long communications lines (for example, PSTN of cities and larger). The small-scaled optic SCS employ plug&jack connectors or simply connectors to splice optical fibers. Further we shall examine the connectors in detail.

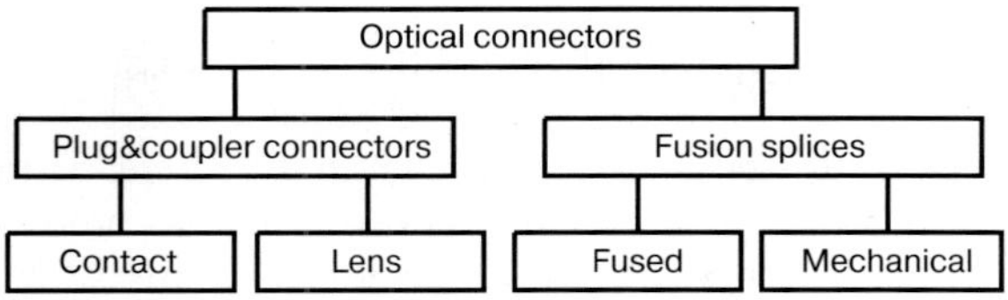

Fig. 4.13. Basic options for optical connectors

4.2.1 Optical Connector Functions and Basic Requirements

Optical connectors, which are sometimes plug&jack connectors, are intended to connect patch and equipment cords to termination hardware in telecommunications closets, telecommunications outlets of workplaces, and network equipment.

The basic functions of the optical connectors are:

- fiber entrance into splicing point with the specified bend radius;
- fiber protection against external mechanical and climatic exposures;
- fiber fastening in centering system.

The products examined in this section meet the following basic technical requirements:

- minimal and high attenuation of backward dispersion;
- long-term stability and reproducibility of parameters;
- high mechanical durability and minimal dimensions and weight;
- easy cable mounting;
- easy connection and disconnection processes;
- terminals with convex end surfaces;
- preliminary special treatment of terminals.

The both normative documents (TIA/EIA-568-A and ISO/IEC-11801) contain standard requirements for optical connectors. The standards give only general positions and specify what type of connector is allowable for application in SCS optical subsystems, the basic transmission parameters of various types of connectors, the requirements for connector durability, and the procedures for connection of optical connectors. The standard requirements for the limiting values of attenuation, reflection losses, and durability of optical SCS connectors are listed in Table 4.11.

Table 4.11. Basic characteristics of optical SCS connectors according to ISO/IEC 11801

Parameter	Multimode	Single-mode
Attenuation, dB	≤0.5	≤0.5
Return loss, dB	≤–20	≤–26
Connection–disconnection cycles	500	500

In SCS, according to actual standards, it is only possible to use two types of optical connectors—SC and ST. SC connectors only apply in newly created SCS. In existing SCS with ST connectors, it is possible to use them in the future, particularly for the build-up of an SCS. For the connection of network equipment to SCS using other types of connectors, it is possible to use equipment cords with SC connectors at one end of the cord and another type of connector at the opposite end. It is also possible to use adapters between SC connectors and other types of connectors (Section 4.5).

Connectors should be with the letter A or B. An A-marked plug always terminates outlets bearing the same symbol, and vice versa. A duplex SC plug should have different marks on its two pieces according to standard, and, if it is looked at from the ferrules so that the keys are up, the left plug is always marked with the letter A, and right plug with B. The mark on an in-line outlet has a special feature. Its different sides have different marks (Fig. 4.16). This mark shows the direction of an optical signal. The A-marked plug always is a transmitter, while the A-marked outlet is a receiver, and vice versa. The same is true for network equipment, where the connector marked with an A is the input for the optical receiver, and that marked with a B is the output for optical transmitter.

Today, the majority of connectors are designed for connection with two optical fibers; these are called multichannel connectors. Such connectors provide simultaneous splicing of two or more pairs of fibers. Multichannel connectors make up an increasing fraction of the total number of connectors. Tight connectors are used in critical conditions (high humidity, vapors of aggressive materials, etc.). There are also so-called hybrid connectors that allow splicing of optical fibers and electric conductors.

There are lens and contact variants of optical connectors. Lens connectors (Fig. 4.14) were widely used in the early stages of optical communications engineering, which focused on lenses or their analogues. With the aid of a lens, the light from the transmitting fiber is transformed into a large-diameter parallel beam and is then focused by a second element on the core of receiving fiber. The key advantage of the described solution is its smaller sensitivity to axial and side shifts of the spliced fibers.

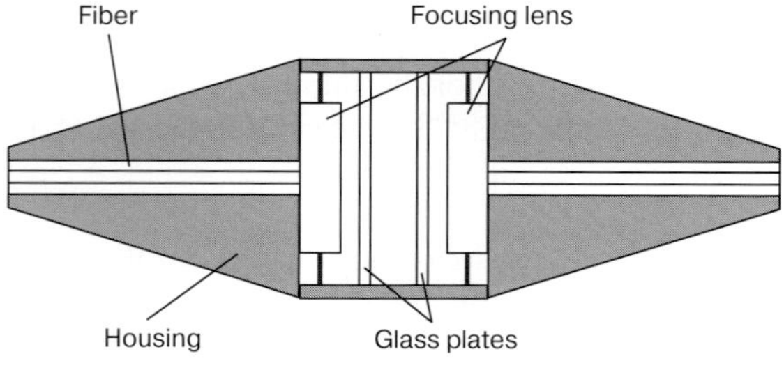

Fig. 4.14. Lens optical connectors

Contact connectors (Fig. 4.15) suppose end-to-end connection of fibers, and moreover, parallelism of their axes and the minimum possible distance between their end surfaces are monitored. This connector architecture is smaller and lighter, and it gives lower signal attenuation by eliminating lens losses and Fresnel reflection losses. For this reason, the overwhelming majority of modern connectors are contact connectors.

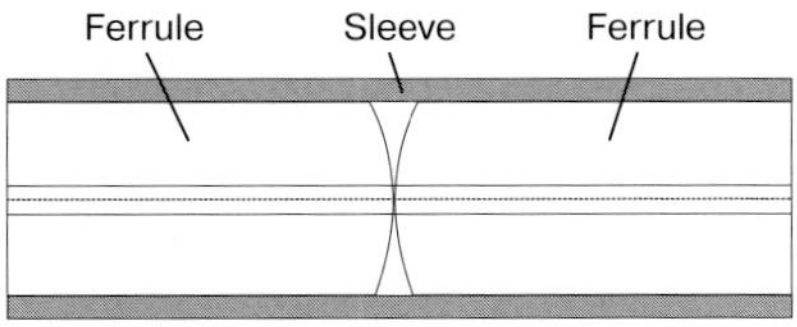

Fig. 4.15. Optical contact-type connector

A plug ferrule serves as the basis for the majority of contact connector structures. This ferrule is inserted into an adjusting element as a thimble, and the connector contains two basic components: a plug and coupler.

The majority of the produced connectors use a balanced scheme, i.e. both spliced fibers are fastened by similar plugs, which are then inserted from two sides into a connecting coupler that is equipped with a special centering device. There is also a small group of optical connectors that contain only two elements: a plug and a coupler. Such connectors are called non-balanced.

To fix a plug in a coupler, a bayonet element (a so-called ST connector) and a latch may be used. The latch can be internal (a SC connector) or external with a lever (LC connectors, E-2000). Polyhedral or round coupling nuts (FC and SMA connectors) are also utilized. The connection of terminal active equipment to an optical cable is performed by the same way. The equipment is supplied with the reciprocal part of the optical outlet.

The connectors are manufactured both in multimode and in single-mode versions. A single-mode connector possesses a structure that is similar to that of a multimode connector; it differs basically by more rigid tolerance to the geometrical size of the plug ferrule and the centering elements of the coupler. These features reduce the splicing losses of single-mode fibers to within allowable limits. For example, a standard diameter of the ferrule aperture of a plug for single-mode optical fibers is 126+1/–0 μm, whereas in ferrules of plugs for multimode fibers, this parameter has a value of 127+2/–0 μm.

Many multimode connectors have plugs of several options for installation on fibers with various diameters of claddings (125, 140, 280 μm etc.). Their designs differ from each other only in diameter of the ferrule aperture.

The working temperature range of the majority of optical connectors designs is -40–+85 °C, which coincides with the working temperature range of the majority of outdoor cable structures. The key parameters of some types of optical connectors are listed in Table 4.12, and the connection scheme is shown in Fig. 4.16.

Table 4.12. Key parameters of optical connectors

Connector type	Ferrule material	Fastener	Average attenuation, dB, at wavelength 1300 nm	
			Multimode	Single-mode
FC	Ceramic	Captive nut	0.2	0.3
MIC	Ceramic	Latch	0.3	0.4
SC	Ceramic	Latch	0.2	0.25
SMA	Steel	Captive nut	1.0	–
ST	Ceramic	Bayonet	0.25	0.3
E-2000	Cupronickel	Latch	0.2	0.25

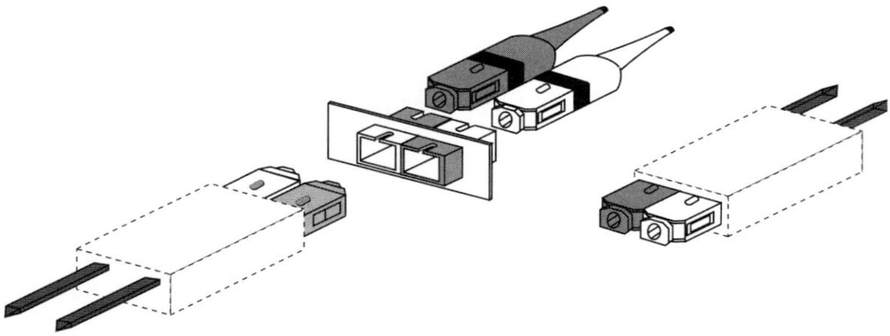

Fig. 4.16. Connection scheme of optical connectors

4.2.2 Parameters of Optical Connectors

Insertion loss

Losses in optical connectors are determined by many factors, which generally may be divided into the following groups:

- internal factors determined by tolerances for geometrical sizes of optical fibers;
- external factors determined by the quality of manufacturing of separate elements of a connector and its technological tolerances;
- losses caused by reflection and dispersion;
- losses caused by pollution.

Among the primary *internal factors* that cause losses in optical connectors, we can enumerate eccentricity and noncircular structure of the core, as well as the differences in diameters, numerical apertures, and refraction profiles of spliced optical fibers. The need to take into account eccentricity and noncircular structure of the core appeared at the early stages of the engineering development of optical communications lines. Today, these factors are minimal due to the state-of-the-art level of the fiber production industry [82]. Thus, for example, when the core noncircular coefficient value is not greater than 5%, insertion losses do not exceed 0.1 dB.

Losses caused by differences in diameters of spliced optical fibers most frequently occur when multimode hardware is used, because the standards enable the use of two types of fibers with core diameters of 50 and 62.5 μm in SCS. We note that these losses occur only at transitions from large-diameter fibers to smaller ones. When fibers with equal nominal diameters are spliced, insertion losses occur due to the core diameter tolerance.

Losses caused by difference of numerical apertures generally arise because of the presence of industrial tolerances for this parameter.

Losses caused by *external factors* include losses due to presence of an air layer between the end faces of spliced fibers, radial and angular displacement of fibers, and wedging of fiber face surfaces in connectors. The losses are caused by industrial tolerances for the geometrical sizes of the separate components of optical connectors, which perform the centering of spliced fibers.

When there is an air layer between the face surfaces of spliced fibers, F*resnel losses* caused by partial reflection of the light beam in the air–glass interface occur.

Return loss

There are air layers with varying thickness and size in any optical connection between the face surfaces of spliced fibers. These layers are caused by inevitable errors and tolerances during manufacturing. Fresnel reflections arise in glass–air–glass interfaces, which results in back-reflection of the light beam and return losses in the signal.

The back-reflected beam has a negative influence on high-speed laser optical transmitters because its return back into the resonator causes strong distortions of the transmitted signal. Basically, any inhomogeneity of an optical fiber may cause return losses, however, the optical connectors are the principal cause. Based on that, the return loss value should be very strictly controlled when creating optical communications lines. The return loss can be defined by a return loss coefficient, which is defined as the ratio of the power of the back-reflected light to the power of the input signal. Because this value is very small, it is usually measured in logarithmic units.

The standards TIA/EIA-568A and ISO/IEC 11801 describe solutions for multimode optical connectors, which possess return loss coefficients not worse than –20 dB, as the parameter value must not be worse than –26 dB. Actually, the last value is not acceptable for many applications, and a number of methods have been elaborated in order to decrease the value. Based on the coefficient of return loss, single-mode connectors can be divided into classes:

- PC <-30 dB
- Super PC (SPC) <-40 dB
- Ultra PC (UPC) <-50 dB
- Angled PC (APC) <-60 dB

For minimization of return loss it is mandatory to have a physical contact (PC), in which the glass of the fiber core is tightly pressed to the core glass of another fiber. The presence of physical contact is especially important for single-mode connectors. Ferrules with a flat form of the end surface, which were often used in connectors designed before 1985, fail to meet this condition (Fig. 4.17a). A number of engineering and technological solutions were developed in order to achieve physical contact; an incomplete list of these solutions includes:

- pressure springs that press end faces of ferrules to each other when the plugs are inserted into an outlet;
- ferrules with convex face surfaces (radius of rounding is 10–15 mm, Fig. 4.17b);
- special face surface treatments.

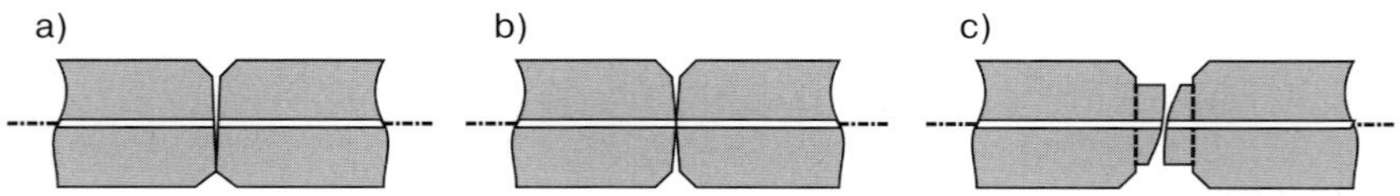

Fig. 4.17. End surfaces of ferrules: a) flat; b) convex PC, Super PC, Ultra PC; c) Angled PC

Pre-angled (approximately 8°) end-face ferrules (Fig. 4.17c) can be the most effective at minimizing return loss, although they are the most technically complicated and, consequently, the most expensive ferrules.

We should also note one more engineering solution that was widely used in the early stage of fiber-optic communications development. To minimize return losses in connectors, a transparent immersion liquid with a refractive index approximately the same as that of the glass is dripped in the contact area of the optical fibers. This solution makes the use of connectors inconvenient, and, because of improvements in ferrule manufacturing technology, it is rarely used. The immersion gel is applied only to some types of mechanical connectors and mechanical splices, that is, elements with minimal splicing and disconnection cycles.

4.2.3 Design Features of Optical Connectors

The structure of an optical connector includes the following basic modules and components:

- ferrule or other element for fiber fastening;
- element for the alignment of spliced fibers;
- housing equipped with elements to protect against turning and wrong termination;
- elements that fasten to the reinforcing coatings of fibers and cable;
- boot;
- protective cap.

Depending on the optical connector design, some of the above elements may be omitted.

Ferrules of optical connector plugs

The basic component of the majority of plugs is an axisymmetric ferrule with a central aperture to fit in the fiber end. To reduce losses at the splicing point, the fiber endface must be trimmed and polished to the endface of the ferrule. The most connectors are designed to fit cylindrical ferrules with a diameter of 2.5 mm. There are also noncylindrical ferrules of other diameters. The endface of the ferrule is designed with a facet, which facilitates installation of a plug in the coupler. Moreover, when the connector is assembled, there is a free space between edges of the ferrules where dirt particles can penetrate; therefore the endface surfaces of ferrules are tightly pressed.

Ferrules can be divided into monoblock and composite types by their design. Monoblock ferrules (Fig. 4.18) are made of the same material, as a rule, such as ceramics, metal, plastic, and sometimes glass. Aluminum and zirconium oxides are frequently used for the manufacture of ceramic ferrules. Aluminum oxide is cheaper than zirconium oxide, however, zirconium oxide ferrules possess better mechanical stability and strength properties than aluminum ferrules. Ceramic ferrules possess better durability and stability when applied in a wide range of temperatures than those made of other materials. There is an additional property of ceramic ferrules—they enable more rigid tolerances on geometrical parameters. This results in lower insertion losses (0.2–0,3 dB, Table 4.12).

Plastic is used for the manufacture of ferrules in order to reduce the connector cost, but it possesses somewhat poorer properties for stability and loss. Some types of connectors have a metal ferrule made of stainless steel and have characteristics between those of ceramic and plastic ferrules. Glass ferrules are used when the plug is bonded to the fiber by UV-cured glue.

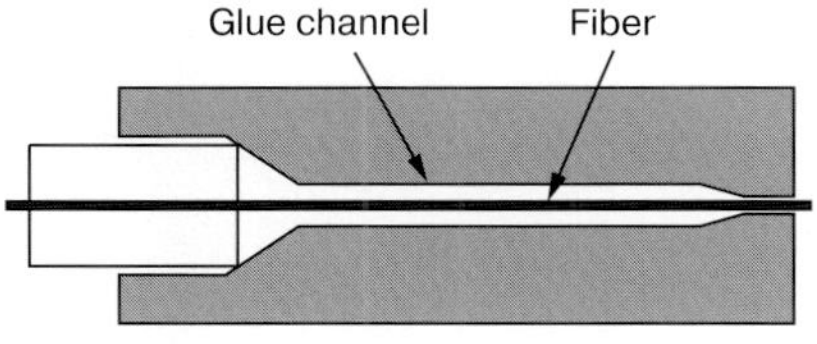

Fig. 4.18. Monoblock ferrule

The SCS standards specify rather strict requirements for the compliance of optical connectors with the given values of insertion and return losses at not less than 500 connection–disconnection cycles. Hence, most of the plugs used for the implementation of optical subsystems employ ceramic ferrules.

Compound or composite ferrules (Fig. 4.19) are sometimes used. There are following options here: the E-2000 connector has a ceramic ferrule with a cupronickel insertion. A similar idea is used in the plug&coupler connectors, such as “List-Bulava”, designed in the former USSR in the mid-1980s. This is a ferrule made as a glass capillary that is sealed in an outer alignment metal tube. In some types of connectors, such as the SMA-906, the ferrule is made of metal covered with a ceramic alignment tube. This type of ferrule differs from the previously mentioned ones by its somewhat short length. Use of complex composite structures is justified where wear-resistant outer coverings are needed to achieve high durability of connections. They are also used where, because of an insufficient level of technology (especially at the early stages of fiber-optic communications engineering development), it is not possible to achieve high accuracy of central channel manufacturing in order to fix fibers in firm materials. Finally, this type of ferrule is also used where the application of a multilayer ferrule with a relatively soft internal part allows for additional alignment of the fiber by means of technological hardware and decreases insertion losses.

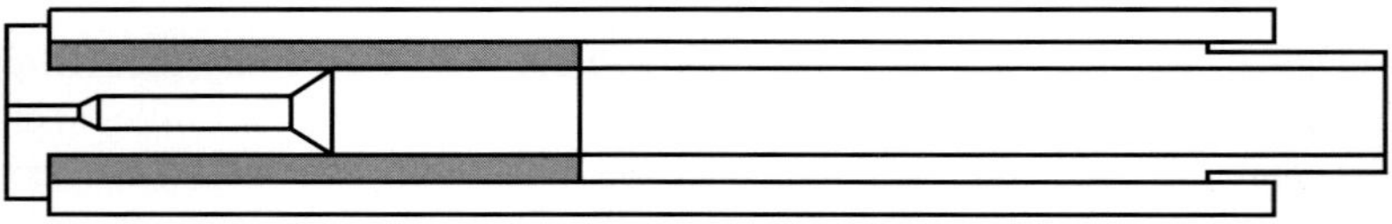

Fig. 4.19. Composite ferrule

We comment on the latter statement in detail. The alignment procedure can be carried out in two steps; multimode connectors employ only the first step. At the first stage (passive alignment) when inputting a fiber into the channel prior glue hardening, one presses the endface of the soft insertion of a composite ferrule by means of a circular stamp with a triangular working element. Due to plastic deformation of the internal material, it tightly covers the fiber end and reduces the core residual eccentricity up to the degree allowable by manufacturing standards on fiber core and cladding eccentricity, that is, up to 2 μm (Fig. 4.20).

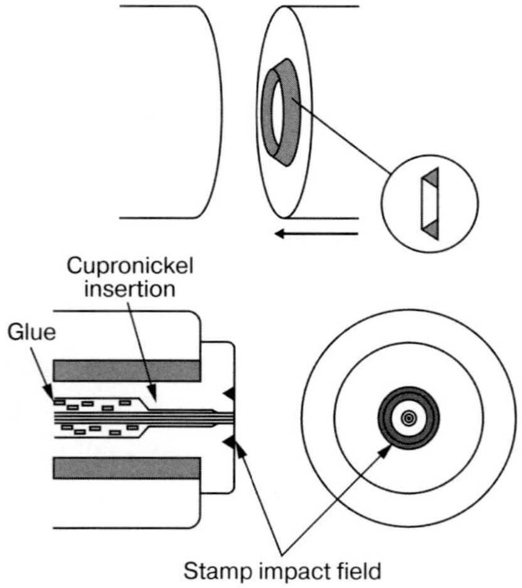

Fig. 4.20. Passive alignment layout

At the second stage, once the glue has hardened and the ferrule has been finished, the stamp looks like a section with an angle equal to 120°, which is preliminarily oriented in order to minimize the residual deviation of fiber and ferrule axes (Fig. 4.21). The standard eccentricity value of the cladding (core) of up-to-date fibers is 0.8 μm. Once the active alignment is done, the core–ferrule eccentricity is no more than 0.5 μm, which corresponds to average losses 0.12 dB [83].

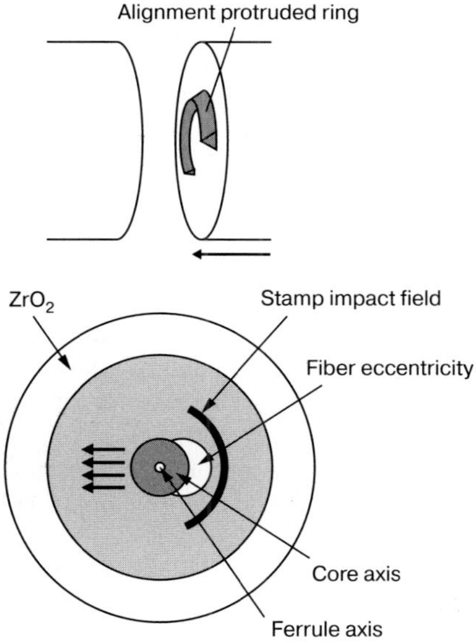

Fig. 4.21. Active alignment layout

Protective elements against ferrule rotation and wrong termination of plugs

One of the necessary conditions to obtain low insertion and return losses and to improve the stability of these parameters during the whole maintenance period of a cabling system is the presence of physical contact in spliced fibers. However, such contact gives a mechanical interaction between the fibers at the moment of connection and disconnection, which results in damage to their endface surfaces and in deterioration of their parameters. The risk of damage increases greatly if, at the time of installation or disconnection of a connector, the fibers rotate relative to each other. To prevent such damage, modern connectors must be equipped with turning protection elements (Fig. 4.22). The practical solution to this problem may be achieved in the following ways:

- equip the plug with a guide lug that slides into a slot or groove on the coupler housing during installation;

- use the push–pull principle to connect plugs with cylindrical ferrules into couplers;
- use noncylindrically and nonconically shaped ferrules that connect only by linear movement.

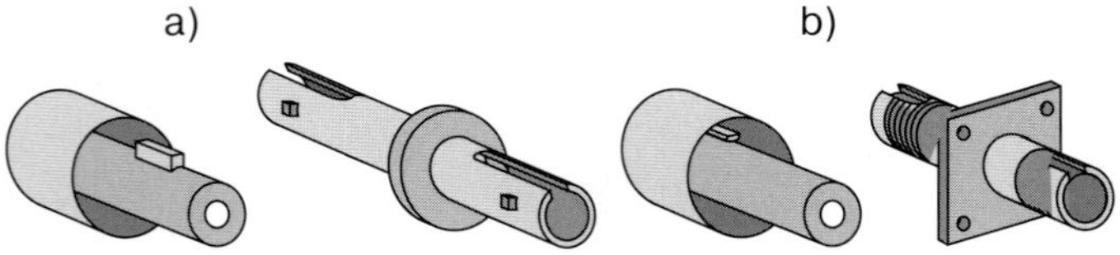

Fig. 4.22. Ferrule protective elements against rotation: *a)* with radial lug and slot; *b)* with axial keyed-lug and groove

The first method is mainly used for single plugs. Their connection to a coupler is made with common or bayonet nuts (ST, FC, SMA, and other types of connectors). The other two methods are widely used in modern connector designs.

Protection against wrong termination can be divided into *passive* and *active measures* and is realized by means of different hardware. There are several passive protective options. The first option uses various color markings and legends to provide visual control of connection accuracy. The second option uses multipurpose ports (for example, different functional sections or subsystems). These ports are based on various types of connectors.

Active protective measures are based on various elements of mechanical blocking to prevent wrong plug termination to the coupler. One may use the following techniques:

- asymmetrical plug houses;
- guide lugs on lineally terminated plugs;
- various insertions and frames, including mobile ones to fit in both a plug and a coupler.

One may sometimes combine active and passive protective measures. For example, blocking frames of adapters can be made of colored plastic.

Cable fastening elements and techniques

Optic plugs are usually adapted to installation on cables for cords in protective hoses of 2.5–3.0 mm in diameter. If a plug is installed on a buffered (0.9 mm) fiber, the fiber is covered with a 2.5–3.0-mm diameter tubular adapter in order to ensure the given fiber bend radius at the entrance point. Some constructions are equipped with a rubber boot for the same purpose. If there is no such adapter in the delivery kit, it can be replaced by a short piece of the protective hose of a cable for cords.

To improve the maintenance reliability of plugs when mounting them on cable for cords in hose of 2–3 m, an internal ferrule is used in the structure of modern connectors. This 3–6-mm long ferrule has a supporting collar that is put on the fiber buffer (0.9 mm) and fits tightly in the hose.

Plugs of many connectors are designed for only a certain type of fiber (for example, in a buffer of 0.9 mm). There are also more expensive universal solutions, which use only a part of the components in any specific case during assembly.

When gluing a plug on a fiber in tight buffer of 0.25 mm, it is recommended to re-cover it with a secondary buffer with outer diameter of 0.9 mm. There are a number of techniques. For example, cambric tubing with outer diameter 0.9 mm from the kit D-181755 produced by Lucent Technologies provides a reliable fiber protection to external mechanical exposures. For the same purposes, Mohawk manufactures the Field Breakout Kit. It comprises a metal tube with six (M90272) or twelve (M90273) cambrics of 0.9 mm in diameter; the metal tube is terminated to an outdoor cabling module with a crimper. Kits of series 91.B0610–91.B0640 by Mod-Tap consist of a support, cover, and terminal assembly. They are designed for 4, 6, 8, and 12 fibers and are an improvement over the above-described products, as they have color-coded cambrics with an outer diameter of 0.9 mm that can be fixed on the module tube housing without crimping, however, they possess poor dimensions and weight. Some plugs of the bonded connectors by AMP are equipped with an inline plastic thimble, which is fixed by a crimping sleeve during assembly and provides a reliable entry of fibers in tight buffers of 0.25 mm.

During installation of a plug on a cable for cords, the fastenings must possess high mechanical durability. The boot is designed depending on the plug fastening technique. The basic solutions are presented in Fig. 4.23.

The plugs in the first category have a wide cone-shaped metal boot, which is crimped during assembly. The cable buffers and hoses are bonded and fixed by the friction force of the crimped boot (Fig. 4.23a). The advantages of this design are easy assembly (it is convenient for less-experienced personnel) and the opportunity to considerably reduce dimensions. The main drawback is poor resistance to plucking axial mechanical exposures that accompany the fastening technique to buffers.

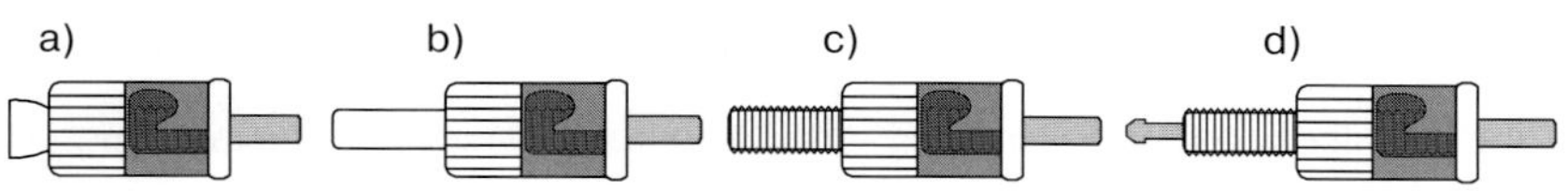

Fig. 4.23. Boots of plugs for fastening to protective coverings of cables for cords

The second (more widespread) option applies a cylindrical boot of rather small diameter and a crimping sleeve (Fig. 4.23b). The reinforcing Kevlar threads of cables for cords are laid along the boot during plug mounting; a metal sleeve slides over this assembly and is crimped. The reinforcing threads provide resistance to plucks that dramatically reduces the possibility of connection rupture. The additional increase in mechanical strength of patch cords terminated in plugs of some connectors is achieved by utilization of boots with ridges or grooved surfaces, which improves the reliability of the Kevlar thread fastening. The sleeve is crimped not only on the boot, but also on the outer jacket of cables for cords to provide the same properties (Fig. 4.23c). The strength of the jacket of cables for cords fastening can be improved by the insertion of a small-diameter cylindrical lug under the protective jacket on the boot end (Fig. 4.23d).

It is necessary to note that in some types of group connectors the mechanical strength of the plug fastening is provided only by fastening to the protective jackets of cables for cords. Such fastening can be created with the help of both a crimping ring and a split terminal.

Plug boots

A boot with a length of about 3–5 cm defines the fiber bend radius at the plug entrance. This boot is manufactured of rubber or soft polymeric material. The flexibility of this element in modern designs is frequently increased by means of perpendicularly oriented splits. A special element is inserted into the boot in plug model 943 manufactured by Amphenol. This element provides a cable bend at 90° maintaining a given bend radius. The last feature is very useful when cables terminate to ports of optical shelf racks and some models of wall enclosures with protective shutters. The boot also provides for color coding of plugs, for example, when the design of the connector does not support a duplex plug. In this case, boots or plugs with a boot of various colors can be ordered separately.

Optical couplers

Optical couplers are mounted on the faceplates of telecommunications outlets, wall enclosures or distributing shelves. The coupler consists of the housing with fastening elements on the faceplate and an internal centering device. The centering device aligns the plug ferrules that are inserted into the coupler relative to each other. There are some different design options for this element. The centering device is mostly manufactured as a two-piece sleeve made of ceramics or phosphorous bronze and inserted in the coupler housing fast or flexible. In the couplers without an alignment ferrule the centering device is implemented without application of a centering sleeve (Section 4.2.6).

The coupler housing is designed with fastening elements of the switching center (a screw thread on a nut, a square, rectangular, rhomboid, or round flange with two, three, or four apertures sized for M2 screws, a latch and the additional details for fixing a plug in the working position (groove, lugs of a bayonet fastener, elements contacting with a latch). Use of a latch is typical for couplers in plastic housings. In some models of small wall enclosures, the installation of SC couplers with plastic housings is performed using a groove where the plastic housing flange is inserted. Sometimes for the expansion of functionalities, two types of restrainers are applied simultaneously, for example, a latch and a flange with apertures.

There are two kinds of couplers—multimode and single-mode, and the differences between them are generally the materials of the housing (metal or plastic) and the centering device (bronze or ceramics).

Optical couplers may be supplied with directing lugs for protection from plug ferrule slippage at the moment of connection; according to the working standards they should be mounted so that directing grooves for the given lug are directed to one side. Duplex couplers with slits oriented on the opposite sides [84], which are commercially available, are nonstandard products, however, they may be used to allow additional mechanical coding and blocking of certain ports.

Color-coding is used for single-mode and multimode couplers with plastic housings, particularly in SC couplers. Single-mode and multimode couplers with the metal housings (for example, ST) in most cases differ from each other only by the marking legends on the packing and the color of the protective cap. The most widely-known exception to this rule in Russia is ST-couplers produced by Lucent Technologies. The company's trademark and abbreviation SM and MM for single-mode and multimode variants, respectively, are printed on their housings using the punching method.

Protective caps and covers

Protective caps are practically mandatory elements for plugs and couplers of many types of optical connectors. They are used to protect the ferrules or face surfaces of plugs and the female contact of couplers in nonworking conditions from dust and dirt.

There are two basic designs of plug protective caps. They can enclose either the front part of the plug housing or only its centering ferrule. The second option is often used for plugs, whose ferrules protrude from the housing (for example, ST and DIN). The housing is a separate detail; in MIC connectors it is supplied with a sword knot, and it hangs on the cable for cords when the plug is connected. Some types of plugs designed by specific manufacturers have caps covering both the plug endface and ferrules.

Rubber and polymeric materials are used for manufacturing of plug protective caps. Sometimes caps are painted in various colors, serving as a color-coding element for various types of products. The material of the cap shall be sufficiently rigid to prevent collapse at the moment of cap removal from the plug, causing dust particles to accumulate on the ferrule endface.

A number of modern constructions of connectors have protective covers that are components of both the plug and the coupler. The solution makes the use of protective caps unnecessary. For ferrule-free connectors (Section 4.2.6) a protective cap is mandatory.

The caps of couplers provide similar functions. Multimode couplers usually have black or red protective caps, but single-mode couplers have yellow protective caps. In couplers of series 954 produced by Amphenol, a spring-loaded outer cover is used instead of a cap, but couplers of series E-2000 produced by Diamond employ an internal cover. SC couplers produced by Alcoa Fujikura have a protective cover designed as an external adapter that is installed on the housing. Presence of the cover, which automatically closes when disconnecting the plug, is especially important when powerful long-wave laser oscillators are used in network optical transmitters. This cover protects the eyes of service personnel from radiation emitted by the laser.

The rubber caps cover also couplers of the network optic interfaces. Hewlett Packard employs safe plastic insertions to be fastened to lugs of bayonet connectors in its transceivers for this purpose. In SC couplers a rubber or plastic insertion that is sometimes supplied with a lug for easy installation and removal serves as a dust-protective element. The protection of FC couplers is implemented by a threaded plastic or metal cap screwed on the coupler.

4.2.4 Basic Types of SCS Optical Connectors

SC-type connectors

An SC connector (subscriber connector, but sometimes it is informally called a stick-and-click connector) was developed by the telecommunications corporation NTT for subscriber multipurpose devices (Fig. 4.24). Today, it is approved by the international standard IEC-874-13. Working editions of the standards determine it as the basic type of a coupler for application in SCS. It may be implemented in simplex and duplex options. Its construction was designed as a plastic housing to protect the ferrule and provide smooth push–pull connection–disconnection.

Fig. 4.24. SC plug

The overwhelming majority of SC plugs is supplied with ceramic ferrules; there are also individual samples of these products with ferrules made of stainless steel. The SC ferrule is deep within the plug housing to protect it from pollution. The push–pull movement at connection–disconnection makes the connector especially useful for application in 19-in shelves, since it allows for an increased density of ports because of the compact position of couplers. The latch is unlatched only when pulled out of the housing, which increases the service reliability. On the top of the plug housing there is a keyed-lug to prevent it from termination into the coupler. SC connectors provide good stability of parameters (they survive over 500 connection/disconnection cycles), as the ferrules do not rotate relative each other at connection–disconnection. As can be seen from the Table 4.12, this connector has the best attenuation.

Two methods are used to compose a duplex connector from a simplex one. The first utilizes fasteners on the plug housing that interact when assembled. The second method applies an external fastener, which may be designed as a holder consisting of two symmetric pieces with openings for plug housings. It can also be an H-shaped detail, with side slots to fit the plugs. For example, fastener 2A1 produced by Lucent Technologies, which is supplied with symbol marks (letters A and B), is designed as an H-shaped detail. The distance between the axes of ferrules in a duplex connector is 12.7 mm.

The large plastic housings of SC plugs and couplers the use of effective color marking in addition to symbol marks. According to the standard TIA/EIA-568-A, the housings of single-mode and multimode variants of SC connectors are manufactured in blue and gray (or beige). There is also a single-mode SC with a green hosing and a skewed ferrule to reduce the return loss. Some kinds of SC connectors with nonstandard coloring of plug&coupler housings are widely distributed. For example, a multimode plug and coupler delivered by Prospect Technologies (Russia) has a black housing, and a multimode SC coupler made by Methode has a white jacket.

ST connectors

Optical ST-type connectors (straight-tip connectors, sometimes informally called stick-and-twist connectors) were developed by AT&T Bell Laboratories (presently Lucent Technologies) in 1985 to replace biconical connectors. Prior to the development of the SC connector, the biconical connector was the most widely used in SCS optic subsystems and local networks. Today, the connector structure is determined by the international standard IEC 874-10, which prescribes the presence of a ceramic ferrule with a diameter of 2.5 mm and a convex endface surface (Fig. 4.25).

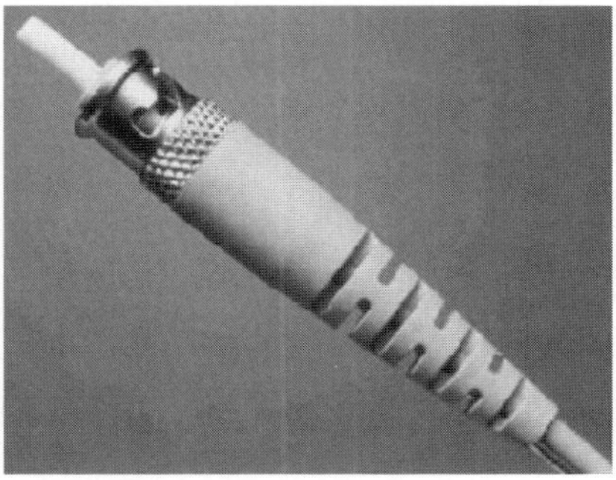

Fig. 4.25. ST plug

The plug terminates into the coupler by a spring-loaded element with a quarter-turn. Therefore, the ST connector sometimes is called the BFOC (bayonet fiber-optical connector).

There are several options for ST-connectors, which differ from each other basically by the form and material of the bayonet fastener, and also by the fastening method of buffers and coatings to the plug housing.

Lucent Technologies has developed three plugs options for this connector: ST, STII and STII+ plugs, which are fully compatible with respect to termination slots of the coupler and have insignificant construction variances that evolutionarily improve their performance. Thus, in particular, a ST plug bayonet nut has an axially open groove, but in both of the last options, a bridge closes the groove (see Fig. 4.26). The key feature of plugs produced by Lucent Technologies is that a crimping tool is not necessary when reinforcing 0.9-mm diameter fibers with buffers.

a)

b)

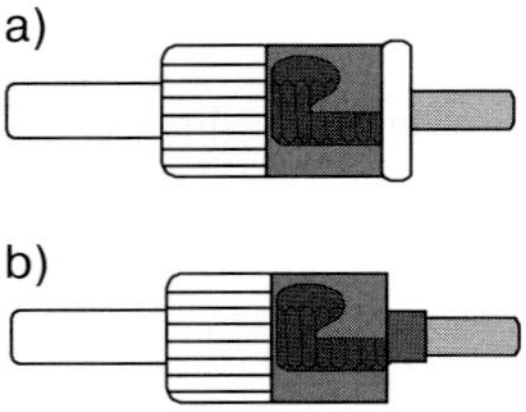

Fig. 4.26. Bayonet nuts for ST plug: *a)* closed groove; *b)* open groove

The metallic housing of an ST plug&coupler provides high mechanical strength, however, it also complicates coding and identification. Only one company produces a few samples of this product with a nut of the bayonet fastener made of either golden or silvery metal, depending on the product design (company Brugg). Sometimes the housings of couplers have pressed-in letters SM and MM to indicate single-mode and multimode options, respectively. Some companies offer ST plugs with plastic boots in various colors, and rings, sleeves, and other similar products are also widely used, which are not the regular marking elements.

The design of a ST connector does not allow it to be formed into a duplex plug, so the majority of manufacturers produce simplex ST couplers. Only Alcatel offers duplex ST couplers in an overall housing.

Low cost and simplicity of installation and connection are the advantages of the ST-connector, but it possesses the following drawbacks:

- forward protrusion of the ferrule increases susceptibility to damage;
- lack of a duplex option makes termination of twin cords inconvenient and can result in wrong termination;
- lack of color or other factory marks complicates its identification;
- turning force on termination causes friction of the plug ferrules that may lead to scratches on the polished surfaces and ultimately increase attenuation after multiple connections and disconnections;

- fastening with a bayonet nut does not provide the stability of parameters necessary for some applications in environments where vibrations are present.

ST plugs are designed with a special lug fitted in the coupler slot for partial protection of the ferrules against friction during termination.

4.2.5 Other Types of Optical Connectors

FC connectors

FC-type connectors (Fig. 4.27) are specified by the international standard IEC 874-7. They are used mainly in a single-mode applications. They are widely distributed in various telecommunications systems for PSTN. In order to get low attenuation and minimize return loss, the connector ferrule is made with a rounded end (at the same time, tolerances for the geometrical sizes are quite strict). The first variant of the plug design had a ferrule with a flat endface that did not enable users to obtain good performance. The next ferrule with a rounded endface that provided physical contact of spliced fibers was called FC–PC (PC: physical contact) in order to distinguish it from its evolutionary prototypes. Today, FC connectors with flat ferrules are not manufactured; therefore, the designations FC and FC–PC are identical.

The connector provides reliable protection of the ceramic ferrules against pollution, and use of a female swivel-fastening nut provides a tight connection and vibration resistance. Its main drawbacks are its large dimensions and installation inconvenience, since it is necessary to make several turns of the fastening nut during connection/disconnection.

There are two basic options for FC couplers: an SF coupler with a square pad and screwed mounting with two M2 screws, and an RF-type coupler with a round pad and nut mounting.

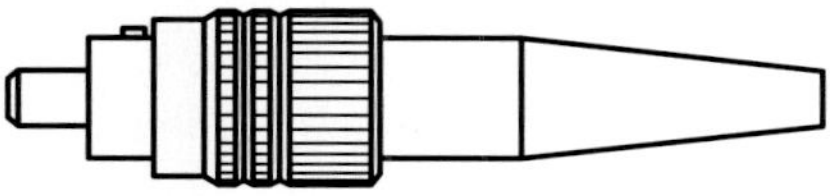

Fig. 4.27. FC plug

MIC connectors

The duplex MIC connector (medium-interface connector)[4] was developed specially for FDDI networks. The plug is made of plastic (Fig. 4.28), is equipped with a latched fastener, and has an asymmetrical form to prevent wrong termination into the coupler (mechanical blocking). The use of replaceable keys as color insertions also enables coding of the MIC connector for connection to only one port: A,

[4] The MIC manufactured by the company AMP is called FSD (fixed shroud duplex).

B, M, or S in FDDI networks. MIC connectors have average values of insertion loss when compared with other types of connectors (Table 4.12).

The cable is fixed in the plug housing by a crimping ring or a plastic fastener. Multimode and single-mode versions of the MIC connector are available. In the mid-1990s, an improved version of a MIC connector was developed. It differs from the previous model by its reduced housing length and the use of a 45°-angled boot for cable entrance. From the designer's point of view, these innovations have increased the ease of operation with optical cabling systems at high densities of FDDI network ports.

The push–pull connection–disconnection of an MIC connector and its good protection of the ferrule–plug housing connection from dust contribute to its advantages. Its large dimensions, complexity of installation, and high cost are the drawbacks of this construction. The MIC connectors are widely used in FDDI equipment. They are also allowed for use in 100BaseFX equipment; however, the authors of this book do not know of any examples of commercial 100BaseFX systems supplied with connectors of this type.

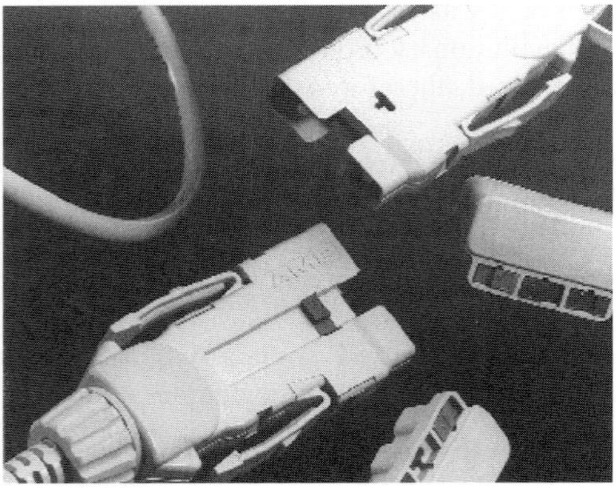

Fig. 4.28. MIC plug

SMA connectors

The SMA or FSMA (fiber subminiature assembly) connectors (Fig. 4.29) were developed at the late 1970s by Amphenol, and at present are specified by the international standard IEC-874-2.

Fig. 4.29. SMA plug

The connector represents a successful attempt to apply the principles worked out during the development of SMA electrical plug&jack connectors to a new field. The SMA plug has a 3.175-mm (1/8-in) diameter metal ferrule with a flat endface that does not guarantee physical contact of the spliced fibers. There are two kinds of such connectors, SMA-905 (FSMA-I) and SMA-906 (FSMA-II), which differ from each other by the form of the ferrule ends. The plug is fixed to the coupler by a hexagonal or, more rarely, round female swivel nut.

The connector is not equipped with a guide pin to prevent ferrule rotation during installation, which negatively influences insertion losses and long-term stability. In addition to LAN and SCS, connectors of this type are widely applied in industrial systems as well as in medical and military engineering. We note that the protection of spliced fibers may be gradually increased up to the level IP-65 [85] by means of special facilities. Today, the design of the FSMA connector is out-of-date, but its couplers are used in Ethernet active equipment and modems with data transmission rates not exceeding 2 Mbps, which are produced mainly by American companies.

DIN connectors

DIN connectors, which are sometimes called LSA and LSB connectors (abbreviations from the German term Lichtwellenleiter Steckerverbinder, meaning optical plug, variant A or B), are regulated by German standards DIN 47256 and DIN 47255, and by the international standard IEC-874-6. They are implemented by means of a 2.5-mm diameter ceramic ferrule, which has a rounded endface surface to provide a physical contact. The plug is terminated in the coupler by use of a round female swivel nut and a knurled upper surface. The connector is basically used in single-mode applications and has very small dimensions. Today it is considered out-of-date, and it is progressively excluded from practical use. It is seldom used in SCS engineering, except when cable system equipment is made in German-speaking countries or will be installed in these countries. These connectors are also widely used in Australia.

4.2.6 High-density Connectors

The general drawbacks of the duplex SC, which is prescribed by working editions of the SCS standards for use in optical subsystems, are its large dimensions and the resulting inability to achieve the density of ports both on patch panels and network equipment that is equivalent to such values in electrical solutions. Some designs that have density not less than that of electrical modular connectors and that have been developed to commercial production levels and are or more or less introduced in wide engineering practice are be examined below. All the examined products have a common feature: their coupler in the duplex version (or its replacing element) has a slot that fully complies with that of the modular jack and is interchangeable (both outlets have an identical form). Along with an increase in the port density, this feature also enables designers to standardize the installation of elements to modular panels, similar to the case for electrical solutions. All investigations in this area can be divided into four basic directions:

- application of ferrules with a diameter reduced to 1.25 mm to decrease the dimensions of other plug&coupler components;
- use of a principle of traditional design modernization and miniaturization of separate components to increase the installation density;
- application of the designs and features that were elaborated during the creation of group or multichannel connectors;
- rejection of centering ferrule.

We should note one more important feature of the new generation of connectors: the average cost is 30% less (at the beginning of 2000) than the cost of traditional designs due to technological improvements and materials savings [86].

Decreased-diameter ferrules

By mid-1999 the most widely known representative of the pilot high-density connectors was the LC connector (link control) developed by Lucent Technologies in 1997 (and by other sources in 1996 [87]). This connector is produced in both single-mode and multimode versions. It uses a ceramic ferrule with a reduced diameter of 1.25 mm and a plastic housing equipped with an external lever-type latch to fit in a coupler (Fig. 4.30). The connector is produced in both simplex and duplex options.

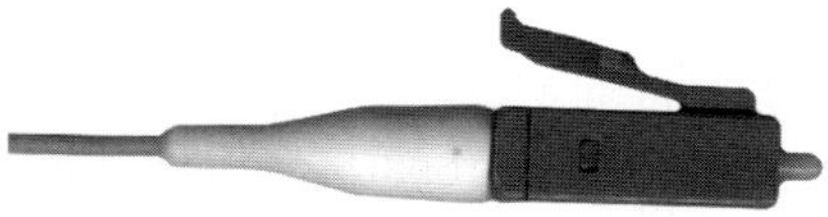

Fig. 4.30. LC plug

According to existing and prospective editions of SCS standards, the optical connector designers guarantee up to 500 cycles of connections and disconnections without deterioration of loss coefficients. This is accomplished by using ceramic ferrules and by the push–pull principle of connecting the plug into the coupler.

The LC plug is bonded by epoxy resin. The design of the plug enables its installation on a buffered fiber (0.9 mm) and on patch cords in 2.4-mm loose tubes. The LC plug can be installed in field conditions on 900-mm fiber, but it can only be fabric-bonded on cables in 2.4-mm loose tubes when manufacturing patch cords because of its small dimensions.

The basic characteristics of LC connectors are listed in Table. 4.13.

Another version of this connector is the MU connector by NTT. This product can be considered a compact variant of the SC connector. Similar to the previous version, the MU connector has a housing with an internal latch (based on the push–pull principle). It is approximately half the size of the SC connector due to the decreased diameter of the ferrule and the miniaturization of other components

[88]. Both simplex and duplex options of the connector are commercially available.

Table 4.13. Basic characteristics of single-mode and multimode LC connectors

Parameter	SM	MM
Average loss, dB	0.1	0.1
Root-mean-square deviation (rejection) of loss, dB	0.07	0.1
Coefficient of reflection, dB	-50	-20
Maximum variation of losses after 500 connection/disconnection cycles, dB	0.2	0.2
Maximum variation of losses within temperatures -40–+75°C, dB	0.1	0.3
Ferrule material	Ceramic	Ceramic

Compact connectors with 2.5-mm diameter ferrules

The second approach is based on the use of a 2.5-mm diameter ferrule (which was previously applied) in a connector. Improvements in dimensions and weight are achieved by tight integration and by miniaturization of some elements of the housing. The most widely known connectors of this type are the E-2000, SC-Compact and FJ connectors.

The company Diamond has designed the E-2000 connector (Europe, 2000). It was widely distributed in some European countries (Switzerland, Germany). The ferrule is made as a composite in form of a cupronickel cylinder, which is tightly covered with ceramic sleeve. The plug is latched in the coupler by an external lever-latch. The connector may be either simplex or duplex. The duplex connector is produced in traditional (duplex, distance between the axes of the ferrules is 12.7 mm), compact (compact duplex, distance between the axes is 6.4 mm), and vertical (low-profile duplex, with stacked plugs at an angle of 180°) design options. A special latch is used to make one duplex plug from two single ones; the duplex coupler is compatible with a standard modular jack by slots, but only in its compact option. The E-2000 connector differs from earlier solutions by possible effective color coding (today, the standard coding includes eight colors) and mechanical blocking when the replaceable mounting frame of an outlet is used, and also by a protective cover integrated in the construction. The protective cover opens automatically at the moment of installation in the coupler and reliably protects the ferrule from dust.

The SC-Compact connector produced by Reichle & De Massari is a successful example of the improvement of a widely commercially applicable product in order to obtain new properties. The prototype of this connector was the well-known SC connector. However, by the removal of external fastening elements and a new mandrel design, Reichle & De Massari engineers reduced the distance between the ferrule axes from 12.7 mm to 7.5 mm. This allows the connector to fit in slots of the modular one. We note that the vertical option of a SC-duplex plug produced by Honda Tsushin Kogyo has 8.5 mm between the ferrules axes. The slots of this coupler are very similar to those of a modular jack, however, these two outlets are not mutually replaceable.

One more type of connectors, which applies the similar design approach, is a High Density SC Connector by the company 3M. This connector differs from a standard density by overall dimensions of the plug with reduced cross-section up to 6,0x7,2 mm compared with 7,4x9,0 mm for the prototype. The solution is most advantageous when a quad outlet is used for connection. This design provides the distance from one coupler center to the other one center of about 7 mm, i.e. this connector provides the density of ports nearly equal to those of electric circuits, however, it does not support the backward compatibility.

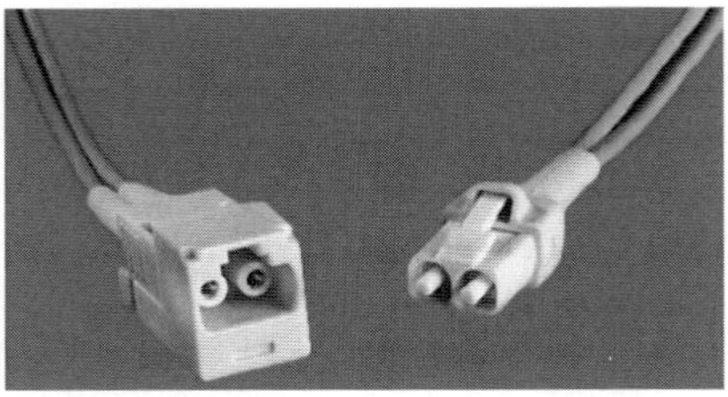

Fig. 4.31. Opti-Jack connector by Panduit

In 1996, Panduit offered a FJ (fiber jack) or Opti-Jack-type connector (Fig. 4.31). This product was designed for use in PAN-NET structured cabling system, and it is produced only in duplex variant. Like other connectors, the core of this connector is the ceramic ferrule with a diameter of 2.5 mm; but because of its more dense configuration and, in particular, the reduction of the distance between the axes of the ferrules to 6.4 mm (0.25 inches), the coupler dimensions are reduced to those of an electrical modular jack. A lever type latch fixes the plug in the coupler. A dome-shaped cap closes the latch-lever in order to improve the operational conditions. This design enables field assembly. For this purpose the original gluing technology utilizing two-component anaerobic glue was developed. Cleaning of dust particles from ferrule endface surfaces may be needed during day-to-date operation and is provided by the demountable design of the coupler: its components are fastened to each other by latches.

The FJ connector differs from other designs by its coupler, which is not a separate element, but is always combined with one of the plugs. The classical coupler for connectors of this type [89] was developed in 1998, and it is intended and used only for measuring.

At first, the FJ connector was manufactured only in the multimode variant in a beige housing. In 1998 a single-mode variant of the FJ connector was produced in a blue housing.

Group connectors

Examples of the third approach to connector design are the numerous designs of multichannel or group connectors [90]. The highest quality products of this group enable splicing of up to 18 fibers simultaneously, that is, their density is 9 times higher than that of electrical modular connectors. These products are quite frequently designed as a reduced or simplified variant of the "large" group connector, which was developed for use in telecommunications applications. The common distinctive feature of all the designs reviewed below is the push–pull principle of plugging into the coupler without using threaded or bayonet fasteners.

SCDC and SCQC connectors are promoted by a consortium, which includes Siecor, Siemens, and IBM. Their distinctive feature is the application of an external housing of a traditional simplex SC plug. This approach reduced the development time and gave partial unification with already-existing products. The new feature is use of a centering element, which is very similar to a usual ferrule and has two (SCDC) or four (SCQC) channels to fix spliced fibers.

The Mini-mass termination (mini-MT) connectors developed by Siecor and the MT-RJ connectors developed by a consortium of companies including AMP[5], Siecor, Hewlett Packard, USConec, and Fujikura also use the principle of partial unification. These products include an identical centering element with a nearly rectangular cross-section designed for two or four optical fibers. The difference between these designs is that the MT-RJ connector has a plug-fixing element, which is friendly for SCS users and similar to a lever-latch of an electrical modular plug. We especially note that the MT-RJ connector is one of the key elements of the Solarum optical fiber cabling system produced by AMP.

The Mini-MPO optical connector made by Berg Electronics has the highest capacity among the prospective types of optical connectors designed for SCS. This connector enables simultaneous splicing of up to 18 fibers.

Connectors without centering ferrules

The ferrule of an optical connector plug is an expensive precision detail, and the process of reinforcing fibers by the centering ferrule is a complex and long operation (see Section 2.3.3 of [7]). To eliminate these drawbacks, two kinds of ferrule-free connectors have been designed to center fibers during connection by other means. The common distinctive features of connectors of this group are:

- the fiber protrudes from the holder by some millimeters, the endface of a fiber is sheared and prepared for splicing by special installation hardware;
- the obligatory presence of a spring-loaded cover, which closes fibers in the nonoperating condition;
- the installation of a plug or coupler is possible only with use of a toolkit supplied by the manufacturer.

The Optoclip II connector manufactured by Huber+Suhner is designed on the basis of a widespread balanced circuit. It is based on the use of a single plug, which may be joined with another plug to obtain a duplex variant, if needed. The prior fiber alignment at their connection is performed with use of a cone-shaped guide, the final alignment is performed with use of a system of three balls positioned at an angle of 120° from each other, one of which can move in the vertical direction.

[5] The abbreviation is a registered trademark of the American corporation 3M.

Alternately, the VF-45 optical connector (sometimes also called VG-45) by 3M, is implemented by means of a V-shaped groove. This connector is designed for simultaneous reinforcing of two fibers of a ribbon cable by one plug. The end fibers are terminated in the coupler at an angle of 45° to ensure precise centering of optical fibers in guiding flutes and to obtain physical contact between spliced fibers in the mounted plug. This feature slightly reduces the common length of the spliced connector. An original feature of the plug is that, unlike the overwhelming majority of other designs, in the process of installation the protective cover is shifted sideways, instead of upwards.

The problem of cleaning the endface surface of spliced fibers was originally solved in VF-45 connectors (Fig. 4.32). This is a rather difficult problem for products without centering ferrules. A special washing device cleans the fibers since a great deal of the cleaning liquid passes through the connector coupler.

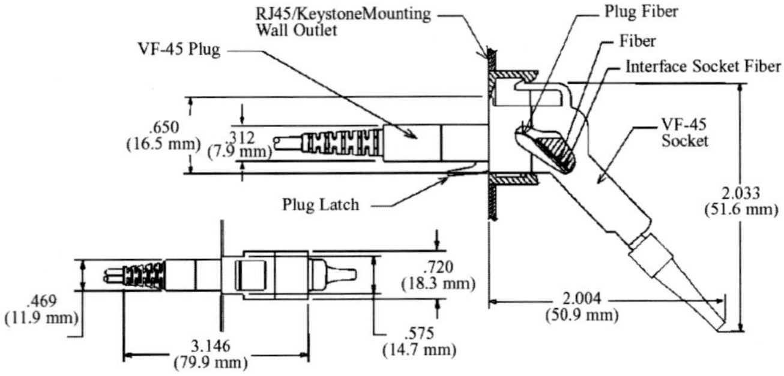

Fig. 4.32. VF-45 Plug&coupler

At first, neither of the examined connectors accurately aligned the fibers, which is necessary for single-mode applications; therefore, they were initially only used only in multimode designs. Single-mode applications have for these connectors appeared only in the late 1990s. To obtain the necessary level of return reflection, the endface surface of a fiber is canted at an angle 9° when the fiber is processed in a shearing device during the installation of a coupler.

It is also worth mentioning that the problem of color-coding is solved differently in these products. In the Optoclip II product line, the differently colored plastic housing is used for coding; in VF-45 differently colored shutter serves for coding of multimode and single-mode options.

Various types of optical connectors used by some manufacturers are listed in Table 4.14.

Table 4.14. Types of optical connectors produced by various companies

Company	SCS Type	LC	E-2000	MT-RJ	MTP	LX.5
ADC Telecommunications, USA	Enterprise		•			•
AMP, USA	NetConnect (Solarum)			•		
BTR Telecom, Germany	Opdat		•			
Corning, USA	Corning Cable Systems			•		
IBM, USA	ACS		•	•		
Lucent Technologies, USA	SYSTIMAX	•				
Molex, USA	Molex Premise Netwoks	•		•	•	
Ortronics, USA	GigaMo GigaMo+	•		•	•	
RiT Technologies, Israel	Smart			•		
Siemon, USA	Siemon Cabling System	•		•		

4.3 Telecommunications Patching Equipment

4.3.1 Design Features and Connection Variants

Optical patching equipment consists of a thin-walled plastic or metal housing with external fixing elements. The couplers of optical connectors are installed on the side or endface surfaces of this external fixing. Various organizers, fasteners, and other auxiliary elements are installed inside. Optical patching devices are intended for:

- connecting the fibers of various SCS segments to each other using patch cords;
- connecting the network equipment to SCS by equipment cords and adapters;
- permanently linking (splicing) fibers of various backbone or horizontal cables inside its housing.

The typical design elements of a patching device are:

- *Housing*: The housing includes external fastening elements, a demountable, fold-down, or movable cover (it may be translucent), or hinged doors for access during installation and maintenance. The housing must ensure access to the fibers and optical connectors and reliably protect the optical fibers against external mechanical exposures, extraneous particles, and dust inside the connectors.
- *Panel with optical coupler:* According to requirements of the working standards, the basic type of connector for SCS and, correspondingly, for patch panels, is an SC connector. As only twin patch cords are used for connections, the panel shall be equipped with duplex SC couplers. The apertures of uninstalled couplers must be covered with end caps for protection against the penetration

of extraneous particles inside the housing. The construction of the panel allows convenient connection of optical plugs to the couplers from both inside and outside. According to the working SCS standards, the couplers of the demountable optical connectors installed on the panels should be mounted so that their guiding grooves are pointed on one side. Both horizontal and vertical (one above another) installation of couplers in a pair is allowed.

- *Marking elements of optical ports:* The same marking elements as those used in various electric panels are primarily used. We note only two features that follow directly from the specificity of this equipment. The first marking method by insertions with icons (an exception to this rule is shelves by Siemon) was not widely applied. The second, glued labels and replaceable legends placed on the external surface of the panel, is widely applied due to the specific design of shelves and wall enclosures as well as the use of protective screens to cover couplers with installed plugs.
- *Fiber organizer:* The basic purpose of this element is storage of a technological stock of fiber length with a minimally allowable bend radius. Structurally, it can be made as a cylinder with tabs on its sides, a pallet with upward-turned edges, two-piece rings, and two-piece rings installed on a cross-shaped frame.
- *Permanent connector organizer*: This organizer is intended for fastening the pipes of protective sleeves for fused fiber connections or enclosures of mechanical splices. Optical fiber and permanent connector organizers are frequently combined. This combined element is called a splice-plate. The capacity of one splice-plate usually does not exceed 16 fibers with the appropriate number of connectors. If it is necessary to store many optical fibers, the splice-plates, as a rule, are installed in a row or, most often, one above the other.
- *Cable fastener:* This fastener is used for fastening a backbone or horizontal cable at the entrance of the enclosure. The fastener may be designed as clip, latch, collet, perforated slat, or tab. The solutions used to mount ties are sometimes applied in combination with other fastening methods.

The internal wiring scheme of termination hardware depends on the technology used for the installation of optical plugs on a cable. In practice, three basic wiring versions (Fig. 4.33) are applied.

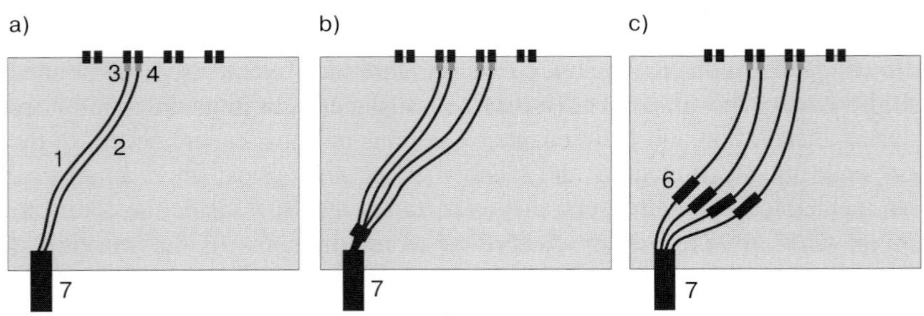

Fig. 4.33. Standard internal wiring of optical devices

In the primary version (Fig. 4.33a), an additional intermediate connection is not applied. The optical plugs 3 and 4 are installed on the optical fibers 1 and 2 of the cable 7. Then, these optical plugs are connected to the internal part of couplers. This version is typical for glued and mechanical plug installation [7], and for application of a preterminated assembly (Section 4.4.2.1). The same cabling is used when mini-pigtails are applied [7], however, an additional fiber fusion connection of optical fibers is included in the enclosure. A more complicated version (Fig. 4.33b) is typical for preterminated assemblies with a group connector 5. To wire optical fibers in couplers, a short combined cord is used. This cord remains inside the enclosure and is inaccessible to users. Finally, welding and mechanical splice installations (Fig. 4.33c) are widely applied in the creation of SCS optical fiber subsystems. The application of pigtails with lengths of about 120 cm is typical. The use of both technologies is necessarily accompanied by the introduction of an additional permanent connection in the optical signal transmission channel. The elements 6 that provide this connection (protective sleeves and housings of mechanical splices) are necessarily fixed inside the housing of the device. The considerable volume of these elements in comparison with the fiber element fixation often makes it necessary to fix them in organizers.

Depending on the design of a housing and the method of its fastening, one can define some optical connecting devices, each of which may have various designs. These devices are described below in more detail.

4.3.2 Racks

Racks are used when the number of optical ports in a cabling system exceeds several hundred. Their usual design is a supporting frame where the standard and additional equipment that is optimized for work with a large number of cables and patch cords is installed.

In the process of rack design, special attention was paid to increase the configuration density and to improve the simplicity and ease of work with patch cords. The known designs in this area are based on pull-out and fold-down cassettes; to reduce the dimensions, the cassette in its working position after installation in the holder may fold down at an angle of approximately 30°.

An example of equipment of this type is the NGF rack produced by ADC. Its basic element is a pull-out block with optical connectors of various types. Twelve blocks are installed in an FMDF frame in one or two vertical columns with a height of 2.1 m. The central panel located between the columns is used to arrange a redundant length of patch cords. The fibers are arranged on the panel by means of 15 round lugs located one above the other. The redundant length of cord makes a loop, which is placed on one of these lugs. This approach allows connection with fixed cords (lengths of 5.5 m). Instead of modules with optical connectors, installation of other passive optical components (for example, splitters and multipliers) into the rack is allowed, which essentially increases the rack functions.

4.3.3 Nineteen-inch Patching Equipment

Traditional shelves

Shelves are designed for installation in 19-in constructions (Fig. 4.34). The brackets are used for installation. Sometimes they are designed as a movable components providing smooth or discrete depth regulation during installation. The same effect is achieved when brackets of various lengths are applied.

The front panel where the couplers are mounted may be designed as an integrated component of the housing or as a demountable element fastened by swivel latches or screws. In this way, a shelf can be adapted for a certain situation since a particular optical connector can be chosen.

A modular design is used when couplers are installed on replaceable insertions that are part of similar elements of wall enclosures. This design provides a lower density of ports, however, it enables installation of various types of couplers in one shelf, which can be used for mechanical blocking of wrong connections.

Fig. 4.34. Nineteen-inch shelf

To increase the number of shelf ports without increasing the dimensions of the shelf and to simplify the identification of the shelf ports, sometimes pair grouping of couplers with a vertical arrangement of pairs, and zigzagged front panels with a pair of optical couplers on each platform (Fig. 4.35) are used.

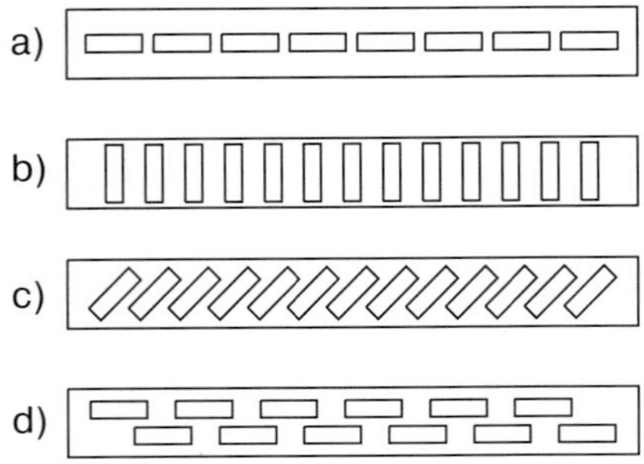

Fig. 4.35. Optical coupler placement on the front panel of shelves: *a)* horizontal; *b)* vertical; *c)* angled; *d)* in two rows with displacement

An additional advantage of the zigzagged panels is a smaller bend radius of a patch cord. Less frequently, the couplers are placed in two levels with a small horizontal displacement of the top line of couplers relative to the bottom one (Siemon, shelf FCP 2), or in one row, but at an angle of 45° to the lengthwise axis. An increase in shelf capacity is also achieved by increasing its height and by using a multilevel coupler arrangement.

RiT Technologies also applies the functionality of the PatchView system (Section 13.1.3.1) to optical constructions. For this purpose, each duplex SC coupler in the SMART F/O 96O shelf is supplied with an LED, the shelf is connected to the scanner, and Smart Jumper cords with an additional electrical conductor, which has a diameter of 26 AWG, are used for the transmission of identifying signals.

The cable usually enters the housing from the rear side, however, some shelves have a backbone cable entrance from the left, right, or back side of the housing. This feature allows for functional flexibility of the cables. The OR-625MMC series of shelves produced by Ortronics has two rectangular cable entrances in the cover of the housing.

Optical shelves manufactured in the early and mid-1990s had a direct cable entrance whose plane coincided with a plane of the side or (more often) rear wall of the housing. This design led to some problems, especially in the termination of outdoor cables, which arose from the difficulty in supporting the adjusted entrance in the limited space of 19-in constructions. To eliminate this disadvantage, designs that provide a cable entrance at a certain angle to the longitudinal axis of the shelf are widely used in up-to-date designs. Technically, it may be implemented as a lug, as a triangular groove on the rear side of the housing, or as an aperture in the rear wall behind which a cleat (Fig. 4.36) is installed at an angle of 45°. The first two solutions are preferable, because they provide excellent air-tightness of the shelf inner space.

a) b) c)

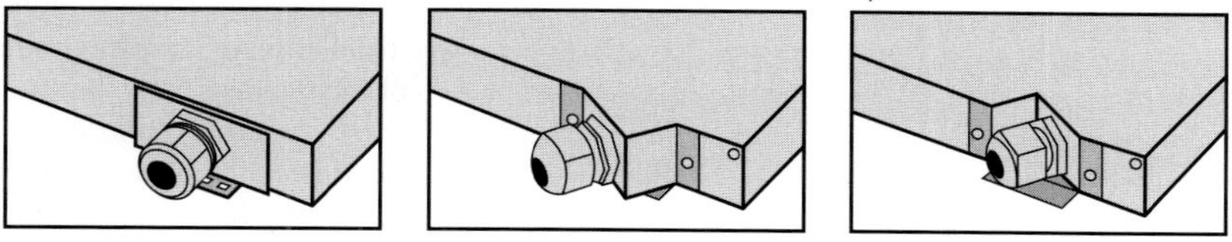

Fig. 4.36. Entrance placement in optical shelves: *a)* right; *b)* with outer angle; *c)* with inner angle

A design with one entrance used as a central aperture in the rear wall was widely distributed in the very late 1990s. In this design, the cable was arranged horizontally along the rear wall up to the aperture and fixed by hoops in several places, but only tubes of modules enter the shelf housing [91]. The cable and module entrance may be covered by a blind cover (for example, in shelves manufactured by Volokonno-Opticheskaya Technika, Russia). This solution provides reliable cable fastening by distributed fixing, but it is inconvenient for installation in housings with widths of 600 mm and, consequently, it is not widely used.

As a rule, for easy installation and maintenance, some types of shelves are supplied with rails, which allow the shelves to slide forward and sometimes to swing down approximately at 30°. Alternatively, the housing of a shelf may be installed on the central or side axis of the external enclosure with a hinge for possible access into the inner space. Use of this solution is strongly limited due to complications of both linear and angular relocation of the housing in cases where practically rigid outdoor cables are used.

A horizontal organizer is usually installed under or in front of a shelf for arrangement of excessive patch cords. In addition to the traditional design in the form of rings or a pallet, swivel organizers are widely used. These organizers are also made as shelves. Inside the housing of this element there are several ring lugs on which the extra length of patch cords can be wound. The rotation axis is on the side, as is the entrance aperture for cords to prevent the cables from pulling out. Swivel organizers allow the extra length of patch cords to be kept inside the housing, which essentially improves aesthetic characteristics and convenience of shelf maintenance. It is used to advantage in SCS with advanced optical subsystems, which demand the application of many cords. Brief information on the shelves of some SCS equipment manufacturers is given in Table 4.15.

Table 4.15. Optical shelves

Manufacturer	Type	Number and type of couplers	Comments
Lucent Technologies	600	12 or 24 ST or SC	Transparent top movable cover, replaceable front panel with hinged latches
Panduit	FRME	24 or 48 ST	Installation of couplers on replaceable 6-position front panels
Siemon	FCP-DWR-(X)	12 or 24 ST or SC	Transparent top cover, angled installation of couplers on replaceable insertions, built-in shelf organizer for cords
RiT Technologies	SMART F/O 96	96 SC	Shelf is equipped with PatchView system
	R3203050	12 ST	Angled coupler mounting

Other kinds of 19-in optical equipment

Some kinds of optical equipment that are not widely used in the construction of optical SCS subsystems but that are used for solving various specific problems are considered below.

So-called dismountable blocks of 19-in patch panels (3.3.2.3) are frequently used in electrical patch panels. Insertions with optical connector couplers and electrical modules can be inserted in slots on these panels. The optical connectors are used for the organization of workplace telecommunications outlets. This solution enables flexible configuration of the panel and creates panels that are supplied with optical couplers. Practically, however, this design is rarely used due to its inconvenient installation (an organizer for optical fibers and connectors is not used). To eliminate this drawback, Panduit has designed a fiber spool cylinder organizer, which is hung on the rear panel by means of a special bracket.

Alcatel uses another method to increase the functional flexibility of dismountable blocks. A universal front panel of the Omega series, type ACS-202.125, with slots to fit in a 6-port optical fiber module (Fig. 4.37) is offered.

The module has a front wall with optical couplers and a rear horizontally placed (when active) shelf with elements used to install a splice-plate and fastener. The entire assembly may be considered as an open-frame (without a cover) shelf. It is cheaper than other options, but it fails to protect cables and fibers from mechanical impact and dust particles.

The global fiber panel is free from this disadvantage. It is produced by the RW Data Cabling Sciences, and has the same structural layout. After installation of the modules with optical couplers in the front panel, individual protective covers protect the cable holder and splice-plate.

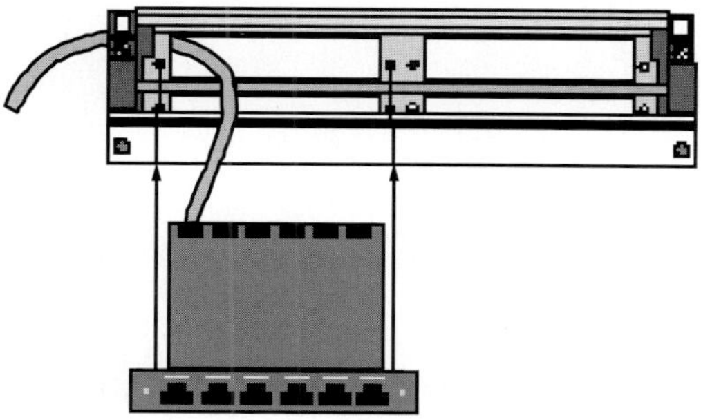

Fig. 4.37. Alcatel patch panel with optical module

Termination hardware can be implemented not only as a 19-in shelf, but also in in-line enclosures. Examples of this approach are products 404 866 and 404 867 (with maximum capacity of 24 and 48 connections, respectively) manufactured by Datwyler. Fibers splicing in such shelves is carried out via two-piece optical connectors. To install the couplers, the internal space of the shelf is split into two parts by a longitudinal wall.

Other in-line enclosure options for installation in 19-in constructions are implemented by the following design. Housings of wall enclosures LIU100 and LIU200 manufactured by Lucent Technologies can be oppositely installed on mounting plate 742A. The panels with two-piece connector couplers are fastened only in one of the housings.

The use of the configurations considered above instead of the more common 19-in shelves provides the following advantages:

- lower cost because of the decrease in the number of couplers and face panels and the absence of patch cords;

- decrease in the total losses in a channel from the reduction in two-piece connectors;
- some improvement in mass and dimensions.

4.3.4 Wall Enclosures

Wall enclosures are usually used as terminal and connecting hardware for cables with small fiber bundles (as a rule, no more than 24). They are most convenient in situations when the network and patching equipment is mounted without the use of closed constructions such as housings. A typical wall enclosure is shown in Fig. 4.38.

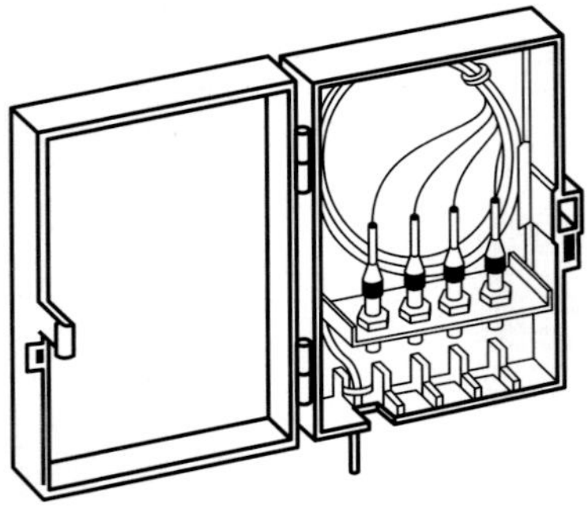

Fig. 4.38. Wall enclosure

Wall enclosures are also quite often used near the entrance facility to transition from an outdoor cable to an indoor one. In this situation, the enclosure may have no optical couplers and may function as a common in-line enclosure. Some manufacturers produce wall enclosures in metal protective housings with locked doors. This allows installing them in premises with free access.

Structurally, the enclosure is a thin-walled metal or plastic housing inside which organizers and cable-fastening elements are placed. Optical couplers are usually located in one or two rows on the right lateral surface of the housing (on the right hand of the expert who establishes the connection). When the number of couplers is between 4 and 6, they are sometimes angled down for protection against pollution. The openings to install the couplers are factory- or field-drilled in the housing. In the latter case, the use of a special template is recommended to obtain a good-looking enclosure. In the overwhelming majority of enclosures produced by foreign manufactures, replaceable insertions with six slots each are applied to fit in the couplers. Two grip fasteners fix the insertions. Sometimes these insertions are unified to make it possible to use them in modular shelves.

To provide mechanical protection of a patch cord connected to a coupler, widespread designs with internal walls to fit the couplers or with additional external folding protective screens are used. The protective screen is made of metal or transparent plastic and is frequently supplied with an internal lock. If it is necessary to install several enclosures one above the other without right or left displacement relative to each other, designs in which the left part of the housing contains a place for in-line cable transition to the upper enclosure may be selected. One of the disadvantages of high-density wall enclosures is the difficulty in accessing one of the sides of the coupler during connection and reconnection. To eliminate this drawback, Ortronics introduced a surface-mounted fiber cabinet with a fold-down mounting frame, which is designed similar to three-section wall 19-in cabinets. A horizontally swiveling central section ensures the access inside the cabinet. The kinds of optical wall enclosures of some SCS manufacturers are listed in Table 4.16.

Table 4.16. Optical wall enclosures

Manufacturer	Type	Number and type of couplers	Comments
Lucent Technologies	100A3 200A3	12×ST, 6×D-ST, 6×MIC 24×ST, 12×D-ST 12×MIC	Enclosures are equipped with 2- and 4-face panels with 6 and 3 slots for various couplers, unused slots can be covered with blind covers
Panduit	FWHE	24 or 48 ST	Enclosures have auxiliary protective wall and door to cover couplers
Siemon	SWIC-(XX)	12, 24, 48×ST and SC	Angled insertion modules installation with 1 or 2 pairs of optical couplers, couplers are covered by a hinged door
Molex	M.O.R.E.	12, 24×ST, FC and SC	Enclosures are equipped with two or four 6-port modular face panels with slots to fit in various couplers, the couplers are covered by a hinged fold-down door with a lock

4.3.5 Optical Modules

The 19-in construction in various designs is rather large. The efforts to eliminate this drawback have resulted in several new versions of installation equipment. New kinds of optical patching products were developed especially for this equipment. The optical module, which can be considered a compact shelf with a one-sided fastening, is shown in Fig. 4.39.

This kind of product is a part of the INFRA+ cabling system. It is designated for latched installation in various distributing frames. The module is intended for termination of a two-fiber cable. If it is necessary to increase the number of serviced optical ports, the modules are stacked one above another. Access to the module in this configuration is made after horizontal hinged folding of its housing.

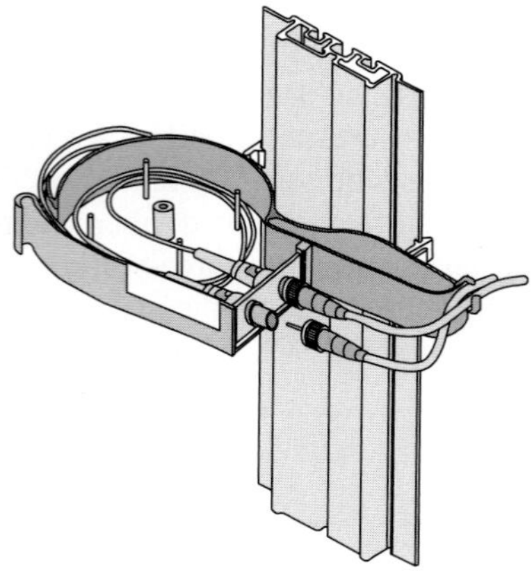

Fig. 4.39. Optical module

4.3.6 Optical Multiuser Couplers and Consolidation Points

Optical multiuser telecommunications outlets (MUTO) and consolidation points refer to a rather new type of the optical patching equipment, which appeared only in the late 1990s because of the rapid growth of open offices. Products of this type apply the fiber-to-the-desk concept to the open office business. In some cases, this kind of equipment is used in addition to multimedia outlets (Section 5.2.3). These outlets and consolidation points are made as closed housings, they have no fastening elements in 19-in construction, and they possess the required aesthetic appearance for applications inside the office. These products are intended for installation on columns, walls, and other parts of buildings. The maximum capacity of known examples of these outlets is limited to 12 ports, so the outlet is functionally analogous to a wall enclosure within the specific application area. Examples of 12-port multiuser outlets and consolidation points include AMP products 406818–406820 and 406771, respectively.

4.3.7. Telecommunications Outlets

Optical telecommunication outlets implement the functions of an SCS user interface element, and they are installed in workplaces. A horizontal cable is connected to the outlets for their connection with the floor distributor. A traditional telecommunications outlet looks like a simple insertion. It consists of a housing with one or several optical couplers mounted on it (Fig. 4.40).

In practice, there are outlet options for right and angled installation, as well as installation with lugs or grooves. Installation of outlets with downward-facing guide slots is preferable. During installation it is necessary to use wiring circuits such that the left coupler is marked with the symbol A. According to the normative documents applied from the middle of 1999, the basic kind of reviewed couplers are SC modules. In the described solutions, the use of telecommunications outlets with ST-connector couplers is allowed in SCS.

Fig. 4.40. Optical telecommunications outlets in the form of insertion blocks

The design of the telecommunications outlet housing and the method of its installation should provide an opportunity to obtain a bend radius of a horizontal cable not less than 30 mm. In all other respects, the construction of an optical telecommunications outlet does not differ from the design of its electrical analogue.

Multimedia outlets are widely used to install the optical ports of cable systems. Multimedia outlets are described in Section 5.2.3 in more detail. Their basic advantage is they allow the arbitrary combination of electrical and optical ports depending on the required configuration of a workplace. Their large dimensions easily solve the problem of obtaining the required fiber bend radius.

The majority of first-generation constructions of telecommunications outlets are made as insertions intended for installation in a raceway with a decorative panel. The optical couplers are mounted on this insertion. The common drawbacks of this design are the inconvenient field installation of connectors and the impossibility of using termination technology, which is based on the use of pigtails. In addition to multimedia outlets, two-level designs have recently become widely distributed. This design accommodates an organizer for extra fiber length on the first level, which usually consists of an installation box to be installed inside a raceway or a mounting box. The second upper level consists of an outlet with optical couplers. To increase the maintenance reliability, an additional guiding element, which controls the required fiber bend radius, is sometimes installed near the couplers.

Among the designs based on a mounting box that can be classified as "closed" solutions, there are also half-opened options (Fig. 4.41).

Their common feature is the presence of a cylinder-type optical fiber organizer on the rear side of the outlet. This cylinder may be placed perpendicular to the planes of the outlet on a plastic bracket (Panduit fiber spool series CFS2XX, Fig. 4.41a), or parallel to it (coupler type 2-966936-5 designed by AMR, Fig. 4.41b). The cylinder-type organizer has enough capacity to store up to 12 m of buffered fibers (0.9 mm), and up to 2 m of standard duplex cable for cords. The first version differs from the others by its universality. It can be used in patch panels along with outlets. The construction offered by AMR favorably differs from its analogue by its rather small dimensions.

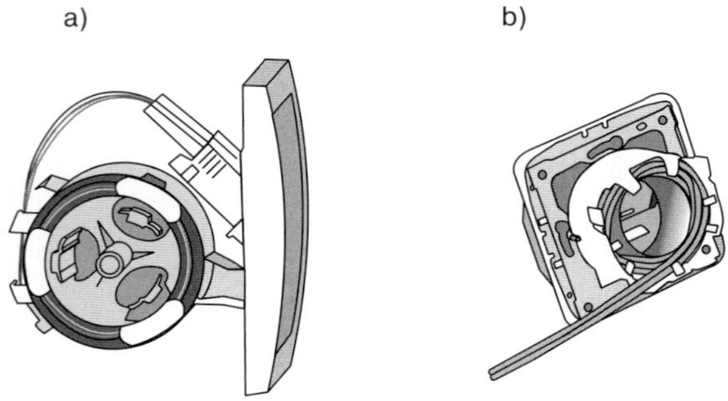

Fig. 4.41. Spool organizers for telecommunications outlets: *a)* with perpendicular spool orientation; *b)* with parallel spool orientation

4.4 Terminated Optical Fiber Cable Products

4.4.1 Patch Cords and Equipment Cords

Similarly to electrical patch cords, optical patch cords are designed for manual replacement of various SCS cable segments. The patch cord consists of a piece of cable for cords with optical plugs installed at its ends. In the overwhelming majority of cases, duplex cords are applied for the reasons of convenient utilization in SCS. A simplex cord is rarely applied in practice. Its main application is the connection of some kinds of measuring equipment.

The working editions of the SCS standards define an SC connector as the basic optical connector. The plugs of SC connectors should be installed on a cable for cords so that both ends of an optical cable terminate plugs with different marks: A–B or B–A (Fig. 4.42).

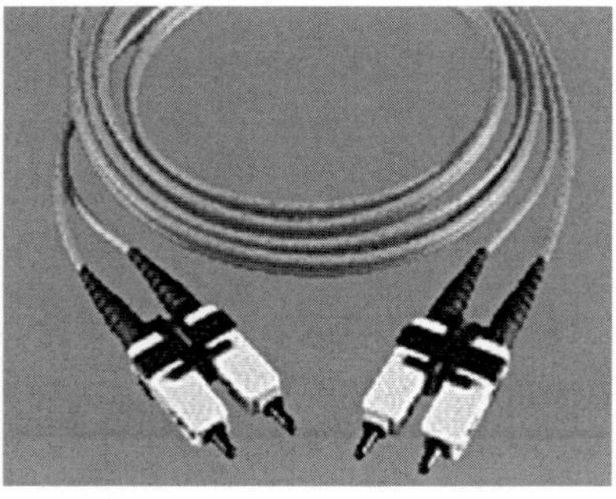

Fig. 4.42. Patch cord with SC plugs

If the patch cord is reinforced with ST plugs without any factory marks, the marking should be made by any other way according to the same rules. Colored polymeric boots are mainly used for marking. For example, ST plugs manufactured by Siemon may be completed with black or beige boots. More rarely, markings may be based on clamping nuts made of colored metal (for example, white and yellow for ST plugs manufactured by Brugg. When necessary, self-laminated labels or replaceable legends may be applied for marking.

The optical equipment cords are a part of the network equipment subsystem. They consist of a piece of a duplex cable for cords. Duplex SC plug for connection to the telecommunications outlet of a workplace or to the patching equipment in a telecommunications closet is installed on one end of this cord. The second end of the cord accommodates one duplex or two simplex optical plugs of the type required for connection to the network equipment. For example, for connection to the FDDI concentrator the equipment cord on the second end should accommodate MIC plug (Fig. 4.43).

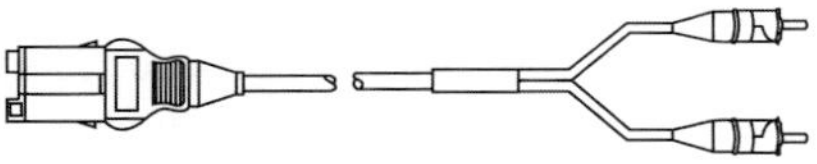

Fig. 4.43. Equipment cord ST-MIC

As a rule, standard lengths of commercially manufactured equipment and patch cords are 3, 5, 10 and 15 m; longer cords or cords with lengths differing from the standard ones are custom-ordered. Alphanumeric indexes are used to mark and label the cords. These indexes contain information on fiber type, plugs, and cable length as codes. The marking and coding methods are considered in detail in [7]. In most cases, the manufacturers pack the cords in transparent plastic packages. The label with the cord passport is supplied. This passport contains some basic parameters of the cord, in particular, actual attenuation and return loss coefficient (the last value is required for single-mode cords).

4.4.2 Preterminated Cable Products

Preterminated assemblies

Preterminated assemblies or cables allow the construction of cable lines without field welding and fiber termination, which saves cost and installation time. A preterminated assembly looks like a piece of multifiber optical cable that is factory-terminated with optical plugs; it is already checked and supplied with special termination hardware. The hardware usually consists of a sealed ferrule made of a metal or plastic tube to fit plugs with an extra fiber length inside. When a preterminated assembly is used, the work of installation is reduced to cable drawing, removal of a reinforcing ferrule, and connection of plugs to couplers of a shelf or wall enclosure.

The surface of a ferrule may be smooth or fluted. Using a ferrule with a fluted surface increases its flexibility, however, the flute lugs snag the protruded parts of cable channels. To solve this problem, a protective cover made from a thin elastic material may be put on the fluted ferrule. The front end of the ferrule is equipped with an eyelet or eyebolt for lacing. Despite its high cost, the last version is preferable because it prevents possible cable twisting during its lacing.

To simplify the installation in a shelf or in a wall enclosure, one may use a two-piece protective hardware. After lacing is completed, the head part is deleted and the back part remains on the cable. It is utilized as a reinforcing element to fit the clamp of a cable holder. Brugg uses this solution in a series of fiber quick assemblies.

It is a good idea to use a group connector for cable termination of a preterminated assembly, as it reduces the length and diameter of the protective ferrule. In this case a short combined cord is used for wiring of optical fibers on couplers of enclosures or shelves.

Preterminated assembly may be manufactured on-site from a common patch cord of the required length. Molex produces a special kit for this purpose. The kit includes a soft plastic ferrule, which fits optical ferrules and is equipped with a lacing ring and fastening elements of cable for cords. The assembly is usually delivered on a spool. When both cable ends are terminated, they are freely accessible by winding them together with the termination hardware in a section of the spool, which is separated by means of an internal wall. Some companies (Lucent Technologies, ADC, Fibertron, Molex) produce preterminated assemblies with one of the cable ends to be inserted in a shelf and terminated to couplers; they are known as preterminated shelves. The shelf may be placed at one or both ends of the cable. To prevent penetration of dust during lacing of the cable in the cable channel, Suhner delivers these products with the shelves in protective covers.

Preterminated assemblies are also very convenient for the organization of temporary connections. In this case, the assembly cable is short (up to 300 m). The cable is delivered spooled on a cylinder with a diameter of about 70 cm. The cylinder has adjusting legs and a handle for cable reel-out (the construction provided by Brugg).

Application of preterminated assemblies and shelves is advisable for routes up to 2 km; exact knowledge of the length is required. One or both cable ends may be terminated, as specified in the order. The preterminated assembly may also serve as the central element for special-purpose structured optical-fiber cabling systems. We note the FLine system by Kerpen as an example of this approach.

Repair cable insertions

Repair cable insertions are used for fast restoration of communications in the case of backbone cable damage. This product is a piece of a reeled cable with length of 100–300 m. The cable is equipped with enclosures on both ends. The enclosure may fit optical couplers or simply extra fiber for various insertion options. The coil is supplied with handles and an adjusting frame on rollers to allow for easy movement of the cable. The enclosures are packed inside the coil so that they do not protrude from its edge to allow for easy transportation. The connection of backbone cable fibers depends on the construction of the terminal enclosures and can be arranged with the help of mechanical splices or quick adapters.

4.4.3 Monolithic Distributing Panels

Monolithic distributing panels or minipanels are a new type of terminated cable products that were launched for commercial manufacturing in the very late 1990s. By idea and design, these panels are similar to preterminated shelves; however, they have an essential difference. The panel housing is monolithic and, since it is distinct from the shelf, the user and assembler have no access inside it during installation and operation. This monolithic design of the housing essentially reduces its dimensions. The plug of a group connector terminates the panel cable.

An optical pathway may be constructed according to two basic schemes. The first is called the "economical scheme" [92]. It is based on the application of two panels with cables of various lengths. Thus, the cable length of one of the panels is fully determined by the length of a certain route. In the "flexible scheme" the panel cables have equal and quite small length (about 1 m), and the usual preterminated assemblies are used for their connection. Hence, the economical and flexible schemes predominantly differ by the number of plug&coupler connectors in the pathway: one connector is used in the first case, and two connectors are used in the latter one.

4.5. Adapters

Optical adapters as well as equipment cords are used in the network equipment connection subsystem and are formally beyond the scope of the standards. Basically, these products provide the connection between the plugs of different types. Adapters are applied for replacement when the necessary equipment cords are missing. Usually adapter outlets and FM connectors are used for this purpose.

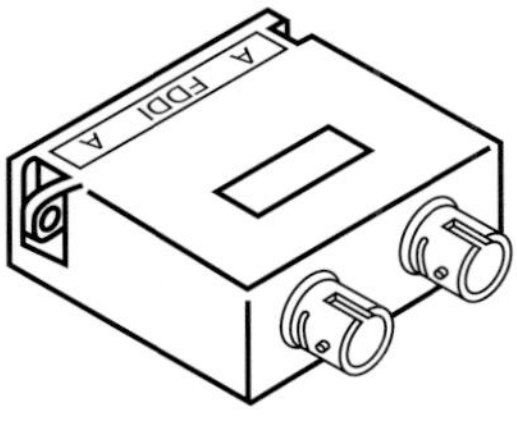

Fig. 4.44. MIC-2ST adapter

Adapters are used to connect plugs of two various types. This device consists of an in-line coupler with a common centering ferrule and couplers of two different types on both sides. Figure 4.44 shows an example of such an adapter connecting MIC and ST plugs.

The FM connector (female–male, Fig. 4.45) connects one type of plug to another type of coupler. They consist of a plug with the back part made as a coupler to connect the ferrule of the other plug type.

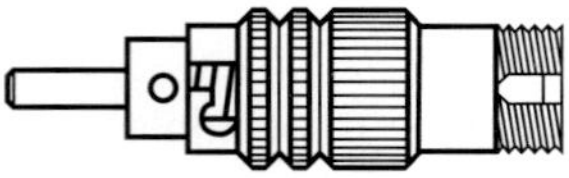

Fig. 4.45. FM connector

Some kinds of the network equipment have very high photodetector sensitivity, which results in its overloading in the short lines that are common for structured cabling systems. In this situation, optical attenuators are used to provide serviceability of the equipment. There are several constructions of these devices. It is convenient to use attenuator-outlets and FM adapters with built-in attenuators in SCS. These kinds of devices may be fixed or flexible with smooth adjustment of insertion attenuation. Because of its special functions, the FM adapter with attenuator may have a coupler and a plug of one type of connector. When there is an alternative for SCS pathway construction, it is preferable to use attenuators based on FM adapters.

4.6 In-line Enclosures

In-line enclosures are applied in campus backbone cabling subsystems to splice segments of outdoor cables because there are limitations on the length of outdoor cable constructive segments. Therefore, long external optical-fiber backbones consist of several cable segments (usually having lengths of 1–4 km) that are spliced in in-line enclosures. In-line enclosures are also used to protect splicing points of damaged or broken outdoor cables. Unlike connecting shelves and wall enclosures, in-line enclosures have two basic design features: they have no couplers for connection of outdoor cords, and they have sealing glands, fillings, cup-type seals, and other elements that provide hermetic sealing of the internal volume.

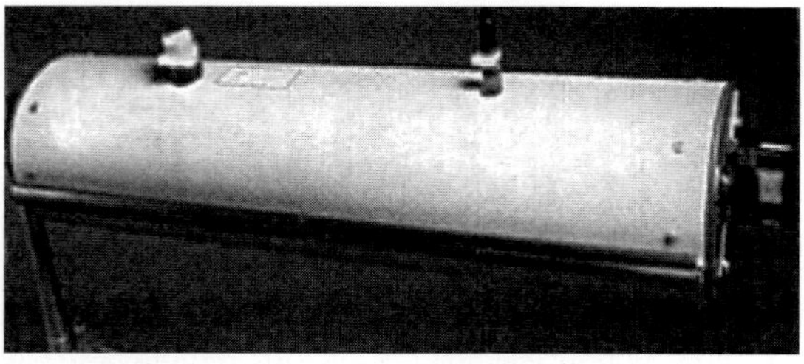

Fig. 4.46. In-line enclosure

In-line enclosures survive year-round maintenance under severe conditions, such as in collectors and manholes of the cable ductwork, buried in the ground, at the bottom of water reservoirs, and on aerial line posts. The design of all products examined in this section should provide sealing when frozen by ice.

An in-line enclosure consists of a cylindrical or parallelepiped-shaped housing with rounded edges. When the housing has a parallelepiped form, the mechanical strength of the housing is sometimes increased by bracing struts on its external surface. There are also disk-shaped housings, but they are very rarely used. The housing includes a tray with cassettes for the accommodation of optical fusion splices and elements that hermetically seal the internal volume and provide continuity of the armored and reinforcing elements of the cable.

The enclosure is hermetically sealed by cold or hot methods with the use of a filling mixture, thermal transfer pipes, fillings and cap-type seals, and special mastics and sealing stripes (for example, VM stripes produced by 3M). Sealing of a special inert gel mixture additionally supports the internal sealing of enclosures manufactured by Raychem. In some designs, because of the application of high-quality fillings and cap-type seals as well as special forms of pressured surfaces, it is possible to obtain the required sealing level by mechanical methods only (tightening by bolts or clamps), without the use of hermetic materials and the thermal transfer process. This essentially speeds up and simplifies the work, especially if it requires multiple assemblies and disassemblies.

Depending on the number and the arrangement of cable entrance facilities, the following kinds of enclosures can be classified as *direct* (entrances from different sides of the housing), *dead-end* (or cup-like, entrances from the one side of the housing), *in-line* (two entrances), and *splitting* (three or more entrances) enclosures.

During SCS construction, the splicing of separate lengths of cables is seldom performed because of the rather small extension of cable pathways. Therefore, only some SCS manufacturers, for example, Lucent Technologies and HellermannTyton, include in-line enclosures in their official catalogs.

4.7 Blolite System

The Blolite system was developed by BICC in the late 1980s and consists of a successful adaptation of a well-known cable engineering method of channel pneumopreparation to the SCS area. According to the concept of the Blolite system, optical cables are not routed along the channel, but rather along empty cable channels (Bloducts) where the optical fiber is blown by compressed air (Fig. 4.47). The basic difference between pneumopreparation and the traditional method is that the auxiliary parachute is not applied during fiber routing in the Blolite System.

Special optical fibers of three basic types are designed for use in the Blolite System: 9/125, 50/125, and 62.5/125. These fibers differ from the usual ones by the presence of an additional external housing made of an antistatic material. The special structure of the top surface of this material increases the efficiency of the optical fiber movement in the airflow. To increase the fiber flexibility, the external diameter of its secondary protective coating is reduced approximately to 500 μm. The optical fiber enables direct installation of optical plugs without application of welding or mechanical splices.

Cable channels have diameters of 5 or 8 mm (internal diameters of 3.5 and 6 mm). Up to seven channels can be placed in this overall jacket (Fig. 4.47a). Bloducts having diameters of 5 mm are used for lines with lengths not exceeding 500 m; channels having diameters of 8.5 mm enable the creation of lines with lengths up to 1000 m. The maximum length of vertical lines is 300 m. Bloducts may have common or special water-resistant covers, which enables their connection between buildings. The Blotwist cables can be used for routing inside a building with up to four buffered or unbuffered category 5 electrical modules and one cable channel tube (Fig. 4.47b). These modules may be designed in UTP (series C5U), STP (C5F), and S/STP (C5S) variants. To ease identification, their external jackets have various colors.

a) b)

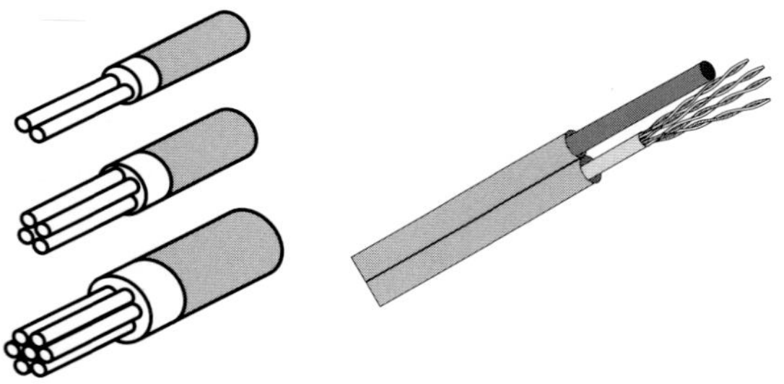

Fig. 4.47. Blolite system elements: *a)* Bloducts; *b)* Blotwist cable

Up to 4 optical fibers can be laid in one Bloduct (in the future, 8 or even 12 fibers may be laid); the routing speed is 40 m/min, and one line may have up to 300 bends with bend radius of approximately 25 mm. The additional routing of lines does not necessarily require a shutdown of the cabling system.

According to the Blolite concept, during the installation of a cabling system relatively low-cost cable channels are laid instead of optical cables. The number of optical fibers connecting various points is defined by the current needs. The replacement of one type of fiber by fiber of another type and the repair of damaged optical fibers is considerably simplified since the replaced fiber is simply blown out of the channel.

Some accessories and special equipment have been developed in order to provide development and maintenance of cable channel systems. This equipment includes splicing elements of cable channels with the same or different diameters, plugs, and in-line enclosures for indoor and outdoor installation. The fibers routed in the cable channel are carried out with the help of the BloCentre set, which includes an electric air compressor and an air drier. The air drier can be connected to a compressed air cylinder. The compressor and drier are packed in two suitcases for delivery.

After routing, the cable channels are checked using special test devices. To guarantee the absence of bends, pinches, and other defects of the same kind, a polymeric ball is blown in the channel. A special catcher installed at the end of the channel catches the ball. The absence of holes and ruptures is checked by measuring the constancy of the pressure created in the channel (maximum pressure: 10 bar).

According to the rating made by the developers of the system, use of this system reduces the time and funds necessary to realize a cabling system by 65%. Furthermore, the repair of damaged sections is approximately 40% less expensive than that of common lines.

4.8 Conclusions

Technical, mechanical, and operational characteristics of optical fiber products of various purposes fully comply with state-of-the-art requirements for structured cabling systems. This enables implementation of optical subsystems under specific project requirements without special restrictions. The level of standardization of optical fiber components and their production enable the use of standard approaches, technical solutions, and key rules of construction in the process of implementation of SCS optical and electrical subsystems.

Optical fiber cable products for SCS and PSTN are unified to a considerable degree. This provides an essential improvement over some actual standardized parameters of SCS cables (for example, attenuation) and decreases the cost of cable elements.

The use of secondary optical fiber protective coverings of various types during optical cable development allows adjusting the cable design to its application area. As a result, a basic type of optical cable is designed for each of the three SCS subsystems.

Today, multimode cables are generally used in SCS. The growth of single-mode solutions is mostly caused by the need to support super high-speed network interfaces, such as Gigabit Ethernet.

The majority of SCS multimode cables and patch cords use 62.5/125 fiber and, consequently, these fibers are recommended for the installation of optical subsystems. When it is necessary to transmit Gigabit Ethernet signals over large distances, it is preferable to use 300–500 m cables of a special design based on high-quality multimode optical fibers.

The basic connector used in SCS is the SC-type connector. In the near future, one or several types of connectors with increased density and form similar to that of a modular electrical connector will be added to the list of connectors allowed for SCS applications. This will considerably increase the standardization level of electrical and optical connecting hardware and simplify the integration of electrical and optical networks.

The platform of the patching equipment for SCS optical subsystems is 19-inch shelf. The presence of serial devices such as wall enclosures and racks enables the installation of equipment with respect to the technical requirements of a specific project. The specificity of construction of SCS optical networks lines and, in particular, the increase in popularity of "fiber-to-the-desk" solutions in open offices determines the requirements for some types of optical patching equipment (for example, the user's outlet and the consolidation point), which is absent in PSTN.

The presence of a range of terminated cable products that are more advanced than electrical subsystems ensures the quick construction of SCS optical communications lines. It also ensures communications recovery in an emergency without complex and expensive specialized technological hardware

5 Auxiliary Components

5.1 Installation Equipment

5.1.1 Nineteen-inch Racks and Cabinets

One critical problem that may occur during the design of a structured cable system (SCS) is the need for compact housing of equipment, which often includes network devices for different applications (most often, hubs, multiplexers, routers, and servers) in the constrained space of telecommunications closets and equipment rooms. At the same time, this housing must provide easy access during construction and operation. Nineteen-inch installation equipment is specially designed to satisfy these needs. The classification of 19-in equipment is given in Figure 5.1. Along with 19-in equipment, there are also a number of 17-, 23-, 24-in and metric (with 535-mm mounting width) options. Since 19-in equipment is unpopular in Russia, it is assumed that the use of 19-in equipment will be further reviewed. We note that 19-in equipment can be easily mounted into 23-inch cabinets by means of different height-adjustable adapters.

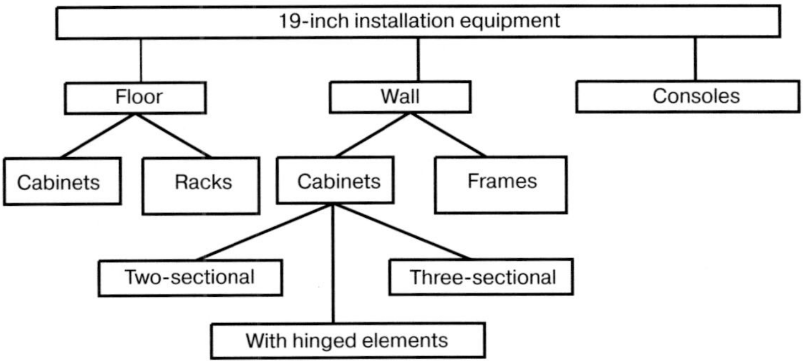

Fig. 5.1. General classification of 19-in equipment

The principal dimensions of 19-in racks and cabinets were initially certified by the American standard ANSI/EIA RS-310D. This document initially defined only the size and height of 19-in front panels as well as the position of the mounting holes. The key regulations were adopted in 1969 by the international standard ISO/IEC-297 with nearly no changes.

This type of equipment is produced by companies that specialize in metal working (Rittal, Wilsher & Quick, Shroff, and others). The equipment is sold to the system integrators for SCS mounting. At present, 19-in equipment is actively integrated in accessories delivered by SCS producers (Siemon, Ackermann, Mod-Tap, ITco., and others).

Dimensions of 19-in equipment

Mounting rails with holes for mounting different passive and active devices are mandatory elements for 19-in equipment. The space between internal and external guide edges, and also between their center points is certified [93, 94].

The working height of mounting equipment is measured as a standard unit (*U*). The unit defines a minimum height of any device suitable for mounting in 19-in assemblies and is equal to 1.75 in. (44.45 mm). For some solutions this parameter is abbreviated as RMS (rack mount space) and RU (rack unit). The German abbreviations HE (Hohen Einheit) and HU (Hohen Unit) appear in the German technical literature. We should also point out that some producers recommend using a negative allowance of around 0.8 mm when selecting the height of separate elements of hardware to fit in 19-in hardware, particularly front panels, to increase the ease of hardware mounting [95].

Two or three mounting rail holes are positioned per each height unit, and as a rule, are square-shaped. Thus, according to the international standard IEC-297-1, two holes per each height unit are certified. American-manufactured cabinets, which conform to the standard EIA RS-310-C, are supplied with additional holes that are positioned between the main holes on the same axis (Fig. 5.2, right).

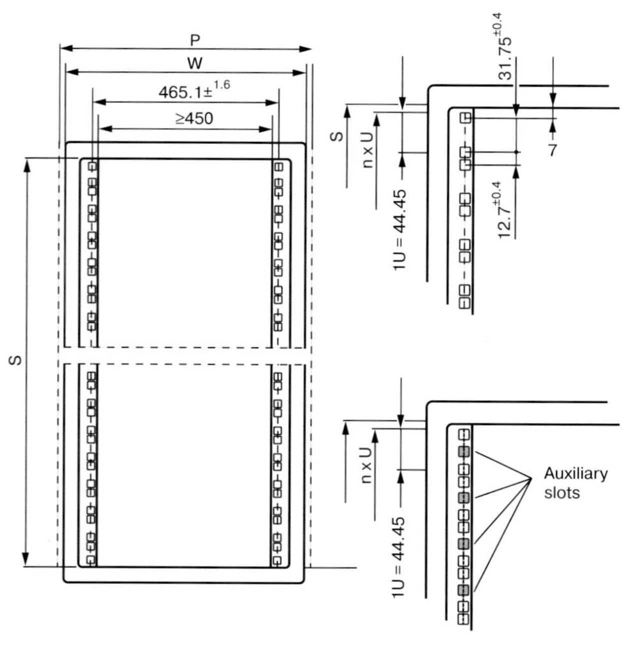

Fig. 5.2. Master size of 19-in rack

These normative documents also regulate the external dimensions of 19-in cabinets. The dimensions are presented in Tables 5.1–5.2. The parameter *S* in the Table 5.1 defines the minimal number of anchor locations in the cabinet. The height *H* includes legs. The cabinet width *P* is the maximum allowable.

Table 5.1. Nineteen-inch cabinet height

Rack height, [mm]	Unit number, S>n×U
800	13
1000	18
1200	22
1400	27
1600	31
1800	36
2000	40
2200	45

Table 5.2. Nineteen-inch cabinet width and depth

Rack width, W<P, [mm]	Rack depth, W<P, [mm]
550	400
600	600
700	800
800	900
900	

The most useful cabinet sizes in practice are 600×600 mm and 600×800 mm. The size 800×800 mm is rarely used.

Two versions of the mounting slots on the hardware are available, open and blind. Their dimensions are also certified as well (Fig. 5.3.).

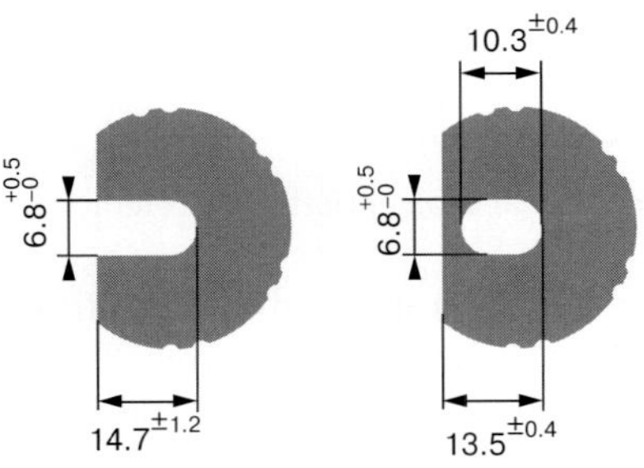

Fig. 5.3. Dimensions of mounting holes to fix hardware in 19-in rack

Auxiliary features of 19-in hardware

The fit-in hardware is mounted to the guide rails by fastening screws and nuts. The square nuts are equipped with external hinged tag assemblies by which they are anchored in the holes of the mounting rails. M5 and M6 screws with straight and Philips-slot heads, respectively, thread into the nut. The nuts are chromium-plated for appearance and are equipped with metallic or plastic washers of appropriate profile and surface finish.

Metallic parts of 19-in mounting hardware are preferably colored various tints of gray. Small adjustable components are sometimes chromium-plated. If required, the producer may deliver components in any other customized color from the (RAL Color System) catalog. However, this increases the end product price by 5–30% (depending on the delivered lot size and selected color). Rarely, various colored elements such as emblems and company logotypes are used to improve the aesthetic value of the product; this is done predominantly for finishing front doors.

Special attention is paid to the continuity of the painting materials for cabinet coloring to ensure efficient functioning of the telecommunications equipment grounding system. For this reason, the previously widely available varnishing and anodizing techniques for finishing of cabinet panels are rarely used any more.

Fig. 5.4. Design features of cabinet frames for traditional cabinet (left) and Comrack (right)

Mounting hardware of the Comrack series (Schroff) and hardware identical to it are widely used in European countries. This hardware differs from traditional cabinets by the presence of H-shaped top and bottom components, and also supporting profiles with slots (Fig. 5.4). The use of specially designed top and bottom profiles increases the number of side channels for cable routing. Three side-by-side mountable bases enable the formation of three 19-in assemblies with three pairs of rails and only two pairs of bases and tops (Fig. 5.5). The use of slots instead of traditional holes for square anchoring nuts enables fine height adjustment of mounting hardware. The nut is fixed in the slot by a leaf spring.

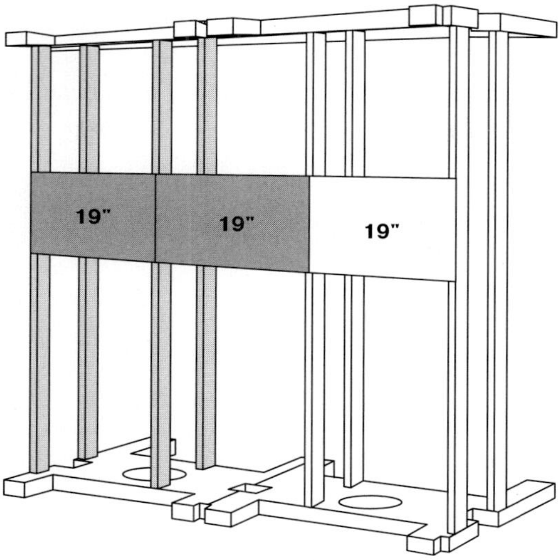

Fig. 5.5. Side-by-side installation of several Comracks

5.1.2 Mounting Cabinets

The mounting cabinet is a primary mounting element for installation of SCS equipment, and consists of closed 19-in hardware. The hardware base components are the frame and guide rails.

The frame is commonly made of steel and is used to fasten other elements and to make the housing rigid. The frames are assembled or welded. Welded frames provide more rigid cabinet housings than assembled ones, but cabinets with housing assemblies allow easier transportation of their components.

The frame houses external components of the housing and mounting rails of steel or aluminum profiles. Depending on the fastening technique, the rails may be adjustable or fixed. The adjustable rails enable control of their depth in the cabinet. Fasteners can be used for fine adjustment. If an adjustment with a discrete step is required, fixing bolts that fit in the holes of the horizontal plate are used. Frequently, the mounting rails must be relocated from standard positions when different kinds of add-on devices are to be connected to the connectors of the cabinet-mounted equipment.

In addition, the cabinets restrict unauthorized access to the cabinet equipment and effectively protect the installed cabinet equipment from dust, contaminants, foreign particles, and moisture according to the IP classification (Section 14.4). The cabinets also screen against external electromagnetic fields, and provide a pleasing appearance.

Cabinets can be floor or wall mounted.

Floor-mounted cabinets

The floor-mounted cabinets are installed on the floor (Fig. 5.6). The standard size of the cabinets is:

- Rail heights: 21–48*U* with discrete steps being defined by increments of *U*;
- Width: 600 or, rarely, 800 mm;
- Depth: 600 or 800 mm.

Floor cabinets include mounting rails, a frame with a base, a top cover, side panels, and front and rear doors.

Fig. 5.6. Floor-mounted cabinet with a glass front door

As a rule, floor-mounted cabinets are produced with four rails, which provide fastening of the installed equipment at four points. Some companies deliver customized cabinets only with one pair of rails on the front. The floor-mounted cabinets are very tall, which complicates the height adjustment of the equipment. To eliminate this drawback some producers mark the rails with height markings.

The mounting frame is usually equipped with an additional perforated plate in the middle, which provides rigidity and serves as a supporting or fixing element at cable routing and installation of auxiliary equipment. In some cabinets these plates are removable with mounting height adjustment.

The general drawback of traditional 19-in cabinets is that access from the side and rear to the mounted equipment inside is complicated in the case when several cabinets are mounted side-by-side. There are two approaches to eliminate this problem. The first is based on using a hinged front pair of mounting rails (an assembly with a hinged frame). The frame equipped with special swivels can be opened a full 180°. The second approach supposes an active mounting frame, which slides back and forth along telescopic rails. In cabinets for large installations (33*U* and higher) only one part of the19-in rails may be hinged or made as a pull-out block. The two parts of the cabinet with different types of rails may be separated by a built-in shelf, which is an integral part of the frame (the Intelligent series of cabinets by Elgadphon). The benefit of the construction with a hinged frame is its somewhat higher maximum installation capacity. The frame makes it possible to install 200 kg of equipment versus 50 kg of equipment with the pull-out frame.

The base, or base unit, has an opening in the middle, which is often closer to the rear side. This opening provides for airflow, and for power and communications cables from under the raised floor. Rittel produces cabinets with openings covered by removable, adjustable three-section panels, which enable selection of the required opening size for cable pass-through. Prospect Technologies (Russia) produces cabinet bases with cable lead-ins covered by compact covers, which are formed during manufacturing by use of the partial manufacturing method, and fixed to the base by metallic ribbons. On cable entry, the covers are broken out through ribbons. The number of the tapped covers is customized to allow entry of various sizes of cable bundles.

The cabinet bases of some manufacturers are made as frames that are covered by a removable plate. If it is necessary to make holes in the plate, their number and shape as well as the installation of dust-protective textile fabric filters are customized for a specific order.

Some types of cabinets have bases that are also equipped with side openings for cable access, which are sealed by covers when not in use. To provide for cooling efficiency, external vertical base panels may have holes, which can be sealed by filter inserts to reduce the dust content of the cabinet. For small bundles of cables, it is efficient to use brush guards on entry points to protect the connections from dust and other contaminants. The bottom rear of the Compaq cabinet houses an additional cutout for cable entry.

The base unit is installed on supporting legs with adjustable control levers. To provide for a level cabinet installation on an irregular floor, some types of legs allow for a vertical deflection of their supporting elements at an angle up to 35° due to the available inner swivel. The legs are equipped with an underlining made of rubber or nylon. The fixed legs are fastened to the base by screws or glue. The use of self-adjustable rolls with supporting elements in the form of single or double discs, or a hemispheric ball (Fig. 5.7) enable relocation of the cabinet within a building, which is sometimes is useful.

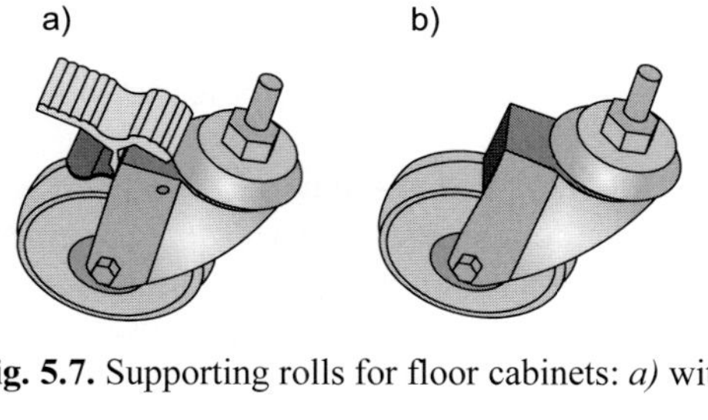

Fig. 5.7. Supporting rolls for floor cabinets: *a)* with catching mechanism; *b)* traditional

The drawback of disc-type legs is the accidental travel of the cabinet on site. To eliminate this, some companies produce legs equipped with catching mechanisms that are locked after the cabinet is moved [96].

The top cover is made of steel and often has oblique edges to somewhat increase the effective cabinet volume. It may be solid or have openings, cut-outs, or anchoring positions to fix different equipment and other similar elements. Cable entries on the top covers of floor cabinets are quite rare, as the existing office building architecture and practice of SCS arrangement make cable routing predominantly via the base unit frame.

The space between rails and sidewalls is often used for cable management. Usually manufacturers design their cabinet floors so that 82% of the floor cabinet space serves for equipment housing, cable entry, and other purposes [97]. The cabinet-mounted equipment can weigh up to 500 kg. The cabinet base is specially designed for mounting heavy equipment such as servers and power supplies, and tolerates loads up to 300 kg.

Normally, this type of cabinets are made for installation on the floor. Reichle & De-Massari has designed solutions for premises with raised floors. These cabinets differ by special height-adjustable elements, which are anchored on the main floor and enable leveling of the bottom of the cabinet to the surface of the raised floor. The type of cabinet allows cable entry from under the raised floor, and compared with similar products, is somewhat lower due to the lack of a visible base unit.

Doors and sidewalls. The sidewalls are manufactured of steel with corrosion-resistant paint and are removable, since they are fixed to the frame with screws, latches, and captive locking fasteners. In large cabinets the last two variants are applied together. At the same time, the locks of fasteners and the front door must accept the same key. The removable sidewalls provide convenient access to equipment and cables for installation and maintenance. To room large bundles of cables, the cabinet is sometimes equipped with sidewalls of 50-mm and 100-mm depth, which protrude from the base unit (Rittel QuickCabinet series of cabinets).

The use of hinge-mounted sidewalls in floor cabinets is quite rare. Cabinets of the Ultima Access Cabinet series by Wilsher & Quick have sidewalls made of two pieces that are hinged on a central supporting rod. This provides very convenient access to the side parts of cabinet-mounted equipment, vertical organizers and cords, and vertical power supply distributor. Cabinets of the TS8 series by Rittal have hinged sidewalls, but are continuous.

When several cabinets are mounted in a row, the adjacent sidewalls are removed to provide continuity of the cabinet inner spaces. To provide continuity and rigidity of such a construction, their frames contain supporting holes for bolt anchoring.

The cabinet door is usually opened a full 180°, and is sometimes equipped with a special captive locking fastener. The rear door is usually made of galvanized steel, but the front door can be manufactured of steel or shock-resistant glass.

The glass front door is a low-cost solution, and enables visual access to front panels when the cabinet is closed. There are two assemblies of glass doors: a full glass door with openings for hinges, lock, and handle, and a door with a metallic frame. The frame functions as a load-bearing support for the door and also fits the above-mentioned components.

The use of a steel door provides more reliable protection of the installed equipment and ensures shielding of the cabinet inner space against electromagnetic emission. For efficient shielding a metallic two-piece skirt is used to ensure reliable round galvanic continuity of the closed door and the cabinet housing.

For easy maintenance the door hinges are removable, or if required, can be easily and without any tools moved from one side of the cabinet to the other and spring-latched by one person. A less popular construction is available without a spring-latch, but with a stub element, which functions as the axis. The solution is not popular due to the risk of losing the stub.

The majority of cabinets have a door, which is designed as one right-angled panel. There are also two-piece and two-leaf front doors. Each piece of these doors is individually hung on a pair of hinges and can be opened independently of the other. They differ from each other by positioning of wings. Two-piece doors have wings, which are located one above the other and cover the entire cabinet width, but two-leaf doors are positioned side-by-side and have a half-width. It is advisable to use two-piece doors when access to the part of the equipment mounted in the cabinet (for example, the server) is very rare, as compared with access to other devices. Wilsher & Quick produce this type of cabinets with the top door made of glass and the bottom one of steel. The hinged two-piece doors do not require any construction changes to the cabinet. It is advisable to use them in space-constrained equipment rooms. Unlike the two-piece solution, the two-leaf door arrangement may be applied both for front and rear cabinet doors.

The front and rear doors have the same structure and are sometimes exchangeable. The rear door may be replaced by a removable cover, which is designed like those on the sidewall (Wilsher & Quick). The front and rear doors have the same height. There are also rear doors of less height, which are called compact doors [96]. If they are used, the rest of the space is sealed by a special panel. The panel can be equipped with a cut-out to ease the entry of large volume cable bundles into the cabinet.

There are also doors with tapered profiles used by Elgadphon (Israel) in their Intelligent series cabinets. It is considered that complicated manufacturing methods and cabinet designs are compensated in full by more convenient access to the side panels of the equipment and cable routing channels. In addition, according to many specialists' points of view, this type of door configuration contributes to the aesthetic appearance of the cabinet.

In order to prevent unauthorized access to the installed equipment, the sidewalls and doors may be equipped by a captive locking fastener, which requires the use of a common wrench or binding-post wrench, or other tool with a square-, triangular- or otherwise-formed working element. Often cylindrical secret locks are used. It is possible to use electromechanical locks with centralized controls or to unlock the door by means of chip and magnetic cards (products in the 7200 series by Rittal). Door handles may be of various shapes (Fig. 5.8) and are often supplied with locks.

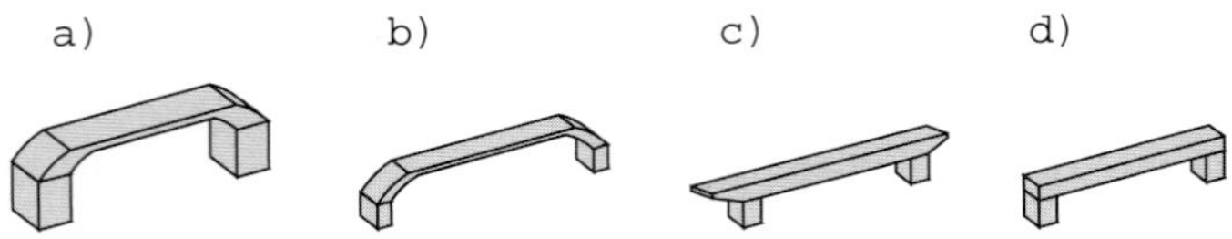

Fig. 5.8. Handles for 19-in cabinets

Temperature conditions for network equipment. When designing the cabinet, special attention is paid to the cooling of installed network equipment. Cooling may be provided by using active and passive solutions. Active or dynamic solutions suppose the use of axial or radial fans (Fig. 5.9). Passive or static solutions (Fig. 5.10) are based on the use of various openings to ease natural air circulation and heat venting due to convection. In brief, the solutions include the following:

perforation in the form of a system of vertical openings 2–3-cm long in the bottom and top of sidewalls and along the edges or bevels of the top cover as well. To control air circulation strength, the vent openings may be covered by louvers of a specific length;

- top covers with removable hatches, which may also be additionally perforated;
- top covers with one or two central square or rectangular louvers, whose plates may be raised some centimeters above the top;
- so-called server cabinets have frames of the front glass door and rear door that are equipped with a system of vent openings and slots;
- some server cabinets have a deep front door with two vertical vent openings in its sides.

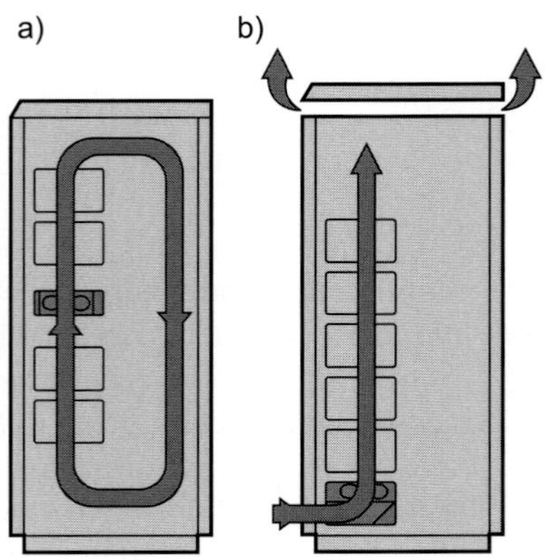

Fig. 5.9. Options for active cooling of network equipment in a 19-in cabinet: *a)* closed circuit; *b)* open circuit.

In practice, passive and active solutions are often mixed together. Thus, covers of some cabinets have openings for warm air outlets, along with locations to install fans, if required. Air circulation inside the cabinet may be increased by the use of so-called vent shelves (Section 5.1.4).

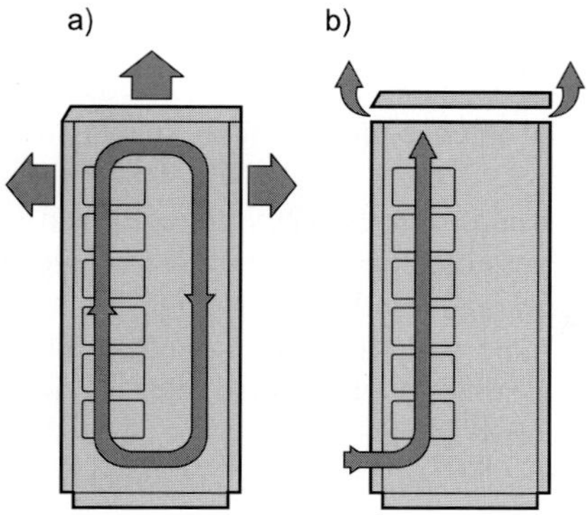

Fig. 5.10. Options for passive cooling of network equipment in a 19-in cabinet: *a)* in a hermetic cabinet; *b)* with vent openings

The active solutions to provide the required temperature conditions inside the cabinet also include an air conditioner installed on the hermetic cabinet housing (Fig. 5.11). The rear wall of the cabinet is the most typical installation position for the air conditioner. In some cases, a side installation of the air conditioner is also used (Rittal).

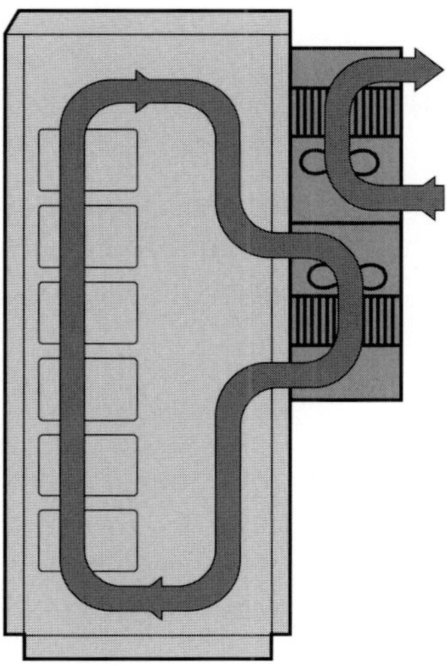

Fig. 5.11. Cooling of network equipment by means of an air conditioner

Special modifications of floor cabinets. The majority of products reviewed in this section refer to the so-called mixed or universal constructions, which are used for the mounting and maintenance of active and passive public equipment in telecommunications closets and equipment rooms. Additionally, there are a few specialized cabinets in the market that were initially aimed at a specific application and differ in design from the standard solutions.

The so-called *server cabinets* differ from standard cabinets by their reinforced base and 19-in rails. Since this type of cabinet houses a large number of devices with considerable power consumption, it is often equipped with special forced cooling systems.

Industrial cabinets are installed in industrial premises. They are equipped with a number of protective solutions against possible impacts (protective level not lower than IP43). As a rule, they are nonseparable, with steel welded frames and simple designs. The glass doors have a frame structure with sponge rubber or polyurethane foam sealing to ensure the hermetic level required. The air is supplied inside the cabinet through a dust-protective filter.

Office cabinets differ from their analogues by their superior appearance. As a rule, they are painted in light colors with glass-front doors without frames. Since it is necessary to transport these cabinets through narrow doorways, their frames are demountable. Aluminum alloys are used for the manufacture of separate details. The protective level of the cabinets is not higher than IP20, that is, it provides only contact protection. Custom colored 19-in rails, which are visible through the front door, are often important design elements.

External cabinets are used in small quantities in SCS when mounting external system equipment, and differ from their analogues for internal mounting mainly by different construction elements for inner space protection to environmental impacts. The elements include all-welded housings, double walls, screw fastening of removable panels, and use of inner conditioners.

Wall cabinets

Wall cabinets are intended for mounting on walls in telecommunications closets and equipment rooms (Fig. 5.12). While they use the same fastening techniques as floor cabinets, they are shorter and lack rear doors. These cabinets are divided into two-piece and three-piece units.

Fig. 5.12. One of the possible designs of wall cabinet

The primary elements of three-piece cabinets are a rotary section with mounting rails and a front door. The base and housing are made of steel. The base has a depth of about 10–15 cm and is wall-mounted. The rotary section is hinged to the base and may swing out horizontally to provide access to the rear part of the installed equipment. In the majority of wall cabinets, the rotary section is fastened by a continuous-type hinge, which gives the rotation direction. In cabinets of the XL series by Legrand, the fastening is hinged with possible installation on either side of the base in order to change rotation direction depending on local conditions. Like floor cabinets, the door is made of glass or steel.

The primary design elements of the two-piece cabinets are a housing and a front door. The cabinets are lower cost than the three-piece ones due to their simple design, however, they are less convenient to maintain due to complicated equipment mounting procedures and inconvenient access to their rear. These drawbacks may be eliminated by making the elements of the cabinet housing removable or by using rotary 19-in cabinet elements.

The first option is usually arranged by means of removable or hinged side panels. Rittal has developed an interesting solution in their QuickBox series of cabinets. In these cabinets the mounting frame is U-shaped, which after wall-mounting accommodates equipment. Once the equipment is mounted on the frame, it is covered by an external housing with a glass front door. The same design is used by Contour (Russia).

Rotating elements present two options. The first was suggested by Wilsher & Quick in their TF series of cabinets. Side panels of the housing and 19-in frame are hinged. The hinges are designed to allow the side panels to swing out horizontally and forward during equipment installation and maintenance. CNZ Neubauer has designed KISS series cabinets with L-shaped hinged frames and vertical orientation of 19-in rails. In addition to more convenient maintenance, the solution enables the maximum use of the inner cabinet space since LAN hubs are mounted in parallel to the door.

The vast majority of this type of cabinets is intended to accommodate all the equipment, which leads to a considerable increase in their dimensions (the cabinet's depth is defined by the depth of the largest equipment unit, since panels do not differ in depth). Legrand has designed the Almural series of cabinets, which use a combination of 137-mm housings for panels and 33207-type shelves for hub installation in order to decrease the dimensions of the cabinet.

There is an extended opening, which is positioned close to the rear wall and covered by a screwed protective cover, in the bottom and top panels of two-piece cabinet housings that provides air circulation and cable entry. In three-piece cabinets the same openings are located in the base. Cables enter the cabinet through grommets for improved protection of its inner space. Some of the DK series of two-piece cabinets by Rittal have side panels with viewing windows to control active equipment.

These cabinets are often used outside telecommunications closets and equipment rooms, therefore their doors are equipped with high-security locks. In three-piece cabinets, the latch that holds the hinged piece on the base is also equipped with a lock. Cabinets by AESP are designed with fastening of the hinged piece by a long bolt on the frame. The design provides superior protection against unauthorized access (only one lock is used).

To protect equipment from dust, the cabinet door is equipped with rubber or textile seals.

Wall cabinets are most often fastened directly on walls, for which four or five slots for anchoring screws are made on the cabinet rear wall. A paper template is sometimes included in the cabinet mounting kit for easy layout of slots. Fastening brackets can be used when it is necessary to route cables behind the rear wall.

Fans are used for forced air circulation in the inner space of the cabinet. It is preferable to use a single fan instead of a fan shelf to save space. In addition, it is quite rare that high power consumption active network equipment is installed in wall cabinets. In this situation, the fan shelf produces excessive air circulation; it is quite enough to install a single module. The latter is installed on the sidewall of the housing in space close to the 19-in rails. There are openings for anchoring bolts and air circulation (quite often together with low-density textile filters) on the wall.

The standard dimensions of wall cabinets are:

- rail height: 3–15*U* with discrete steps to be specified by the producer company in increments of *U*;
- width: 600 mm;
- depth: 250–450 mm.

The standard color for cabinets is gray. The company Mod Tap offers cabinets in green and beige, which are intended for installation of various SCS subsystems equipment.

We note that when designing SCS, often compact wall enclosures with less width than the 19-in cabinets are used. Their main application is mounting of consolidated and transition point equipment, various switches, sensors, and fire and burglar alarm system controllers. Such enclosures can be both wall mounted or mounted directly in ducts. They are usually produced of high-impact plastic and are often equipped with a keyed clear front door.

5.1.3 Alternative 19-in Installation Equipment

Racks

Open racks used as mounting elements (Fig. 5.13) are a low-cost alternative to mounting racks and are used when:

- there is need for ready access to equipment during operation;
- the required cooling efficiency fails cannot be achieved in a closed rack.

The open racks consist of one or two rows of guide rails, which can be equipped with bottom braces in the form of short plates or small angle-plates for additional strength. The base and rails are made of steel or aluminum alloys. In cases where two versions are available, steel racks are recommended for mounting heavy equipment (multiplexers, and multiport hubs).

Like floor racks, open racks with two pairs of rails are available preassembled or as separate components. In the latter version, it is possible to change the space between the front and rear pairs of rails in some solutions.

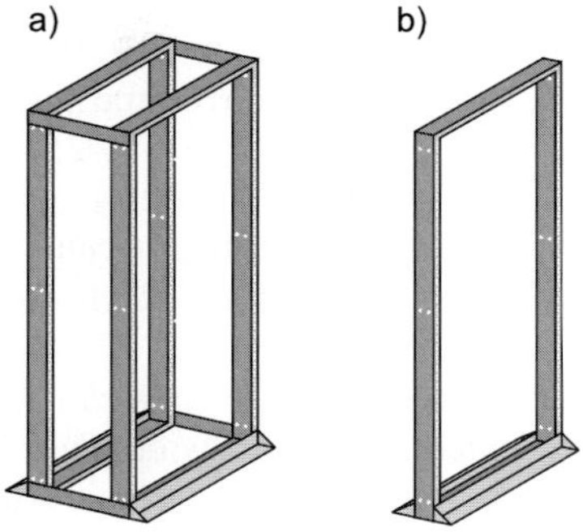

Fig. 5.13. Racks: *a)* two plane; *b)* one plane.

Most open racks are intended for floor mounting and have sufficiently broad bases to provide the required steadiness. It is also possible to mount racks on supporting rolls. To allow easy access to the rear and side equipment surfaces during mounting and maintenance, the front pair of rails can be hinged to swing out horizontally, similar the case for closed racks. This feature is very convenient when installing several racks side-by-side.

For steadiness producers recommend fastening racks to walls and/or floors through appropriate openings. In practice, this recommendation is sometimes not followed, or may only be followed for mounting of heavy equipment at a considerable height above the floor. There are also some racks without broad bases, which must be fixed by regular fastening elements.

Mounting frames

If open racks are the functional analogues of floor racks in a certain sense, then mounting frames (wall adapters or brackets) mainly perform the same functions as two-piece wall racks. They consist of a C-shape construction, which is mounted by its long panel to the wall (Fig. 5.14).

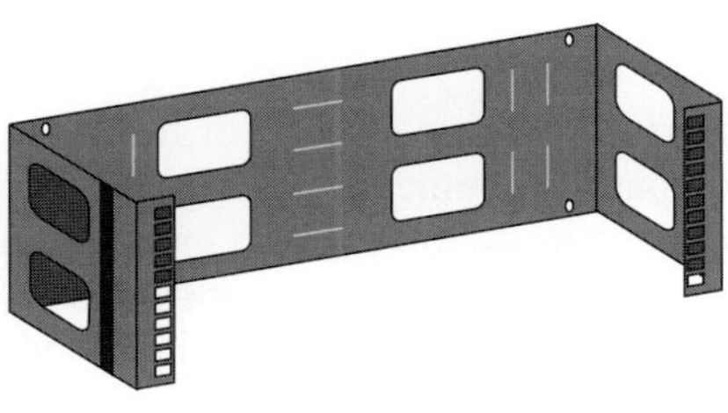

Fig. 5.14. Mounting frame

The height of the rack is usually not more than 10U, however the Mod-Top system has frames with heights of42U. They are used when it is necessary to install small amounts of equipment in 19-in hardware, but the use of open racks or cabinets is not possible or is not advisable for some reason. To ensure protection of the inner space of the frame, it may be used with a removable top cover and bottom plate.

The majority of producers offer brackets of the same depth (the most widely used is the 250-mm gauge). There are also options with two different values of this parameter. The sidewall may accommodate ring managers for patch cords, and thus has openings for anchoring screws.

To ease the mounting and maintenance of equipment that is installed on the frame, two main solutions are applied. The first, which is the most widely used in practice, is to install the left rail on a vertical continuous-type hinge. This enables the rail to horizontally swing out while the equipment is installed. This frame almost coincides with three-piece cabinets by its design and functionality. In the so-called bottom-hinged wall-mounted brackets by Ortronics, the working part of the 19-in rail is made as a separate V-type detail and is fixed to the base with two screws. The fastening method enables users to unscrew two upper screws and swing the modular panel forward, providing access to connectors on its rear during mounting and maintenance.

Some types of mounting frames include elements for the mounting of adapters that can then be mounted in a standard 19-in rack.

Wall brackets

The wall brackets are often used when a telecommunications closet or equipment room, because of its constrained space, cannot accommodate any other type of mounting equipment or its use is not advisable for some reason. The wall bracket consists of a rectangular structure, which is directly wall mounted and is equipped with accessories for fastening of various equipments. Because of the small depth of the frame, it fails to provide proper access to the rear of installed devices, and it does not allow mounting of most active network equipment types. These drawbacks essentially reduce the utility of such frames when building SCS.

Mounting consoles

The mounting console is a specialized piece of furniture, usually of modular design, that is designed to accommodate computer equipment. Consoles enable arrangement of the workplaces of the system administrator, the duty operator, and other specialists. They are oriented for use in control, operator, and similar work areas where personnel spend all or most of their time. They are also often used for mounting of equipment in telecommunications closets and equipment rooms, especially in large-area premises.

The key element of the mounting console is a bracket with fastening holes, which accommodates shelves, boxes, plates, and cabinets to house various computers or other equipment. Because the console is modular, its configuration is easily adapted to local conditions. Separate elements of consoles may be equipped with glass or metal doors and have 19-in mounting rails. The 19-in guides with positive or negative tilt angles are widely spaced to improve visual control of the installed equipment.

Some mounting consoles may be equipped with external frames, which accommodate wall panels and glass doors. At the same time, in the rack option it is possible to install separate parts of a console (most often, racks to accommodate servers), as well as the whole console.

Flexible equipment racks

A flexible equipment racks is usually made as an open frame with one pair of 19-in guides, which is installed on rolls. The racks are usually made to accommodate testing diagnostic equipment that is used for measurement. They are often equipped with integral shelves for additional frame rigidity. Fastening corners with adjustable openings are useful when it is necessary to change the mounting angle of a shelf. Traditionally, these racks were widely used for telephone exchanges. They have not yet become popular in SCS engineering and are used only in backbone networks. Two examples are the RiLab by Rittal and the Metramobil rack series by Knurr.

5.1.4 Equipment and Accessories for 19-in Racks

There are many auxiliary devices to improve the effectiveness and maintenance of 19-in hardware. For review purposes, they can be divided in several groups.

Shelves, winged pallets, and fastening L-bars

Shelves, winged pallets and fastening L-bars function as mounting elements for equipment that is not equipped with regular in 19-in hardware fastening elements (some types of multiplexers, monitors, printers, computer, and server system blocks). Shelves are mostly widely used. The maximum equipment weight for shelf mounting is 50–100 kg. The mounting surface of the shelf may be flat or may have openings. The latter design reduces the weight and improves the cooling conditions of active devices. The shelves can be fixed or extendible. The extendible design is provided by telescoping bars and has a considerably lower weight-carrying capacity than does a fixed shelf. However, they are very convenient for the installation of such equipment as keyboards and printers.

Some types of extendible shelves are equipped with auxiliary side-folding/pull-forward panels for mouse pads (Rittal type PC4610/4620 and PC4000/4621) or two-story racks (type PSS by W7Q). Extendible shelves are also available with integrated keyboards, which minimize the height required for keyboard housing by from 1 to 2*U*. The two-story structures are designed for housing printers. The top of the product is used to accommodate the printer, the bottom for paper. The printer may be placed parallel or perpendicular to the closed door of the rack.

Shelves for open racks may protrude forward relative to the mounting rails (the space is not limited by the door) or can be equipped with removable auxiliary plates, which extend the shelf length, if required. This design is sometimes called a shelf with double fastening or a double shelf, as traditional shelves for racks have been called shelves with edged fastening or single shelves (Fig. 5.15). The mounting depth of a fixed shelf is adjustable by a number of openings on its sides.

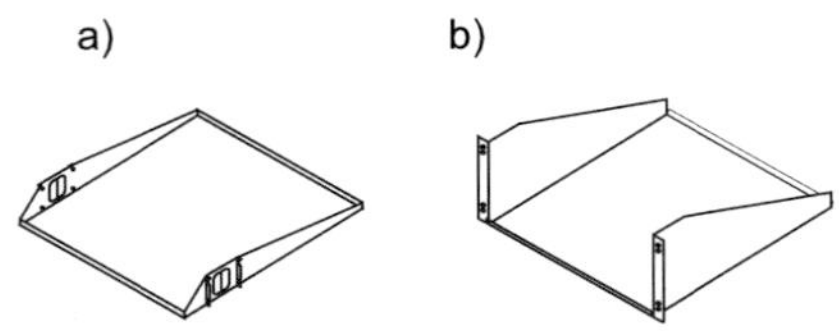

Fig. 5.15. Shelves for 19-in racks: *a)* with center fastening; *b)* with edge fastening

Desk-type shelves are equipped with an axis swivel and a fixing brace. In the off-mode position, when the brace is removed they swing out vertically on the swivel. The second option is brace-free with two supports on the edges. In this case, the shelf turns upwards in the off-mode. Besides mounting rails, desk-type shelves may be also housed on the rear of the rack door. Desk-type shelves are not widely distributed in SCS practice.

Mounting L-bars are made of steel and are quite often chromium-plated for superior aesthetic appearance. They are smaller, however, and are not as convenient for installation of equipment of a certain width. They are anchored at one or two points in specific slots of different shapes.

Other mounting structures, such as winged pallets, which are located in the bottom of the rack, are used for mounting heavy equipment such as servers and power supplies. They have a maximum load-carrying capacity of 300 kg.

Power-supply distributors

Power-supply distributors or outlet socket blocks are used to connect active and auxiliary rack-mountable equipment (vent shelves and lights) to one power input circuit. They consist of outlet socket blocks with fastening accessories for mounting on the rack frame (usually by screws). As a rule, they are equipped with a master switch with an optical voltage indicator and automatic circuit-breaker, as well as an auxiliary network filter and varistor protection system against high burst noises (reaction times of several tens of nanoseconds). The efficiency of noise suppression by the filter depends on the frequency and may be up to 70 dB or higher in some models.

The vast majority of distributors consist of one type of outlet modules. Other solutions with various modules are very rare in practice (for example, products by the company ZPAS [96]). The length of distributor's three-wire cable does not usually exceed 2–2.5 m. The selection of a plug type is usually customized.

The distributors are divided into vertical and horizontal versions. The vertical option is a high-density device (versions with up to 10 outlet modules are known to the authors) and is fastened vertically on one of the rear rails. They are convenient to use in high density racks, where active equipment is installed with power on the rear of its housing. The vertical distributor, which is intended for open racks, possesses superior design and geometry and is located from the outer side of the rack rail.

Horizontal distributors, sometimes called outlet panels, are usually installed on the front rails of racks and cabinets; however, they can be also fixed on the rear pair of rails. They are specially designed to fit in small-density hardware for the power supply of small amounts of active equipment, if required. Horizontal distributors include not more than five outlet connection modules and use 1*U* of height when assembled.

In most horizontal distributors, the outlet connection modules are placed on the front panel. In the product MCCP5919 by Hubbell, only the master switch is on the front panel. This solution combines the benefits of the horizontal and vertical options, in particular, the convenient power cord plug-off as in the horizontal distributor, and an improved appearance as in vertical one (power cables are not visible through the glass door).

To provide an efficient work of network equipment, the power cables of distributors should not be routed closer than 10 cm from telecommunications cables. They can fit in a separate organizer, if necessary.

Grounding equipment

The existing operating rules for electrical equipment and metal hardware for different applications requires mandatory grounding of all elements made of conducting materials. The grounding system protects working personnel from electrical shock and increases the efficiency of the rack inner space shielding. The grounding equipment is provided as a kit for 19-in hardware and consists of the following components:

- *inner grounding bar*, usually in form of a copper plate with cross-sectional area of 0.5 cm^2. The grounding bar has slots with M5 threads for mounting of inner grounding distributors and is fastened on the inner side of the rack sidewall horizontally or vertically;
- *inner grounding distributors*, which consist of thick isolated or bare copper round or flat leads with screw terminals on the ends. All basic components of the rack (base with a frame, sidewalls, front and rear doors, and top cover) as well as the housings of mounted equipment are securely connected to the inner grounding bar. When an unassembled rack is delivered, the inner grounding distributors must be installed directly after assembly of the housing.

The distributors may be components of some rack elements of unassembled racks, or may be delivered as a set of separate leads. In some cases the rack base may serve as an inner grounding bar. The grounding kit also includes bolts to securely join separate elements and to connect equipment grounding leads and regular building grounding rings.

We note that the auxiliary accessories of some racks include self-adhesive indicator tapes with metal-plating. The tape adds to the auxiliary balancing of potentials and increases the efficiency of the rack inner space shielding.

Patch cord organizers, jumper and cable managers

Constructions for closed racks. Patch cord organizers or cable management panels are intended for neat management of patch and other cords at their termination to the equipment mounted in a cabinet or open rack (Fig. 5.16). They are quite often used as elements for neat management of cables that are housed in 19-in hardware. They usually consist of a plate with several split ring hangers or nearly closed loops (usually single loops, rarely double loops) to hide excess patch cord lengths (Fig. 5.16a, ring-type manager). The hangers may be made of plastic, and in this case, they often have round-shaped cross sections. The metal versions are made of flat sheet metal.

Besides the hangers, smaller auxiliary plates, which are located along the plate to manage the lateral entry of cords, are also sometimes used (Fig. 5.16c). Large fixing elements in these combined panels are sometimes called hoops, small ones are referred to as clips.

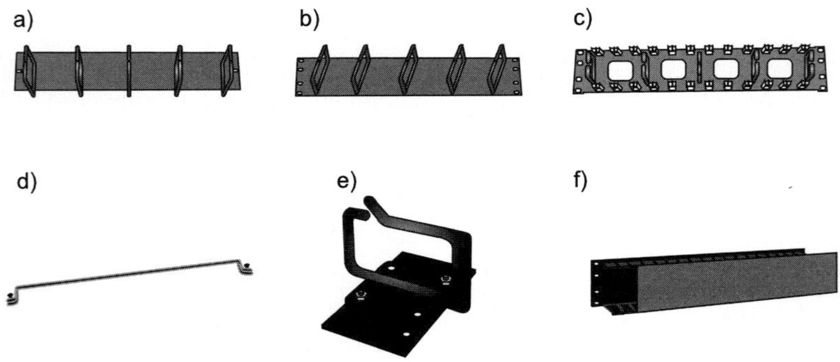

Fig. 5.16. Patch cord managers: *a)* classic; *b)* tilted; *c)* with hoops and clips; *d)* cable holder; *e)* single hoop; *f)* slot-type

The main fastening elements in the majority of models with installed organizers are vertical. In Siemon organizers WM-144-5-A and WM-145-5A and Rittal organizer 7255.035, these elements are tilted at 45° (Fig. 5.16b).

There are also organizers with vertical hoops in the central part and with single or double hoops, which are positioned along the edges and oriented horizontally. The latter hoops serve for vertical management of patch cords.

Panduit has panels of the CPP series with organizers in form of a comb with spring tabs to interlace patch cables. This solution differs from solutions with rings and clips by allowing more precise anchoring of the cable in a desired point.

A so-called cable management support bar by Ortronics is designed as a U-bracket, which is fastened on the faceplate of a modular panel (Fig. 5.16d). This element is used for frequent patching of cords, whose cables are fixed on the bracket by plastic hoops. The product is not as convenient for patching procedures, but it is light and small. A supporting organizer is also available from the same company; it differs by available clips for fixing patch cables. This organizer enables the use of fewer plastic hoops. A supporting organizer by AMP is called a cable support bar. It differs from its analogue, which is reviewed here, by the fact that it may be delivered as a 2-in (product 406042-1) or 5-in (product 557548-1) option.

The supporting organizer by Ortronics is placed at about half the height of the panel and is made as a wire bracket. However, the similar SpaceMaker bracket by Leviton has a flat structure and is placed at the bottom of a shelf or panel in the ready-to-work position. The latter option allows users to secure cable bundles by hook and loop straps.

In general, supporting organizers save useful space in 19-in racks (mounting heights maximum are reduced by half). The principal drawback is inconvenience when frequent patching of cords is required. Most organizers have a permanent structure. The Bud Industries products CM-2225 and CM-2227 have a modular design. They are designed with plate-like bases with heights of 1 or 2*U* and slots to fit in split rings, if required.

All the above-reviewed products provide for efficient operation of cords. However, they do not make it possible to arrange a neat connection field (and avoid the "cable noodle" effect). The so-called slot-type organizers are free from this drawback (Fig. 5.16f). They are designed as U-shaped details. The sidewalls have slots for lacing patch cables. A cover, which snaps on after the cables have been arranged on the base, provides the required neat appearance.

Fiber-optic SCS subsystems widely use organizers in the form of shelves to hide excess lengths of patch and cross-connect cords. The functional analogue of these solutions for electrical systems is the 4710-501 Telesafe shelf. Unlike optical shelves, this shelf is free from inner organizers. The active part protrudes forward along telescoping guide rails, but does not swing out horizontally on hinges. The top cover of the housing is equipped with a system of short parallel slots for patch cable pass-through.

Horizontal and vertical versions of organizers are widely used. Most of the horizontal organizers are 1*U* or 2*U* high. They are usually mounted between modular panels. Vertical organizers installed next to 110 panels have the same height as cross-connect towers. Very tall vertical organizers (up to 40*U*) that can be installed on mounting rails or rack hinges instead of doors are available. They can rotate for easy routing of cables.

The general drawback of vertical organizers is that they take a lot of space in space-constrained racks. Organizers in the form of single (or, more rarely, double) cable rings can be mounted on a base in order to eliminate this drawback. The cable rings are fixed on mounting rails by adhesive tape or by bolts in any desired point (Fig. 5.16e).

Almost the same hardware as for electrical cable management is used for fiber-optic cords. However, optical cords are sensitive to bends. Therefore, in order to maintain the required bend radius, sometimes single organizers in the form of 90° tubes with several parallel grooves on the top surfaces are used for patch cord management. The groove radius is selected by taking into consideration the allowable bend radius of a patch cable (Table. 4.9).

Constructions for open racks. The same organizers are used for open racks. We note only two differences. The first refers to vertical options of these products. Thus, an organizer in form of a long narrow pallet has become widely popular. It has a U-shaped cross-sectional shape, and in the vertical position is mounted with bolts between two racks. Bars, U-shaped brackets, and various L-shaped gauge brackets function as cable guides and are fixed by screws or latches on mounting rails. These elements are usually mounted on the front or on the rear side of the mounting rails (for example, Panduit organizer D31). The organizer may be mounted on a single rack, if required. A cover snaps over some models for aesthetic purposes. Intercabinet organizers in the form of 100–150-mm high U-elements that are covered by removable covers are used less frequently.

The second difference is that it is possible to use a so-called upper organizer, which is installed on upper part of rack and provides neat cable management. Such organizers are used if stationary cable trays in telecommunications closets and equipment rooms are not available.

Organizer rings, and wire distribution spools with heads that prevent the cable from sliding off are used to organize cross-connect wire for 66 panels. The organizer rings are equipped with latches and are installed on mounting frames. The wire distribution spools are fixed on the wall by a captive wood screw. They are mounted in 19-in racks by captive machine screws.

Cable managers. Cable managers are of almost the same design as patch cord organizers. There are only two principal distinctions to note. The first is that these elements are poorly finished, since they are hidden behind rack sidewalls after mounting. There is a design that consists of flat or perforated U-brackets, but this option is very rare for patch cord organizers. The brackets are installed on the rear of the modular panel to retain single cables or their bundles using different fastening techniques, predominantly by means of plastic hoops and loops of different designs. Such brackets may be integral or removable elements of a panel. Loops are sometimes delivered as separate components, and in this case, are most often installed on the perforated transverse bars of lateral rack frames. Sometimes cable managers incorporate bottom pallets to restrict cable bend radii.

Shielded panels regularly include rear brackets that serve as cable managers and grounding bars for shields of cables that are connected to the panel. Cables are fixed to the bracket individually for maximum transfer impedance and complete shield coverage. Collars and fasteners of different designs are used for the procedure. In order to make it easier to position cables, the brackets are sometimes equipped with protruded tabs; T-tabs prevent the fastening collar from sliding off.

The functional analogue to the patch cord organizer in form of single hangers is the so-called cable bracket, which looks like an L-plate with anchoring openings on the short side. The long side of the bracket has openings to fix plastic hoops.

Some types of racks may be used for mounting ladder-type cable managers. These racks include a number of individual perforated cross-bars, which are mounted in the rear of the rack and are displaced by height and width with respect to each other. When assembled they look like stairs. The cross-bars are perforated to anchor single cables or bundles with plastic hoops or sticky bands. This design uses a great deal of space inside the rack, however, it provides more convenient cable routing to panels from any direction and without dramatic bends. This design also gives easy routing of auxiliary cables.

With sufficient inner rack space, one can use spiral cable managers to route low volumes of cables or cables that are sensitive to small-radius bends to panels. They consist of a spiral made from a tube of a hard material for the accommodation of cables. Since it is very rigid, it maintains a minor bend radius.

Fan systems

The fan system is mandatory for effective cooling of active equipment (Fig. 5.17). There are single, integral, and group fans. The first group is mounted on lateral or upper panels, the second type is designed as a vent shelf that incorporates, at maximum, nine individual vent modules (the most useful designs accommodate from one to four fans). The shelf is 1–2*U* high and is secured on 19-in rails. The majority of shelves have a slot in the top covers of the housings for air release. There are also designs with air release slots in the lateral panels of the housing.

a) b)

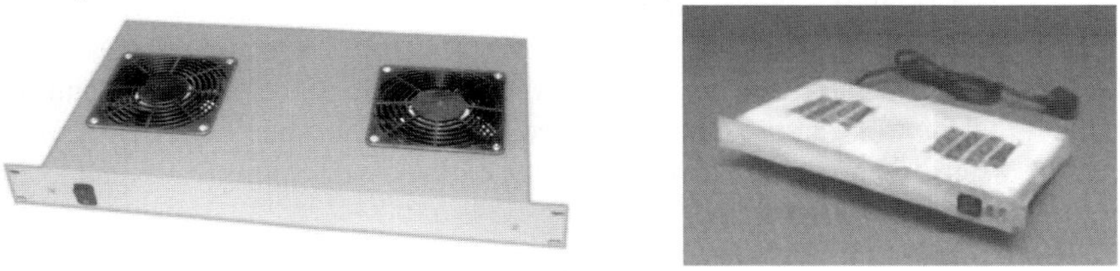

Fig. 5.17. Vent equipment: *a)* ventilator block for mounting to the upper cover of rack; *b)* ventilator shelf

Performance of a vent module depends on its engine capacity and is within 50–150 m^3/hour. As a rule, vent devices are designed with noise suppressors. The standard value of acoustic noise is about 40 dB

Manufacturers recommend that one vent device or shelf for input ventilation should be mounted in the bottom of the rack [98]. In practice, such device placement gives no advantage for more efficient cooling of equipment, therefore, vent shelves are often secured directly under active devices. Ceiling vent models are also used. They are located in rack cover hutches and function as air exhaust units.

There are options for inner and outer mounting. It is preferable to mount a vent shelf in a rack cover, as it reduces the height of the rack.

Since the end of 1990s, designs that include vent shelves with integral controllers to control the cooling capacity depending on the specific local conditions have become available. The controller fits in a shelf or it is a part of a sensor, as discussed in Section 5.1.4. The cooling capacity can be controlled by changing the impeller rotation speed, or by shutting down part of the modules on a multivent shelf.

Auxiliary accessories

Manufacturers of 19-in racks offer a wide range of accessories for easy maintenance of rack-mounted equipment.

Internal lighting units for racks represent lights of a specific design that allow easy and neat installation and maintenance of the equipment. It is preferable to use incandescent lamps as current spikes, which are used for fluorescent lighting, sometimes lead to the breakdown of network equipment. The lights can be mounted on rack covers by screws or magnets; they can also be incorporated in a pull-out shelf. The main advantage is their possible adjustment for optimal lighting. The devices are equipped with standard manual switches. Some manufacturers equip racks with terminal switches for automatic illumination when the doors are opened.

Racks may be equipped with various inner sensors to control parameters such as temperature, humidity, power status, and burglar alarm system status. There are two designs of sensors: as 19-in shelves with different modules in which sensor elements fit (produced by Schroff and Rittal, and ZPAS), or as individual controllers. The last solution is promoted, for example, by RiT Technologies, and saves on rack height. Its drawback is that it is complicated for reading LEDs and the status of other indicators. Grips, buttons, slide and changeover switches, and other elements to control parameters are housed on the front panel. The sensitive element of a sensor can be connected to the base block with a long cord and secured in the square slot of a 19-in rail.

Sensor readings are used both to activate various execution units (ventilators, acoustic and optical alarms[1], and fire-fighting systems), and are sent to the local or central console of the system administrator via an RS-232 interface, modem, or local Ethernet. The last operation runs by SNMP protocol. Local indication is made on a LED or liquid-crystal display. Such solutions are very popular due to the rapid growth of local networks and maintenance of most equipment areas for state-of-the-art SCS without need for the permanent presence of working personnel.

There are some climate-control solutions for racks that are installed in nonheated areas. Legrand has designed a heater that is incorporated in a rack.

---

[1] They are used predominantly for fire alarms and for unauthorized access to the rack.

It is much more convenient to operate and maintain equipment installed in a mounting rack when the rack is completed with operator documents with configuration descriptions, final changes, diskettes with the files for network equipment, and an installation tool kit. There are special casings, pockets, and boxes in racks to house all the supplies. Quite often, the boxes are supplied with telescoping guides and locks for convenience. Pockets and casings are made of metal or plastic, and are hung up or glued on the lateral walls of racks. Racks with steel front doors are usually equipped with pockets from the inner side.

Some mounting positions in 19-in racks are left free (for example, when fewer devices are required, or when it is necessary to provide quick cooling of network equipment). Racks with glass front doors have neat inner appearances when the mounting positions on the front pair of 19-in rails are not used and are covered by blind panels or filler panels. Such panels, as a rule, are 1 or $2U$ high, are made of steel or aluminum, and are secured with screws.

There is a wide range of adapters for mounting of narrow equipment; these adapters are included in the accessory kits of wide racks. The adapters are designed as securing L-bars or U-details. In addition, they are sometimes used to secure equipment on the rear pair of rack mounting rails.

Sealing kits for doors, lateral walls, bases, and covers help to increase the protection of the inner space, if required. Thus, for example, Ackermann kits 27414BR and 27414BP provide for the IP54 protection level of standard racks.

Some manufacturers of installation equipment offer cast metal units to serve as ballast, which are located on the base of mounting rack in order to displace its center of gravity down and increase steadiness. Application of such units is mandatory when mounting heavy equipment such as high-capacity servers and power supplies on pull-out shelves.

5.2 Surface Cable Raceways

5.2.1 Application and Features of Wall Raceways

Main specifications

Wall raceways serve for the routing of communications and power cables for various needs, and for the installation of receptacles. Surface raceways are used when:

- it is not possible or advisable to route cables otherwise;
- there is a need to protect cables from mechanical impacts, water, and other liquids;
- there are strong aesthetic requirements for office area interiors.

Being an electrical device, the cable raceway features a number of parameters and must meet a certain range of requirements. The aesthetic appearance of surface raceways is a very important factor, as they are primarily specified for office areas. The area of use of raceways also sets strong requirements for their fire safety. Their design considerations include all statements that refer to SCS cable products (Chapter 9). Additionally, the aerodynamic criterion must be taken into consideration, that is, the raceway susceptibility to drafts, fresh air intakes, and other factors that contribute to flame spreading.

For long-life, the raceway has to be impact-resistant and UV-proof and must take into consideration the application area.

Electrical parameters imply breakdown voltage stress of up to 240 kV/cm and higher. This enables the efficient protection of equipment and personnel in emergencies. Sometimes the raceway maximum sustained stress is given instead of the breakdown voltage stress parameter. The transfer impedance between sectors is also specified for metal and metal-plated raceways.

Types of raceways

The raceways consist of hollow closed chutes of various cross-sectional areas, which are mandatorily equipped with removable or, at least, lift-up covers[2], and serve for mounting on any flat main or false vertical surface. The most useful are those with rectangular cross sections (Fig. 5.18).

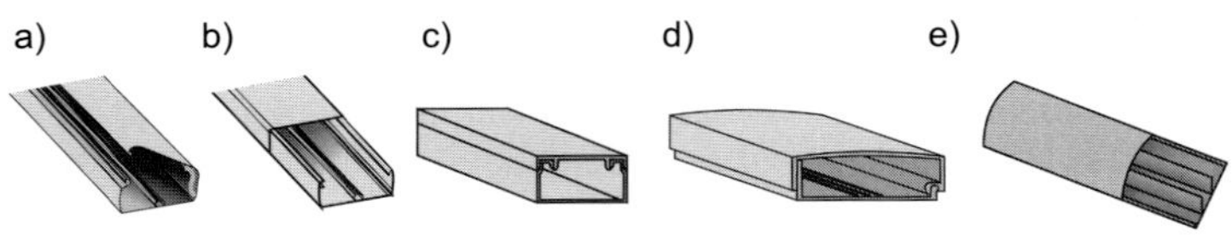

Fig. 5.18. Rectangular raceways: *a)* one-piece design; *b)* two-piece with U-cover; *c)* two-piece with flat cover; *d)* assembly with open cover; *e)* floor raceway

In addition, there are also tapered, triangular[3], and half-round raceways and plinths. The raceway can be manufactured as:

- *One-piece*, which consists of a solid plastic piece; one of its ribs has an opening with fastening slots. The plastic is thin on the opposite side and can bend for high flexibility (Fig. 5.18a). One of the sides can be lifted up and can function as a cover; the cover snaps over by a common or two-sided latch (Fig. 5.19). The maximum raceway gauge does not exceed 38×24 cm.
- *Two-piece*, which consists of two components: a base and a cover. Covers are U-shaped (Fig. 5.18b) for round covering, and flat (Fig. 5.18c). The U-cover is secured to the base by lateral latches. The latches can be placed both on the upper and bottom of the base lateral walls. The flat cover attaches by anchoring lugs in slots on bends of the base sidewalls. The first fastening option is specific for small raceways (up to 60×16), the second is for large raceways;

[2] There are also a limited number of cover-free raceways. However, they are inconvenient during maintenance and possess poor aesthetic values, and therefore are not widely used.

[3] For installation in corners and under carpets.

- *Assembled*, with open covers (Fig. 5.18d). The cover here is an L-shaped separate element, however, during assembly it fits by its roll-shaped lug in an appropriate slot on the base. If it is necessary to access the inner raceway space, the lateral latch snaps out and the cover is hinged to open horizontally or down. The assembled raceway differs from the one-piece one by a longer life and more convenient maintenance (the cover opens at a very large angle); however, it is more expensive.

We note that some types of two-piece raceways, which serve as plinths and fit tightly by their bottom parts to the floor, may have one top-latched L-cover. The cover surface tightly covers the raceway inner space, since it is made from flexible material. Some covers are equipped with an auxiliary visor to accommodate a carpet edge.

It is usually easy to remove covers to access the inner space. However, in the AXIS series of raceways made by Rehau, the procedure can be completed only by means of a special tool, which provides for efficient protection of cables from unauthorized access during cable routing in offices and residential areas.

In some cases, it can be difficult to use a standard raceway because of irregular surfaces. Exposed brickwork is a very common example of such a surface. For this purpose, the Panduit recommends the use of several metal brackets for wall mounting with screws, instead of the base. Cables are managed inside the brackets and then covered by U-covers.

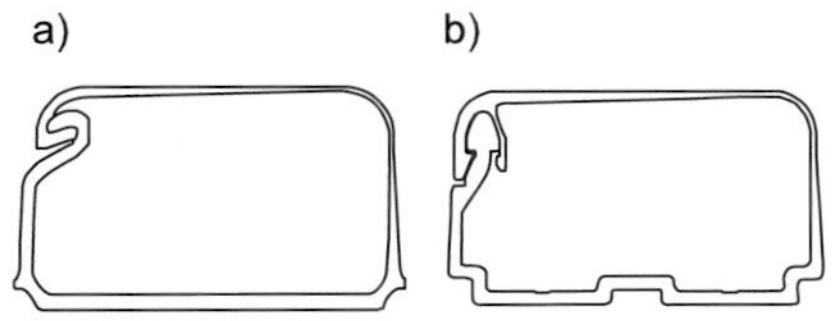

Fig. 5.19. Latches for one-piece raceways: *a)* single; *b)* double

Cable raceways are almost always installed directly on walls with mechanical securing; the retainer type is selected depending on the wall material (Section 11.5). Small raceways (up to 40×16 mm) may be glued on flat surfaces. Glued fastening may be done in two ways. In the first, the raceway is fixed by adhesive tape; a protective film on its adhesive surface is removed prior to gluing. In the second method, glue is put on the bottom of the raceway prior to gluing. It is preferable to have the bottom corrugated to increase strength. Some types of 19-mm and 25-mm wide raceways can be secured to metal furniture by magnet strips (Panduit). Various retaining brackets and elbows can be used for mounting under false ceilings, irregular walls, and other similar conditions.

The rectangular raceways differ by their cross-sectional sizes, which, as a rule are specified in dimensions. Manufacturers usually have standard dimensions from the minor cross section (for example, 14×7 mm) to large (for example, 250×60 mm). We do not know of any standards within the field. The raceway size is selected according to the cable volume being routed and the receptacle mounting technique. An approximate correspondence between the raceway dimensions and its capacity when using "half face" and "along profile" techniques for receptacle mounting (Section 5.2.3) is given in Table 5.4. We note that some SCS manufacturers (for example, AESP) forbid using raceways with minor cross-connects for their systems, as they fail to provide a given cable bend radius when bypassing bends.

Table 5.3. Standard capacity of surface raceways with full capacity

Duct gauge, mm	40×12	60×16	75×20	100×50
Number of 4-pair UTP cables	12	16	24	80

The inner space of large raceways (not less than 40×16 mm) is divided into two or more compartments, which makes their use more convenient. The need to section the inner space also comes from the norms specifying shared cable routing for different purposes, for example, power and communication. Lateral spaces normally serve for the routing of power and communication cables, since the central part is used for mounting of different receptacles. Since power cables take up less volume in standard office applications than communications cables, sometimes an asymmetric inner separation is used. For example, in the T70 raceway by Panduit, power section has a cross-sectional area of 587 mm^2, as the area of the communication section of 2013 mm^2.

In most cases, the inner space dividers, which are integral parts of raceways, are used for raceways with cross sections up to 60×16 mm. Large raceways are equipped with removable dividers that can be inserted in slots on the bottom. Depending on the design of raceways with regular dividers, those with the same cross-sectional area may have a different number of dividers, which should be considered when ordering. Usually, fixed dividers have a solid structure. In some raceways of the INKA series by Thorsman, dividers have slots every 200 mm, to minimize the weight of the end product and to ease cable entry to outlet connection modules. In large raceways with regular or auxiliary dividers often a longitudinal cover made in several parts is used for more convenient maintenance.

The lateral walls of traditional large raceways are usually not rigid enough, which leads to some weakness in the direction perpendicular to the routing axis. Two approaches eliminate this drawback. The first is based on the considerable reduction of sidewall height (or even elimination of it), and the use of an L-cover, which covers the lateral part of the raceway. The second approach is used by the company Mita. They use double lateral walls with inner cross-dividers in their raceways. Rigidity is also achieved by the use of chamfered raceway edges with diminished heights of the external walls.

Mita produces original transformer trunking in sizes up to 25×16 mm. The raceway is flat on delivery to the site. Once it is wall mounted, its lateral wings are raised to form the sidewalls. Thorsman uses the same solution, however, the raceway is coiled in a cardboard box with carrying handles. Large lengths (12 and 15 m) of the product guarantee minimum joints during installation, since the coil length allows the installer to cut a piece of the required length. Usually, the raceway is fixed to the main surface by wood screws or similar retainers. If such mounting technique seems to be unsuitable for some reason, then fastening brackets of different shapes and designs are used. Raceways by Rehau have predrilled tapped slots in their bases for easy mounting with wood screws.

Materials and color

Raceways are mainly manufactured from impact-resistant plastic and metals (aluminum, or galvanized or stainless steel). PVC is widely used for plastic raceway manufacturing, because it is inexpensive, is easily processed, and can easily achieve aesthetically superior products. The key drawback of PVC (its halogen content) is eliminated in halogen-free plastic, however, such polymers are more expensive [99].

The key advantages of plastic raceways are lower mass and higher flexibility, which enables them to cover irregular surfaces tightly during mounting. Metal raceways also provide grounding in addition to mechanical cable protection as an outer screen for routed cables. There are also plastic raceways with aluminum film-covered inner walls and grounding facilities. The grounding facilities often consist of a short flexible bar with ring contacts on the ends for screws.

Experimental results have revealed [100] that the use of an aluminum raceway reduces the cross-talk power by about 25 dB, and the use of metallic powder painting on plastic raceways reduces cross-talk power by 10 dB within the frequency range 100 MHz–1 GHz. The steel raceway has intermediate shielding characteristics between those of aluminum and powder-painted plastic ones. The shielding efficiency depends strongly on the mounting quality and the transfer impedance value between raceway compartments. To achieve a low value of the latter parameter, it is recommended that auxiliary uninsulated grounded wire be routed inside the raceway (the analogue of a shielded cable drain wire). Some companies offer kits with grounding wires for their metallic products to connect compartments.

The external surface of plastic raceways can be matte or glossy. The glossy finish is less exposed to dust, and since it reflects light, it gives a slight tint of the environmental interior color; however, matte raceways cost less. The majority of manufacturers produce raceways in several basic colors. The most widely used raceways are white. In some cases, it is possible to use raceways of other colors to adapt to local conditions. In order to improve the aesthetics of raceways with latched flat covers, the company Legrand snaps colored inserts into mounting slots on the cover. Recently, solutions with attractively covered outer surfaces and base protection by self-adhesive polymer tapes that are removed after mounting have become very popular (for example, products by Rehau and Unex).

Metallic raceways can be also painted in one of the classic colors listed in producer catalogs, or in any customized color by special order. Painting enamels or anodic coating techniques are used. Usually, custom-colored painted raceways cost twice as much and take 2.5–3 times as long to produce. As a rule, it is also possible to order a special color for plastic raceways. This is a high-cost solution and takes time. Since plastic is a smooth material, outer scratches, which inevitably appear during maintenance, are not as visible on a plastic raceway as they are on a painted metallic one. When selecting materials for plastic raceways and colors for metallic raceways, it is necessary to prove their UV-stability, since this parameter affects the aesthetic appearance of raceways over long time periods.

The cost of metal raceways is higher than that of plastic ones. Metal raceways are recommended for use when high mechanical and electromagnetic cable protection is required.

Raceways are usually delivered in 2-m packages. Exceptions in mid-1999 are 3-m raceways by Panduit and the Perimeters series of raceways by Siemon (package lengths of 1.8, 2.4, and 3 m). The 3-m sections are less convenient for transportation and carrying, however, their use allows, in most cases, for vertical parts to be made without intermediate joints and to save accessories. One standard package contains 20–100 m of raceway, depending on the section length and the size of its cross section.

The raceway inner surface is marked by index that includes the type, digital code, name of certification organization, and product standard.

5.2.2 Standard Kit Components

Manufacturers offer a number of standard components for each size of raceway. These elements considerably extend the range of possibilities for routing and installation as well as improve the aesthetic appearance of mounted raceways. Some of the standard kit components are shown in Fig. 5.20.

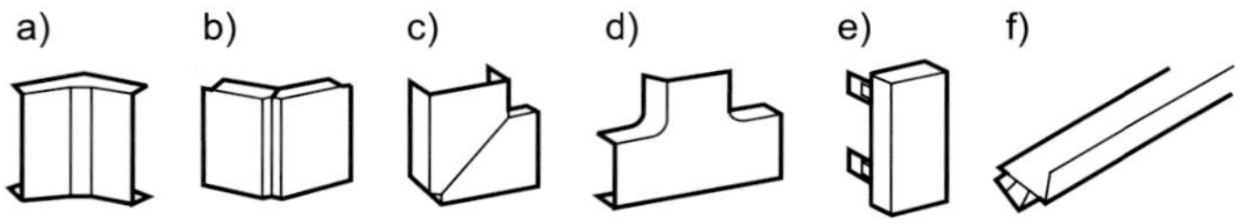

Fig. 5.20. Standard kit components for raceways: *a)* inside elbow fitting; *b)* outside elbow fitting; *c)* flat elbow fitting; *d)* T-fitting; *e)* end cap; *f)* spacing fitting

Inside elbow fittings (Fig. 5.20a) are used to cover inner wall joints. In the majority of cases, the element looks like a single detail; sometimes it can be divided, as can many of the elements discussed below, into a base and a cover.

Outside elbow fittings (Fig. 5.20b) are used for protruded wall joints. The elbow is designed as a cover over the joint of two raceways, or is installed as a raceway angle part. Inside or outside elbows in the form of covers may have fixed or flexible openings. In the first variant, the elbow consists of a one-piece plastic piece with a full fixed opening of wings at 45°, 60°, 90°, 120°, or 135°. The wings are hinged at the flexible opening. The hinges allow for an opening between 10° and 170°. The last property increases the cost of the end product, but is very valuable for Russian conditions, since, in practice, the wall seams often make angles other than right angles. In addition, the flexible opening enables the use of the same piece as an inside or outside elbow for small raceways, depending on conditions.

Flat elbow fittings (Fig. 5.20c) are used for raceway right bends on flat walls. In cases when cables are routed in raceways with large minimum allowable bend radii (mainly, optical ones), then a special elbow design with auxiliary extrusions into the inner or outside fields (Fig. 5.21) are used.

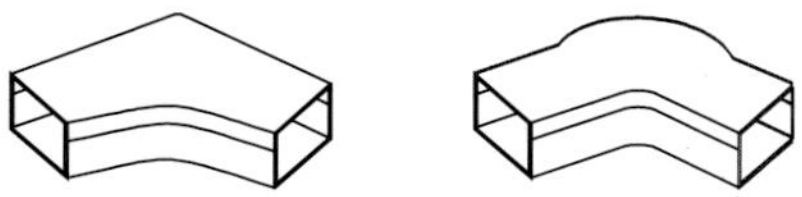

Fig. 5.21. Examples of designs of elbow fittings for cables routing with greater allowable bend radii

T-fittings (Fig. 5.20d) provide raceway branching at 90°. In the most popular versions, all three raceways have the same dimensions. There are also individual samples where the tapped raceway is smaller. T-joints may either protrude above the raceway by means of a protruded cover or they may not protrude above the raceway. Despite its somewhat poor aesthetic appearance, the first solution enables installers to avoid cutting off the raceway sidewall, as the tapped cables are routed above it.

Cross-joint fittings are used for cross-points of two raceways of similar or different sizes at a right angle.

Adapter fittings are used for raceways joints with different cross sections. Such elements usually serve for raceway transition from one size to another, for sizes produced by one company. There can be straight-through adapters, combined with (outside or face) inside elbows, as well as T-adapters. They are often used to adapt to intrasystem cable channels; however, some companies produce separate elements for this need.

End caps (Fig. 5.20e) cover the raceway end cut and are retained by the friction force of special tabs that fit between extrusions on its bottom and the walls. Sometimes the end cap is screwed. When it is applied as a plinth, it often has an asymmetric profile. In this case, the end caps are divided into left and right versions.

Cover clips are used to cover joints where two pieces of raceway come together. They cover the joints and even seamed raceways and their covers. In metallic raceways, they also minimize the transfer impedance between individual sections. They can be combined with inside splicing units for insertion inside the raceway to tighten its lateral walls (often in special slots) and precisely even sections. The inside splicing unit may be in the form of a plastic detail (Panduit) or a metallic shank (Rehau).

Dividing fittings (Fig. 5.20f) are removable elements that divide the inner space in a large raceway (usually 75×20 mm and larger) into separate compartments, which are used for cable routing, and a central void to fit the outlet connection module. Single walls are the most popular solutions, but there are also double-wall versions.

Blind sealing covers cover raceway entries to walls, false ceilings, and similar structures. Sometimes producers offer a number of modifications of symmetric and asymmetric profiles. Each is used for raceway entry into a specific opening (in corners, on wall joints, and others).

Holders or cable *retainers* are covers of various widths and lengths that fit into raceway cover slots or its separate sections under the main cover. They prevent cables from falling out of fully packed routing voids when the covers are removed. It is sometimes made as a special opening in the base slots, where the flat cover protrudes by its mounting tab. The majority of retainers are made of one piece. There are also retainers that consist of several identical elements that are coupled side-by-side to fit in the cover slots or the upper part of the divider.

Auxiliary components include sound-absorbing plates or braids for acoustic insulation of passages between premises, fire-resistant inserts, protective caps for screw heads, wall tube insulators, and edge covers for cable jacket protection from mechanical damage during routing.

The kit components are delivered in boxes or packs. The package provides service information and codes for automation control. The number of separate elements in a box or pack is defined by their dimensions, and ranges from 1 to 20 pieces.

5.2.3 Requirements for Outlet Mounting in Work Areas

The standard kit for raceways mandatorily includes a number of elements for mounting of different outlets: modular, coaxial, optical, and electrical, which are all called *outlet connection modules*. The design of the outlet provides a finished look to the raceway. All these modules are completed in one style with the same fastening method. The outlet may be installed in the inner raceway space, on the raceway, or next to the raceway.

Each of the installation options meets requirements, which are reviewed below. General qualitative characteristics of the options are given in Table 5.5.

Table 5.4. General qualitative characteristics of outlet installation options

Parameter	In-duct	On-duct	Next to duct
Mounting complexity	Low	Average	High
Cost	Low	Low	High
Aesthetic value	High	Low	Average
Required duct capacity	High	Low	Low

We note that raceway manufacturers sometimes design specific adapters to install some types of widely used outlet connection modules of another company into their own fasteners, in order to extend the functional possibilities of their products. Moreover, some manufacturers of SCS equipment who do not produce raceways certify products of other companies for compatibility with their own products, and recommend their use during cable system construction. The most well known example of such integration in Russia is the alliance of Lucent Technologies with Legrand and Thorsman.

We must also point out two aspects of raceways. First, besides outlets, raceways can also accommodate lighting on–off switches, regulators, tumblers, sensors, and other similar components, including those equipped with optical indicators, without using any auxiliary hardware. This standardization simplifies the mounting procedure of different outlets and extends the functional potential and application areas of raceways. Second, it is very popular to use the so-called one-post options for the installation of outlet connection modules. Multipost solutions are used for installation of several modules in one place. They differ by the number of mounting slots. The maximum number of such slots is 6; however, the most popular are the 2- and 3-post options.

Outlet installation in raceway inner space

A general view of a raceway with an installed outlet is shown in Fig. 5.22. An installation kit includes a mounting unit, an outlet connection module (with adapter, if required), a module holder, and a faceplate.

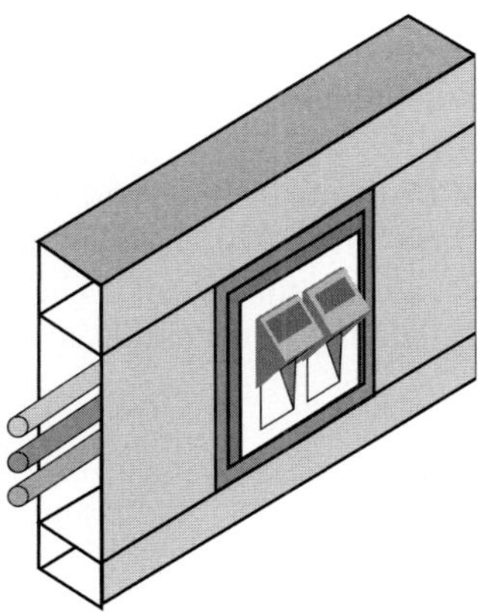

Fig. 5.22. Outlet installation in raceways

The mounting unit consists of a plastic housing, which is open from the faceplate side, and fasteners to mount inside the raceway. There are two fastening options. The first is when the unit is mounted in raceway cover slots or slots of inner space partitions, and is usually based on latches. The second option allows fastening on appropriate extrusions (for example, on DIN plates) of the raceway bottom and is made by hinged clips. The raceway housing has slots to fit in the fastening holder, cut-outs for cable entry (which are often covered by removable covers), and also two openings for holder screws. The distance between these openings is 60 mm for the European option, and 93 mm for the American one. Some models of mounting units (for example, the FB175 by Marshall) have a regular inner protective screen.

The module holder consists of metal plate with several shaped cuts and openings to enable mounting inside the mounting unit. The central cut accommodates the plastic latches of the outlet connection module. As a rule, a multisection raceway is used for outlet mounting in the inner space. The central section is used only for outlet mounting. The power and communication cables are routed in lateral sections.

The faceplate covers the module fastener and functions as a decorative cover. Usually, the faceplates are latched or screwed. Fastening by a two-piece stub, where the faceplate is retained only by friction, is a very rare solution. Sometimes faceplates also have mounting slots and cuts to accommodate the outlet connection module, which enables the installer to avoid the use of a separate fastening holder.

Outlets installed in the inner space of a raceway have a neat and sleek appearance, however, this requires a large raceway since once the mounting unit is installed, it overlaps a part of the raceway inner space. Two solutions were developed to eliminate this drawback. The first uses faceplates with angled connection modules. It enables the use of mounting units with less depth. The second supposes replacing the mounting unit by small holders, which may be supplied with removable lateral screens.

Outlet installation in raceways

Outlet mounting on a raceway is made by means of a mounting frame and an outlet connection module (Fig. 5.23). The mounting frame is a plastic base with two installation slots and a cut to accommodate the outlet connection module. The outlet connection module is latched or screwed to the mounting frame and is covered with the faceplate.

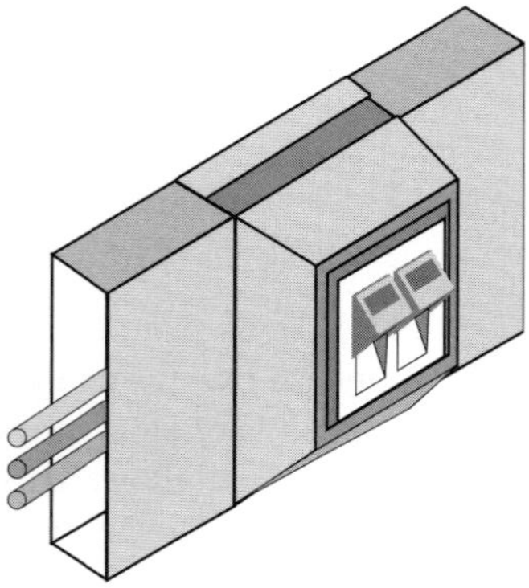

Fig. 5.23. Outlet installation on raceways

The reviewed outlet mounting technique enables the use of raceways with smaller cross sections than those in previous options. However, the outlets protrude above the raceway, are sensitive to mechanical impacts, and are the least aesthetically pleasing. This mounting technique is often called "half-face fastening.

The solution, which was discussed in Section 5.2.3, is based on the use of faceplates with angled outlet connection modules. It may be considered an intermediate option, which combines the main techniques of outlet fastening in the inner space of the raceway.

Outlet installation next to raceways

Installation of outlets next to raceways (Fig. 5.24) combines the advantages of the two previous options, however, it is used for small raceways. This method is also implemented by mounting frames and outlet connection modules. The mounting frame is a base that is mounted on a load-bearing surface (for example, a wall or furniture) next to the raceway, and has a cut for the outlet connection module and slots for screws. In the majority of cases, it is supplied with a supplementary cover, which covers the cable exit point from the raceway to the outlet connection module. The frame is fixed with screws or two-sided adhesive tape so that the cover overlaps the raceway. The outlet connection modules fit in the upper frame opening. The frame is used in horizontal and vertical locations, depending on the raceway orientation. The latter is most often used.

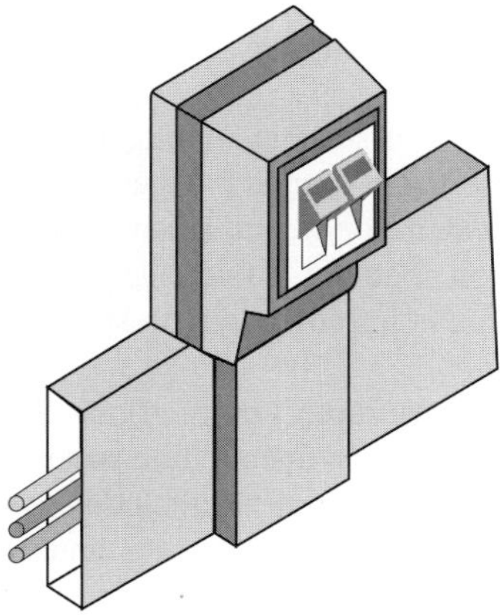

Fig. 5.24. Outlet installation next to raceway

Outlets installed with low-profile fastening do not protrude as much above the wall surface, possess good aesthetics, and enable the full use of the raceway inner space for cable routing. Their drawback is their time-consuming mounting (it requires the installer to drill a minimum of two additional holes in the wall), and also their use of mounting frames.

Combined solutions

Combined solutions are used for outlet mounting next to raceways. In addition, they include features of other installation techniques. Two products are available. Outer telecommunications outlet housings, which enable next-to-raceway mounting (usually for 6 and 12 slots for different module type) have become popular. The housing consists of two parts. The bottom part functions as the base, the top part covers the base after its mounting. The designs visually differ from frame solutions since they protrude less beside the raceway (since its base partly protrudes on the raceway), but considerably above its surface.

Panduit has designed a so-called WorkStation Outlet Center to be mounted next to raceways. The product serves as a standard workstation and has slots for two power and two telecommunications outlets. The power outlets are mounted outside the raceway, the telecommunications outlets are mounted in the cover of the raceway opening. Since the installation of large power outlets occurs outside the raceway, this configuration has reduced dimensions.

Multimedia outlets

A multimedia outlet is a small plastic or, more rarely, metal box with fastening slots for electrical and optical outlets for different connectors. The multimedia outlet is named such because it is suitable for both electrical and optical cables. The mandatory element of the outlet design is an internal standard or additional fiber manager, which serves to wind 1 m of standard reserve fiber following a given bend radius. Sometimes the fiber manager is equipped with an additional organizer for mechanical splices or fusion splice sleeves.

The outlet housing is typically white, but other colors are available. The most widely used models have a parallelepiped shape with rounded edges of some sides. There are also housings of a more complicated polyhedral prismatic form, for example, the CT multimedia outlets by Siemon. Usually, cable entry is at the rear and bottom of the housing, through appropriate holes covered with spring end caps or caps that are removable during installation.

The multimedia outlet is mounted on a wall or (very rarely) on the floor. Some types of these outlets are provided with magnets and can be installed on metal surfaces. As a rule, fastening screws and two-sided adhesive tape are included in the delivery kit. The magnetic holders must be ordered separately. Cables enter the outlet via the raceway or an inner wall cable duct. There are also models that are mounted directly on raceways by the low-profile approach, but without the use of a separate mounting frame due to available raceway entries on the narrow lateral surfaces.

Outlets for optical and electrical connectors are usually installed on replaceable inserts on the lateral surfaces of the housing. Flat insert units slide to their positions along guides. The angled inserts are fixed to the bottom by screws. Some models house one or two outlets directly in protective covers (for example, BRACOC by Hubbell). Besides individual insert units, there are also removable front panels with a maximum of four slots to accommodate different types of outlets. The standard capacity of the most popular multimedia outlets is 4–8 ports, that is, they can serve a group of users and function as MUTO outlets, if required. General information on multimedia outlets of some manufacturers is given in Table 5.5.

Table 5.5. Technical specifications of multimedia outlets

Manufacturer	Type	Dimensions, mm	Number of outlets	Outlet type
AMP	559274-1	170×146×38	6	ST, SC, FC, modular, BNC
Hubbell	BRACOC	133×133×38	4	ST, BNC, IBM, modular
Lucent Technologies	40A1	175×142×41	8	ST, SC, MIC, BNC, modular
Mod-Tap	17-5229-02 17.B143G	197×159×57	8	ST, SC, BNC, modular
Panduit	CBXF6 CBXF12	170×120×25 170×170×46	6 12	ST, SC, BNC, modular, RCA
Siemon	CT-MMO-(XX)	200×200×57	24	ST, SC, BNC, RCA

5.2.4 Workplace Connection Elements for Large Premises

Workplace connection elements for large premises consist of floor and under-floor boxes, columns, and outlet panels. We also refer here to enclosures for consolidation point mounting. They functionally complete raceways and enable them to essentially extend the task area to be solved during SCS construction by standard facilities. In addition, manufacturers of raceways include these elements in their production plans to maintain design compatibility. These elements do not require auxiliary facilities during installation. The interest in these products in Russia is defined by the growing need for open office environment creation.

Under-floor boxes

Under-floor boxes serve for the connection of workstations, which in large premises are remote from walls and power supply systems, to SCS. They are also used when horizontal cabling is routed along under floor ducts. Besides different outlets, some specific kinds of active equipment can be also mounted in boxes [101].

The under-floor box includes a housing and a cover that is put on raised floor posts instead of one of the panels or is simply inserted by its flange into an opening in the panel. The majority of models are rectangular, but there are also round versions (for example, GRAF-9 products by Ackermann).

When the box is produced by the same manufacturer as the raised floor, it is equipped with elements for mounting in a variety of plates or posts of the raised floor. Power and telecommunications outlets are installed vertically opposite each other in the box or are fit in a horizontal panel. Some boxes are designed with combined horizontal and vertical outlet mounting. The outlet connection modules are installed by their kit fasteners. Sealed cable lead-ins are used for introducing horizontal and power cables into the bottom or sidewalls of the housing. Box covers are equipped with grommets for protection against dust and other contaminants.

The cover has a smooth bend of 4–5 cm for the exit of patch cords and power cables, or a small vertically opened door with a holder in half-open position (at an angle 30°). The box cover can painted any color, but its surface is suitable for gluing of carpets or linoleum.

Box design options are shown in Fig. 5.25.

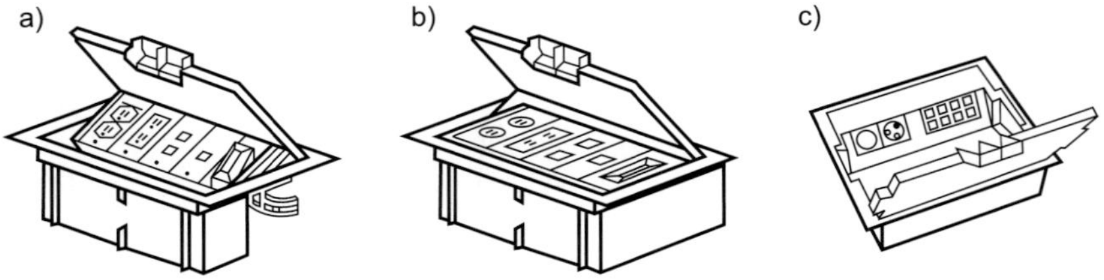

Fig. 5.25. Box design options: *a)* cover and frame fixed together; *b)* independent cover; *c)* opposite mounting

They basically differ from each other by installation of outlet connection modules. They can be mounted on angled panels (Fig. 5.25a, the cover and frame are fixed together), located horizontally (Fig. 5.25b, the cover is independent from the panel), or inside opposite each other (Fig. 5.25c).

Floor boxes and pedestals

Floor boxes, minipedestals, or simply pedestals (Fig. 5.26) are mounted on raised surfaces without carpets and introduce computer, power, and telephone outlets to workplaces [102]. Unlike under-floor boxes, the majority of raised boxes are used for a maximum of two workplaces, and therefore are smaller. The box housing is made of aluminum or plastic. Plastic housings usually look like truncated pyramids, aluminum housings have a round punched cover.

Fig. 5.26. Floor box

The front panel with outlets is vertically angled to ease the termination of work area cables. Besides the traditional position of modular jacks, there are also boxes with jacks angled at 90°. Use of such asymmetric designs has not become widespread. The solution has been designed, apparently, in order to make it possible for the user to visually control the latch when the patch cord snaps out.

The operating conditions of pedestals lead to specially designed elements for mechanical protection of their front panels with modular jacks. There are two known solutions, which are often supplementary to each other. The first is based on using a protruded cover cornice, the other uses a protective bow.

The company Deutsche Electraplan produces high-capacity pedestals. The selection of power and telecommunications outlets in these models is made by insertion of removable modular insertion units. The 622141 XXX series of products is a two-tier solution, which enables users to increase the housing capacity up to 16 ports. Manufacturers usually offer several housings in different colors for superior aesthetic appearance. Power and communication cables enter from the bottom via grommets. The box protrudes above the floor, and therefore, is inconvenient during operation. Floor boxes are produced by a rather small number of companies.

The pedestal is rarely used in SCS. It has no visible differences from a standard floor box. It is recommended for use in office premises with stationary furniture, and also in large auditoriums, since it is very convenient for connecting laptop computers during presentations, conferences, and other events.

Columns

Vertical columns also serve for the connection of workplaces to SCS in large areas. These products are manufactured in two basic options: in the form of a short column that protrudes up to 0.6 m from the floor, or as a post that runs continuously from the floor to the ceiling.

The short column (Fig. 5.27) has a rectangular or square cross section; triangular cross sections are very rare. One can use an inside permanent or removable partition wall to increase the mechanical strength. For extra capacity one may tightly join two short columns by their rear permanent panels.

Fig. 5.27. Floor-mounted short columns

A short column is mounted on a mounting base. Telecommunications and power cables are introduced to the column from under the raised floor or by means of a floor-mounted box.

Short columns are sometimes used in auditoriums of educational centers, where they are tightly installed to the side panels of tables. This design allows tables to be easily moved and replaced, if required.

The other type of columns known to the authors are made of anodized aluminum and built according to two basic schemes. The first uses beams with rectangular or square cross sections, as well as T-shaped beams (columns of the Prestige series by MK Electric [103]). The second uses a support with hung elements. Various design options are available. For example, Thorsman uses two parallel load-bearing tubes of small diameter, but Nordic Aluminum applies a single vertical post. The base of the INTEGRATION 45 series columns by INFRA+ is a vertical raceway in which one or two columns with slots for telecommunications and power outlets join in the bottom as joint bars.

One of the problems that arises when using columns is securing it between the floor and the ceiling. The problem can be solved in two ways. In the first, the column consists of two parts and has a telescoping design (an option from Legrand). The bottom part is installed on a rubber center pad or other elastic material with high friction coefficient; the upper part is ceiling mounted by a mounting bracket. A jackscrew is used to make the construction sufficiently rigid. It is mounted on the support of the bottom part under the active upper one. The second approach is used by Nordic Aluminum, and is based on a continuous column and a jackscrew with broad travel of the working element. Sometimes auxiliary holders are also used to fasten the column to the desktop. The entrance of the column to the false ceiling is usually covered by a plate for a neat appearance.

Outlet panel

The outlet panel is made as a short part of a raceway with a rather large cross-sectional area with lateral covers, and can be equipped with several power and telecommunications outlets as well as switches. The device is placed under the desktop and is fixed by special holders. The panel is suitable for furniture with voids for cable routing. It can be successfully combined with columns. The use of a flexible corrugated tube for cable routing is recommended in the latter solution.

There are also outlet panels for vertical mounting, which are fixed in the on-position close to the sidewall of tables. Their design is identical to that of floor columns, the only difference is their base-free design.

Housings for consolidation point equipment

Housings for consolidation point equipment are intended for mounting on walls or columns in open office environments. They consist of a box that is specially designed for open mounting with an appropriate external appearance. The cover of the housing is equipped by a lock or a locking bar. The central part of the cover can be made of transparent plastic, which enables users to visually control the cable position. Usually, cross-connected 110 blocks of different designs with supplementary organizers serve for electrical cable management. There are also housings on the market for mounting of 66 blocks.

Depending on the dimensions and the mounting techniques, one can find vertical and horizontal options for housings (for example, products CPEH-(XX) and CPEV-(XX) by Siemon).

5.2.5 Alternative Raceways

Raceways for fiber-optic cable routing

Special raceways for routing of fiber-optic cables appeared at the end of the 1990s because of the rapid growth of fiber-optic equipment as a part of SCS. They are used for local cable routing in telecommunications closets and equipment rooms. They are mainly intended for overhead mounting, and arrangement of chutes to shelves and active equipment that are mounted in open racks. The raceways accommodate patch cords and riser cables for vertical routing. In some cases, manufacturers prohibit the use of raceways for routing of cables other than fiber-optic ones.

The products are specially designated for the space division of optical cables from other cable types, maintenance of minimal allowable bend radii, stretching force constraints, and easier cable routing and operation with optical patch cords [104].

Unlike the standard raceways, those reviewed here feature the following:

- the presence of only rectangular cross sections and substantially fewer sizes;
- the use of special facilities and solutions to constrain the minimum bend radius of 30-mm long routed cables;
- a wide range of vertical chutes with appropriate accessories for mounting of intrarack connecting cords;
- brightly painted raceways (orange or yellow).

The biggest differences between these raceway types are defined by their designations. Raceways are usually made of PVC and polycarbonate [105].

Splicing of separate sections and mounting of accessories is done by special fasteners for easy assembly and mounting of raceways in work areas. Hanging the raceways on ceilings and similar structures is completed by means of brackets of various designs and gauges.

The range of accessories for raceways is similar to that for standard raceways and includes elbows, T-adapters and other types of adapters, as well as end covers. A snake-shaped centering adapter is useful to easily by-pass vertical posts (it is often used due to the peculiarities of raceway application). Elbows and adapters made of clear plastic allow visual control of the proper optical cable routing (Panduit).

Optical cables are more sensitive to the maintenance of minimum bend radius and compression strength values than are electrical cables. Therefore, raceways for optical cables are quite often equipped with partition walls in bends, taps, and other similar accessories.

According to the information available in mid-1999, raceways for routing of fiber-optic cables are offered both by SCS equipment manufacturers and by manufacturers of accessories for optical cables. They are delivered to the market as end-to-end solutions (raceway, accessories, fasteners, and tools) under individual trademarks (for example, Lightway by Siemon[4], PANRACEWAY by Panduit, FIST by Raychem, and Lightguide fiber-optic protection system by Hellermann Tyton).

Conduits for mounting under raised floors and behind false ceilings

Besides wall raceways, there are also specially designed conduits for general purposes for mounting under raised floors and behind false ceilings. These products do not differ from decorative raceways, except for a less neat appearance and, possibly, higher mechanical strength as defined by their maintenance conditions. The main features are specified by the frequent need for two-level mounting of the conduits and for the introduction of cables to decorative wall raceways and other elements for the connection of workplaces in large premises. It is also possible to use them for the arrangement of vertical conduits.

The products are made of metal or plastic and possess approximately the same dimensions, mass, and strength characteristics. The key advantage of plastic conduits is that there is no need to ground them, since this is a time-consuming procedure because of the peculiarities of their mounting locations.

The conduits in this Section differ from their office analogues by the lack of sectioning of the inner space and small dimension-types. It is also possible to use a perforated bottom that serves to fasten cables and cable bundles by plastic hoops, which is especially useful for the arrangement of vertical conduits.

Special adapter devices are used for two-level routing. They are selected by taking into consideration the conduit height with a cover. The transition can be fixed or hinged. During assembly of conduits for mounting under raised floors, their separate sections are placed on supports. Separate conduit sections for mounting behind false ceilings are placed into the traverses of suspended supports or wall brackets. Large radii bends are bypassed by means of elbow fittings with housings of corrugated materials.

Mounting under raised floors and behind false ceilings requires that the installer follow strict fire-preventive norms. Therefore, a standard kit for conduits includes a set of different adapters for metal sleeves and similar products, and fireproof mats and other facilities for the installation of protective end caps.

Besides conduits for wall or ceiling fastening, there are so-called floor raceways. Their main function is cable protection against mechanical impacts when

[4] This product is manufactured by Warren & Brown Technologies and is called Lightpaths.

routing in pedestrian underpasses. The products, as a rule, are of low capacity and have a semicircular or nearly semicircular cross sections, with inside reinforcing ribs to increase compression strength (Fig. 5.18c). Floor raceways are manufactured of metal or plastic and are colored differently.

Wiring ducts or slotted trunking for routing outside office areas is the functional analogue of decorative raceways [106]. It possesses less exterior finish, and its sidewalls are designed as combs with fingers that provide free cable access (Fig. 5.28).

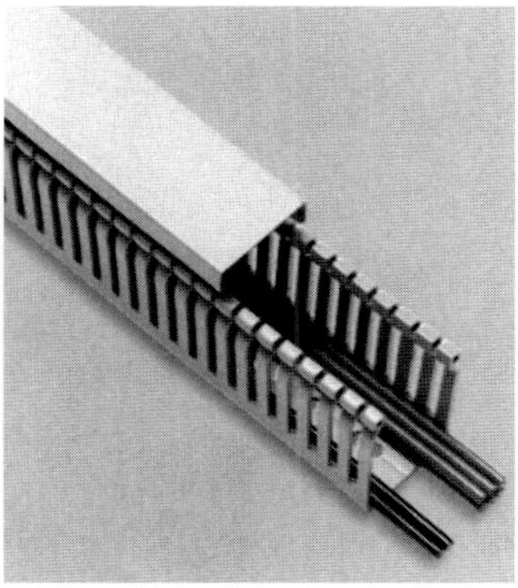

Fig. 5.28. Wiring duct

The fingers can be easily broken out by a special tool or pliers in order to extend the pass-through, if required. This option, along with a somewhat reduced mass, simplifies the splitting procedure for one or several cables or a bundle. Once the cables are routed, a cover included in the kit covers the duct.

5.3 Conclusions

The use of the auxiliary components in SCS allows designers to achieve superior aesthetics for offices and work areas for the entire cable system lifetime. The products contribute to convenient maintenance, enhance mechanical protection, and restrict unauthorized access to cable and cross-connect elements. When special requirements must be met, these components provide additional electromagnetic shielding of the passive components of cable systems as well as active network equipment. There is a wide range of constructs and raceways with appropriate accessories on the market, which makes it possible to solve all tasks arising during SCS construction.

The mounting equipment used in SCS is manufactured to comply with standards that ensure the interchangeability of accessories of different manufacturers. There are almost no standards for raceways, which sets high requirements for the product suppliers.

Various sensors and electro-mechanical execution units with the transmission of signals (via SNMP protocols) to a central control board and back have become widely used as a part of mounting equipment.

The development of a technical platform for decorative raceways is aimed at both the creation of specialized products as raceways for fiber-optic cables, and the extension of their application areas. We would like to note examples of the latter are backbone cable pathways under raised floors, behind false ceilings, and in service hallways in office buildings, as well as in vertical conduits.

6 SCS for Non-office Buildings and SOHO Sector

Computer technology for the most diversified purposes, as well as other electronic and electro-technical devices with managing controllers, are used increasingly in all spheres of modern life. The functioning efficiency of such devices grows when integrated into a uniform system, which is carried out with the help of various networks. They can be constructed by various principles, for example, by overlaying a telephone network, by means of infrared interfaces, on the basis of radio channels, or even with the usual electric wiring. However, in a majority of cases the bases of their infrastructure are communication cables, specially allocated for this task, and the appropriate switching equipment. During the last decade, the cable layout in office-type buildings (business centers, banks, public authorities, etc.) has always been constructed in the form of structured cabling systems. The various aspects of realizing such systems as technical objects are described in great detail and are set out in a number of familiar national and international standards.

The necessity of organizing information for transmission networks is becoming more and more important for non-office buildings. Typical examples of such premises are residential areas (beginning with an individual apartment or cottage and finishing with a micro-district in an urban area or a modern settlement in a rural area), as well as hospitals and hotels in the public building sector. The growing complexity of low-current cable layout in such buildings strongly requires applicable solutions that are based on structured principles. This chapter describes SCS that are mainly installed in non-office buildings and for which some criteria (often based upon formal criteria) do not correspond in full to current editions of the standards for SCS. Similar systems in the small office/home office (SOHO) sector and small firms are also mentioned in this chapter, whose distinctive attribute is the small number of communication ports (less than 50). The principles and technical solutions used in this sphere can also be successfully carried out for the organization of cable layout in allocated and remote premises away from the main office in which some of the employees of any average or large company work. The given examples do not include cable systems for industrial buildings, which are constructed on the basis of specific industrial enterprise requirements.

6.1 SCS Design for Non-office Buildings

6.1.1 Features of the Field of Application

The general features of a cable system in non-office buildings and an SCS in a typical office building appear rather similar because of their economic character and in view of their function. In order to find solutions during the creation of such projects, it is convenient and economically favorable to use approaches that are used when constructing "classical" SCS. This is described, in particular, in working editions of the standards TIA/EIA-568-A and ISO/IEC 11801, which recommend using as much as possible principles and solutions during construction that are based on office SCS. However, the specifics of non-office SCS in the field of application make a sufficiently strong impact on the basic principles used for their realization and on construction principles. The general features of such cable systems are defined by the following circumstances:

- Selection means for SCS creation are limited (a significant number of SCS in non-office buildings are realized in the residential sector and are paid from the user's personal budget), which emphasizes the first component in the rigid requirements of the "price-productivity" parameter;
- In many cases, the small relative installation density of users' information outlets and correspondingly significant role (position) of a system's number of ports is not higher than several tens;
- A large number of applications, in comparison with offices, are simultaneously serviced by a system, even today, and in a combination with a wider spectral range of transmitted signals. The last characteristic is caused, mainly, by a combined operation of radio and cable TV.

In regards to the general distinctive SCS attributes of non-office buildings, we shall note:

- Equipment and solutions that are used for cabling outside office buildings are promoted in the market as a single product under its own trademark;
- Minimization of equipment nomenclature, which is necessary for the realization of a functioning system (deliberate orientation to a principle of the "thin catalogue");
- Maximum wide application of SCS standards as the essential base for office-building SCS in combination with the use of the construction modular principle;
- Production of specialized hardware products with uniform design and with their functional analogues based on a "large" system;
- Quite often solutions are exploited to transmit TV- and broadband signals without regard to the current standards, which prohibit to do so for traditional SCS;
- Relatively wide-spread use of active equipment, which was specifically designed or adjusted for such systems, as a standard element of those systems.

- High popularity of technical means that provide a ready system installation that fails to require use of special instruments and makes minimum requirements of professional skills of the user as a designer and an installer of SCS.

6.1.2 Functional Capabilities and Requirements of SCS

Functionality of non-office SCS strongly depends on that market segment, which addresses its needs to specific technical solutions.

The cable system of a small building or remote premise with a workgroup is normally focused on satisfying classical office needs and primarily on the functioning of a mini-automatic telephone exchange and LAN. The element base with all hardware products used for its construction, provides characteristics that correspond to the appropriate category 5e, though the requirement to transmit Gigabit Ethernet and ATM 622 signals at small offices is improbable in the foreseeable future. Nevertheless, the advantage of the use of standard components essentially exceeds profit on some cost reduction in cases where of out-of-date components are used.

The cable layout, which when installed in houses or in the large multiroom luxury apartments, is intended for wide spectrum signal transmission and distribution to separate rooms for the different digital and analog appliances. Therefore the cable layout requirements are as follows:

- Support of multichannel TV signal transmission, including high definition TV, which is transmitted from satellite and/or transmitting antennas, and also from an external TV signals transmitted over cables;
- Provision for the operation of video cameras, including door intercommunication systems, security facilities, and supervision for children by displaying received images on a control monitor;
- Control signal transmission for illumination system devices, air conditioning, satellite TV receivers, radio and cable TV receivers as well as for household broadcasting systems;
- Support of functioning of modern audio–video systems and other devices united in the single definition "home cinema";
- Ability to organize such cable layouts in house or home office room;
- Provision of convenient Internet access;
- Support operation of fire and security systems;
- Make it easy to relocate telephone sets, faxes, and other similar devices in user-friendly locations.

The given fields of application are the most complex for the designers of a cable system, since this cable system is inseparably combined with rather small volumes of digital information flows and the need for the provision of high-quality transmission of broadband analog signal.

The home network Frend concept, developed in Russia is deployed in small urban districts or towns on a basis of full-size SCS and provides the general types of "normal" local network service to its residents subject to the specific requirements of its users. A subscriber of this network is provided with access to electronic bulletin boards, information about various municipal services, on-line payment of household bills, Internet services, and network games, and printing via network printers with delivery of printouts to mailboxes. The proposed solution has the important advantage, according to the designers, that the connections of the separate users are carried out for a price that is less than the price of usual Internet access at higher and various levels of given service.

Additionally, the Frend network can serve as the basis for the construction of a house information readout system for electric power, water, and gas meters, since many advanced models of these devices are equipped with interfaces based on modular connector outlets.

6.2 Construction principles

6.2.1 Switching nodes and switching centers

The basic principles of *switching nodes* and *switching centers* in "classical" SCS organization, as is well known, are defined by the TIA/EIA-569 standard, and consist of a technical room, in which the various equipment of a cable system is installed and the network devices of various purposes are mounted. Because of the relatively small dimensions of non-office-type SCS, the assignment of a separate room for cross- and equipment rooms is seldom done; in practice "open" solutions are used more frequently.

Because of the small dimensions of the cable systems, in comparison with those of an office, the cable system in a house or multiroom apartment is constructed on the "simple star" method, which realizes a principle of the centralized administration. The main items of such networks (analogous to equipment rooms for "full size" SCS) are predominantly installed in an antechamber, ground floor, storeroom, or in a cellar and are carried out in the form of wall assemblies of this or that version. In case of necessity it can be placed in any suitable place. Similar solutions are frequently used at offices of small firms, and in most cases, closed or semiclosed assemblies are used, which are mounted in a remote corners of a room and, whenever possible, in places closed from unauthorized persons.

The cable layout of the Frend system is an example of non-office-type SCS, for which the necessity of special room assignment arises. It occurs because some hundreds of subscribers can be connected to this network. The Frend central switching node installs near the central dispatch office of the district. It is a classical equipment room, in which, along with switching equipment of a cable system, servers of various service supports, cable TV system receivers, modems for connection to dedicated Internet lines, as well as workstations of technical personnel are installed. The local switching nodes are located in back rooms of the dwellings and are structurally carried out in the form of usual 19-in wall cabinets containing the passive and active network equipment. Floor-switching equipment, which services groups of apartments on one entryway, is carried out as blocks such as 66 blocks, which are installed on free places of electric boards.

Except for large housing systems, the special technical rooms are assigned in large public buildings and buildings of non-office types (hospital, hotels, cultural institutions). This done because of the significant number of SCS ports that serve similar buildings (an average-size hotel has some hundreds or more information outlets), and also from the need to interact with numerous external telecommunications services (radio, cable TV, broadcasting, etc.).

6.2.2 Element base

At present, there are numerous specialized devices and solutions in cable layout constructions for non-office systems. These devices are supplied in the form of functionally completed systems, which are promoted to a market under independent brands (Table 6.1) and are mainly aimed at applications in the domestic sector.

Table 6.1. Cable systems for SOHO and non-offices

Company	Title	Cable type	State mounting construction	Active equipment
Brand-Rex, Great Britain	FlexiLAN	UTP cat 5	10-in housing	–
Corning, USA	Home Way	Hybrid S-UTP cat 5 + coax	Not available, 19-in cabinet or chassis is recommended	–
Elgadphon, Israel	Elgadphon SOHO	UTP cat 5, STP cat 5	10-in cabinet	–
Krone, Germany	MiniLAN	S-UTP cat 5, STP cat 5	10-in cabinet	–
Legrand, France	VDI	UTP and STP cat 5	10-in cabinet	–
Network Systems & Services, Great Britain	Midi-LAN	UTP and STP cat 5	10-in cabinet	Ethernet hub, alarm system
NORDX/CDT, Canada	RUN (Residental Universal Network)	UTP cat 3, UTP cat 5, coax	10-in cabinet	Ethernet hub
Ortronics, USA	In House	UTP cat 5, coax	10-in cabinet	Ethernet hub, TV amplifier–splitter
Siemon, USA	Home Cabling System	UTP cat 5, coax	17-in chassis	–
Telesafe, Norway	WTP (Wideband Twisted Pair)	S/STP	19- and 10-in cabinet	TV amplifier–splitter
IT, Russia	IT-SCS-mini	UTP cat 5	10-in cabinet	–

The construction of structured cable systems of non-office types based on the standard element base for "large" SCS is not excluded. It is natural that in the system structure, the separate components, which are specially developed for a given application field and for the system structure of a specific SCS, can be also used.

Cable hardware products

The cable types for organization of cable layouts in non-office SCS are mainly defined by the application field for which a given specific technical solution was initially aimed. Thus, in cable systems intended for SOHO sector networks, because of their significantly smaller volumes of transmitted information than in corporate settings, even in modern situations cables of category 3 are very popular. In those cases when the cable layout is intended for installation in the residential sector, there are currently two approaches, which have approximately equal popularity. The first is based on the use of so-called multimedia cables, such as S/STP with multilayer common shields and characteristics, which are certified up to frequencies that considerably exceed the operating frequencies even of category 7 projects (for example, the upper frequency of fairly popular cables in Western Europe, such as SOHO 860 MultiPurpose by NEK/CDT and MegaLine 8 by Kerpen, is 1 GHz). Because of the considerable cost of such hardware products, in order to increase the economic efficiency of the technical solutions carried out on this basis, the cable-sharing principle is widely used. The second approach is based on the use of category 3 symmetrical cables to provide operation of telephone and computer networks, whereas coaxial cables are intended for TV-signal transmission. Therefore, it can be seen that in both cases, to meet the specific requirements of non-office applications, solutions that exceed the limits of the working standards for office-type SCS are used. The need for their used is caused by the tendency to find less expensive alternatives to optical-fiber elements, whose price prevents its wide use in non-office cable systems at the present time.

It is practically rear, that backbone cabling is used for multiport non-office systems because of its small dimensions. When it is necessary, the principle of its realization strongly depends on the place of network deployment. Thus, according to the Frend network concept, upon its realization in a small residential district, the subsystems of external paths are mainly constructed by the use of fiber-optic cables. When a network is deployed in a residential area, in most cases the subsystems are organized using symmetrical electrical cables, because of the need for transmission of local automatic telephone exchange signals as well as to minimize the total cost of the project. At the same time, common cable modems are used to transmit digital information. The connection of local switching nodes to on-floor centers (analogues of an internal path subsystems of common SCS) is made by multipair cables with crimping on 66-type panels. The cable layout for apartments is carried out by common 4-pair horizontal cables.

Switching equipment

The switching equipment structure of non-office SCS basically reiterates the technical solutions used at similar equipment constructions for "large" SCS. Thus, in particular, all types of panels are used: panels with 66-type and 110-type modular connectors outlets, the first of which occupies a leading position. In order to connect telephones and other equipment, which does not place high requirements upon the path capacity, distribution blocks are used. An example is the so-called *LSA Plus distribution block* from the V.D.I. system by Legrand.

Note also the basic features in this field that directly follow from the specificity of non-office SCS applications. The first is the increased use of connector types that differ from the abovementioned connector types, including applications that are not allowed in "classical" SCS by working and future editions of the standards. Thus, in particular, in the RUN system by NORDX/CDT, the list of regular products consists of RG6U coaxial connectors for cable TV (including splitters), and in the PL system (Section 6.4.1), I-adapters with modular connector outlets carry out the functions of elements for patch cord connection. The second feature is that the patch *panels* are often designed using removable units, and in these units, as well as in the panels of telecommunication outlets, the same insert units with elements of the appropriate connectors are installed. In order to decrease the assembly width, in some cases the panels are installed in a vertical position, which facilitates the use of the required bend radius for patch cords.

Except for switching panels in the assembly of central points of home networks, various specialized adapters executed in the form of panels are frequently installed. Thus, for example, ISDN adapters and passive splitters of TV signals made as panels are used.

The small sizes of cable systems in the SOHO sector and the accordingly small dimensions of the switching-distribution equipment used for their realization also determine the considerably smaller length of patch cords. Thus, for example, in a RUN system the minimum length of a standard cord is merely 8.9 cm, and the structure of the In House system by Ortronics includes 7.6 cm cords (in the majority of office SCS this value is not less than 60 cm). Hence, there are even cords that are inches long in some systems instead of traditional lengths of feet or meters.

When the transmission path of television signals supported by a cable system is constructed using coaxial cables, the cords of various lengths with coaxial connectors are included in the list of regular SCS equipment.

Telecommunications outlets

In residential sector, cable installation is traditionally carried out under the *hidden scheme*. Therefore the equipment structure for construction of home networks includes a wide spectrum of various faceplates. They are intended for fastening on common mounting boxes and are supplied with standard slots in order to install the connectors of various types. At the same time, the list of these hardware products, which are frequently built with unified plastic holders, along with the traditional components, includes non-typical RCA and F-type outlet modules for office SCS and some other modules that are primarily focused on the support of TV and broadcasting channels as well as voice equipment signals. In practice, there are also specialized telephone outlet applications in accordance with the national standards. When more than two posts are located on one such telephone outlet, it is sometimes called a multimedia panel. Modular multipost plates are the other version of the specialized information outlets for installation in home networks and ensure simultaneous access to several applications because of internal switches or splitters, which allow the use of an L-connection.

The creation of specialized patch panels for use in the domestic sector is also possible. An example of such a plate is the MX-WP-SS by Siemon. Depending on its realization, this plate is intended for installation a common or keystone-outlet of a standard modular connector and differs from similar products for offices by the availability of two plastic holders used to hang a wall-mounted telephone set. At the same time, one end of the patch cord is linked to a socket on the bottom surface of the telephone set and is not seen by the user, which provides a very pleasing appearance.

There are many original specialized samples of *passive components* developed for this application field and that are components of signal transmission paths, along with a common element base for use in workstations in non-office networks. For example, we note the Norwegian company Telesafe designs used in WTP cable system structure. The system components list includes an 6610-759-type installation cord, in which, in accordance with the cable-sharing principle, the grouping of individual twisted-pairs of S/STP cable is carried out in three individual outlet modules in a common patch panel. In order to make a connection to a coaxial TV connector, a *baluns* is integrated in the patch cord and the junction with a horizontal cable is made through a CMG high-frequency multimedia connector with the form of a standard modular connector. The use of four-post multimedia outlet panels such as 6800-2010 gives significant convenience upon installation. This panel also realizes the principle of cable sharing, but on the basis of another design. There is a linear band-split filter of Krone contacts in this panel. The distribution of individual S/STP cable pairs to the outlet modules with Ethernet grouping, telephone, and television connectors is made on current-carrying paths of the internal printed-circuit board. The panel is also equipped with a baluns in order to switch from twisted-pair to coaxial cable; the baluns is connected to an unbalanced television connector. In addition, a radio broadcasting outlet is placed near the TV outlet, which is electrically separated from the radio broadcasting outlet by a low-band frequency filter and operates in case of radio signal transmission together with TV signals by a frequency multiplexing principle. The use of this solution provides the capability to obtain WTP signal transmission parameters in the system paths, which considerably exceed the known category 7 requirements, despite the presence of three additional plug&jack connections.

The In House system patch panels, which are not related to the transmission signal path, are an example of the specialized custom hardware products. They differ by a vertical arrangement of outlets and have a mounting plate of rather complex spatial form. The use of patch panels allows, primarily, an increase in the bend radius of patch cords and, second, organization of the internal space that is protected from external actions, in order to carry out the cable layout at the outlets in the small space of mounting constructions.

Raceways

The hidden character of cabling that is typical for modern residential premises makes the use of raceways rare for cable layout construction; decorative conduits are not included in component structures of popular domestic-purpose SCS. Nevertheless, there are individual samples of such raceways that are focused on the residential sector. In particular, the German company Rehau offers decorative raceway with separate sections intended for central steam or water-heating pipes.

In those cases when the use of the hidden scheme is impossible or inexpedient for some reason, decorative raceway made in the form of floor or ceiling moldings are often used, along with the traditional hardware with rectangular cross section. These hardware products mainly differ from common conduits for offices by their triangular cross section. Telecommunication outlets are usually fastened along a profile due to the small size of the conduit.

6.3 Mounting Constructions

6.3.1 Design Options

The common solutions in for mounting constructions are, first, the practically obligatory inclusion of a specialized mounting construction in the structure of the components structure, and second, the almost general refusal of 19-in mounting constructions as too large for the given application in favor of smaller equipment. For example, the standard cabinet width of the IT-SCS-Mini system is 310 mm, and the panel width of the so-called (Command Center) Home Cabling System by Siemon is 531 mm (17 in). Moreover, because of the small number of serviced ports in the popular solutions, the wall-mounted versions of mounting constructions are exclusively used.

Because of the considerably variable installation conditions that can occur in cable system construction in the SOHO sector, there is a great deal of variation in mounting constructions in practice (Fig. 6.1).

Closed and open mounting constructions are the reduced models of the common 19-in equipment (cabinets and wall frames). The mounting constructions are referred to as semiclosed equipment, whose system part is openly installed outside the cabinet or outside the similar mounting equipment. Two basic ways of installing the equipment outside are possible, depending on the equipment type: switching or network equipment. Finally, a combined solution represents the installation base on which the protective housing is hung after placing the devices.

In those situations when of 19-in mounting constructions must be used for some reason, the technical solutions can be used that ensure minimization of the overall dimensions of the mounting construction, particularly their depths. The given task is solved in traditional constructions:

- By changing the hub installation method (the hub planar to the wall with the modular connector outlets turned sideways or upwards);
- By installation it in a shelf outside the cabinet (semiclosed design).

There is also a significant number of non-traditional mounting constructions, which are described in detail in Section 6.3.3.

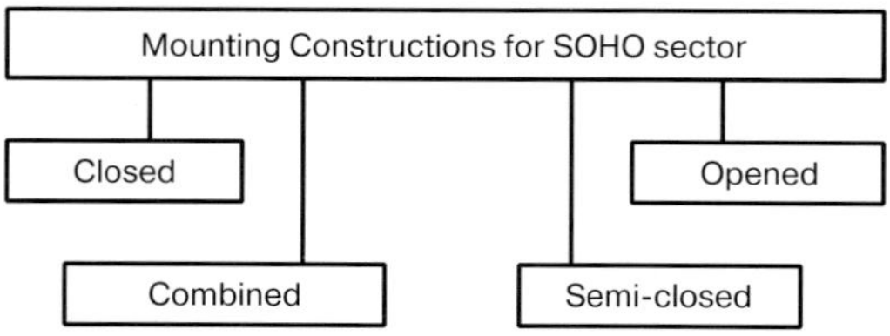

Fig. 6.1. Mounting constructions for SOHO

6.3.2 Cabinets

Design features

The most widespread installation element for non-office SCS construction is wall cabinets, which use the closed scheme of rack construction. In contrast to cabinets from "large" SCS, the given cabinets have smaller width (Fig. 6.2).

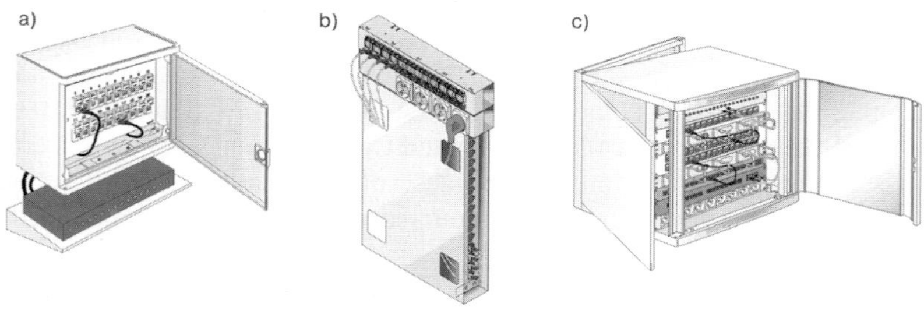

Fig. 6.2. Mounting cabinets: *a)* with adjustable shelf; *b)* case; *c)*10-inch classic

The half-format (10 in or 19/2 in) from the standard width is very popular in Russia. As is the case for for "large" cabinets as well, these cabinets are produced by companies who specialize in manufacturing of equipment for installation and by the SCS manufacturers. The advantage of this solution is also the availability of a wide spectrum of accessories (shelves, organizers, the main power supply distributors, etc.). Thus, because of its small overall dimensions and use of thinner steel sheet (the common value is about 1 mm) and therefore more economical use of metals, and the decreased requirements for manufacturing accuracy, the cost of such cabinets is 60–80% less than for 19-in cabinets of similar heights.

Based on basic technical features of 10-in cabinets we note:

- Most are built in two sections;
- Only one pair of guiding rails is used with the capability to change their installation depth;
- Other mounting solution is possible along with the common guide rails with screws and square nuts.

Because of the small power (no more than several tens of Watts) of active equipment installed in such cabinets, the use of compulsory cooling devices is excessive. In order to eliminate heat, it is possible to place several ventilating holes in the sidewalls of the cabinets.

The cabinets can be various heights. In practice, as shown in Table 6.2, the most popular are the mounting constructions that consist of 6 and 8 units, which allow service to 16–20 ports of a cable system.

In case it is necessary to increase the number of ports serviced, two cabinets can be installed one above another (for example, the IT SCS-mini system) or near each other (for example, the Molex USN-32 system). Further cascading is inexpedient, and it is more useful to use the common 19-in mounting constructions.

In practice, except for the cascading principle, solutions based on including mounting constructions of several heights (usually three heights) in the system structure gained a certain popularity. This provides high flexibility of coordination with the requirements of a specific non-office SCS. In particular, NORDX/CDT, Ortronics, Ackermann, and some other companies use to such an approach.

An example of a mounting construction in which the installation of switching equipment is realized without fastening rails is the Armural cabinet from the V.D.I. system structure by Legrand. There are four 12-mm diameter vertical round poles in this cabinet for installation of patch panels and telephone plinths. Because of its small depth (it is only 137 mm), this design is called a board instead of a cabinet in the company's technical specifications.

The simplification and price-reduction requirement has an effect on the design of mounting constructions for application in non-office SCS. Thus, in particular, this requirement is one of the reasons for the widespread use of combined methods in the realization of mounting construction. For example, the FlexiLAN cabinet system by Brand-Rex does not have a back wall and is made in the form of a case with a front door. It is hung on the mounting base for the installation of panels. The classical type of cabinet can be made without a back wall. In those cases when for some reason additional protection of internal space is required, the cabinet can be completed with a removable back wall. This element is included in complete systems (for example, the IT-SCS-mini) or, if necessary, it is ordered separately and is established in place before installation (a solution by the Czech company Zerak). The cabinet produced by IT from the IT-SCS-mini system structure is made under the "double-design" scheme, and it has an absolutely symmetrical construction, In order to change the direction of door opening, this cabinet is simply turned over by 180°.

Table 6.2. Mounting cabinets and accessories for SOHO networks

Company	Product type	Mounting construction	Modular panel	110-type cross panel
Elgadphon, Israel	Elgadphon SOHO	10-in cabinet, 6U, 3 sections	UTP and STP, 8 ports, 1.25U	–
IT, Russia	IT-SCS-mini	10-in cabinet, 8U, 2 sections	STP, 10 ports, patching panel, 1U	50 pairs
Kontur, Russia	–	10-in cabinet, 6U, 2 sections	UTP, 8 ports, 1U	50 pairs
Krone, Germany	MiniLAN	10-in cabinet, 8U, 2 sections	UTP and STP, 8 ports, 1U	–
Rittal, Germany	–	10-in cabinet, 6U, 2 sections	–	–
Molex, USA	USN-32[a]	10-in cabinet, 2 sections	UTP and STP, 16 ports, modular, 2U	–
Legrand, France	V.D.I.	10-in cabinet, 2 sections, without installation rails	UTP and STP, 20 ports, 2U	–
Brand-Rex, Great Britain	FlexiLAN	10-in housing		
Network Systems & Services, Great Britain	Midi-LAN	10-in cabinet	8 ports, UTP and STP, patching panel, 1U	–

Company	Patch cord organizer	Shelf	Power supply panel	Ancillary equipment
Elgadphon, Israel	Panel, 0.75U	0.75U, depth 159 mm	–	
IT, Russia	Circular, 3 rings	1U, depth 170 mm	2 outlets, switch	Cords 0.4 m
Kontur, Russia	Circular, 4 rings	1U, depth 170 mm	2 outlets, switch	
Krone, Germany	Circular	1U	–	Cords 0.3 and 0.5 m, gag
Rittal, Germany	–	–	–	–
Molex, USA	–	1U, depth 161 mm	–	–
Legrand, France	–	–	–	Cords 0.6 m, telephone plinths
Brand-Rex, Great Britain				
Network Systems & Services, Great Britain	Circular, 3 rings	1U, depth 100 mm	–	Telephone and ISDN panels, gags 1 and 2U

[a]USN: Uniwersalna szafka nascienna, universal wall-mounted cabinet

Note also that in the given systems because of the clear features of deployment places, it is not as necessary to protect the installed equipment from unauthorized persons. Therefore open wall frames (Home Cabling System by Siemon) or light cabinets without locking high-security front doors (RUN system by NODRX/CDT) are frequently used in these systems for equipment installation. Even the variant with a heightened security patented lock (USN-32 system by Molex) is not excluded. The need for the use of locking closed mounting constructions arises, mainly, upon installation of equipment in corridors or in small offices.

Accessories

The cabinets for SCS realization in the SOHO sector are completed with an advanced set of accessories, which consists of various patch panels, patch cord organizers, shelves, and main power supply distributors.

The realization of patch panels with modular connector outlets is carried out under two different schemes. The first is the fixed scheme, from the point of view of separate ports and is rarely used in practice. The patch panels of Elgadphon are realized under the classical fixed scheme. The so-called modular inserts for 6–8 ports of the NORDX/CDT RUN system are another example of this version. Moreover, except for patch panels in the form of modular inserts, various adapters, splitters, and other similar hardware products are also produced.

The flexible scheme of patch panel organization has become much more popular. This scheme allows the flexible coordination of system capacity with the actual requirements of the user and allows the use of a uniform type of outlet module in all systems . Thus, depending on the adopted construction principle in SCS, both modular constructions (Molex inserts such as Euromod) and board constructions (IT-SCS-mini) can be used.

Patch panels with modular connectors are supplied with common and shielded outlets. The configuration density allows use of 8– 10 ports per one panel of height 1*U*; the most popular solution is 8-ports. Along with 8-contact universal outlets, it is also possible to use the panels with 6-contact outlets for telephones (for example, the V.D.I. system by Legrand).

The panels with 110-type connectors are intended primarily for telephone applications. The well-known solutions for SOHO sectors are only available in Russian products (IT and Kontur). Note also that because of the significant overall dimensions of a 110connector plug, IT recommends the use of installation cords with modular connector plugs for switching. The idle cord end is crimped on the outlet block with 110-type contacts.

The shelves are made with edge point-to-point fastening, their height is 1*U*, and they are intended for installation of low-port hubs, routers, and other similar equipment. The working surface width of such shelves is 17 cm. Shelves are produced by D-Link, 3Com, Allied Telesyn, and Cisco. In order to provide additional fastening, holes for fastening screws are sometimes placed on the working surfaces of the shelves. In some cases, the shelf is supplied with a rear working surface that bends upward for more convenience.

The power supply panels are made under the horizontal scheme with a front outlet arrangement and are furnished with two outlet modules and a button switch.

There are two types of cross-cord organizers—slotted and ring. In the latter case, the organizers are made with three or four rings and installed under patch panels by usual methods.

6.3.3 Other Construction Types

In addition to cabinets and wall frames, other construction types are also used. For example, the so-called compact is applied in In House system. This hardware product is a 30.5×36.8×10.2-cm steel case with a demountable cover. It is mounted on a wall and is intended primarily for equipment installation of TV and radio broadcasting appliances, which are located in the nearest rooms.

The basic mounting construction element of the Czech cable system SOLARIX SOHO is a plain installation panel (with 12, 18, 24, 36 or 54 ports) with modular connector outlets. In addition to telecommunications outlets, there are slot-type organizers and a power supply outlet on the central part of this panel. In the right bottom part of the mounting construction is the place for fastening the LAN low-port hub in a vertical position. In case it is necessary to protect the equipment against nonauthorized access, the panel can be installed in the cabinet without mounting guides and with a locking door.

The wall-mounted boxes of ABCplus (France) are an example of a nontraditional way of mounting construction development. This element realizes the semi-closed scheme of construction and operates as a carrying frame and patch cord organizer. The modular connector outlets are essentially placed outside the closed part of the box, and the hub is placed planar to a wall and its port connectors are placed sideways. The length surplus of the patch cords is put inside the box between an external wall and the top cover of the hub. In order to lead these patch cords out, a large square opening is provided on the front wall. It can be that a drawback to the compactness of the solution is some inconvenience of termination cord connection to the hub, especially for a significant number of active ports.

6.4 Realization Features

6.4.1 Cable Layout

The SCS in the residential sector (both in a separate apartments, and in the whole complex) are usually realized as individual projects at the construction stage or major building repairs. SCS installation is in most cases carried out by a specialized construction and installation company. Therefore we shall describe here only those features that are frequently used in practice for SCS installation in offices of small companies. There are very popular modular solutions in this application field, for which two basic versions are characteristic: availability of complete cable system installation sets and preterminated detail kits. Their use allows rapid deployment of small SCS by the users or technicians of the automation department. The advantage of the first solution (Section 3.5.1) is the lower cost of initial components. However, practical realization of cable layout on this basis requires certain qualifications and skills of the users as SCS installers.

The advantage of the second solution is its extreme simplicity for practical realization (all work is reduced to three basic operations: outlet and panel installation, preterminated cable cabling, and their connection). In addition, with its use the danger of incorrect installation is considerably reduced and it removes the need to read user's guides, to view training videos or compact discs, or to test the assembled system.

The manufacturers of equipment based on preterminated hardware products can consciously break the provisions of the standards in the field of allowable sockets number in a path (including plug&sockets). For example, note the PYPLine system by AMP. The system is based on the use of original multipair cords with a short branch from a multipair connector. The modular connector plug is mounted on a cable branch, which is connected to the outlet, and the multipair connectors are used to increase the cable length and to connect to the following outlet. The allowable number of sockets in a path is logically described, first of all, by the small probability of the use of cable lines of over 30-40 m in a SOHO sector, and second, by use of applications with rates more than 10–16 Mbps in this area.

The distinctive feature of the PL system by the Hsing Chau Industrial Co. (Taiwan) is its complete lack of horizontal cables, as all cable layout is exclusively carried out with patch cords. The minimization of the number of plug&sockets in a signal path (there are no patching cords in the system and terminating cords are used) is the confidence in obtaining characteristics of category 5 based on NEXT. The connection of the patch panels and telecommunications outlet at workstations (both elements are realized on the basis of I-adapters) are carried out by cords with one-wire (monolithic) conductors (in fact, it is the preterminated cable), which operate as horizontal cables. Such an approach is necessary to decrease attenuation.

In some cases, the installation location and features of the transmitted information type cause the designers to breach the particular provisions of the normative technical documents, which standardize not only the cable type, number of connectors, etc., but also the construction principles of the SCS horizontal subsystems. Thus, for example, in order to compensate for high attenuation and to eliminate strong amplitude–frequency distortions in broadband multichannel television signals, which occur in standard S/STP cables at the top part of the frequency range (more than 500 MHz), an additional corrective amplifier is used in the WTP cable system. This device is used for paths more than 60–75-m long (the distance depends on the cable type used and on basic amplifier–distributor type in a technical room). It allows, along with transmission path splitting, an increase in pathlengths up to 90 m. The amplifier and the panels are installed in a 10-in cabinet, and all necessary connections are carried out by patch cords. If this solution can be considered as a peculiar analogue of a crossover point or a consolidated point of “classical” SCS, then the use of active equipment and the opportunity to carry out the administration functions in it are outside the working standards.

6.4.2 Active Equipment Installation

The more frequent use of specialized active equipment, which is included in the cable system structure as standard equipment, in comparison with "large" systems is a distinctive feature of non-office SCS. It also concerns equipment by third-party companies that have SCS manufacturer identification numbers and their various warranties. The offered hardware products provide functioning of a LAN, of various types of TV signal distributed on separate sockets, as well as reception of wire and radio broadcast signals (Table 6.3).

Table 6.3. Active equipment by various companies

Company	SCS type	LAN fiber-optic equipment		Radio equipment	TV equipment	LAN	Monitoring systems
		Installed	Noninstalled				
AESP, USA	Signa-Max	–	–	–	–	•	–
Alcatel, France	Alcatel Cabling Systems	•	•	•	–	•	•
AMP, USA	NetConnect	–	•	–	•	•	–
Datwyler, Germany	UNILan	–	–	–	•	–	–
INFRA+, France	IBCS	–	•	–	•	–	–
Lucent Technologies, USA	SYSTIMAX	–	–	•	–	–	–
Network Systems & Services, Great Britain	Midi-LAN	–	•	–	•	•	–
Ortronics, USA	GigaMo, GigaMo+	–	•	–	–	•	–
RiT Technologies, Israel	Smart	–	•	–	–	•	•
Telesafe, Norway	WTP	–	–	–	•	–	–

A version of LAN equipment installed in a mounting construction is shown in Fig. 6.3.

Hubs and other LAN equipment for use in home networks, because of the limited number of appliances connected to a network, have usually small number of ports. Therefore they are frequently mounted in special transition panels, which allow to their installation in mounting constructions (for example, so-called 5- and

8-port LAN modules of the RUN system with Ethernet hubs of the 10Base-T system).

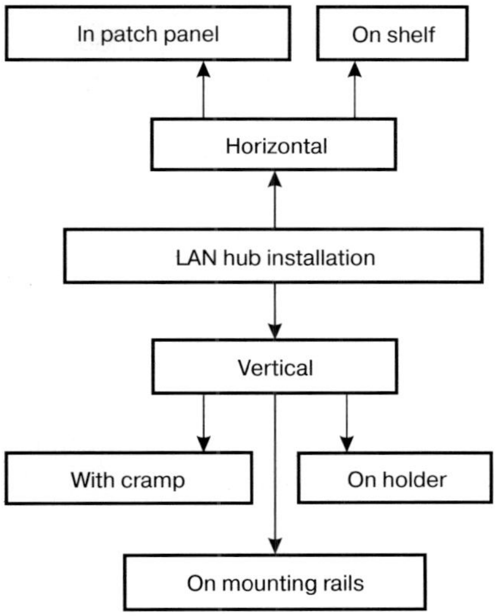

Fig. 6.3. Installation options for LAN devices in mounting construction

In those situations when the SCS-based local network is realized with the help of the hub that is not included in the delivered equipment set of a cable system, it is usually installed on a shelf. The shelf can be in a cabinet (Molex designs) or it can be installed under a cabinet (Armural cabinets by Legrand). As a rule, hubs may be installed in many ways in practice. In those situations when the customer wants the LAN network equipment of a specific manufacturer, integrated fastening elements can be provided. For example, the shelves of an Elgadphon SOHO system may contain holes for fastening screws for an Allied Telesyn microhub. It is possible to use the holder intended for a certain network equipment type, except for a common shelf. The 195402-0016 type of Midi-LAN systems by Network Systems and Services (NS&S) that was designed for a Bay router is an example of such a holder.

The solutions based on vertical installation of hubs are commonly used. The basic advantage of this solution is the decrease in the weight of the construction at the expense of decreasing its depth with the corresponding cost reduction. The installation is made directly with the help of special gripping or on the mounting base that carries out the adapter functions. In the latter case, fastening is done by the use of adhesive tape or even magnetic holders.

The hub, installed in a vertical position, is mostly turned with modular connector outlets directed sideways or upwards. The last version, despite its universal expansion, is less preferable because of the danger of extraneous objects and dust ingress in outlets. This version is allowable only for closed constructions such as cabinets. An example of a design based on the side installation of the hub in the operating position is KISS-type cabinets by the German company CNZ Neubauer.

For greater operation convenience, the frames of this cabinet are L-shaped (one of planes is used for the hub installation and the second for the panel installation in a vertical position) and can be turned.

As described above, optical transmission paths are rarely used in SCS with small number of ports. One of the few exceptions is the connection to the basic network of the small dedicated workgroup. This task is usually solved with the help of a wall-mounted socket. Another possible variant is the use of a custom design, such as an open shelf with a front wall by ITT and Zerak companies. A medium converter is installed on the shelf, and its electrical output is connected by a mounting cord to the electrical outlet in the front wall (Fig. 6.4). At the same time, depending on the converter design, the outlets of the electrical and optical connectors can be placed in parallel or transverse to the front wall. The front wall protects optical cables from possible damage during routine operation. The drawback of such an approach is the absence of an organizer of the optical waveguides and splices. Use of external cables is unlikely for the case, and it is better to choose a more simple construction. The splice-plate installation on a back wall of USN-32 cabinet is one more solution of an optical communication channel creation in "small" SCS and this solution is made by Molex company.

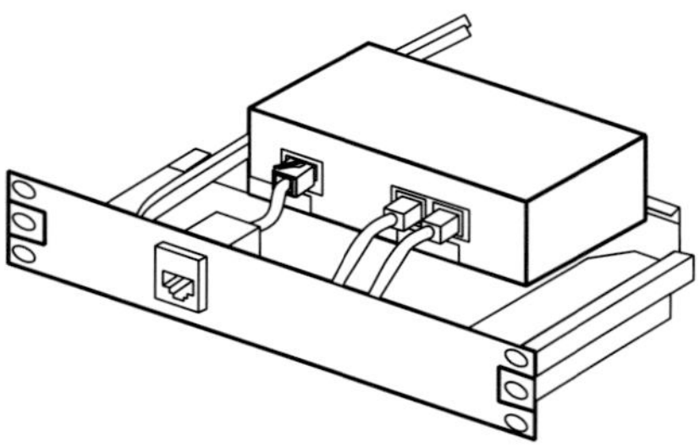

Fig. 6.4. Converter in 10-inch cabinet

6.5 Conclusions

The use of structured cable layout in non-office public buildings, small offices, and residences provides functioning of the basic modern telecommunications services, as well as new forward-looking communications systems. Moreover, in combination with administration flexibility and simple adaptation for specific requirements, the cable layout gives other similar advantages.

The equipment for non-office SCS construction is produced by a wide range of the manufacturers, therefore many independent designs of low-current cable layouts are available.

The popular systems are realized under the basic principles of full-scale "classical" SCS construction. At the same time, the specific features of this field and the small number of ports have an effect on the approaches and solutions used.

Among the most frequently occurring differences we note the admissibility of coaxial cables and a lack of rigid observance of some provisions of the TIA/EIA-568-A standards and ISO/IEC 11801 standard for construction of horizontal cable layouts.

There are small differences between “small” and “large” SCS in the installed equipment field, pleasing appearances can be obtained by omitting standard 19-in cabinets and frames in favor of designs with reduced dimensions or custom designs with a common set of accessories.

7 Specialized Active Network Equipment for SCS

The layout of an SCS inside a building or inside a group of buildings became a business long ago. Nevertheless, SCS usage for the general body of companies that work in the information technology market is often the first stage in the creation of a customer's enterprise infrastructure. In that situation, delivering and mounting the fair quantity of different types of equipment begins immediately after mounting the cable system for the specific project. Previously, equipment selection, on the basis of which higher levels of information and computing systems are built and the system is adapted to the conditions of the specific project, were the prerogative of system integrators. Recently, specialized companies that manufacture separate components of SCS as well as systems as a whole often began to deliver active network devices. Their interest in this direction is conditioned not only by the sales growth volume (at the expense of the spread of nomenclature for the proposed equipment) but also by the important factor of giving clients and partners a complete package from one company. It also gives them market advantages over competitors.

As a whole, rules and principles of SCS creation, as a technical object, are described thoroughly enough in a number of active and prospective American and international standards. SCS, a complex technical system with tens of thousands of elements in its composition, has sharply defined the composition of component interfaces and structures. In principle, there are no power supplies and active equipment among them. Different SCS manufacturers and equipment manufacturers use a different approach to the problem of the introduction of active network units into their system composition as a market product. These units are not standard from the point of view of well-known native normative and technical documents. Many manufacturers follow “the rules” carefully and make no mention of any specific active network units in their catalogs at all. Other companies with the above-mentioned advantages more or less introduce different active units into their SCS (Table 6.3). Sometimes these units are divided into a certain subsystem with their own trademark. The manufacturers that follow this approach are such leaders in the SCS market as Lucent Technologies (WaveLAN radio system), Ortronics (TransOptix system of fiber-optic media converters), Alcatel (FiberCon lower-level fiber-optic units), and some others.

In practice, a great variety of additional equipment is offered to users by the SCS manufacturers because of the universal routing of up-to-date cable systems. There is a well-known suggestion in this field to introduce into official catalogs radio communications units, TV and radio broadcasting units, different intruder and fire alarm sensors, LAN units, and so on (Table 6.3). It stands to reason that in such situations, it is more expedient to use equipment of the lower level. First, this equipment directly uses SCS resources and is a "logical" extension of cable system for users and installers. Second, the specialty of the application field is that it is no need for these devices to achieve particularly high and superior parameters. But the typical nature and the high volume of output of the element base that is allocated inside the equipment makes it possible to sufficiently minimize the price of the finished product.

7.1 LAN Units Using Fiber-optic Interfaces

The last several years of SCS creation were marked by the rapid growth of fiber-optic usage, including all levels of realization of very different projects. The objective prerequisites of situations in this field are the well-known advantages of fiber-optic communication lines, such as high capacity of signal transmission route, long range, galvanic isolation of message source and receiver, and lack of sensitivity to electromagnetic interference. A significant decrease in the cost of fiber-optic elements and the introduction of mounting methods into wide practice that use inexpensive equipment, as well as the ability of systems to be mounted by installers of medium skill are also factors. Taking into account all these items, fiber-optic lines are nearly always used to organize backbone subsystems and, in some cases, their use is also economically sound when creating horizontal parts of cable systems. At this, in most cases, the high capacity of optical methods leads to their use for providing LAN functions. Up-to-date LAN apparatus differs from the other kinds of network equipment by its very high requirements for this parameter.

Currently, four basic approaches are used to create a terminal part of cable layout on the basis of optical fibers. According to these approaches, the optical fiber cable goes straight to the user's work area or terminates near a user's desk. It can be the usual horizontal SCS subsystem with a centralized optical architecture, according to TSB-72 (Section 1.4.2), with the concepts of fiber-to-the-desk (FTTD) and fiber-to-the-office (FTTO) that are often called fiber-to-the-room (Fig.7.1).

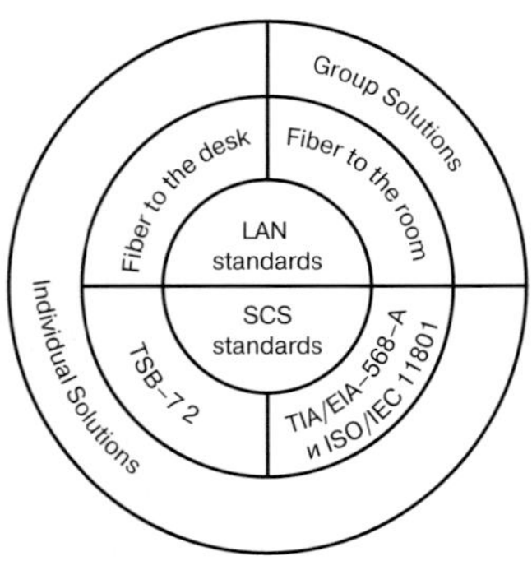

Fig. 7.1. Approaches to construction of fiber optical lines at the lower layer of enterprise informational infrastructure cabling

Due to similarity of the main ideas that were used create these structures; the technical solutions seem similar to each other by the principles of their realization. The most significant difference between them is that, in general, the horizontal subsystem and centralized optical SCS architectures have strong and fairly rigid quantity limitations to the main route parameters that are determined by national and international SCS standards without reference to any network interface. FTTD and FTTO concepts are more the principles of LAN creation with reference to the corresponding standards for fiber-optic interfaces of network equipment and are not described in any normative or technical documents that are known to the authors. It should be noted that, regardless of initial prerequisites, all equipment that is discussed further can use SCS routes for information transmission because the parameters that are determined by SCS and LAN standards are similar enough and do not differ in principle.

7.1.1 Installation Equipment

General principles of the creation of installation units

The installation units enable the realization of FTTO principles in practice. The typical distinguishing features of a given network equipment variant are:

- installation directly in the office premises with compulsory use of decorative boxes or other functionally similar mounting elements;
- orientation to use at the lower LAN level and compulsory availability of a user's interface based on the standard outlets for network equipment (usually, modular outlets; more rarely, AUI outlets) to connect a workstation and another similar units (for example, printers or scanners). That is, units that do not make high demands, in general, on the network resource volume;
- presentation of a standard communication channel that has a maximum bit rate in the range 10–100 Mbps to a single user or to a small group, in this case, more often used for a capacity-splitting mode;
- use of numerous circuits and construction decisions that are unique, in some cases, for active LAN network equipment. These factors allow the full adaptation of these units to the specific conditions of their installation and operation.

As early as the middle of the 1990s, these units ceased to be novel in the network equipment market. Currently, the serial production of these units has been launched and the units are very popular in German-speaking countries [107]. The units, in general, are oriented to work in the composition of Ethernet networks. The solutions for Fast Ethernet networks appeared later and were seldom distributed, but their market share is growing with the increasing popularity of this network standard in all branches of industry. The list of Ethernet units of both types being offered, in general, coincides with the nomenclature of the usual lower-level network units. It includes transceivers, repeaters, media converters, and mini- or microhubs of different kinds, but the most popular is the last version of the equipment (Table 7.1). Token Ring hubs with automatic bit-rate selection options of 4–16 Mbps are units for functional support of other applications. The varieties of equipment that are distributed more or less correspond to the numbers of units of this network standard that are produced in world practice. At the end of the 1990s, technical information from some foreign companies about finalizing the ATM hub development appeared, but any mention of their appearance in wide commercial distribution is not known at this time. In particular, in the official catalog of the Telegartner Company for 2000 there are no technical specifications for ATM hubs, no part numbers for ordering units, and they are available only by special request.

Table 7.1. Installation devices with fiber optical interfaces by various manufacturers

Manufacturer[a]	Ethernet media converter	Fast Ethernet media converter	Ethernet transceiver	Ethernet hub
Ackermann, Germany	–	–	–	•
Alcatel, France	–	–	–	•
Dafur, Germany	–	–	–	•
FiberCraft, Germany	–	•	–	•
Microsens, Germany	•	–	•	•
Telegartner, Germany	–	–	–	•
UBF, Germany	–	–	–	•
Manufacturer[a]	**Fast Ethernet hub**	**Fast Ethernet microswitch**	**Token Ring hub**	**Optical port connector type**
Ackermann, Germany	–	–	–	ST
Alcatel, France	•	–	•	ST, SC, MT-RJ, VF-45
Dafur, Germany	•	–	•	ST, SC, MT-RJ
FiberCraft, Germany	•	•	–	E-2000
Microsens, Germany	•	–	•	ST, SC
Telegartner, Germany	•	–	•	ST, SC
UBF, Germany	–	–	•	ST

[a]In this case, production is assumed to be the output of units at a company's own production facilities or equipment made by other companies as OEM products.

It should be noted, that there is a limited number of installation devices that contain up-link modules based on UTP.

They are used only to build up LAN ports when there is a shortage of lines, and to cascade several devices, first of all, microconcentrators. Their limited application and range make them less popular.

Full-scale production of installation units is made by SCS manufacturers (mainly, European companies, among them Alcatel, Telegartner, and Achermann) and independent companies. We will not discuss the differences among these manufacturers, since the installation units operate primarily over the optical routes of the building cable layout that is, in general, SCS. Besides, current practice shows the great popularity of introducing installation equipment into SCS as OEM units.

The installation units differ from their desk or 19-in analogues in common usage by some construction features. As a rule, these differences only minimally affect the functional characteristics of the components of the local network, but they allow the full adaptation of these units to the mounting and maintenance conditions that are specified from the point of view of network equipment usage practice. Their installation at the state work area is made by the same principles as the installation of the usual internal information or power SCS outlets. The fastening on the sidewalls or on the button of the ducting as well as on the element of functional analogue to it (decorative column, under-floor ducting) is used directly or through an adapter. According to the manufacturers recommendations, the mounting time and the time to put it into operation shall not exceed 5–10 min. Moreover, the time is only for the installation of the unit at the work area, power supply cable connection, fiber-optic cable connection, as well as the connection of other cords.

A high level of aesthetic characteristics is one of the paramount demands that are made on the equipment directly installed in office premises. To realize this condition, external decorative elements of the installation units are usually designed by taking into account the manufacturer's style of any specific of SCS installation component (ducting, different kinds of columns, and under-floor boxes). The user has often the opportunity to make a choice from several versions of a design. The base module of a unit has a similar construction for all types and is adapted to a specific manufacturer's style by using detachable plastic decorative covers, mending plates, and other analogous elements of different colors. The decorative elements can be fastened by screws or by latches. The nameplate and a protective title label often cover the flat side of the screw head.

Installation units are designed to only minimally differ from the usual SCS outlets. Sometimes the manufacturer of the installation units makes them in two main versions; they are installed into raceways or under-floor boxes of one or some specific manufacturers of such equipment. Despite the increase in assortment of devices, the manufacturer takes an advantage by producing them for installation without using additional adapters.

In a great number of cases, the installation units give normal use of the network devices to maintenance personnel and to the users. Naturally, it is fully provided for by the support of the most complex multiport representatives of installation group units (by hubs and switches). Particularly, these units, more or less, realize the following functions:

- support the automatic electric port switch-off from an internal bus when this port fails;
- provide maintenance by SNMP protocols using a graphic interface;
- change the inside tuning by introducing the corresponding constants into the ROM (read-only memory) through the COM-port;
- have a microswitch to reset the apparatus circuits.

However, the hub, as well as any other installation unit, is usually supplied with a number of colored LED indicators (for example, receive, transmit, collision, and power) that make it possible to determine with high accuracy the current state of both the separate ports of the unit and of the network segment as a whole. To increase the volume and effectiveness of information, the LEDs can operate in continuous and pulse modes. In the most cases, the LEDs are located on the decorative cover of the unit; if there is a grouping of a pair of hub outlets with a fixed configuration, the LEDs can be located between the groups of the modular outlets (see Fig.7.2).

Fig. 7.2. Microhubs with four ports for installation in raceways: left-horizontal option, right-vertical option

Features of optical ports

In principle, the optical port connector outlets of any installation unit are located in the lower part of its case and are not available for the user in the normal maintenance mode — that is the main difference in this kind of equipment. On the basis of the reasons that follow, there are two main versions of this port, which can be a built-in fixed module or a changeable module. The second approach is the most popular. In general, the practical realization of the up-link port as a hinged module or unpackaged module that can be inserted, that is, a changeable unit, is widely distributed during the design of network equipment. The most significant differences in the field under consideration are:

- the up-link module, in a great number of cases, has only a fiber-optic interface (single module samples with UTP ports are known, and BNC ports are distributed widely; AUI and MII ports are no longer used in this apparatus);
- the up-link module is not made in the form of an unpackaged slide-in unit, but normally has a fully closed case that is typical for hinged microtransceivers.

In operation, the module fits tightly to the housing of an installation device that makes the whole construction very compact. Several pin contacts that are divided into two equal groups and that are symmetrically located relative to their long axis are provided for on the flat side of the module case, along with a spring contact to provide continuous shielding. For example, in the units made by Microsens, this module has 32 pin contacts. These pin contacts are used to power the unit and to connect the unit to the electrical circuits of the main unit. Such a construction feature makes it possible to install the module, depending on local situation, in one of two or four (depending on manufacturer) positions with discrete of angular offsets of 180° or 90°, respectively. For this purpose, the hub case has two or four equal lead-in outlets with contacts. These lead-in outlets do not protrude from the hub. By selecting the angle of installation, it is possible to obtain the proper bend radius of the optical cable inside the limited space of decorative ducting, on-floor columns, and under-floor boxes. It should be mentioned separately that there are no additional elements to mechanically fix the module in the operating position, that is, the module is supported in its static position only by pin friction in the outlet.

The fixed port installation is used, mainly, in those situations when the developer is trying to minimize the mounting depth of the unit by decreasing its maintenance parameters and by increasing its sizes of that part of the unit that is larger then the ducting. To illustrate such a situation, let's look at the technical specifications of the FiberCraft G014-10 and G014-100 microhubs. Both units have similar types of connectors (E2000 low-profile duplex to minimize its height), but in the second case, the use of the detached version of the module increases the minimum mounting depth from 32 to 48 mm, that is, one and one-half times. Also, it should be noted that the most manufacturers select the minimum value of this installation unit parameter in a fixed configuration that is equal to32–35 mm, regardless of the outlet type used. This is due to the sizes of the E2-type mounting boxes that are widely distributed in Europe. The adaptation of units with fixed configurations to the specific conditions of maintenance is being done in some constructions since it is possible to easily dismount the optical interface board, turn it by 180°, and fix it in this position.

As a rule, the user or system integrator can select two or more types of up-link modules or types of fixed configuration units with the main outlet types of optical SCS connectors: SC and ST. The manufacturers' conservatism is not excluded from this field and, in some cases they fail to support SC outlets at all, probably to minimize the price. It may also stem from their unwillingness to change the production process, taking into account the fact that this connector is not available to the user. Through mid-2000, the other types of connectors were used more rarely, although developments have begun to appear based on the SFF-perspective (small form factor) products (MT-RJ, VF-45, and others, Table 7.2).

By its information transmission characteristics, the optical ports of the well-known installation units have no noticeable differences from the up-link ports of usual network devices. So, the nameplate range of the most popular multimode transceivers is 2000 m, which is a tribute more to the LAN standards than to the sensitivity of the receiver. The range of 412 m that is given for Fast Ethernet hubs is not determined by the power parameters of the optoelectronic elements, but is determined by the well-known limits of the protocols over the diameter of the network. It should be noted that there are only very broad proposals for lower-level networks of single-mode variants and that single mode transceivers operate at a wavelength of 850 nm (the main technical specifications of typical units of this class are shown in Table 7.2). Moreover, some manufacturers specify this spectral range of single-mode units as the main one, and the spectral range of 1310 nm is considered an additional one because of its higher cost. The use of such a nonstandard and illogical solution, at first sight, is easy explained by the lower prices of shorter-wavelength laser emitters (Section 4.1.4), and by the lack of the need to get longer parts, which is caused by the field of application.

Table 7.2. Installation units of the Microsens Company: technical specifications of a single-mode fiber-optic up-link module

Wavelength, [nm]	850
Transmitter power, [dBm]	-20
Receiver sensitivity, [dBm]	-32.5
Range of standard single-mode fiber in accordance with technical standard G.652, [m]	2000

Installation microhubs

The data in Table 7.1 suggest that the most popular versions of the installation units are the Ethernet and Fast Ethernet hubs, since they are included on the standard lists of all manufacturers. These units, often enough called mini- or microhubs, functionally are the usual hubs with several ports of type 10 Base-T or 100 Base-TX (see Fig.7.3). Naturally, they have fixed configurations with the outlets of modular connectors for user's ports and optical ports as built-in or detachable up-link modules to provide connection to higher-level equipment.

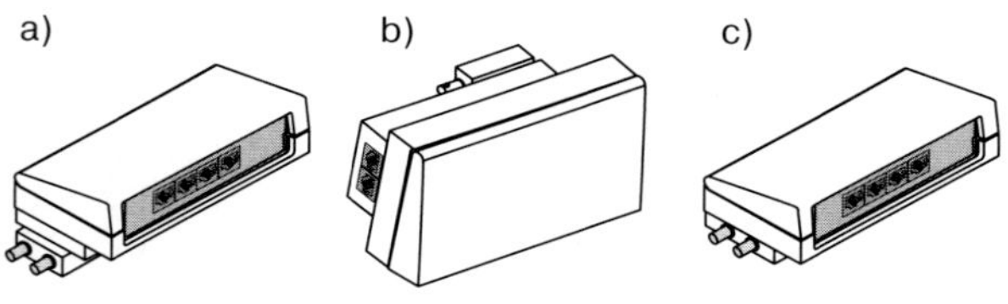

Fig. 7.3. Installation microconcentrators: *a)* with removable module of fiber optical uplink-port; *b)* with additional electrical ports; *c*) with fixed fiber optical port

Due to standardization, units supporting different network standards very often have cases of the same type. Since Fast Ethernet hubs have greater power consumption, they can be provided with perforation holes in the case. These cooling perforation holes make it easy to achieve normal temperatures of the active components of the unit's electronic circuits, and except for marking signs, are, in fact, the only visual difference from lower-speed prototypes.

The outlets of the electrical ports that are provided to connect a user's workstation and other network units, such as printers and scanners, are located in most models in the upper part of the case and can be fulfilled with straight or angular mounting. Two arrangements are used. In the first, a single row of outlets is available; in the second, the outlets are grouped by pairs, with their locations in the separate independent units identical to the outer fulfillment. The advantage of the first arrangement is its great density of ports and the ability to increase their quantity in case of corresponding order without radically remaking the unit. The advantage of the second model is its increased mounting flexibility, since the outlet blocks, depending on fulfillment, are located above each other (so-called vertical version) or side-by-side (horizontal version).

Outlet blocks in which the electrical circuits of the easily detachable upper side are connected to the circuits of the stationary base by short flexible ribbon cables are widely used. This makes it ease to change the outlet orientation of the modular electrical connectors of the user's ports (depending on the specific mounting conditions) by block replacement and to use the same type of the hub for installation in the vertical or horizontal part of the ducting.

Outlets that are grouped by pairs are found in units of the FiberShare series by Alcatel. In these units, the electrical port outlets are placed in groups of three or four in one square module in order to minimize the size and for use with the newest elements. This slightly decreases the time for microhub adaptation to the specific conditions of its installation.

In widely used models the number of the electrical ports varies from two to eight; the most popular is the four-port hub. Most probably, this is explained by the efficiency of such units for servicing a group of four work desks in the open offices that are widely distributed abroad. Usually, the manufacturers of this kind of equipment state that it is possible, within some limits, to change the number of electrical ports by special order in cases of microhub fulfillment based on a single row. Also, it is possible to add a backbone optical port by two internal electrical ports with standard range of 100 m. These ports, which are modular connector outlets that are analogous to optical ones, are not available to users in the normal maintenance mode of operation, but are provided primarily to connect two ports of a remote information outlet to the microhub. Sometimes such features are presented in the name of the unit, for example, the 6+2 Port Ethernet Installation Hub made by UBF (six user ports and two expansion ports). The ports, in contrast to optical ports except in their installation place and assigning, have no differences in their telecommunication parameters from external ports. Also, it should be noted that the limited sizes of the microhub cases do not to allow choices in devices where the configuration is fixed by the optical ports.

In some cases, the installation units also provide the ability to cascade. The necessary condition of this functional support is the presence of the "linear" electrical ports, which are in the references in this field, invisible for the user part of the unit. Particularly, Alcatel FiberShare TP3 units support the cascade function.

Because of the minimal power consumption of modern elements that are used for installation units, typically there is no problem with normal temperatures, even for operation in the limited space of decorative ducting. Exceptions include microswitches and some early models of Ethernet hubs that use miniature forced cooling units, or where the case perforation that was mentioned above provides cooling.

Features of power supplies for installation units

Currently, in LAN both internal and external direct current power supply units are used with the electronic components of installation units. External power supplies are usually supplied for use with media converters, repeaters, and other single-port units; two variants can be found. The most widely distributed is the "installation type" of power supplies, that is, those that are installed into adjacent mounting boxes and closed with decorative covers. In this case, to improve mass and size factors and to lower cost, a single supply can be used in the group mode, that is, several units can be connected to it (usually two units). The hinged power supply design, usually an external power adapter, is rarely used due to its inconvenience (the danger of accidentally switching it off and the need for a separate outlet). For example, this scheme is used in the Alcatel 3-port FiberShare 3+ microhub. The built-in power supply is more typical for different types of hubs, that is, for multiple port devices. A connector block on the case is used to put the power mains and the ground wire to units with built-in power supplies. There, an additional plastic fixer is provided for the power cable. Only FiberCraft uses a 30-cm long power cable integrated into the case in some of its products.

To increase the functional flexibility and maintenance reliability of power supplies, the power supply circuits are designed to provide normal functioning of the installation unit when the power varies within certain limits (usually, the range is from 85–100 V to 240–265 V. Sometimes the range is divided into two subranges, with nominal voltages of 110 V and 220 V that are selected by a switch). This also solves the problem of connecting the unit to power mains that have different nominal voltages. In most cases, the power supply is constantly connected to the mains, although sometimes a small-size switch is provided for and installed by special order (Telegartner Company).

Among the other technical decisions that are made during the design of microhub power supplies, one should note the need for silent operation. The main point is that when the network units are not in the operation mode, the power supply is switched to low-consumption mode by turning off part of the electronic circuits. In this case, the power consumption is decreased to 20–25% of its nominal value. For example, the 4-port ISABEL 100 Fast Ethernet microhub made by Dafuer has a power consumption of 1 W versus 5 W when it is under a full load. This mode of operation is used because the unit is switched off from the local network a large part of the day (13–14 h a day, for example, from 6 o'clock p.m. until 8 o'clock a.m.), since the workstations of users are switched off. The data indicate that the main benefit is achieved not only by power economy but also by decreasing the heat load to the electronic components of the circuits, which leads to an increase in maintenance reliability.

Economic aspects of using installation units

When realizing a specific project, the technical and economic advantages of using installation units, particularly the multiport versions of these units (microhubs and microswitches of different kinds) are determined by several circumstances. First of all, these units surpass the 100-m limit of the length of twisted pairs at bit rates of more than 10 Mbps. Thus, their use gives designers more freedom for building cable systems. In particular, centralized optical SCS architectures can be designed according to TSB-72, and small workgroups, up to 2000-m remote, can be connected to the network. Second, since their use minimizes the number of optical interfaces as well as optical cable consumption, the capital expenditures to arrange the communication are minimized [108]. Third, during network updates there is no need to change LAN network adapters with UTP interfaces that were bought earlier. This means, in turn, that the transfer process to full optical networks becomes smoother and is not so sensitive to budgets. Finally, the use of installation microhubs decreases the number of ports for passive and active equipment in installations, which decreases the cost of their construction.

Some manufacturers of installation units insist that the usage of microhubs instead of traditional network adapters and converters with optical interfaces makes it possible to save 75% over other alternatives. However, this is not true in practice, because of the significant price of the installation unit. Because of the small size of installation units, it is necessary to use miniature-mains power supplies inside the units. Nevertheless, the cost is cut by 50% for full microhub capacity usage.

Wide-spread introduction of the reviewed installation units is limited in spite of their economical advantages, in general, for technical reasons:

- the units require local power supplies, which leads to a lack of communication when the main is damaged;
- the installation units can be mounted only by internal schemes inside large cross-section ducting or in under-floor boxes, which leads to a noticeable price increase for SCS realization (this is the case for Russia);
- the ports that are created by the units serve only LAN of single type, that is, they are not universal, which is a distinctive part of SCS.

The last disadvantage, to a great extent, is offset by a great variety of hubs and other unique network devices of different systems that are made and that are changeable on-site, as well as by the readiness of manufacturers to modernize their products within broad limits according to the requirements of customer specifications. In addition, the first examples of microswitches that support the autonegotiation mechanism over electrical user ports are appearing in the market. In connection with computer telephony development, it may be possible to make versions that transfer telephone operation through the installation units as the second main user of the SCS resources.

7.1.2 Non-installation Units

Non-installation-type units differ from installation ones only by a single main feature that determines their subsequent construction features: these units, in principle, are not provided for mounting in ducting, in floor columns, or in under-floor boxes. Therefore their linear optical ports are easily accessed by the user or system administrator in the maintenance mode. Switching, measurements, technical maintenance, and other analogous operations are simpler than in installation units.

In respect to LAN creation principles, the nearly all non-installation units realize the fiber-to-the-desk concept. There are single samples of microhubs with optical up-link ports whose use allows for the creation of a network based on the fiber-to-the-office principle.

The list of the manufacturers of non-installation-type active LAN equipment to be used in SCS composition to a great extent coincides with that for installation units. The same is true for the standards supported: most of the equipment is provided to work in Ethernet and Fast Ethernet networks; the units using Token Ring standards are seldom used. The manufacture of equipment for AS-400 systems is even less popular, and proposals for ATM devices are not known at all. The aggregate group of non-installation units is analogous to that of installation ones and repeats the nomenclature of the usual lower-level LAN networks. Hence, significant differences are seen even at the production output levels of different kinds of equipment: the most popular among users and, correspondingly, manufacturers are the media converters rather than microhubs.

In contrast to the installation units, everything that is considered below is oriented to the realization of the fiber-to-the-desk concept. It allows designers to make a transfer from a basic optical cable layout (basic cabling), when the active network equipment continues to use "copper" network interfaces, to a completely optical SCS subsystem (complete cabling). In addition, the use of such units makes it possible to more effectively realize the so-called collapsed backbone of an enterprise that, in combination with a large range, creates improved conditions to solve the problems of switching of information flows and supporting the functioning of multimedia systems.

In general, the great number of non-installation types of units may be divided into the units that are used directly at the users' work areas and the units that are installed on the technical premises.

Units for users' work areas

LAN serial network equipment with fiber-optic linear interfaces may be classified by different criteria. In this specific case, it is convenient to consider the form of construction fulfillment, as the basis of the classification of LAN network equipment that is part of SCS and that is primarily oriented for use at the users' work areas. That is, the equipment is classified by whether it is packaged, hinged, or unpackaged.

Packaged units are those units (in this context) that have closed cases with outlets for electrical and optical interfaces, indication elements, and management, as well as the outlet for the mains, if necessary. Usually, such units are provided for use on the desk or for external connection to the workstation, including use of their outlets and other construction elements (by the hinged scheme). In mid-2000, these units are the primary state components of optical SCS developed by the original equipment manufacturers of optical connectors. Currently, the most advanced devices of this type are offered by FiberCraft and 3M (Table 7.3). The manufacturers of classical SCS, in practice, rarely include these devices in their catalogs (for example, AMP).

Table 7.3. LAN fiber-optic equipment with original optical connectors

Manufacturer	SCS type	Connector type	Standards supported	Equipment structure
FiberCraft, Germany	Flex OS	E-2000	10Base-FL, 100Base-LX	Network adapters, media converters 802.3 and 802.3u, microhubs, switches
3M, USA	Volition	VF-45	10Base-FL, 100Base-SX, 100Base-LX, 1000Base-X, 802.5j	Media converters 802.3, 802.3u, and 802.5j, network adapters, hubs, switches
Alcatel, France	Alcatel Cabling System	MT-RJ	10Base-FL, 100Base-SX, 802.5j	Media converters 802.3, 802.3u, and 802.5j

The hinged units are the least popular with designers and manufacturers. These units also have completely closed cases, therefore they can be considered as a variety of the packaged units. The units are separated into a special subclass because they are hinged directly at the SCS outlet. An example of a typical product of this type is the by INFRA+ (France) media converter of 7932 NC. The unit is a part of the company's IBCS cable system and supports 10Base-T and 100Base-TX Standards by its electrical ports (depending on modification). An external detachable mains adapter connected to the adjacent outlet provides DC power supply. An external cord with the standard two-pin plug provides DC power. The reviewed solution represents itself a real mini-complex, and is primarily interesting because of its closed integration of active and passive parts that was designed during the development of the construction. In particular, the outlet module of the RJ OPTO-type (product 7220) where the converter is installed, has a modular connector outlet as well as an SC duplex optical connector outlet. Because the converter is heavy, mechanical stability is provided on mounting by the additional fixation of its case with two screws at the main fastening screwed elements of the outlet module. The converter is provided with an internal Y-adapter with two outlets at the front panel. When the unit is connected to an optical port, the plug of this adapter fits into the outlet of the main modular connector of the RJ OPTO module. The logical continuation of the complex from the SCS side is the MNC880FO hybrid cable, which contains not only the electrical elements with 4-twisting, but also the fiber waveguide.

Unpackaged LAN units are the most popular variety of the equipment used at the users' work areas. The main reason for the wide introduction of different unpackaged versions of network equipment with fiber-optic interfaces is, perhaps, the increased popularity of workstations with integrated network adapters. In this case, the use of traditional hinged transceivers and external media converters is not convenient, mainly from the maintenance point of view, and for financial reasons it is often just impossible (Table 7.4.).

Table 7.4. Modular converters "with vertical scheme" for mounting in equipment rooms

Manufacturer	SCS type	Equipment type	Optical interfaces supported
AMP, USA	NetConnect	Modular Converter System	10Base-FL, 100Base-SX, 100Base-LX
Ortronics, USA	GigaMo, GigaMo+	TransOptix	10Base-FL, 100Base-SX
3M, USA	Volition	VOL-0213	10Base-FL, 100Base-SX, 100Base-LX, 802.5j

Manufacturer	Optical connector type	Number of Slots	Height	Number of power supplies
AMP, USA	ST, SC, MT-RJ	12	2U	2
Ortronics, USA	ST, SC	18	2U	1
3M, USA	VF-45	21	3U	2

The most natural versions of active devices that are used at the user's workplace in the unpackaged variant are extension board. The logical and clear reason for using an unpackaged construction is that, in practice, in the majority of cases, different adapters fail to engage one-half of the slots of the typical workstation that is connected to the usual SCS information outlet.

Traditionally, in unpackaged LAN units one can find the main functional modules of a half-length printed board on which electronic and optoelectronic components are mounted, as well as a front panel with different interface outlets, miniature LEDs, and other elements that are necessary for maintenance. The unpackaged devices that are widely offered in the beginning of 2000 can be installed into ISA, EISA, VLB, MCA, and PC1 bus outlets.

The list of unpackaged LAN equipment with fiber-optic interfaces produced by SCS manufacturers, as well as the techniques for its creation, includes LAN network adapters, media converters, and microhubs. Besides the different functional assignments, they also differ from one another by their different use of workstation technical resources (Table 7.5).

Table 7.5. Workstation resources used by different unpackaged LAN units with fiber-optic linear ports

Workstation resource	Setup slot	Power supply	Information bus
Network adapter	•	•	•
Media converter	•	•	–
Hub	•	–	–

The first unpackaged unit with a fiber-optic interface oriented at this field of application was a LAN adapter with different standards. Since information exchange is necessary over the internal interface, network adapters are provided for a specific base, the most popular ones from a large list are ISA and PC1 buses.

The popularity of media converters has grown with the development of optical communication techniques, because of:

- the need for a natural and smooth transfer from "copper" to optical systems by using network adapters with twisted-pair interfaces that were purchased earlier;
- cost: when there is a network adapter, a variant based on the converter is less expensive than the installation of a traditional optical fiber adapter.

Examples of unpackaged converters include units of the TransOptix series by Ortronics that are offered for use in their GigaMo and GigaMo+ SCS, as well as the analogous Alcatel FiberCon 1 PC units that are part of the Alcatel SCS cabling systems. The common feature of the Alcatel units is that they are connected into the idle slot of the workstation and use the station as a power supply and mounting construction without direct connection to the system block over the information signal circuits. A short 4- or 2-pair cord with a modular connectors plug provides connection of the network adapter to the converter. The developers of unpackaged converters point that it is unnecessary to change the program tuning of the workstation and that there is a minimum delay time for the transmitted signal, that is, the converter influence is negligible for the maximum allowed diameter of the collision domain. This is provided, together with high-speed element base, by the circuits design for signal transmission between two mediums without leaving the AUI and MII interface level.

Multiport unpackaged units are represented by the DIANA series of units (desktop integrierter adapter fur optische netzwerk anbindungen: adapter integrated into computer for connection to the optical network) by Dafuer (Germany). It is interesting that the units are not formally part of a specific SCS, however, they are designed to solve the problem of network creation on the basis of fiber to the desk. The units are unpackaged Ethernet or Token Ring 4-port hubs with built-in fiber-optic up-link ports. In contrast to the other types of unpackaged network units, the external hinged adapter provides a power supply for the microhub, which makes it possible to operate the network segment with this microhub even when the workstation is turned off.

Units for installation in technical premises

The active SCS fiber-optic units that are provided for installation on technical premises are, in most cases, different modifications of media converters that are versions of the usual general network equipment. Their inclusion in manufacturers' catalogs is mainly to show the full spectrum of options from one manufacturer that is available to the terminal user.

The units that are installed in technical premises are made by several construction schemes. The first is the usual 19-in horizontal shelf construction of height 1*U*, where unpackaged modular converters are installed using the traditional side-by-side location of the electrical and optical outlets on the front panel. Fiber Con N units by Alcatel are an example of such a design. These units are used mainly in small and mid-sized networks, where a high port density is not required. In such situations, the main advantages of this type of units are the ease of transferring from one network standard to another by simply changing modules, and the ability to create a mixed configuration (for example, 2 × Ethernet and 1 × Fast

Ethernet). When port density is not maximized, an additional advantage is the ability to give extended service to the system administrator. So, in the media converter modules of FIBER CON 2GI by Alcatel, small-size DLP switches are provided for each port; their use makes it possible to turn off the Link-mode and to change the assignment of the separate pins of the modular electrical converter. The latter makes it unnecessary to use an inverted cord, which greatly increases the convenience of cable system maintenance.

In practice, media converters and other units with horizontal designs and fixed configurations are used more rarely than modular devices. The optical Flex OS by FiberCraft has 6-port and 12-port Fast Ethernet media converters of the 6MC100- and 12 MC100-types, respectively. We should also mention that an 8-port 2-speed VOL-2008 minihub with a 100-Mbit optical up-link port based on a VF-45 connector is available in the Volition SCS by 3M.

Another design is a card rack with height increased to 2–3*U*, in which the media converter modules are located vertically. A brief list of such devices is shown in Table 7.4. The specific differences between the equipment with this design are the ability to change one converter to another, to connect new converters in current-maintenance mode without turning off the power supply (hot-swapping), as well as the wide use of back-up power supplies to increase the maintenance reliability as a whole. A built-in Internet agent provides remote management and status control.

In this case, the converter is designed to achieve maximum port density. Therefore, besides the corresponding circuitry decisions, the following design measures are always used:

- converters are located in a horizontal case of height 1*U* (sometimes such a decision is called an economic one due to the decrease in size);
- use of fixed configuration;
- optical connectors from the SFF group as well as a pair of TELCO (RJ71) electrical connector outlets are installed; in this case, the optical connectors are located on the front surface of the case and the TELCO connector outlets are mounted on the back wall of the case;
- The number of LEDs is minimized; in some cases, there is no status indication of each port.

It is interesting that TELCO connectors are used in some converters to transmit Fast Ethernet signals; that is, they correspond to the parameters of the 5-th category. In practice, it is not yet often seen.

Comparing the data of Tables 7.4 and 7.6, one can easily determine that unit height is decreased from 2–3*U* to 1*U*, since use of the measures mentioned above increases the port density by approximately 2–3 times over that of modular construction. These designs achieve a noticeable savings for the user since expensive boards with optical interfaces in modular switches of the lower and medium level are not used.

The high port concentration in combination with small sizes of the converter case may make it difficult for the designers to provide the technical conditions that are necessary for the electronic circuits. This problem is acute in the case of equipment that supports the Fast Ethernet standard (even when up-to-date micro-electronic elements with minimal power consumption are used). Thus, for example, the nameplate power dissipation of the FiberCon 2400 24-port converter by Alcatel is 80 W, versus 40 W of the FiberCon 24 of the Ethernet standard. To stabilize the internal temperature parameters, traditional measures are used, one of which is the use of an aluminum alloy case with a perforated upper cover.

Table 7.6. Multiport media converters with fixed configuration for mounting in technical premises

Manufacturer	SCS type	Equipment type	Supported network standards	Optical connector type	Converters quantity
Alcatel, France	Alcatel Cabling System	FIBERCON 24/2400	10Base-T/100Base-TX	MT-RJ	24
3M, USA	Volition	VOL0202	10Base-T	VF-45	24
FiberCraft, Germany	Flex OS	24MC10 Ethernet	10Base-T	E-2000	24

Besides media converters and hubs, switches are also included in the list of equipment. The technical specifications of the switches correspond to those for up-to-date lower-level LAN units. In particular, these devices have an unblocked external architecture and realize the store and forward principle. To connect to the main backbone of the enterprise, an up-link port is provided for and is realized as a changeable module that supports Fast Ethernet and Giga Ethernet standards. The remaining technical specifications of the switches are shown in Table 7.7.

In practically all cases, non-installation units that are oriented to installation inside technical premises interact with the equipment of users' work areas. For example, in the AS/400 system by RiT Technologies, in the optical fiber port that is used to organize the backbone communication channel, one can find repeaters, hubs, and multiplexes.

Table 7.7. Switch parameters from the 3M Volition optical SCS

Model	Standard	Number of ports	Number of MAC-addresses	Bus capacity, [Mbps]	Memory capacity, [Mbyte]
VOL-0215	802.3	24	2k	480	2
VOL-0215	802.3u	16	8k	4800	2

Additional service opportunities

Most network equipment that is mounted in technical premises serves several users simultaneously. This means that the developers of the equipment must provide the necessary level of maintenance reliability both for separate communication channels and for the system as a whole.

Two main methods are used for this purpose in up-to-date telecommunication techniques: the use of high-quality components and redundancy (doubling). Currently, the achieved level of component development makes it possible for developers to give an unlimited warranty for equipment of the group.

In the local network technique, redundancy of the optical routes for signal transmission is supplied by so-called transceivers with redundancy or with internal switches. Such technical decisions are not used in well-known SCS equipment because it has super reliability, and the probability for damage to the cables in the typical conditions of modern office building is too small.

In contrast to this situation, redundant power supply equipment is often supplied when the equipment is installed in technical premises because its failure, in the worst case, will cause many workstations to switch off from the network. Two schemes of redundancy are used. The first is based on the installation of a secondary built-in power supply or is realized in the form of a changeable power supply unit. This design is typical of modular constructions with vertical locations of the converter boards. The second scheme refers to a system with a "flat" installation and is used by Alcatel. It typically uses an external 19-in shelf power supply with the corresponding number of power outputs to serve up to six converters.

To increase the effectiveness of remote control and network management, Alcatel offers the FiberCon M-type management system that works with converters of the FiberCon 24/2400 series. The appearance of this equipment is a reaction of the industry to the increase in complexity of LAN based on SCS as well as the significant volume growth of optical cable designs.

The device has the form of the usual 19-in shelf of height 1*U*. In its LCD display and keyboard to assign the main parameters are located on the front panel. The transmission of management commands and information messages is provided through the local network (in band) using the SNMP protocol or using a V.24 (RS 232) interface. The system allows the management of a maximum of 100 separate converters of the FiberCon 24/2400 series. Moreover, it provides several digital inputs and connectors for the sensing elements of different sensor types (for example, temperature, humidity, and door opening). This situation makes it possible to use the system to create a subsystem for monitoring mounted construction status as well as the technical premises as a whole. The alarm signal that is generated when any of the sensors actuates is sent automatically to an operator's preset IP-address, if necessary. The system is primarily applicable in complex networks with increased reliability requirements, including those mounted in airports, hospitals, banks, power structures, and in other analogous places.

In the majority of cases, the separate workstations that are connected to a LAN by the installation units described in Sect. 7.1.1 are divided into separate work groups by well-known network techniques (by programs, by switches, and by installation of several modules connected to a common bus). It should be noted separately that the original design was proposed by the specialists of Dafuer. The Dafuer SUSI (sharing unit for switched interface) hub is the norm, for the field under consideration. It consists of a 16-slot chassis with two power supplies (to provide the necessary maintenance reliability level) that is mounted into a 19-in construction near the central switch. Four-port modules with three optical 10Base-T ports and one electrical 10Base-T port are provided to mount into the slots. The hub bus connects all modules. The optical ports of the modules are used only to connect workstations or small work groups arranged by installation units. The electrical port is exclusively intended for connection to a switching hub. Composition of one segment includes all optical ports that are in the module with the electrical port in use, as well as in modules with senior numbers up to next module with an operating electrical port. In this case, the optical ports of the latter module are concerned to the next segment. For example, if modules 1 and 4 are connected to the switching hub, then two segments are formed by the hardware: the first has 9 ports (modules 1–3), the second - 39 ports (modules 4–16). All the switching with forming and integration of work groups is done automatically without interference by the system administrator at the moment of switching at the electrical port.

7.2 Wireless Communication Systems for SCS

7.2.1 Radio systems

In general, the radio systems in SCS are used to connect a small dedicated work group of users to a LAN. The systems are used when the use of the wire technical solutions are physically impossible due to certain local conditions or is inexpedient economically. Radio communication are also used for extremely short-term connections. Wireless network creation with overlay to SCS is very convenient at railway station-houses and bus terminals, in conference halls, at stadiums, and other sports arenas, that is, where there are many mobile users who need to be connected quickly, but for a short time, to different kinds of electronic service without large demands for communication channel capacity.

Currently, the equipment that provides such a connection provides an information transmission bit rate that is high enough to work with Ethernet and Token Ring systems (the information bit rate exchange is 2–10 Mbps). The equipment consists of a network board at the ISA bus (for desk workstations) or PCMCIA bus (for notebooks), as well as a radio bridge that is provided with the corresponding antennas. The bridge wire interface is based on the usual modular connector outlet. This device is connected to the SCS with the standard switching cord.

The transceiver operating frequency is selected in the licenceless ranges of 2.4 and 5.8 GHz; that is, the permission of the radio frequency committee is not required to arrange radio networks. For the same reason, the transmitter output power is limited. Some other parameters of this type of equipment are shown in Table 7.8.

When using omnidirectional indoor aerials, the radiation power and receiver sensibility are sufficient to allow communications within the office premises at distances up to 150 m. That corresponds approximately to the typical distances of the horizontal subsystems of standard SCS. When there is need to increase the communications range, one can increase the transmitter output power as well as use high-gain directional antennas. One can also use a flat remote antenna of playing-card size that is connected to the adapter with a short length of cable and is located by the user to provide the best conditions to transmit the signal. Finally, in the PCMCIA network adapter, Alcatel uses an antenna that is placed above the open screen of the notebook. This automatically provides nearly the best functioning conditions in certain places.

Table 7.8. Main technical specifications of radio equipment used in SCS

System	WaveLAN	Wireless
SCS type, company	Systimax, Lucent Technologies, USA	Alcatel Cabling Systems, Alcatel, France
Operating frequency, [GHz]	2.400–2.4835	5.775 (U-NII band)
Transmitter power, [mW]	100	25–50
Radio channel data transmission rate, [Mbps]	2.048	10
Intranet protocol	CSMA/CA	CSMA/CA
Linear signal type	Noise-type, broadband DSSS	–

To defend from unauthorized access to the transmitted information, which is very important in radio communication systems, the following series of measures is used:

- the use of noise-like linear carrying signals;
- the ability to encrypt the transmitted information (equipment from Alcatel has a key length of 40 bit, which can be increased to 128 bit, if necessary);
- the use of additional access control levels for each node.

In mid-1999, Alcatel, Krone, and Lucent Technologies offered this equipment with their SCS. Earlier, radio communication systems for data transmission of Ethernet local networks was part of the NetConnect SCS by AMP, however, in the late 1990s its promotion was stopped.

7.2.2 WaveStar OpticAir Systems of Open Optical Communications

The WaveStar OpticAir system was developed by Lucent Technologies in mid-1999 and is provided to arrange open atmospheric lines of optical communication. In contrast to radio communication systems, which are primarily oriented mainly operate at the lower levels of local networks, the WaveStar OpticAir system is primarily intended to connect two buildings. That is, it is used in the backbone part of the information infrastructure of an enterprise. The maximum capacity of the channel created with it is 10 Gbps using dense wavelength-division multiplexing (DWDM) spectrum multiplexing on four wavelengths with bit rates of 2.5 Gbps on each subchannel. This principle corresponds completely to the projected physical level arrangement of network interfaces of the next 10-G Ethernet generation. The WaveStar OpticAir system is compliant with optical LazrSPEED cables, that is, the maximum length of the connection cables corresponds to the principles of centralized optical architecture and can reach 300 m.

The communication range of the WaveStar OpticAir system, as in any system of open atmospheric laser communication, depends sharply on the environmental conditions; when the weather conditions are good it can reach 5 km. However, system developers do not recommend increasing the distance between the receiver and transmitter more than 1 km from considerations of high stability.

By the developers' data, the WaveStar OpticAir unit can be used in Systimax structured cabling systems [109], which can be seen if its characteristics are analyzed. Nevertheless, the authors do not have any information of its use.

7.3 Units for TV Signal Transmission

Currently, systems that are intended to transmit television (TV) signals are the third, in importance, to consumers of SCS resources after LAN and telephony. However, the role of these systems began to grow quickly the mid-1990s. It depends on both the rapid growth of automation and informatization of the real estate sector within the concept “intelligent building”, and the significant expansion of the use of structured cables in industrial and residential areas. Currently, the cable lines and SCS routes support the use of the following kinds of image transmission applications:

- air, satellite, and cable TV systems;
- alarm systems;
- remote monitoring systems.

During the arrangement of TV signal transmission over SCS electrical cable routes with the corresponding impedance conversion, there is a need to solve two main technical problems: the unbalanced interface of TV units must be matched with the balanced input–output of SCS cable routes, and the required medium coefficients of signal transmission that are necessary for high-quality TV image reproduction must be provided. A baluns is used to match the interfaces of TV units and cable systems of different construction (Section 3.4.2.).

It is necessity to give specific attention to providing the required quality factors of the route because the increase in attenuation of twisted-pair electrical cables with frequency is approximately proportional to the square-root of the frequency, that is, with a steepness of 15 dB per decade (Section 2.1.3.). This leads to significant gain–frequency distortions of the transmitted wideband signal that significantly decrease the image quality of high-frequency signals of the TV spectrum. To eliminate this unwanted effect, two different technical solutions are used. The first is used by Telesafe (Norway) and is based on the installation of an additional correcting amplifier into the route at a distance of 60–75 m. The second, developed by Dätwyler, is based on individual amplifier adjustment depending on the route length. Such an approach is convenient to use, but the upper frequency limit for multichannel TV signals of the Dätwyler UNILan 2000 system is 450 MHz, versus 862 MHz in the Telesafe WTP. In addition, the amplifier is located in the same case as the baluns at the work area (a so-called active baluns), which may present a problem to the developers of remote power supplies.

TV signal transmission over SCS routes can be carried out both individually and by cable sharing. The first approach is realized by baluns and is typical for routes with UTP cables. To provide TV signal transmission in horizontal subsystems, cables with the upper critical frequency parameter of 550–600 MHz (for example, the Lucent Technologies 3071-type baluns, and the AdvanceNet by NEK/CDT) are often used.

The cable-sharing approach is used on routes with S/STP cables in two different versions. An example of the first one is the WTP system by Telesafe (Norway). The system is based on a 6500-703-type 18-port TV signal distribution amplifier (called a TV hub by Telesafe) of with branching factor 1:18. It can be connected to the TV antennas of satellite TV, TV cameras of video monitoring systems, and another similar TV units. The terminal connector of the S/STP cable in the technical premises is a special high frequency connector of the CMG type, but at the users' work areas, a so-called multimedia panel of the 6800-2010-type is used. In the UNILan system by Dätwyler (Germany), a 12-port active patch panel of the UniTV-type with coaxial connector is used in the switch room to connect the sources of the TV signals along with a Telco connector used to connect telephone ports. The external main power supply is assigned to power the TV signal amplifier. The main advantage of the first design is its ability to serve three applications simultaneously (TV, LAN, and telephone); the advantage of the Dätwyler approach (TV and telephone) lies in using standard modular connectors type.

AMP has proposed the so-called Multimedia Distribution System. A broadband uniform distribution (BUD) panel is among the elements of this system. Since outlets of both modular connectors as well as internal wide band amplifiers with interfaces to coaxial cables are used, this panel makes it possible to provide work areas with telephone communications, TV programs, videoconferencing, and other kinds of services concerned with video signal transmission.

7.4 Other Active Equipment

The other varieties of active network equipment that are offered by some SCS manufacturers as part of their systems should be briefly discussed. This equipment is used to solve different information transmission and handling tasks and to provide network functionality. RiT Technologies and Lucent Technologies use *systems of interactive management* by cable systems, which are considered in detail in Section 13.1.

Lower-level *LAN network equipment* is often used in designs for small offices, in the small office/home office (SOHO) sector, and in cable systems for non-office use. The equipment does not have any technical differences from the commonly used equipment except it uses hardware adapters and connector adapters that are easy to install into different kinds of systems.

Information about LAN equipment for the SOHO sector is found in Section 6.4.2. In "large" systems, the use of the commonly assigned network equipment is more the exception than the rule. In particular, in mid-2000 the nomenclature formed a part of the SignaMax SCS by AESP Company. Prior to 1999 active devices were included into the Systimax SCS catalogue of Lucent Technologies without a use of a special code; however, they no longer appear in it.

By the end of the 1990s, the popularity of parameter monitoring units grew sharply, both for mounted designs themselves and for network equipment built into them. Specialized devices that are manufactured by mounting equipment manufacturers (for example, Rittal) and systems that are primarily oriented to solve other tasks can fulfill these functions, but have an excess of computational capabilities for managing controllers. For example, note the PatchView scanner system by RiT Technologies and the FiberCon M management system of media converters by Alcatel.

7.5 Conclusions

The introduction of active network equipment into the structure of SCS greatly increases the potential of cable systems as marketable products. The most popular kinds of active network elements in SCS are LAN units with fiber-optic interfaces as well as systems for TV signal transmission.

Production samples of network equipment are included in the official catalogs of a number of SCS manufacturers, have their own part numbers as passive components, and give users the option to obtain full systems from one manufacturer when realizing lower-level enterprise information and computing networks.

The need to take the specific mounting and maintenance conditions of SCS cable routes into consideration gives rise to a number of original designs and changes in active device construction and makes this field a separate independent direction for network equipment development

8 Grounding of telecommunications closets and equipment rooms

The quality of signal transmission across SCS network equipment depends on the grounding system in telecommunications closets and equipment rooms. It is necessary for:

- preventing personnel from electrical shock;
- protecting network equipment and cable channels from external interference and reducing EMI;
- reliable signal transmission across some kinds of network equipment.

The principles of bonding are described in "Rules for electrical devices" [110]. Shielded cables bonded at one or both ends can be used for protection against external EMI in SCS. The efficiency of shielding depends on the grounding "cleanness", i.e., on the constancy of its potential. A grounding bar is usually used for the grounding. The bar has a non-zero resistance, and therefore streaming down currents sometimes cause fluctuations in its potential. At ineffective bonding, the shielding of twisted pairs may cause a reverse action, that is, the degradation of communication quality.

For low-frequency network equipment, which operates across unbalanced transmission circuits, the ground potential is given as zero. All levels of transmitted signals are changed relative to this value, but the ground itself serves for the grounding of signal circuit currents. It is obvious that fluctuations in the ground potential levels at both ends of a communication channel relative to each other may cause communication interruptions. Moreover, high current amplitude (for example, on renewal of power supply from a city substation after an outage) may lead to damage of network reception and transmission devices. For example, there were cases when all the network adapters bonded to an Ethernet segment by a thin coaxial cable with a grounded shield could fail.

Hence, quality issues for communications must also be taken into consideration, along with electrical safety when designing SCS. In practice, various approaches are applied by electricians, telecommunications personnel, and LAN specialists. It is recommended that increasing the cross-sectional area of conductors reduce the resistance of the main building grounding circuit. It is also recommended that the efficiency of the bonding element that contacts the ground be increased, that is, increases its area, or even artificially permanently moistens the ground. The third design uses auxiliary functional or signal grounding in addition to the normal methods. Its network is connected to the main grounding circuit at only one point, which is close to the building lead-in of the neutral bus from the substation, and/or elements contacting the ground. The fourth method, that communication equipment use galvanic coupling, is not used.

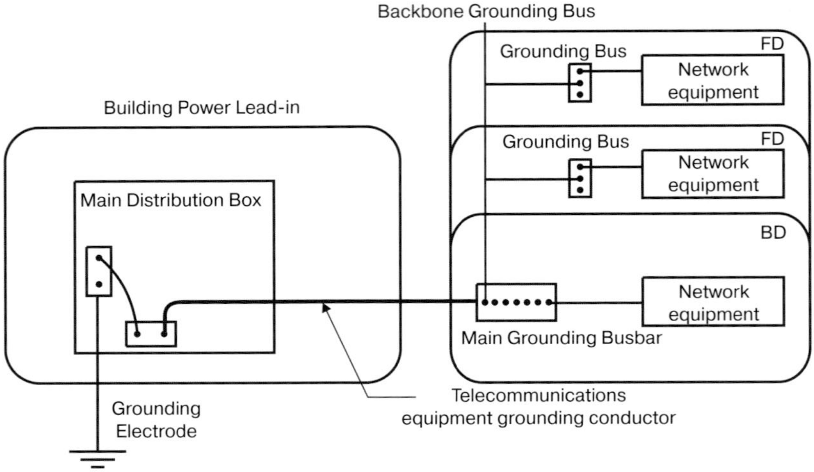

Fig. 8.1. Telecommunications grounding circuit

Experience shows that only the full scope of high-quality grounding measures enables not only reliable protection against electrical shocks, but also high-quality data transmission. An important fact, which must be taken into consideration, is the wide-spread use of grounded network equipment for signal transmission, including cable TV and telecommunications equipment with unbalanced transmission. Moreover, it is important to provide functioning of systems based on shielded cables.

Standard requirements and recommendations of the standard TIA/EIA-607 on grounding in telecommunications closets and equipment rooms are given below. The standard supports the idea that the protective grounding system implies a second telecommunications grounding circuit in telecommunications closets and equipment rooms. Sometimes, it is called the *system grounding circuit* (Fig.8.1), and enables designers to equate the differences in potential between equipment areas.

The use of the telecommunications grounding circuit is important, as the protective grounding circuit is primarily intended for high-voltage power supply systems at a frequency of 50 Hz. Therefore, because of the considerable inductance of its conductors on frequencies above 1 MHz, the efficiency of the system may be sharply degraded [111].

The telecommunications grounding circuit connects the main grounding circuit close to the neutral main lead-in into the building, and/or the grounding electrode systems. Moreover, it may be also be connected to it via grounded metallic parts of the building. The network equipment in telecommunications closets and equipment rooms must be connected to the power supply system through outlets with grounded contacts that are connected with the main grounding circuit. The plug grounding contacts have a galvanic connection with the network equipment housing. Therefore, the telecommunications circuits may have additional connections to the main grounding circuit and operate in parallel to it.

Telecommunications grounding also provides an efficient equalization in ground potential in telecommunications closets and equipment rooms relative to each other. Simultaneously, it ensures a reliable grounding of network equipment, irrespective of the main grounding circuit state.

The telecommunications circuit consists of:

- several grounding bars, one of which is the main;
- backbone and interbackbone buses;
- buses to the main grounding system.

All conductors and flexible buses used for the telecommunications grounding circuits must be made of copper and must be insulated with sections not less than 6 AWG (diameter 4.12 mm, cross-sectional area 133 mm^2). The general rule that is applied for selection of these conductor sections implies that the resistance between the bonded points shall not exceed 1 Ω [61]. All connections of bars with backbone buses are welded.

The main grounding busbar serves for the connection of backbone grounding buses, buses to the main grounding system and nearby network equipment. Moreover, it connects all through-passing parts of cable channels—ducts, trays, and pallets. It is made as a metallic plate not less than 6-mm thick with a minimal width of 100 mm. The length is customized by local conditions. The plate has holes for fastening to the building elements and to the grounding conductors of network equipment (for example, by screw fastening).

The location for the main busbar is selected by taking into consideration two factors: to make it close to the telecommunications campus backbone cables input point to the building and for minimization of connection bus lengths for connection to the main grounding system. The main bar connects to the nearest properly grounded metallic parts of the building, if possible.

Grounding bars are located in telecommunications closets and equipment rooms. They are connected with backbone buses and are intended for connection of grounding conductors of network equipment, racks, and other metallic parts of the cable channels as well. The grounding bar is identical to the main busbar, only its minimum width cannot be less than 50 mm. It is preferred that one or several grounding bars are connected to the nearby properly grounded metallic parts of the building.

Backbone buses connect the telecommunications closet bars with the main busbars. One backbone bus connects several grounding bars located in different telecommunications closets. A large building may have several backbone grounding buses branching from the main grounding bar like a star. Selecting the installation layouts of backbone buses, it is advisable to minimize the buses's lengths, and to make the layouts similar to the structure of the building backbone cabling.

We must also note that use of the building ductwork is not permitted as the backbone grounding buses.

9 Fire safety

9.1 General Information

The state-of-the-art SCS is built in the form of a star and therefore may include dozens of cables hundreds of kilometers long, with a total mass of several tens of tons. The majority of these cables are routed inside the building as a part of the horizontal cabling and building backbone cabling subsystems. Polymer materials are widely used for the insulation of metallic conductors in electrical cables and the protection of fibers in optical ones, and to form the core structure in both constructions. A short list of available materials is given in Table 9.1. Flame impact on the materials may lead to the following negative factors:

- further flame spread along the cable channel (the Bickford's fuse effect);
- secondary places of ignition caused by burning pieces of various jacket materials;
- intense smoke, which hampers evacuation of personnel and fire fighting;
- release of halogen-containing gasses.

Fire-retardant characteristics of cables that comply with the norms are met by using specially selected materials. First, polymer materials have to be fire-retardant. The oxygen index (OI) defines the flammability factor [112]. The value defines a minimum oxygen content in a nitrogen–oxygen flow for which a vertically displaced material sample ignited on top starts burning by itself without any external application of heat. The materials are considered nonflammable if their OIs are higher than 32; values less than 23 classify the material as flammable. Polyethylene, which is used in various cables, is a typical flammable material. Sometimes, a so-called temperature coefficient, which equals the temperature made by burning the material, is indicated.

It is rather complicated to select suitable materials, as existing nonflammable materials whose characteristics make them attractive for cable engineering applications always include halogen content (most often, chlorine or fluorine; bromine, is used very rarely due to medical recommendations). Under high temperatures these chemical agents generate aggressive volatile acids, which badly effect human beings and animals. Halogen-free materials possess low resistance to high temperatures. This is explained by the high chemical activity of halogen ions. Therefore, halogen ions are tightly bound with other jacket atoms and act as a cement material, that is, they hamper the splitting of the polymer splitting into separate components in a fire. Mixtures take an intermediate position between halogen-containing and halogen-free materials. Thus, for example, the fire-resistance of polyethylene can be essentially improved by adding bromine additives.

Table 9.1. Basic polymer materials for SCS cables

Material	Application	Properties
Polyvinyl chloride (PVC)	Jackets	Nonflammable,
Fluorinated polymers (FEP, ETFE)[a]	Insulation of plenum cables	halogen-containing
Polyethylene (PE)	Insulation	Flammable,
Polypropylene (PP)	Insulation, filler	halogen-free
Polystyrene (PES)	Films	
Polyurethane (PUR)	Jackets	

[a]Teflon by Dupont and Neoflon by Daikin

Another additive that is often used to improve cable fire-resistant parameters is aluminum hydroxide. Under high temperatures, the substance splits according to the formula $2Al(OH)_3 \rightarrow Al_2O_3 + 3H_2O - 300$ kJ. The endothermic process implies considerable heat absorption, which dampens the further fire spreading.

The fire safety issues of SCS cables were first seriously examined only at the end of 1980 and the beginning of the 1990s. Today, a number of international standards have been developed. Moreover, many countries have national norms based on cable system fire safety. However, the existing normative documents do not regulate the entire scope of SCS fire safety issues, and research continues within the field. The authors know of only one Russian standard on cable system fire safety [113]. Therefore, this review is based on the requirements of the international normative documents.

9.2 Fire- and Flame-Spreading Resistance

9.2.1 Standards

Upon inflammation, cable systems that cover the whole building can potentially sustain the burning process and cause the flame to spread. Therefore, fire- and flame-spreading resistance is specified by a number of standards:

- The National Electrical Code (NEC) in the USA and Canada, articles 725, 760, 770, 800, and 820. The NEC definitions have been developed by the National Fire Protection Association (NFPA) and are revised every 3 years;
- The International Electrotechnical Committee (IEC) 332 requirements and the HD 405 requirements by the CENELEC (European Committee for Electrotechnical Standardization) in Europe.

We shall review only the NEC requirements. The fire safety requirements for low-current cable systems, that is, those which themselves cannot cause ignition, are given in articles 725, 760, 770, 800 and 820 (Table 9.2).

Table 9.2. NEC articles on low-current cables

Article	Title	Cable indication	Application field
725	Remote control of classes 1, 2, 3, signal transmission and low-current power circuits	CL2 (Class 2 power limited) CL3 (Class3 power limited)	• Industrial control systems • Data transmission
760	Fire alarm systems	FPL (Power-limited fire protective signaling)	• Fire alarm sensors
770	Fiber-optic cables	OFC (Optical fiber, conductive) OFN (Optical fiber, nonconductive)	• LAN
800	Communications circuits	CM (Communications)	• LAN • Telephony • Fire alarm systems
820	TV community antennas and broadcasting	CATV (Community antenna television)	• TV • Video control cameras

The NEC definitions determine four fire safety certification levels for low-current cables. The highest level is the first one; the lowest is the fourth one.

Level 1. Plenum cables, which can be openly routed in plenums without any restrictions. The plenums designate a space between the drop ceiling and overhead floor, which is available for sufficient air drafts to maintain the burning process. A space under the raised floor is sometimes called a plenum, as it is also available for air drafts due to the existence of other plenums, external conduits, or similar structures in the halls.

When manufacturing plenum cables, the manufacturers follow two different approaches [114]. According to the first approach, only low-smoke flame-retardant materials are used to manufacture all insulating covers. The second approach is based on using flame-retardant materials only for outer covers. The drawback of the first option is that it somewhat degrades the cable electrical characteristics and considerably increases its cost. The second approach dramatically decreases the fire resistance when the outer cover is damaged, however, the products are cost-effective solutions due to the low cost of the materials used and the use of standard manufacturing technology. Besides fire- and flame-spreading resistance, plenum cables undergo a mandatory control for smoke levels (test UL 910).

Labeling of cables that are certified as plenum cables includes the index P, for example, the abbreviation CMP means a communications plenum cable. In practice, SCS horizontal cables are mainly routed in plenums in the majority of cases.

We also must note that nonflammable materials such as Teflon, which are suitable for the insulation of individual conductors, are very expensive. Therefore, in order to decrease the cost of cable products, these materials are sometimes used for the insulation of pairs of conductors. The number of such pairs is given in the cable specification and/or designation. For example, a 3+1 cable has one pair of conductors with Teflon insulation or its analogue, and three pairs with standard materials.

As the NEC definitions are oriented to the specific features of American office buildings and Underwriters Laboratories hold the monopoly position, the European cable manufacturers carry out limited tests according to the norm UL 910.

Level 2. Riser cables, which may be routed without any auxiliary protection in building vertical risers. This certification within an area is indicated by adding the index R to the CM cable designation, for example, CMR: communications cable for risers (Table 9.2).

Level 3. General-purpose cables, which may be routed without any restrictions anywhere in a building, except for plenums and vertical risers. They do not have a special marking.

Level 4. Limited-use cables, which have certain restrictions for open routing, for example, their use is allowed only in residential buildings. Routing of limited-use cables is allowed in nonflammable ducts only; or its minimal diameter and etc. are restricted. Certified cables are marked by adding the index X to CM, for example, CMX: limited-use communications cable. Fiber-optic cables and cables for fire alarm systems have no options for limited use.

The certification of cables to one or another level is made once its testing is completed. The NEC definitions foresee UL testing. The Table 9.3 presents UL tests for each level. A cable that successfully passes all the tests is certified.

Table 9.3. UL tests for certification of low-current cables

Level	Certification	UL test
1	CL2P/CL3P, FPLP, OFCP/OFNP, CMP, CATVP	UL 910
2	CL2R/CL3R, FPLR, OFCR/OFNR, CMR, CATVR	UL 1666
3	CL2/CL3, FPL, OFC/OFN, CM, CATV	Vertical tray
4	CL2X/CL3X, CMX, CATVX	VW-1

9.2.2 Marking of Fire-resistant Cables

The fire-resistance information of SCS cables is given in its identification labeling on the jackets. At the time this book was written, there were no normative documents known to the authors that regulate such labeling procedures. Therefore, we note only some general statements that have formed de facto standards.

Information referring to plenum or riser classes is indicated by putting the titles directly on the jackets. Moreover, the labeling may include the following letter indexes: FR, flame retardant; LS, low smoke; NC, non-corrosive; OH, ZH, HF, no (zero) halogen or halogen-free. If the cable simultaneously meets several requirements, its fire-resistance alphabetical index consists of abbreviations that are written one after the other, without any spacing. For example, LSOH means that a cable is low-smoke and halogen-free. We also note that the abbreviation NC is used quite rarely, since the absence of corrosion is directly tied to the use of halogen-free materials in the jacket.

Optical cables for external routing with nonflammable outer jackets are produced in limited quantities in Russia. Cables produced in accordance with the existing norms are specified for the organization of building lead-ins. When using these cables, a changeover into inner cables is not necessary.

9.3 Other Detrimental Agents

9.3.1 Smoke

The smoke from burning cables leads to poor visibility (the optical transmission coefficient decreases by 90% or more). In addition, considerable levels of smoke cause panic and asphyxia among human beings and animals. These factors considerably hamper the evacuation of personnel from buildings and fire-fighting procedures. Therefore, the smoke level produced by burning cable products is strictly controlled and regulated by a number of standards:

- In the USA and Canada by NEC articles 725, 760, 770, 800, and 820;
- In Europe by IEC 1034-2 requirements and by the HD 606.2 S1 document by CENELEC;
- In Russia the smoke levels of various solid substances and materials, including polymers, is regulated by GOST 12.01.044-89 [1].

The smoke-level control procedure, which is known as UL 910 testing, is made by the photometric method in a testing chamber of $3 \times 3 \times 3$ m^3 in compliance with international normative documents (Fig. 9.1.). The heat source is 1 l of spirit composition (90% ethanol, 4% methanol, 6% water).

Fig. 9.1. Testing equipment for the detection of smoke levels

Fans and protective partition walls, which enclose the testing samples, are used to even the levels of exhausted smoke. The optical transmission or optical absorption coefficient variations that are registered by the light shutter installed in the testing chamber serve to determine the level of smoking. The cable passes the test if the optical absorption coefficient does not drop below 60%. In addition to the fire- and flame-spreading resistance, the test measures the peak and general smoke volumes as well as the optical absorption coefficient behavior in time.

The experimental results show that fluoropolymers and low-smoke compounds, unlike PVC, make white smoke that reduces the optical absorption coefficient by not more than 10–15%, that is, they considerably decrease the fire risk.

We must note that, in compliance with the requirements of the NEC articles, it is prohibited to openly route cables that have not passed UL 910 in plenums.

9.3.2 Toxic Asphyxiating Gases

Volatile halogen-containing gases (chromium, bromine, fluorine, and iodine compounds) cause poisoning and asphyxiation of human beings and animals. Compounds of these gases and moisture make corrosive acids (hydrofluoric acid, HF, and hydrochloric acid, HCL), which damage electronics and, for considerable concentrations and long exposures, damage even the structures of buildings.

Statistics indicate that 90–95% of lethal accidents are the result of carbon monoxide poisoning. Any reliable information about poisoning from gaseous halogen-containing compounds was not available in the mid-1990s, at least not in Europe. Therefore, the existing normative documents do not give standards for the toxicity of combustion products, however, the IEC has started working on such a document.

The standards quite strictly limit the levels of halogen-containing agents in gaseous combustion products in order to minimize their corrosive effects. The principal normative document within this field is the standard IEC 754, part 2. Cables that comply with the requirements of the document are zero-halogen cables.

The method for testing of halogen levels is quite simple. The lead sample is burned in a hermetic chamber. The light-end products are evacuated through the waterlock. For analysis by traditional chemical methods, the pH and the conductivity of water samples taken from the waterlock must be controlled. The cable is considered free from corrosive effects on combustion if the pH does not exceed 4.3, and if the electric conductivity is not higher than 10 μS/mm.

Table 9.4 presents the classification of materials by the toxicity values of their combustion products. According to the English norm NES 713 or the equivalent French norm NFC 20454, toxicity is defined by a special dimensionless coefficient. Its value is calculated depending on the chemistry of the light-end materials. It is assumed that human beings can be in atmospheres with coefficients less than 5 for 15 min without any fatal health effects [115]. As a rule, if the cable is designed with respect to the fire safety issues, it meets the requirements of the test NES 713.

Table 9.4. Classification of polymer materials by toxicity of combustion products

Risk class	Ng/m^3 at time of exposure, min			
	5	15	30	60
Very dangerous	Up to 25	Up to 17	Up to 13	Up to 10
Dangerous	25–70	17–50	13–40	10–30
Moderately dangerous	70–210	50–150	40–120	30–90
Low risk	Above 210	Above 150	Above 120	Above 90

The higher cost of halogen-free materials and their low resistance to combustion inhibit the widespread use of halogen-containing cable products. The higher cost is the result of the use of more expensive feedstock and the need to use special extrusion machines for jacketing. However, it is important to note the inferior strength characteristics and, as a result, the poor mass and dimension figures of cables with low halogen contents.

9.3.3 Fire Loads

The fire load is the amount of energy released on combustion of one cable length. The fire load is a quantitative estimate of the cable resistance to flame spreading. The value of this parameter (calculated in mJ/m or kWh/m) essentially depends on the type of material used for cable manufacturing (Table 9.5).

Table 9.5. Fire loads of some insulation materials for cables [116]

Material	Abbreviation	Fire load, [kWh/m]
Polyvinyl chloride	PVC	4.2–6.1
Polycarbonate	PC	2.8
Polyamide	PA	8.9
Polypropylene	PP	8.9
Polyethylene	PE	9.4–13
Polystyrene	PS	10.2–11.7

The authors have learned of only one national normative document that specifies the fire load value. According to the German standard DIN VDE 0108, part 1 (October, 1988), riser cables with halogen-free jackets should have fire load parameters not greater than 14 kWh/m. Other cables, whose constructions include halogen-containing materials (for example, PVC), have values not more than 7 kWh/m.

9.4 Testing of Cable Products

9.4.1 Testing by GOST and IEC

The IEC and GOST 12176-89 [113] standards specify two kinds of fire protection characteristics of cable products. The test IEC 332.1 tests a vertically suspended cable of 600±25-mm length. One or two Bunsen burners with inner nozzle hole diameters of 9 mm are used as igniters. The testing according to the IEC 332.1 proceeds for 60 s. By the end of the test, the cable top should not be damaged by fire, and the combustion of its polymer materials should stop after the burner is turned off.

Stricter testing is performed according to IEC 332.3. In this case, a vertically suspended cable bundle of 3500±50-mm length and not more than 300-mm width is tested. It is fastened by a wire at the center point of a holder in the form of vertical stairs with nine crossbars. The flame source is a band-type propane burner with caloric power of 1.23 mJ/min. The testing time is increased to 20 or 40 min, depending on the category of the tested sample. It is assumed that those samples that have damaged sections not higher than 3 m (by IEC 332.3) or 2.5 m (by GOST 121176-89) pass the test. Both these normative documents clearly specify the design of the testing chambers, the rules for preparation of samples, and the equipment used.

9.4.2 Testing to Comply with UL Standards

Testing of *limited-use constructions* is carried out in compliance with the test UL VW-1 and by means of a setup shown in Fig. 9.2.

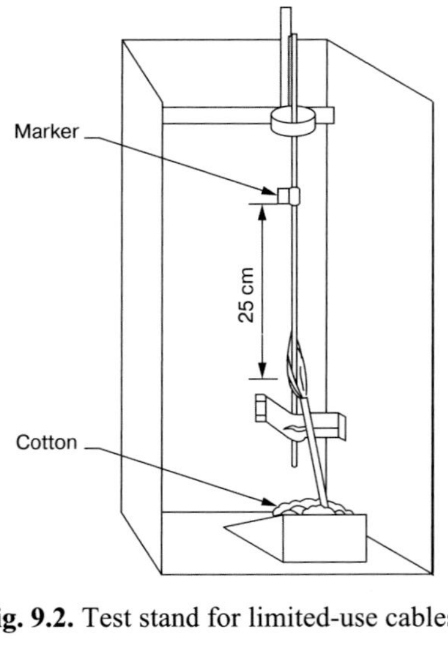

Fig. 9.2. Test stand for limited-use cables

The Torill burner is used, which is the functional analogue of the Bunsen burner. The testing includes several cycles. During each cycle, the flame burns for 15 s, then the burner is turned off for the next 15 s or until the lead stops smoking before 15 s has passed. The process is repeated 5 times.

It is assumed that the sample passes the test when:

- it does not smoke longer than 60 s;
- not more than 25% of the Kraft-paper indicator flag is burned;
- the cotton near the bottom of the burner does not ignite, which indicates the absence of secondary ignition areas caused by dripping of the burning material.

A low-power ignition source is used (caloric power less than 60 mJ/min). The sample is affected by the flame for only 75 s; therefore the test VW-1 is only able to efficiently estimate the fire characteristics of leads with small diameters.

The test stand for *general-use constructions*, which is specified by the vertical tray test, is a vertical cable rack 305-mm wide and 2.4-m high (Fig. 9.3). A 10-in band-type burner with nominal caloric power of 1.27 mJ/min serves as the igniter. The fuel is 1 part propane and 6 parts air. The burner is placed 61 cm above the floor and reaches the temperature of 815°C at a given point.

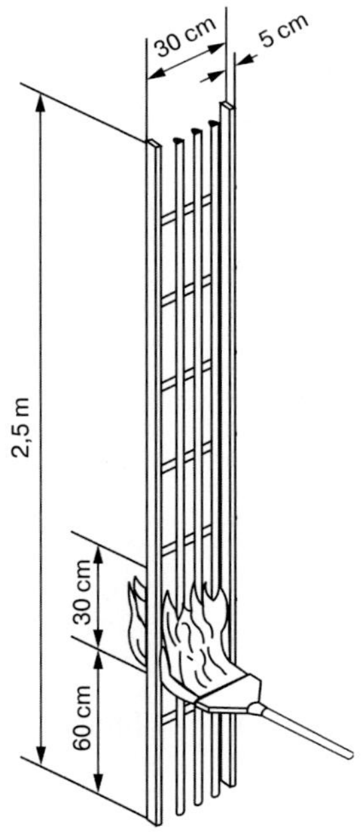

Fig. 9.3. Testing stand for general-use cables

Cables are routed in one row in the central section of the 150-mm wide rack. The distances between the cables must be equal to the half their diameters; the length is 2.4 m (8 ft).

The cables, nozzles, and gas pressure of the burner flames are chosen so that the flame touches 22–30 cm of the sample length. The burner is turned on for 20 min, and the samples are left to burn after it is turned off. The cable passes the test if, after the burner is turned off, the flame dies prior to touching the top of the rack. The vertical tray and IEC 332.3 tests are almost identical in testing approach and estimation of results.

One of the drawbacks of this test is the complexity of using it for small-diameter products, as its principal requirement is to place samples at distances equal to half the diameters of the cables. To eliminate this drawback, the Canadian Standard Association (CSA) has developed a version of the test that requires that cables with outer diameters less than 13 mm be grouped in three-by-three (or more) bundles with minimum diameters not less than 13 mm. However, the test is carried out under the same conditions. The CSA has suggested setting a maximum distance between the cables equal to 15 mm.

The experimental stand for the testing of riser constructions, which imitates routing conditions in risers, is shown in Fig. 9.4. This stand realizes the verification conditions specified by the test UL 1666. The caloric power of the propane burner is 8.7 mJ/min. The test lasts 30 min. The cable passes the test if the height of the burned insulation covers does not exceed 3.7 m. This implies that if the standard height of an office ceiling is 4 m, then riser constructions block the spread of flames between floors via cable risers.

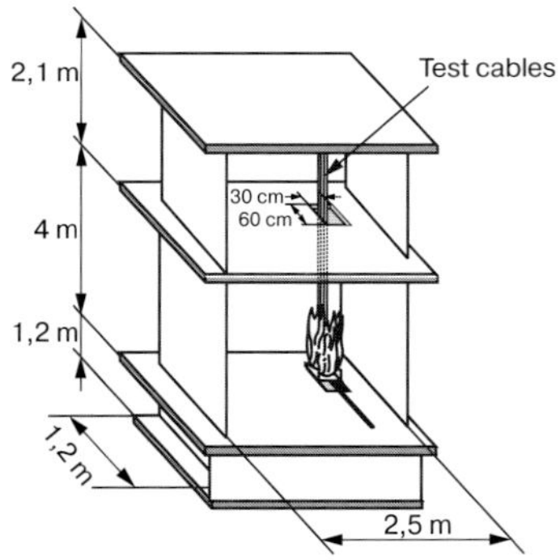

Fig. 9.4. Testing chamber for riser constructions

The testing of plenum-type constructions is carried out in a 25-ft Steiner conduit (Fig. 9.5).

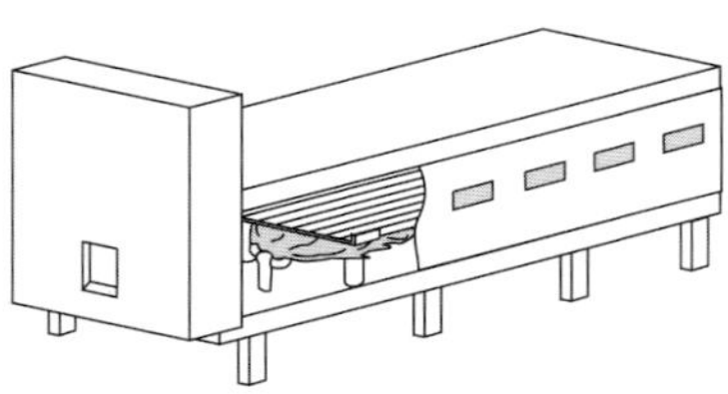

Fig. 9.5. Steiner conduit

The conduit simulates a plenum with inner cross section 45×30 cm, and enables overall fire testing of cables. A 7.3-m long single cable or bundle is routed on the table of the conduit in one layer. Two methane burners with total caloric power of 5.3 mJ/min serve as the igniters. The airflow rate in the conduit is 80 m/min with preheating up to 70°C. The test duration is 10 min. The flame spread may be visually controlled through high-temperature glass windows placed 30-cm apart in the housings. The level of emitted smoke from the burned insulating covers is controlled by sensors in the offtake pipe. The optical transmission coefficient shall not drop by more than 50%.

This testing construction may be referred to as plenum testing if the length of the burned section does not exceed 1.5 m from the end of the 140-cm body of the flame. The walls of the conduit do not allow heat leakage, and the heat is transferred to the entire length of the tested sample, which additionally increases the test accuracy. It has been experimentally verified that this test is passed by cables designed with respect to the fire safety standards.

We must note that the Cardington research laboratory (Great Britain) uses the Fire Research Station, which provides more accurate test verification. The unit consists of a 7.4×5.7-m room, 4-m high, with a 1-m suspended ceiling. The test cable is laid on a table in the conduit. The heat source is a burner with nominal caloric power of 60 mJ/min. The test duration is 30 min.

Cables with jackets made of fluoropolymers, like Teflon, possess minimal flame spread, low smoke, and low carbon monoxide concentration. Smoldering jackets made of low-smoke halogen-free compounds create temperatures above 800°C out of the burner operating area, while the temperatures of fluoropolymers do not exceed 335°C. Moreover, dripping compounds are potential sources of secondary ignition areas.

9.5. Fire Safety when Designing SCS

The principal rules to follow when designing cable systems in compliance with NEC requirements are:

- cables without certification to any of the four levels is prohibited for inner routing in buildings; however, up to 15 m of uncertified cable can be routed inside the building to the point of its connection to a certified cable;
- open routing is made only by means of cables with appropriate certification levels;
- non-plenum cables can be routed in plenums only in high-temperature housings, for example, in metal ducts;
- cables certified to levels 3 and 4 can be routed in risers only in noncombustible ducts and fire-resistant chutes that are equipped with fire-retardant stoppers in floor spaces. The protective elements should have UL certification, but floors should be made of fireproof materials. The fire-resistant stoppers are made by

filling the spaces between the floor and cables with specially hardened fire-retardant compounds (Fig. 9.6);
- in residential areas, CMX cables and CL2X/CL3X cables of remote control systems, that is, certified to level 4 (passed only the VW-1 test), can be routed for common use if their diameters do not exceed 0.25 in (6.3 mm);
- in residential buildings, CATVX cables may be routed as the general-use cables if their diameters do not exceed 0.375 in (9.5 mm);
- in nonresidential buildings, CMX, CL2X/CL3X, and CATVX cables can be routed only in fire-resistant housings. They can protrude from the housings not more than 10 ft (305 cm).

We must also note that many producers of raceways and raised floors include various mats, stoppers, and other similar elements made of noncombustible materials as accessories. Their use is specified by separate standards, whose examination is beyond the scope of this book, but that provide the required fire safety.

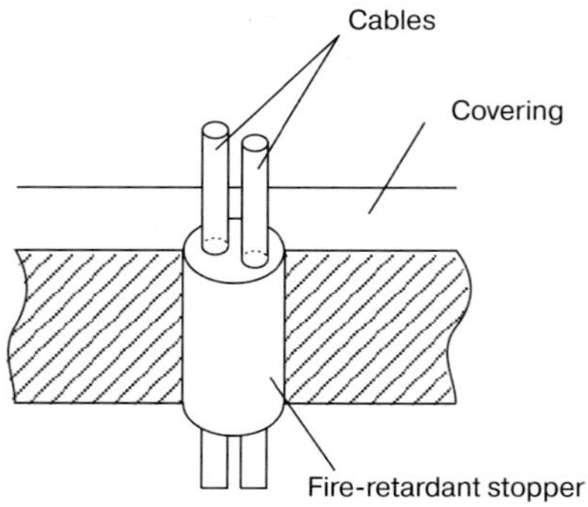

Fig. 9.6. Routing of cables with level 3 and level 4 certifications through floors

9.6 Conclusions

Considerable numbers of cables and other components are routed and mounted inside buildings during the construction of state-of-the-art SCS. These cables and components include organic insulation materials, therefore fire safety issues are crucial in the design and deployment of structured cable systems.

The principal fire risk factors are controlled by international and national standards. The existence of strict quantitative criteria in normative documents of the standards and approved testing approaches enables designers to refer each component to a certain fire safety class.

The fire safety level may be achieved by the use of appropriate insulation materials during the manufacture of cables and other components, and by special architectural or protective construction measures in buildings.

10 SCS Design

10.1 Design Principles

10.1.1 Design Stages

The process of designing telecommunications systems in modern offices using SCS is divided into two fundamental stages: architectural and telecommunication.

The main task of an architectural design stage is the determination of the general structure of the SCS that will be optimum for the complex of technical and economic characteristics during its creation and subsequent operation. This is determined at the time of project development for a new or reconstructed building. During this stage, a cross-room and equipment room, vertical cable riser, methods, and procedures of cable layout both inside and outside a building (cable conduit) are included in the project. The basic initial data for this design stage are (Figure 10.1):

- form, number of floors, architectural, design and other features, geometrical characteristics of a building or their complex as well as adjacent territory characteristics;
- construction and other normative documents to design service rooms for the telecommunications systems and cable routes;
- normative documentation for SCS (the standards);
- additional customer requirements.

At an architectural stage, the work of designing is carried out by the specialized design organizations subject to the requirements of the contractor that will operate this SCS.

The telecommunication design stage sometimes begins upon architectural stage completion, but it is usually carried out after the major civil and installation work ends. During the telecommunication design stage the specific SCS structure is developed and the list of necessary equipment and the technical plans for its arrangement, etc. are made. At This stage of work, companies specialized in the field of SCS creation and system integration are attracted in order to carry out the design. They usually also carry out the major part of installation and initializing operation, which, as a rule, are carried out simultaneously together when interior rooms are finished or as soon as it is completed. The initial data for this stage are (Figure 10.1):

- Building and adjacent territory investigation results or their projection made during the architectural design stage;

- Normative documentation for SCS (the standards);
- Additional customer requirements, for example, quantity and work stations arrangement, number of telecommunication outlets in them, requirements for productivity, reliability, safety, etc.

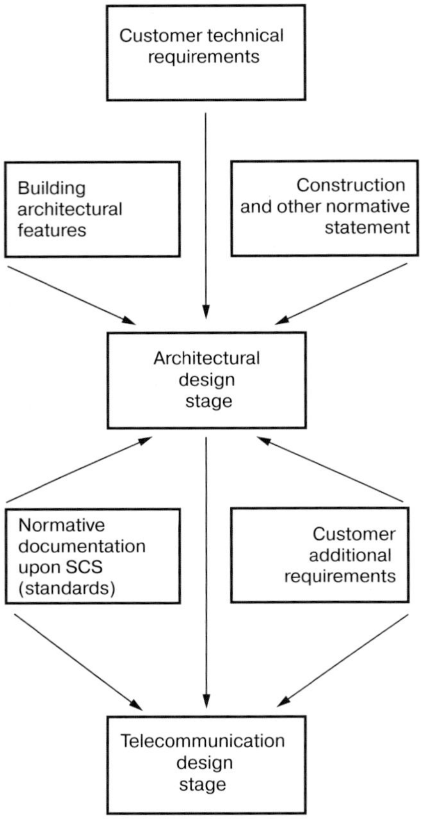

Fig. 10.1. SCS design stages

10.1.2 Stages of SCS Creation

At present, there is no standard that defines the structured cabling system as a technical object and, furthermore, there are no instructions for its design in Russia. Therefore, designing for system realization is conducted with the help of other guidelines. The closest normative document that is frequently used by system integrators for SCS realization is the all-Union State Standard – USS (GOST 34.601-90 [117]). According to this standard, the creation of a system is divided into stages and phases that are listed in Table 10.1.

Table 10.1. Stages and phases of SCS creation

Stage	Phase
1. Form requirements	1.1. Object investigation. 1.2. Form customer's requirements into a system.
2. Technical task	2.1. Development and approval of technical task to create a system.
3. Outline design	3.1. Development of preliminary design solutions for a system and its parts. 3.2. Development of explanatory notes and outline primary design estimate.
4. Technical design	4.1. Design solution development for a system and its parts. 4.2. Documentation development for a system and its parts. 4.3. Development and formulation of documentation upon equipment delivery for a system replenishing.
5. Technical design documentation	5.1. Technical design documentation development for a system and its parts.
6. Enter into operation	6.1. Automation object preparation to enter into operation. 6.2. Personnel training and instruction. 6.3. System integration with delivered items. 6.4. Installation. 6.5. Enter into operation. 6.6. Perform pilot tests. 6.7. Perform pilot operation. 6.8. Perform acceptance tests.
7. System maintenance	7.1. Work performed in accordance with warranty obligations. 7.2. Post-warranty maintenance.

The work of design is carried out in the following stages: the outline design (OD), the technical design (TD), and the technical design documentation (TDD). Furthermore, when a system enters into operation, the operational documentation should be developed. It should take into account corrections that are included in the technical design documentation and that were made during the entry into operation, civil and installation work, pilot operation, and acceptance tests. The operational documentation also includes manuals for system use and maintenance during its operation. The list of documents, which can be included into the structure of technical design and operational documentation, is given in Table 10.2.

Depending on a particular situation during customer interaction, various system creation stages can be omitted, if their omission does not result in a decrease in quality.

The formulation of the textual documentation is conducted according to "Information technology. The methodical instructions" (RD 50-34.698.90). The plans, layouts, and drawings are carried out according to the following standards "Technical Design Documentation System for Construction" (TDDSC) - USS (GOST21.xxx.) In order to prepare technical drawings, automated design engineering systems such as AutoCAD, ArchiCAD, and CADDY can be used.

Table 10.2. List of SCS technical design and operational documentation according GOST 34201-89 [118]

Item no.	Stage	Document description	Availability in structure	
			Design documentation	Operational documentation
1	TT	Technical task	X	–
2	OD, TD	Structural scheme of technical complex	X	–
3	TD	Technical design register	X	–
4		Register purchased units	X	–
5		Explanatory notes to the technical designs	X	–
6		Location plan	X	–
7		Specification of equipment and materials	X	–
8		Calculation of estimated cost	X	–
9	TDD	Register of original holders	X	–
10		Register of operational documentation	–	X
11		Equipment specification	X	–
12		Register of information means	X	–
13		Cost estimate	X	–
14		Technical instruction	–	X
15		User's manual	–	X
16		Operation manual for telephone cable networks	–	X
17		Junction scheme for layout of external cables	X	–
18		Connection scheme for layout of external cables	X	–
19		Table of junctions and connections	X	–
20		Plan of equipment and cable layout arrangements	X	–
21		Overall system description	–	X
22		Program and method of tests	–	X
23		Certificate	–	X
24		Document form	–	X

Initial data for design at architectural and telecommunication stages

The basic design initial data are obtained during the predesign object investigation, and include normal standards and customer requirements. Typically, the technical requirements are frequently placed in an appendix to the invitation for participation in the bidding process. The document, which generalizes the initial data and is the result of mutual activity of the customer and the contractor at the predesign stage, is the so-called technical task (TT) approved by both parties. The TT is made according to the USS standard (GOST 34.602-89 [119]). In order to eliminate any possible mutual misunderstanding during SCS creation, the final system characteristics should be precisely stipulated in the document. The technical design preparation is carried out by the contractor in close contact with the responsible representative of the customer. Where necessary, a third party with appropriate qualifications for preparation of such a document can be involved in formulation of the technical task. During TT development, the design is named with a code according to USS (GOST) 34.201-89 [118].

Outline design

The purpose of work at this stage consists in the development of the preliminary design solutions. The outline design is frequently called the technical offering. The documentation at this stage is general and short (usually 5–10 pages of typewritten text with a minimum number of schematic illustrations, such as SCS common structure and other similar objects). It can contain some variants of the task solutions, a brief analysis of these variants, and recommendations for selection. The technical offer is frequently given to the customer before the conclusion of the official contract for design (for example, during tendering of bids), and therefore is sometimes called the commercial or budgetary offer.

At the outline design stage, the structural SCS scheme and workstation configuration are developed, and the selection of the transmission signal medium and methods of cable layout are also made.

The following documentation can be included in the document structure:

- an explanatory note to the outline design (the document code is P1);
- the structural scheme of the hardware (code C1), which can be included in the P1 document structure;
- cost calculations for system creation (code Б0).

The guidelines in (RD 50-34.698.90[120]) consist of rules for the formulation of these documents.

Technical design

Work at this stage consists of profound development and substantiation of the design solutions for a system as a whole and for its separate parts. The principles of system operation as well as specific tasks and problem solutions, which are concerned with the system creation for a particular object, are to be considered as design solutions. The following documentation is included in the document structure, which is developed during technical design:

- the technical design register (TP code, according to USS (GOST 2.106-96). ESKD – unified system of construction documentation for textual documents);
- explanatory notes to the technical design (P2);

- structural scheme of technical means complex (C1code), which can be included in the document structure (P2);
- specification of equipment and materials;
- calculation of cost estimate (B2).

As a recommendation, it is possible to offer to provide the calculation of cost estimate as a separate document, without filing it in the technical design book.

Technical design documentation development

Work during the technical design documentation development stage consists of the preparation of precise drawings, plans, and tables that will be helpful for installers during system creation. The technical design documentation provides detailed binding of system components to an object, contains the drawings, the tables of junctions and connections, the plans of equipment and cable layout arrangement, as well as other similar papers.

The following documentation is included in the document structure provided at this stage:

- plans for equipment and cable arrangement (the document code is C7);
- tables of junctions and connections (C6);
- installation drawings (SB).

10.2 Architectural Design Stage

10.2.1 Goals and Objectives

The general normative document that regulates the fulfillment of the architectural design stage is the TIA/EIA-569 standard "Standard for commercial buildings upon cable routes and telecommunication cables installation." It was created because construction of an SCS with optimum technical and economic characteristics is impossible in a building for which a number of architectural and planning requirements are not met. The rules of organization are regulated by the standard:

- equipment rooms;
- cross-rooms;
- cable routes of a horizontal subsystem;
- cable routes of an internal backbone subsystem;
- building cable inlet areas of external backbone subsystems;
- cable routes of an external backbone subsystem.

The standard also includes the requirements for power supply, heating, ventilation and air conditioning systems of a building in those part that relate to telecommunications infrastructure construction.

The goal and objective of the architectural design stage is the creation of prerequisites for telecommunications stage fulfillment. Careful study of the design solutions obtained during this stage allows for the achievement of appreciable cost reduction during SCS creation and operation. Moreover, the principles incorporated into the architectural design have a direct impact on cable system reliability and operation safety parameters.

In practice, the design process at the architectural stage is frequently complicated substantially by the situation that many buildings, even for administrative uses, were constructed when there were merely two fundamental cable layout types – power and telephone. Thus, according to those norms, a maximum of one telephone per room was planned, instead of the present situation of one telephone per workstation. Other old buildings were constructed absolutely for other purposes (for example, for dwelling), but not for offices; they were later remodeled. It is generally true that designers encounter problems of insufficient cable route capacity for work carried out in older buildings, which results in the need for new cable riser construction, expansion of horizontal cable ladder capacities on floors, as well as internal remodeling in order to organize the equipment rooms and cross-rooms.

The reconstruction of old buildings, taking into consideration the new requirements, usually reduces these problems to a minimum; however, reconstruction requires significant capital investment and is not always possible. In industrially advanced countries this problem was solved in a drastic way: many old buildings were simply pulled down, and new buildings were constructed in their place. This approach to the problem appears economically more favorable than remodeling. Unfortunately, even many new office buildings recently constructed in our country are very poorly fitted for the construction of telecommunications systems with intensive data exchange.

10.2.2 Equipment Room Design

The equipment room is the space in which the network equipment for collective use (PABX, servers, LAN switches, and network hubs) is installed. Equipment rooms require increased attention by the designers and information system operation services of the building since they contain special equipment. This is related to the normal activity of most organizations, since the owners or the tenants of buildings directly depend on operating access to the information stored in electronic files, as well as on the quality of external and internal telecommunications. The inability to access information or a failure of communications with the outside world can cause large financial losses and adversely affect a company's image, and in the worst case, it can result in the bankruptcy of the company. Therefore, the organization of fire suppression, air conditioning, and access control systems in the equipment rooms is a de facto standard.

Equipment room allocation

During selection of the equipment room location, the following principles should be applied:

- The equipment room should be combined with or moved as near as possible to the building cross-room as in order to minimize the length of the cables connecting them.
- In order to simplify access control procedures, it is necessary to place equipment room near security posts of a building.
- The space for the equipment room should not have public access, since this complicates the access control system.
- There should not be windows in this room, and it should not even closely abut the external walls of a building.
- For equipment rooms located in basements, the risk of flooding by underground water and failures in water pipes and sewer systems should be reduced to a minimum using special construction solutions (for instance, additional drain pipes and appropriate selection of water pipe routes).
- It is not recommended to locate the space for equipment rooms on the top building floors, as this substantially complicates the procedure for installation of external communications cable inlets in the building. Moreover, the top floors suffer more damage than others during fires, and they also flood if the roofs leak.
- It is extremely undesirable to locate the equipment room near the internal constructions of a building: for example, elevator wells, flights of stairs, ventilating chambers, because these constructions limit the future expansion of the equipment room.
- It is forbidden to locate the equipment room near a storeroom of flammable or hazardous chemical materials.
- It is necessary to avoid close placement to power sources of electric or magnetic fields and also equipment, which can cause increased vibration;
- Freight elevators should be near the equipment room.

Equipment room space

The equipment room dimensions are directly determined by the structure of the equipment it contains. If such information is absent, for the design of normal office buildings it is should be 0.7 % of the working space, but not less than 14 m^2. For buildings with a low density of workstations (hotels, hospitals), the area of the equipment room is selected depending on the number of workstations (Table 10.3).

It is necessary to take into account that, in practice, the equipment room is frequently combined with the cross-room on a floor and/or with the internal backbones. Thus, for such space combination, except for equipment of collective use, junctions and network equipment are placed in the equipment room and serve telecommunication outlets of neighboring office space on the same floor.

Table 10.3. Recommended equipment room space for buildings with a low density of workstations

Number of work stations	Equipment room space, m^2
≤100	14
101–400	37
401–800	74
801–1200	111

During the selection of construction solutions, it is necessary to take into consideration that the creation of one large equipment room is less expensive than the creation of several small ones for the same total space.

Environmental conditions in equipment rooms

With the selection of appropriate architectural and planning solutions, and also keeping in mind the engineering systems for the functioning of a building, the following environmental conditions should be kept in equipment rooms.

Air temperature: from 18–24°C measured at 15 m from floor level. The maximum rate of temperature change should not exceed 3°C per hour. When the temperature exceeds the top limit, the majority of network equipment retains its working efficiency, however, this is attended with the accelerated wear and tear of electronic components and results in premature failures.

Air humidity: 30–55% without condensation of moisture, measured at the same height. The maximum rate of humidity change is no more than 6% per hour.

Illumination: not less than 540 lux, measured at 1 m from floor level in a location free from equipment. The light sources should have enough power and to be located so as to ensure uniform illumination of all the equipment room space. The presence of good illumination is especially important during installation. Thus any type of work should be made without the use of additional lighting.

Vibration level: In a 5–22-Hz frequency band the amplitude of vibrations should not exceed 0.12 mm, and in the range 22–500 Hz, the maximum acceleration should not be more than 25 m/s^2.

The intensity of an electric field should not be higher than 3 V/m in the entire frequency spectrum. Contaminants in the air should not exceed the limit values given in Table 10.4.

Additional standard IEC-721 requires that the heat power dissipation of network equipment should be not less than 2.5 kW. The value of this parameter is normalized since the apartment of the equipment room usually has no windows, and the value of dissipated heat allows for the correct design of an air conditioning system.

Table 10.4. Contaminant limits in equipment rooms

Substance	Contaminant limit
Chlorine, ppm	0.01
Hydrogen sulfide, ppm	0.05
Nitric oxide, ppm	0.1
Sulfur dioxide, ppm	0.3
Dust, g m^{-3} day^{-1}	10^{-6}
Hydrocarbons, g m^{-3} day^{-1}	10^{-6}

Requirements for construction and facilities of equipment rooms

The most optimal form of equipment room space is a square or nearly square room with a short wall length of not less than 3 m. The distance between a floor (false floor) and ceiling (false ceiling) should be not less than 2.5 m.

The equipment room floor is designed to maintain a distributed floor load not less than 12 kP and distributed floor point -4.4 kP. In order to organize the cable layout, it is desirable to use false floors with minimum heights of 0.25 m, made of easily removable metal plates. False floors provide cable plait installation in 19-inch mounting constructions from below, taking into account the minimum bend radius of each separate cable. The construction and wall materials are selected with a view for the opportunity of plating them by the use of metal shielding panels and for the fastening of equipment with weight not less than 100 kg to them.

When the network equipment and patch panels are placed in 19-inch mounting constructions, it is extremely desirable to separate cabinets and racks to ensure access to their fronts and backs.

The equipment room entrance is supplied with a metal door, which opens from outside, with dimensions not less than 20×09 m. In order to prevent water penetration from a corridor in the case of water or sewer pipe failures, a baffle is installed in neighboring rooms. The materials and construction of interfloor overlaps, walls, and doors are selected in view of ensuring of fire resistance not less than 45 min.

The equipment room should be equipped with the following systems:

- security-signaling system;
- fire-signaling system;
- fire-suppression system;
- air conditioning and illumination systems, which conform to the requirements of Sect. 10.2.2;
- emergency illumination system;
- protective and telecommunications grounding according to the requirements of Chapter 8, and direct connection to the main grounding bus should be ensured.

Moreover, one or several telephone sets are installed in the equipment room. Practice shows that it is very helpful to have a loudspeaker communications system during the organization of various sorts of preventive maintenance, parameter measurements, and corrections of malfunctions.

The network equipment installed in the equipment room is powered from an uninterruptible power supply (UPS), which, whenever possible, should have two independent connections to an urban electrical network with automatic switching from the general power route to the reserve one. The power supply for security and fire signaling is also provided from two systems.

The cables of public telephone networks and other telecommunications operators are installed in the equipment room. In the presence of external backbone subsystems in SCS, the cables of the external route subsystems are also installed in the equipment room. The cable installation can be made from the cable conduit run, collector, platforms, poles (for aerial suspension), and from other similar structures. In some situations determined by specific local conditions, separate space is assigned for cable inlets. The appropriate junction equipment is installed in it, and the junction equipment is connected with the equipment room using separate internal cables.

Where space allows, it is recommended to provide the following in the equipment room:

- computer or, as a last resort, a table and chair for organization of the system administrator workstation;
- separate bookcases or shelves to store the working and operational documentation, and also metering equipment as well as cable system and network equipment spare parts;
- a wall holder for frequently used connectors and cross-cords;
- carbon-dioxide fire extinguisher.

10.2.3 Cross-room Design

As it was mentioned above, the cross-rooms are divided into external backbone cross-room (EBCR), building cross-rooms (BCR), and floor cross-rooms (FCR). In practice, the EBCR and the BCR are frequently combined with each other, and also with one of the equipment rooms, therefore only the FCR is described below. All cited instructions for it are equally applicable to the rest of cross-rooms.

FCR is a technical room in which the cables of SCS internal backbone subsystems and horizontal subsystem cables are installed. The switching, network, and other auxiliary equipment are mounted in this room. It is not permitted to install equipment in this room that is not related to FCR functions, for example, power distribution panels for the floor power supply.

Like equipment rooms, the cross-rooms require the increased attention of designers and telecommunications infrastructure operation services. However, a failure of the network equipment located in the equipment room results in the complete or partial interruption of the operation of all building information systems, but the failure of equipment located in the FCR usually means an interruption of operation only for workstations that are serviced by this equipment. Therefore, in most cases, construction of the FCR and the FCR equipment have requirements that are less rigid than those for equipment rooms.

Cross-room location

During selection of cross-room locations, the following principles should be applied:

- FCR can be combined with one of the BCR on the same floor.
- FCR should be on each floor of a building. The solution that is frequently used in Russian buildings, based on cross-rooms that also serve neighboring building floors, cannot be called successful, as it fundamentally limits the possibilities for cable system expansion and upgrading.
- FCR should be moved as near as possible to the vertical cable riser on which the cables of the SCS internal backbone subsystems are installed. it is ideal if the cable riser passes directly through FCR.
- In cases when the working space of a floor exceeds 1000 m^2, or if additional cross-rooms are necessary to organize horizontal subsystem cables, which cannot exceed the 90-m limit on cable length, more than one FCR can be installed on each floor.
- In order to minimize the cable length and, accordingly, to decrease the horizontal subsystem cost, it is necessary to locate the FCR as close as possible to the geometric center of the workspace that it services;
- To simplify the access procedure, the location of the cross-room should not have windows, should not be open to the public, or combined with other industrial rooms.
- It is necessary to avoid close location of electrical power sources and/or of magnetic fields, and also equipment, which can cause increased vibration in the cross-room.

If several FCR are planned on a floor, then it is desirable to serve all of them using different vertical cable rises. In this case, the installation of a horizontal cable layout of the SCS internal backbone subsystems can be avoided and, accordingly, the reliability of the FCR also can be ensured. A violation of this condition is allowed to connect part of the FCR to the BCR as a transition through another FCR (Figure 10.2).

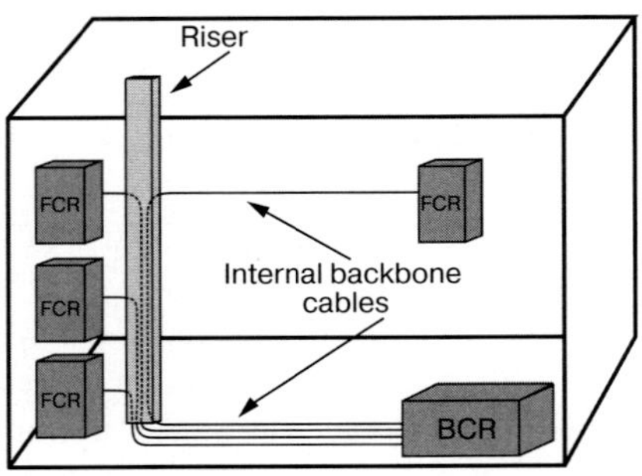

Fig. 10.2. Example of cross-rooms locations

Cross-room space

Usually the workstations are connected to the FCR switching equipment that is located on the same floor. The area of workspace serviced by FCR, according to the standards ISO/IEC 11801 and EN 50173, should not exceed 1000 m^2, that is, one cross-room serves a maximum of 100–250 workstations. The space of the FCR depends on equipment structure that is placed in it, but in any case it should not be less than 6 m^2. If a priori information on the number of workstations served by the cross-room, as well as a priori information on equipment placed in the cross-room is absent, then during the design stage it is possible to use the data from Table 10.5.

Table 10.5. Recommended dimensions of FCR, depending on the workspace served

Served working space, m^2	FCR overall dimensions, m
≤1000	3.0×3.4
≤800	3.0×2.8
≤500	3.0×2.2

Environmental conditions in cross-rooms

The engineering provision systems of the building functions should be designed and realized to provide the following environmental conditions in any cross-room:

Air temperature: 10– 30°C measured at a height of 1.5 m from floor level.

Air humidity: 30–55% without condensation of moisture, measured at the same height.

Illumination: not less than 540 lux, measured at a height of 1 m from floor level in a location that is free from equipment. The principles of realization and selection of quantity, type, and locations of lamps correspond to similar requirements for equipment rooms.

Vibration level: should not be higher than the maximum allowable value for installed equipment in the cross-room.

The intensity of an electric field: should not be higher than 3 V/m the entire frequency spectrum.

Contaminants in air should not exceed the limit values for sanitary norms.

Requirements for construction and equipping cross-rooms

The most optimal form of the cross-room space is a square or nearly square room , in which the length of the short wall is 2 m. The height of this room should be not less than 2.5 m. The construction and wall materials are selected so that they can be plated by metal shielding panels and so that the equipment, whose weight is not less than 100 kg, can be fastened to them.

In the case when the vertical cable riser layout goes directly through the cross-room, it should not have false ceilings or the false floors, and the door must be installed on the wall adjacent to the cable riser. The rest of the requirements for the door and doorway of the cross-room are identical to those for the equipment room. Materials and construction of the interfloor overlaps, walls, and doors are selected to provide fire resistance for not less than 45 minutes.

The cross-room must be equipped with the following systems:

- fire and security signaling;
- ventilation and illumination systems, which conform to the requirements of Section 10.2.3;
- protective and telecommunications grounding, according to the requirements of Chapter 8.

A telecommunications outlet, connected to PABX is provided in the cross-room. The requirements for a power supply system are identical to those for the equipment room. If necessary, the cross-rooms can be additionally equipped with fire suppression, air conditioning, and emergency illumination systems.

10.2.4 Cable Routing in External Backbone Subsystems

The fiber-optic and electrical cables of external backbone subsystems of buildings are installed, in most cases, in the telephone cable conduit run. The basis of the cable conduit run is made of round cement, concrete, or plastic pipes with internal diameter of 100 mm. The conduit run is installed at a depth of 0.4–1.5 m and consists of separate blocks that are hermetically joined. There are sight wells in every 40–100 m of the cable route, on whose walls the consoles are mounted in order to install the cable and couplers.

Recently, the technology of Dura-Line (USA) has been widely used for the creation of cable conduit runs. This technology is based on pipes with 21–33-mm internal diameters and maximum lengths not less than 1750 m. The *Silicore* pipes are delivered on drums or in hanks. The internal pipe surface is covered with a layer of solid grease, which sharply reduces the effort to stretch cables and allows the use of the channel pneumopreparation method in order to tighten the cables. The high durability of the pipe allows it to be installed directly into the ground without the use of additional mechanical crush protection. The system structure includes an advanced accessories set, which facilitates the creation of cable conduit runs.

In order to carry out the layout of cables for external backbone subsystems in industrial facilities, technological constructions with a system of ladders, supporting brackets, and other cable-organizing layout elements are widely used.

The use of aerial cable suspension is not popular mainly due to the complexities of its realization. Generally, aerial cable suspension is used in those situations when cable layout by other means is impossible. In most cases, in order to suspend cables on poles, special suspension or self-bearing cables are used, and the process of their suspension is carried out by the use of special fastening and stretching fittings. It is quite possible to suspend the cables on messenger wires using wrappings or special clamps. The fastening spacing in the latter case is approximately 70 cm.

Sometimes, direct cable layout in the ground is used. It is strongly recommended to carry out the cable layout on a sandy pad in order to provide protection against deformation caused low temperatures during winter.

10.2.5 Cable Routing in Internal Backbone Subsystems

The cable routes of internal backbone subsystems are intended to connect the BCR to FCR, EBCR, and the equipment rooms. Moreover, the external backbone cables are installed on these routes from the building cable inlet to the EBCR or the BCR.

The backbones cables of these subsystems can be installed vertically and horizontally. The constructions for passing of horizontal areas are the same as the constructions used for the arrangement of horizontal subsystems and are frequently used by both types of cables simultaneously. In order to pass vertical areas, cable risers and different well types, specially assigned for this purpose and described in detail below, are usually used.

The dimensions are selected based on the following ratio: a cable riser with area of 8000 mm^2 allows the installation of cables serving 2500 m^2 of workspace. Thus it is necessary to include all the building floors that are served by the cables of a given route in the area considered. Multiplying the received results of the useful area by 3 is recommended in order to provide a reserve for future expansions.

The functions of cable risers for backbone cables can be provided the use of *slots*, *sleeves* and *ducts*. A qualitative comparison of these elements of passing the interfloor overlap is shown in Table 10.6.

Table 10.6. A qualitative comparison of slots, sleeves and ducts

	Brief description	Advantages	Disadvantages
Ducts	Vertically installed along fire-resistant ducts in cross-room walls	• Good protection from water, dust, fire penetration into neighboring floors • Effective cable protection from mechanical damage	• Limited flexibility • Requires large reserves for extension
Sleeves	Short overlapping pieces of pipes made from non-flammable materials, installed vertically along cross-room wall	• Good protection from water, dust, fire penetration into neighboring floors • Easy and inexpensive to install • Simplicity of cable layout	• Provides lower capacity and less-flexible use compared to slots
Slots	Rectangular apertures with borders in the interfloor overlap along cross-room wall	• Flexible use • Good overall dimensions	• Complex to fulfill fire safety norms • Very expensive installation • Distinctly weakens the mechanical durability of overlapping

The slot is an aperture of rectangular form in the interfloor overlap of the cross-room near to one of its walls (Figure 10.3).

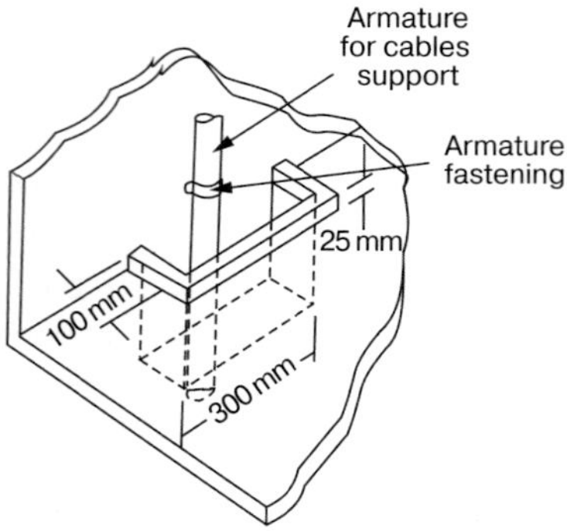

Fig. 10.3. Slot

This element is necessarily equipped with a border, which prevents extraneous objects from falling in it and water penetration. The backbone cables that pass through the slot are installed on a wall of the cross-room with the help of special armature, which is made in the form of clamps. Upon completion of the cable layout, the slot apertures should be closed up by fire-resistant plugs.

The tubular elements of vertical cable risers are sleeves and ducts. The sleeve (Figure 10.4) is a relatively short piece of pipe built in the interfloor overlap.

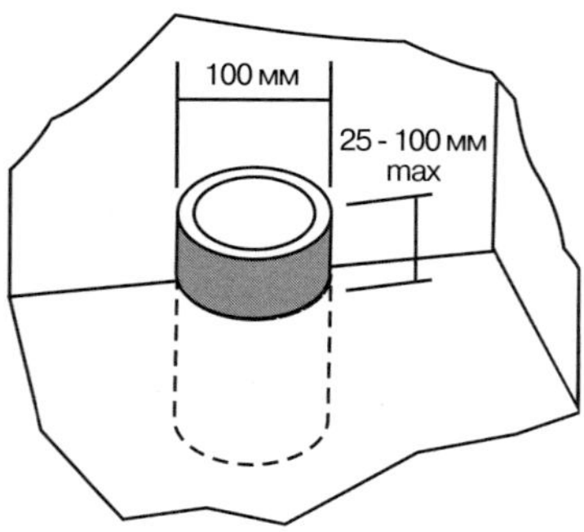

Fig. 10.4. Sleeve

The sleeve ends should jut out from both sides of overlap a minimum of 25 mm. The ducts differ from the sleeves mainly by their greater length. An example of ducts is shown in Fig. 10.5.

The recommended internal diameter of the ducts and sleeves for backbone cable installation is 100 mm. During use of these elements, it is necessary to control the volume of their filling carefully since the limit is exceeded, problems related to the layout of new cables and to the allowable stretching efforts from increased friction can arise. In addition to the duct diameter and the number of installed cables of specific external diameters, the bends and radii are deeply dependent on a value of stretching effort. No duct should have more than two bends with turning angles of more than 90° each.

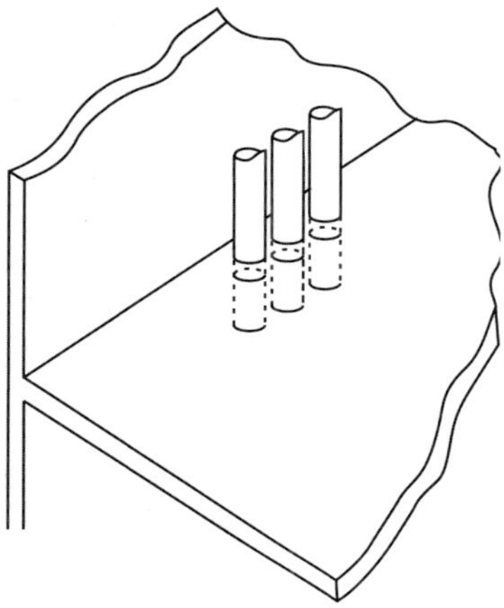

Fig. 10.5. Ducts

The data for filling volume and allowable bend radii of the ducts with the most common dimensions are shown in Table 10.7.

Table 10.7. Filling volume and allowable bend radius of the ducts

Duct diameter "in light", [mm]	Duct space "in light," [mm^2]	Maximum filling space [mm^2] using: One cable	Two cables	Three cables	Minimum cable bend radius with steel strengthening elements, [mm]	Minimum bend radius for other cables, [mm]
20.9	345	183	107	138	210	130
26.6	559	296	173	224	270	160
35.1	973	516	302	389	350	210
40.9	1322	701	410	529	410	250
52.5	2177	1154	675	871	530	320
62.7	3106	1646	963	1242	630	630
77.9	4794	2541	1486	1918	780	780
90.1	6413	3399	1988	2565	900	900
102.3	8268	4382	2563	3307	1020	1020
128.2	12984	6882	4025	5194	1280	1280
154.1	18760	9943	5816	7504	1540	1540

Depending on the dimensions and architectural features of a building, it can contain one or several cable risers. A design with one vertical cable riser and one FCR is used only in those situations when horizontal cables with lengths less than 90 m can be installed from the cross-room to all telecommunications outlets on a floor. In all other cases it is necessary to design several cable risers. Mounting additional FCR without assigning them to cable risers, but connected to the others with the help of vertical backbone cables is an alternative variant. In order to increase the reliability of the internal backbone subsystem it is desirable to install it using two or more spatially arranged cable risers.

For cable installation on the horizontal areas of routes it is possible to use conduits, pipes, ladders, or other constructions that are usually used just for such cases and are described in Section 10.2.6. In the presence of false ceilings on the route, cables can be fastened directly to walls or fixed ceilings by plastic clamps and rawplug anchors, or anchor wedges can be used. In order to install multipair cables, traverses are frequently used. These are L-shaped elements that are fastened to a wall, ceiling, or carrying structure with the direct or U-shaped ends of the supporting part. The distance between traverses should not be more than 1.5 m. This element does not support the entire length of the installed cable, therefore the element is effective only for cables with the increased rigidity.

During the design of cable routes of internal backbone subsystems, special attention should be given to fire safety (Chapter 9). All metal constructions of cable routes, that is, ducts, sleeves, ladders, and conduits, should be reliably grounded.

10.2.6 Cable Routing in Horizontal Subsystems

Cable routing in horizontal subsystems is intended for use with cables installed from FCR to the workstations. There can be also vertical areas that fail to intersect the interfloor overlaps (with the exception of installation methods through overlap, which are extremely rarely used in practice and are considered in Section 10.2.6). The cables of a horizontal subsystem are installed:

- in floor constructions;
- under ceilings;
- in wall channels (cable conduits).

Each of these solutions has some versions that are described and analyzed in more detail in the following sections. Their common requirement is the need to ground all their metal elements, including ducts, ladders, and conduits.

Cable routes in floor constructions

For the hidden cable layout of horizontal subsystems in floors of specially designed buildings, various constructions (Figure 10.6) are provided, which are created during construction, remodeling, or general repair. They include in-floor channels, cellular floors, false floors, and ducts. These elements provide means to hide the cable layout and to provide effective cable protection against mechanical effects. The main parameters, possible uses, advantages, and disadvantages of the constructions are described below. We note that ducts are traditionally the most widely used in our country, despite their inherent disadvantages. The other three elements are rarely used .

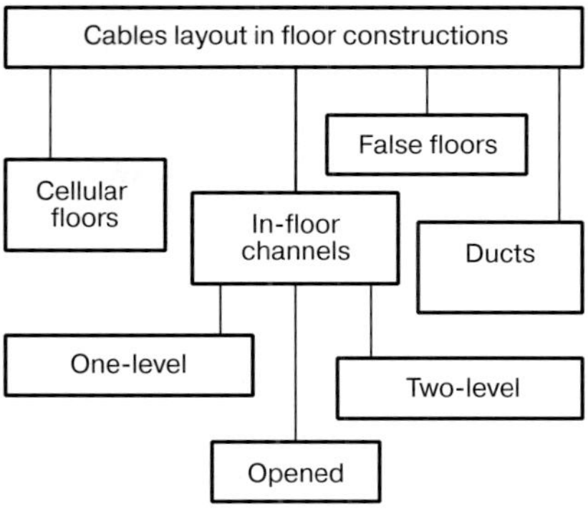

Fig. 10.6. Various constructions of hidden cable layout of horizontal subsystems in floors

In-floor channels. In-floor channels are the specialized metal or plastic constructions with basically rectangular cross sections that are installed in interfloor overlap structures before the "direct filling" of a floor. As constructive elements in-floor channels ensure effective mechanical protection. They reduce the level of external interference and electromagnetic radiation and hide the cable layout. A general disadvantage is their high cost of realization, the need for completion of cable installation before the building construction is complete, the use of special floor boxes in order to provide access to electric and telecommunications outlets, as well as increased floor weight.

The in-floor channels usually form a structure in which the backbone and distributive subsystems can be distinguished. Horizontal cables from the cross-rooms are installed on the backbone channels, and the distributive channels are used for cable branching from the backbone to the workstations. In the intersection of the backbone and distributive channels, the pull boxes are installed. When the channels have a significant length, the same boxes are used. Distances between boxes should not exceed 6 m. The floor boxes with sites for telecommunications and power supply outlets are installed in the workstation locations (Section 5.2.4). The covers of boxes should be on the same level as the surface of the "clean floor." In some cases the distributive channels come from the floor into wall-mounted boxes, whose covers are fixed flush with a wall surface. This method can also be used for combined variants of cable layouts. In this case, the cable is brought from the in-floor channel near a wall and further connected with workstations by means of wall-mounted conduits.

In practice, three types of the in-floor channels are used: single-level, two-level (Figure 10.7), and open (Figure 10.8). In the first type, the backbone and distributive channels are located on the same level as an interfloor overlap surface and differ from each other only by cross-sectional area. They can be installed in buildings with "clean floors" of thickness 63 mm or more.

In a system of two-level in-floor channels, the backbone and distributive channels are installed at different levels. Distributive channels are usually installed above the backbone. In order to organize the two-level channels, the thickness of "clean floor" should not be less than 100 mm.

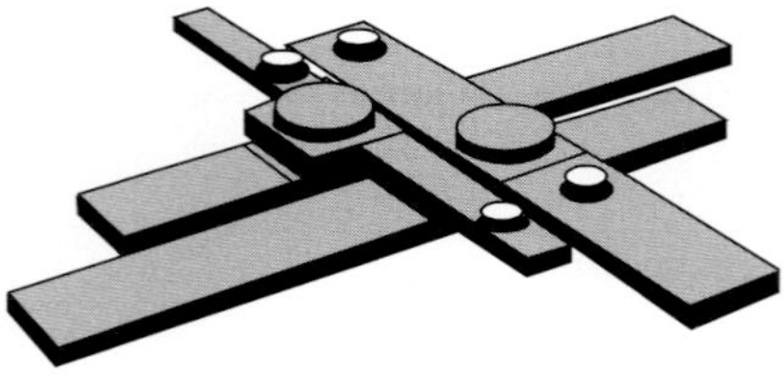

Fig. 10.7. Two-level in-floor channels

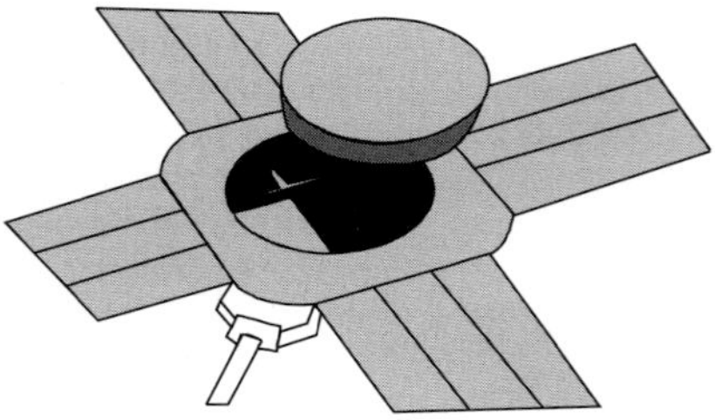

Fig. 10.8. Open in-floor channels

In a system of open channels, which can be considered a “degenerate” type of single-level channels, their top surfaces and the covers of the boxes are located flush on the same level as the “clean floor." They can be installed in buildings with the minimum “clean floor” thickness of 25 mm. This type of channel is the simplest to install and service, but at the same time it only gives a minimum level of cable protection from mechanical failures, water, or nonauthorized access.

Sometimes in-floor channels are divided into a few sections, in which the cables of various designations (for example, information or power supply) are installed.

Every 10 m^2 of the building workspace should be serviced by backbone and distributive channels with cross-sectional area of 650 mm^2. In office buildings, distance between parallel distributive channels should be selected within the limits of 1520–1825 mm, whereas the distance from the channel to an external wall of a building or carrying column should be 450–600 mm. The quantity, cable layout routes, and capacity of the backbone channels are determined from the results of distributive channel design. The distance between parallel backbone channels is usually 18 m. It is possible to generate one or several backbone channel, each of which serves some common channels and provides cable inlets in the cross-room or equipment room.

The terminations of cable channels in the cross-room space are designed to provide convenience of cable outlet and connection to switching equipment panels. Two types of termination that are most frequently used in practice are shown in Figure 10.9.

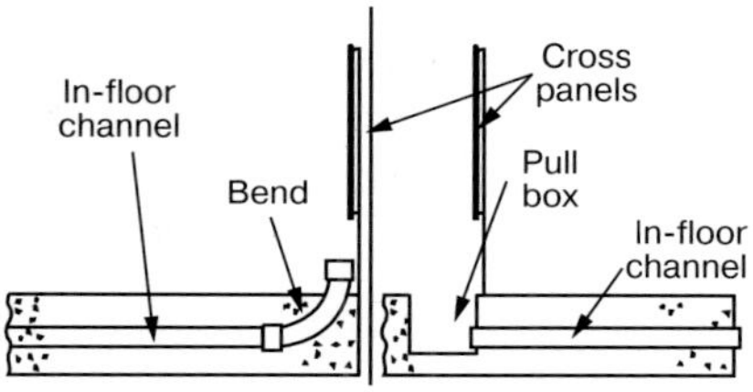

Fig. 10.9. Two types of cable channels termination

Cellular floors. Cellular floors (Figure 10.10) can be considered a type of underground channel; they are a system of continuous cavities in concrete plates of the floor structure. This method of cable layout has practically the same features, advantages, and disadvantages as in-floor channels. In contrast to in-floor channels, this cable layout method has a larger volume than floor cavities, however, it essentially yields the same shielding efficiency from electromagnetic radiation.

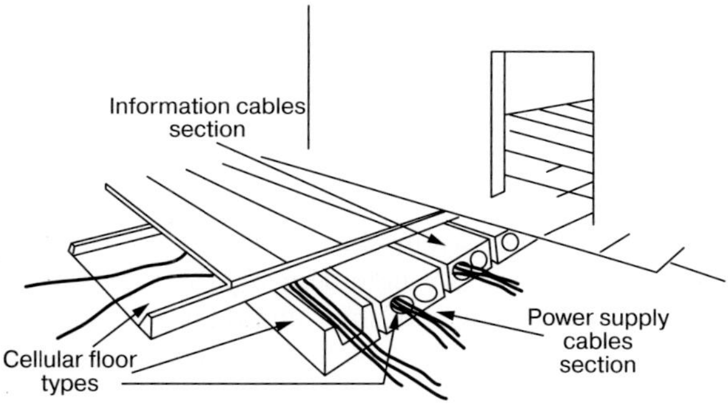

Fig. 10.10. Cellular floors

False floors. Traditional false floors (Figure 10.11) are formed by square tiles of various dimensions installed on metal poles or grates of various heights. Tiles are usually made of cast metal and have linoleum top covering.

The Intercell company offers an original low-profile false floor design in their *Cable Management Flooring System* structure. False floors are assembled by means of 16 pedestals, 55-mm or 85-mm high (depending on the version), integrated into a single unit by planar bottom bridges. The unit is made from one die-formed metal sheet, which is installed on the floor by gluing. The false floor tiles have the following dimensions: 500×500 mm with thickness of 2 mm, and are furnished with side flanges (or additional stiffening ribs) for their installation in fixing rabbets on the pedestals. The bottom surface has a sound-proof covering made of non-woven material. In addition, the tiles are fixed on the pedestals with angular fastening by self-cut screws with a washer to improve attachment reliability. We note the presence of additional accessories of this system:

- in-floor box with a cover (false floor tile), which is installed on this box and has an integrated lid for cable output;
- fireproof in-floor barrier, made of noncombustible material;
- angle bars for the installation of edge areas and doorways.

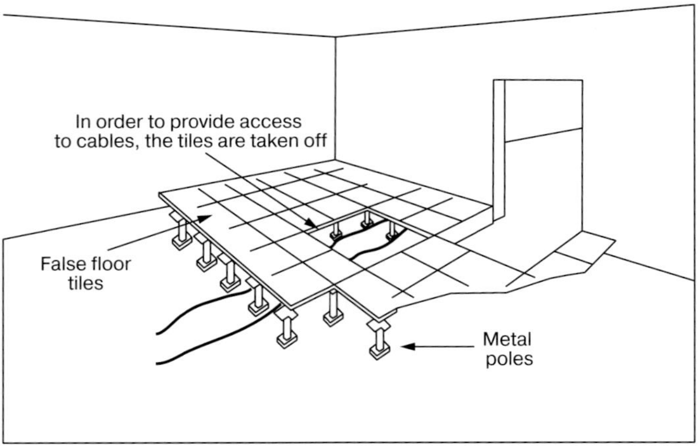

Fig. 10.11. False floors

False floors provide easy access to the cable layout channels, and do not impose limitations either on the number of installed cables, or on the cable layout direction. As an additional advantage, we note their high mechanical durability of 1500 kg/m^2 or more. The disadvantages of this solution are the decrease in room height, the possibility of occurrence of unpleasant acoustical effects, and the necessity for the use of special fireproof cables in the cable layout, as the space under the false floor, in most of cases, is a plenum cavity.

Channels in the form of completely closed metal ladders are frequently used for the cable layout. In this case, the branching of cables to the floor and in-floor boxes of various types for outlet installation is carried out using metal sleeves. This solution essentially allows designers to lower the requirements for the fire safety level of cable hardware products.

Ducts. The network of metal or plastic ducts of various diameters as well as in-floor channels are installed in interfloor overlap structures before the "direct filling" of a floor. They can be divided into two subsystems: backbone and distributive. This approach costs less than the solutions considered above, however, this solution has limited flexibility and small capacity.

A network of ducts is designed to eliminate sections that have more than two bends under a right angle between stretching cables points or intermediate pull boxes. The length of any channel should not be more than 30 m.

During duct installation, it is necessarily take into account the minimum bend radius, and at the same time, the internal bend radius should be not less than 6 internal diameters of the duct. If the ducts have internal diameters of more than

50 mm, then the bend radius is selected to be not smaller than 10 internal diameters of the duct. In case where ducts are used for fiber-optic cable installation, the internal radius of a bend should always be not less than 10 internal diameters. In order to maintain the minimum bend radius of cables, sometimes in practice the duct inlets into the pull box are at a 45° angle instead of the usual 90° angle. The selection of the number of cables according to the dimensions of the ducts is carried out based on the data in Table 10.8.

Table 10.8. Capacity of various diameter ducts

Internal diameter of duct, [mm]	Number of cables and wires at external diameter of cable, [mm]									
	3.3	4.6	5.6	6.1	7.4	7.9	9.4	13.5	15.8	17.8
15.8	3	2	2	2	1	1	1	0	0	0
20.9	6	5	4	3	2	2	1	0	0	0
26.6	8	8	7	6	3	3	2	1	0	0
35.1	16	14	12	10	6	4	3	1	1	1
40.9	20	18	16	15	7	6	4	2	1	1
52.5	30	26	22	20	14	12	7	4	3	2
62.7	45	40	36	30	17	14	12	6	3	3
77.9	70	60	50	40	20	20	17	7	6	6
90.1	–	–	–	–	–	–	22	12	7	6
102.3	–	–	–	–	–	–	30	14	12	7

During installation of ducts, the steel wires are left in them. To avoid damage to the cable sheath, the ducts ends should not have sharp edges or fins. Each duct is marked from both ends by a unique identifier with its length indication.

In those cases when the route length of the duct layout exceeds 30 m, when there are more than two right angle bends on the route between stretching points of cables, or when duct branching is required in any point of the system, then a pull box must be used. The pull box construction should have a lock or a simple latch on its cover, as well as a gasket or other sealant for protection against dust.

The length of a pull box for direct stretching should be not less than 8 internal diameters of the largest duct included in it. The additional requirements for pull boxes for stretching with turns or loops include the following:

- the distance between each duct output and the opposite box side should be not less than 6 internal diameters of the largest duct;
- the distance between duct ends intended for the same cable is selected not less than 6 internal diameters of the largest duct;
- if the duct enters in the bottom of pull box, then the pull box depth should not be less than sum of the internal diameter of the largest duct that came from the bottom and 6 external diameters of the thickest cable.

Under-ceiling cable channels

Cable channel types. In order to carry out the cable layout of an under-ceiling horizontal subsystem in common spaces as well as in rooms and corridors that are equipped with false ceilings, the following types of cable channels are used:

- perforated or solid ladders without top covers;
- cable traverses, which are formed by two lateral longitudinal carrying rails with tubular or wire-carrying elements that are connected with transverse transoms;
- chutes with solid or perforated bottoms;
- closed cable ladders with demountable top covers and perforated or solid bottoms.

Traverses made of wire elements are unique in the considered constructions, because they have high flexibility and are therefore very effective for use in constrained conditions.

All types of channels listed above can have additional accessories: angle bars, channel adapters of various areas, covers, branches, and adapters for pipes.

The effective cross-sectional area of the cable channels is selected at the rate of 650 mm^2 for every 10 m^2 of workspace. The value given is for the case when one workstation has three outlets. In cases where a greater density of arrangement of employees or a larger number of outlets is used, this value should be appropriately corrected.

The fastening of cable channels is carried out under two basic schemes: with the help of side clamps (fastened to a wall) or with the help of vee-type, U-shaped or L-shaped clamps (fastened to a ceiling). The fastening elements should be installed no further than 1500 mm from one another, if another value is not stipulated.

During the channel readiness check before cable layout, special attention should be given to the control of sharp corners and the absence of fins on internal surfaces, which can cause cable sheath damage. for similar reasons it is forbidden to break the cable channels in wall recesses. Upon completion of cable layout, the residual free parts of the recess are enclosed by a fire-resistant material. It is desirable to carry out the layout of power supply cables and SCS cables using different cable channels and to control their routes according to the working norms. The metal elements of the cable channels should be necessarily grounded on a telecommunication grounding bus in the cross-room.

The height of free space between the channel and the main ceiling is selected to be not less than 300 mm, and access to them should not be limited by constructions of other building engineering systems (for example, ventilating conduits).

Similar to the backbone cable layout, it is possible to fasten the horizontal cables bundle directly to a wall or ceiling with the help of plastic clamps and rawplug anchors or anchor wedges. Sometimes the cable traverses are used for this purpose. This element is based on short pieces of T-beam, which are installed to the ceiling by means of two steel rods with threaded ends for fixing nuts. Thus the pieces of the T-beam should be mounted with their edges upwards to prevent the cable overpress effect (Figure 10.12). This requirement should especially be kept in relation to UTP cables of 6 category, which are distinguished by the large sensitivity of electric parameters to external mechanical effects. In case of single wire layout, it is possible to use, as an element of fixing, the self-adhesive platform or the platform with holes for the fixing screw and lug, clips, as well as a special cable holder in the form of press fixing, offered by Rehau company (Figure 10.13)

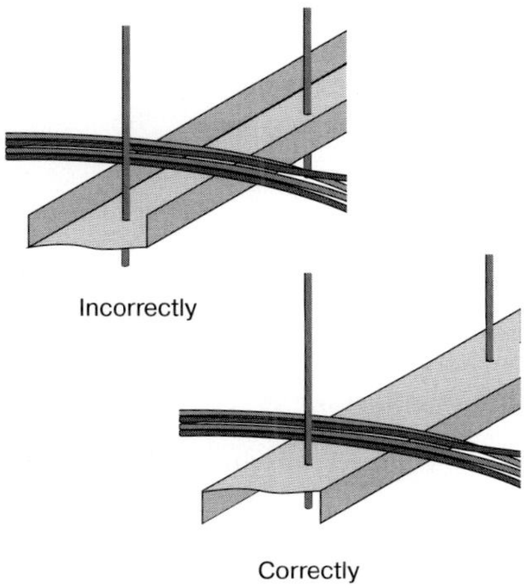

Fig. 10.12. Variants of horizontal cable layout under ceil

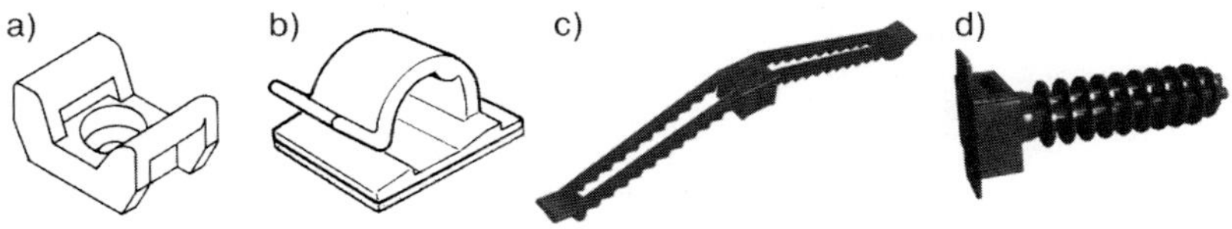

Fig. 10.13. Fixing elements of single wire layout: *a)* platform; *b)* clip; *c)* Rehau cable holder; *d)* dowel.

Methods of cables routing to workstations. During horizontal subsystem cable layout, the following basic requirements should be executed within the false ceiling:

- the false ceilings must have a folding construction and be no more than 3.4-m high from floor level;
- they should have enough free space for the installation of auxiliary constructions and to carry out cables stretching (in particular, the height of the free space between the cable channel and the overlap should not be less than 300 mm);
- provision of channel availability for cable layout and/or a clean wall or ceiling surface for fastening of single cables or bundles;

- during cable layout it is forbidden to fasten the cables or their plaits to the fastening elements of the false ceiling, and it is also forbidden to fasten the cables to cells or rails of the false ceiling.

In order to meet these requirements, the cables can be routed to workstations by four basic methods (Figure 10.14).

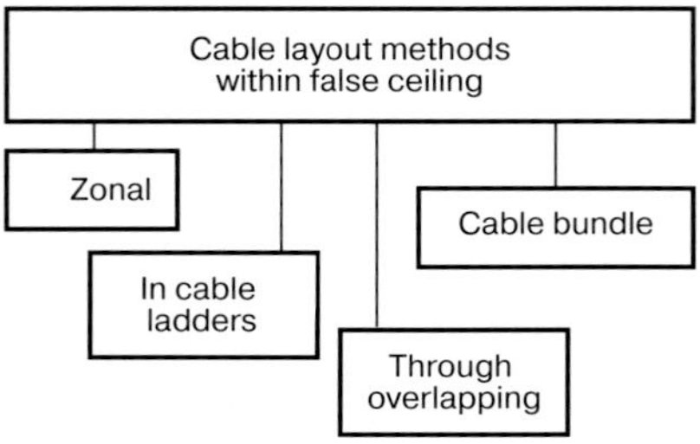

Fig. 10.14. Cable routing basic methods

According to *zone method*, the served area is divided into areas of up to 72 m^2. The telecommunications outlets are installed by one of four methods described in Section 5.2.3, by the use of the vertical conduits or columns, which are mounted with their tops in the space above the false floor. In each area, the branch point is organized by which multipair cable should be installed (Figure 10.15). The horizontal cable is usually installed from the branch point along the shortest distance to the conduit or column and then to the outlet.

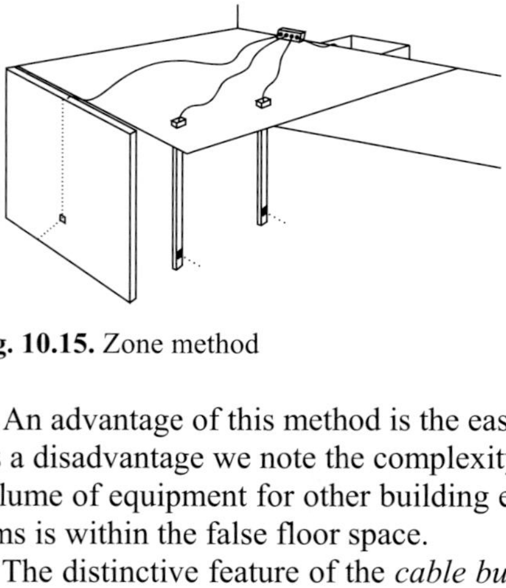

Fig. 10.15. Zone method

An advantage of this method is the ease of its realization and its high flexibility. As a disadvantage we note the complexity of system realization when a significant volume of equipment for other building engineering systems and illumination systems is within the false floor space.

The distinctive feature of the *cable bundle* layout (Figure 10.16) is the forming of cable bunch put in an apartment. The cable bunch is installed along the conduit or column mounting places and fastened to a wall or ceiling. During carrying out the cable bunch layout near the conduit or column, one or several cables are branched from the bunch, which are lowered to the telecommunication outlets then.

The advantage of the considered solution is its high flexibility, the ease of realization, and reduction of daylight lamps interferences when the cable layout is carried out along the walls. As the basic disadvantages are considered the need for plenum-cables use or carrying out the cable bundle layout in protective pipes, otherwise it fails to fulfill norms of fire safety.

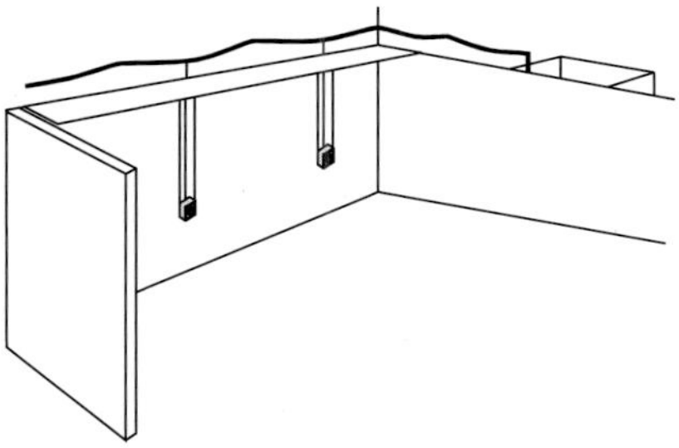

Fig. 10.16. Cable bundle layout

Method of cable layout in ladders (Figure 10.17) assumes, as it follows from the name, presence of cable ladders system in an apartment. The cables bunch is installed on the ladder and, in case of necessity, is fastened with the help of plastic hoops. During carrying out the cable bunch layout near the outlets, as similarly to previous method, one or several cables are branched from the bundle. The advantages of this method are effective protection of cables against mechanical effects and ease of additional cable layout carrying out. As the disadvantage we shall point the following, complexity of realization and increased weight of ceiling constructions therefore given method is generally used in service room and during the cable layout in corridors.

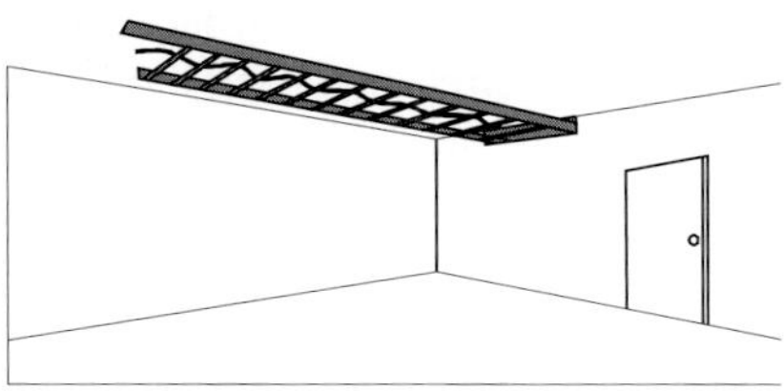

Fig. 10.17. Cable layout in ladders

During carrying out cable layout *"through overlap"* (Figure 10.18), the cable for the workstation is installed from the false ceiling through a hole in interfloor overlap. Due to reduction of building fire resistance and decreasing of overlaps mechanical durability such solution can be applied only in extreme cases, when the cable layout by other ways is impossible because of whatever reasons or related to significant difficulties.

Additionally we shall note, that except for the considered above basic methods, in practice on various sections of route, the combinations of these methods are frequently used, which selection is defined by certain local requirements.

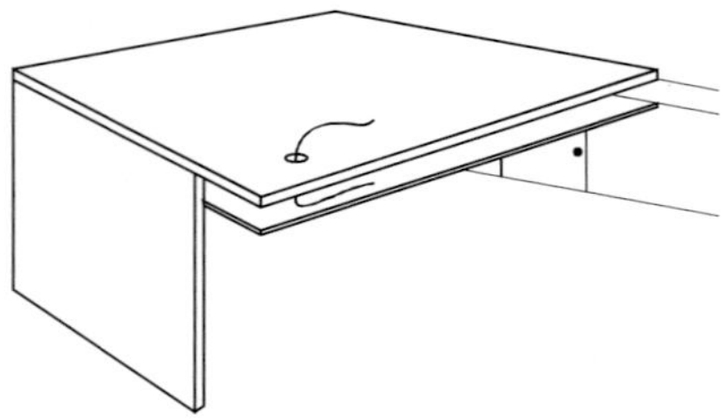

Fig. 10.18. “Through overlap” layout

Cable layout in wall channels

The wall channels are intended for cable layouts to the telecommunication outlets, which are installed on a wall of a room at a convenient height. In some cases, the wall channels are used for horizontal cable bundles in sections that begin at the exit from a cross-room and end directly at the entry to a room with telecommunication outlets.

The following types of the cable wall channels can be used:

- superimposed cable channels, decorative conduits, or plinths (Section 5.2);
- hidden cable channels, which are mounted in the wall thickness so that only telecommunication and/or power supply outlets are on its surface.

Conduit filling by various wires usually does not exceed 30–60 % of their cross-sectional area, the particular value depends on the minimum allowable bend radius of the cables, the method of outlet installation, as well as the need for SCS expansion at the place of conduit installation. In practice, in the absence of a priori information on the number of cables installed in the conduit, the fill factor is usually assumed to be 0.5. In order to determine the required number of decorative conduits, the sum of the cross-sectional areas of all installed cables are divided by the fill factor. In most cases, hidden cable channels are based on flexible plastic tubes of various diameters with parameters determined by the requirements for ducts (Section 10.2.6).

10.3 Telecommunications Design Stage

The calculation of the number of components needed for cable system creation is carried out during the telecommunications design stage. In order to facilitate the design process, it is expedient to use more detailed SCS division into separate subsystems, which differs from the ISO/IEC 11801 standard, and to use equipment that directly interacts with SCS:

1. workstation subsystems;
2. horizontal subsystems;

3. cable system routes;
4. cable subsystems of the equipment;
5. administrative subsystems.

It is simple to see that if the separate subsystems are divided into a classical treelike structure (Figure 1.1), SCS are located at the nodes and the branches of the tree. The design of separate SCS subsystems is carried out consistently. The recommended order of their development coincides with the order in the list above.

The results of the calculations for each subsystem are presented in tabular form. The data from these tables are used as the initial information for the design of the subsequent subsystems. During the final design stage, the equipment specifications are prepared on the basis of these documents.

The table forms can be various, as is convenient for the designer. It is possible to use both paper forms and their electronic variants. In the latter case, the process of final equipment specification preparation is essentially facilitated and accelerated. In the sections dedicated to separate subsystem designs, the recommended table forms are given and can be modified depending on the particular SCS.

10.3.1 Initial Data for Design

Construction solutions

The important part of the initial data structure for cable systems design is the SCS-related information for the construction solutions stipulated by the building project. When the design of the cable system is conducted through the architectural stage, work on the telecommunication stage is somewhat facilitated.

If the SCS design is carried out from the telecommunications stage, then the following initial data are used in order to create the design:

- linear dimensions of the building or floor plans of rooms with indications of the linear dimensions;
- common and/or used area of rooms (available for location of personnel, equipment, and furniture);
- floor height;
- separate floor structures:
  - systems of space allocation: corridors, open, or cellular offices;
  - presence of architecturally allocated areas and their dimensions;
  - ladder locations;
  - presence of technical spaces;
- construction solutions:
  - materials and thickness of walls, partitions;
  - materials and thickness of interfloor overlaps;
  - presence of false ceilings and false floors in corridors and rooms;
  - configuration and location of central heating system radiators;

- location of distributive nodes and vertical risers of the water pipe system, central heating system, sewer system, fire-suppression system, power supply network for high-powered electrical devices, and sources of strong electromagnetic fields;
- cross- and equipment rooms:
  - presence of technical rooms and their readiness for the location of SCS equipment and adjacent systems;
  - dimensions of these room;
- presence and structure of cable conduit runs, ladders, poles, and other similar structures for external cable layout and external cable suspension;
- presence and parameters of building cable inlets:
  - types and capacities of lead-in cables, as well as information about their owners;
  - electrical protection devices use in the cable inlet structure;
  - presence of free channels in the cable inlet and the structure of free channels;
- channels for vertical cable layout areas:
  - types and structures of interfloor overlaps elements;
  - types and structure of cable route elements to link the vertical and horizontal sections;
- channels for horizontal cable layout areas:
  - presence and structure of cable ladders behind false ceilings;
  - presence and structure of cable ducts in floors;
  - presence and filling level of decorative conduits in rooms;
- interior features;
- object power supply:
  - reliability category of power supply;
  - power supply feeder scheme for object;
- grounding:
  - presence of grounding loop;
  - presence of protective zero loop;
  - structure of building grounding system.

The obtained initial data are necessarily controlled in conformity with the architectural and design requirements of the TIA/EIA-569 standard (Section 10.2). If the design necessarily diverges, it is necessary to notify the customer, to prepare the offers and recommendations in order to make alterations to the building construction project, and then to check their performance.

Requirement for the cable system

The following information is necessary during forming of the requirements for SCS:

- types of network equipment that will use SCS for organization of information exchange between:

  - local area networks;
  - telephone networks;
  - safety systems;
  - management systems for the building equipment (lifts, ventilation systems, air conditioning);
  - other systems;
- customer requirements for the telecommunications system characteristics;
- transmission capacity;
- capacity of the internal and external backbone subsystems;
- need for future system expansion;
- customer requirements for the cable layout methods and compatibility with existing interiors;
- customer requirements for compatibility of the equipment that is to be installed in a building;
- other requirements.

Outlet structures in workstations

Outlet block structures in the workstations can include the following:

- telecommunications outlets (TO) connected to symmetrical electrical cables;
- TO connected to fiber-optic cables;
- power supply outlets connected to guaranteed power supply systems;
- power supply outlets connected to household power supply systems.

The presence and number of outlets of each type is determined according to the customer requirements and wishes.

According to the ISO/IEC 11801 standard, it is necessary to install not less than two TO on each workstation. A minimum of one TO should be connected to a cable of category 3 or above. Other outlets are served by cables of category 5 or optical cables. To provide cable system versatility, it is recommended to use TO of category 5. Sometimes, because of local conditions and the specificity of workstations, under the insistence of the customer it is possible to deviate from the recommendations of the standards for the number of outlets and their category. This fact should be noted in the technical documentation with an indication of the reason.

10.3.2 Workstation Subsystem Design

The basic task of this design stage is the development, coordination, and approval of the location plan of telecommunications cable system and power supply outlets, and also the determination of the types and quantity of terminated cords, adapters, connectors, and other similar elements.

The outlet installation locations are marked on the building floor plans. The basic information of outlet installation locations is put in the appropriate columns

(Table 10.9). This table is filled also when the building plans fail to exist or it is impossible to mark exact installation place of outlets on the available building plans. In such situation the table is the fundamental document, describing workstation subsystem and allowing to design a horizontal subsystem.

Table 10.9. Distribution of workstations

Customer: ______________________

Object: ________________________

Building: _______________________

Item no.	Floor	Apartment no.	Number of workstations	Telecommunications outlets			Power-supply outlets		Fastening
				Category 3	Category 5	Fiber optic	Uninterruptible power supply	House hold	method
1	2	3	4	5	6	7	8	9	10

The selection of the telecommunications outlet locations must be based on a uniform distribution of workstations in the room (according to the Construction Norms and Rules (SNiP) 2.09.04-87, Section 3.2, each workstation occupies a minimum 4 m^2 of the workspace; the ISO/IEC 11801 standard (2000) provides 10 m^2, though it stipulates the desirability of minimization of this parameter and the opportunity to apply a national norm). The need to carry out cable layout to a supposed installation location of the outlet and the need for installation of outlets of various types in this location are additionally taken into account.

If the customer requests, it is possible to locate the outlets in accordance with the furniture arrangement plans for workstations. In most cases, this slightly lowers the cable system cost, since the number of outlets decreases. A similar solution implicitly breaks a structural principle, and actually attaches a cable system to the plans for office furniture arrangement, which considerably reduces its flexibility. This approach should be used only in the most extreme cases. The customer should be necessarily warned about the negative consequences of the use of this design in terms of operational flexibility and costs for routine maintenance. The negative consequences can occurred after two or three years of SCS operation or even earlier, for instance, during the first relocation of employees or installation of new furniture.

As a compromise, which slightly decreases the cost of SCS creation and at the same time makes the organization process for new workstations less painful, the following approach can be used. All architectural solutions (cable conduit capacity, number, and overall dimensions of risers) are initially calculated based on the complete cable system capacity and the number of workstations in the first stage is organized in concordance with the actual arrangement of employees. By using this approach, the layout (in the case of necessity) of several additional cable lines is not a labor-intensive operation, can be performed by the operation service employees within a short time period (including days off), and also does not influence the activity of other structural divisions of the organization. We note that all the situations discussed above are correct when the process of laying out additional cable lines is not a permanent operation, that is, the number of installed outlets is first calculated according to the room space.

The Number of terminated cords is selected to be equal to the number of units in a computer network (workstations, network printers, and other similar devices) that will be connected to SCS after the system enters into operation. In order to take into account the need for future LAN expansion and to support the current operation, additional cords (up to 10% more) are put in the Spare Part Accessories (SPA).

Terminated cords for telephone and facsimile machine connection to the telecommunications outlets are usually included in the complete telephone and facsimile apparatus. Therefore they are not taken into account during preparation of the specifications.

The lengths of terminated cords for computer equipment connection are selected depending on the dimensions of the space that is served with a cable system. It is acceptable to use connection cords of one length (2–3 m) for small rooms with uniform outlet distributions. In large rooms, this length can be increased to 8 m. Such situations also arise when the outlets are installed according to the furniture arrangement plans. The use of longer cords contradicts the requirements of the working standards (ISO/IEC 11801, etc.). In any case, up to 10% of the total connection cords should have lengths more than 3 m. In the case of SCS construction in open offices, during selection of connection cord lengths it is possible to be guided by the provisions of Section 1.4.1.

It is not recommended to use self-made cords because of their potentially poorer electrical and operational reliability parameters. Furthermore, the majority of SCS manufacturers fail to certify such solutions.

In cases when it is known beforehand that special active equipment will be connected to some of the outlets of a cable system, it is necessary to provide in the workstation specifications for the availability of appropriate adapters, baluns, and connectors, which carry out the coordination and/or parameter transformations of cables and instrument interfaces.

10.3.3 Horizontal Subsystem Design

The process of horizontal subsystem design is the most difficult and crucial stage of SCS development. The decisions made at this stage determine the technical and economical efficiency of the created system. This is explained by the fact that horizontal subsystems are based on general SCS equipment, both in nomenclature and quantity, and in cost. During design the following activities are carried out:

- connection of separate workstations to the cross-rooms;
- selection of telecommunications outlet types;
- selection of cable types and cable categories, as well as quantities;
- branch-point design (design of branch-point use, if necessary).

The resulting assumptions are added to the data, which are put in Table 10.10.

Table 10.10. Horizontal connections

Customer: ____________________

Object: ______________________

Building: ____________________

Item no.	Cross-room	Number of work-stations	Number of outlet modules			Category 3 cables		Category 5 cables		Fiber-optic cables	
			Cat. 3	Cat. 5	Fiber optic	Type	Quantity	Type	Quantity	Type	Quantity
1	2	3	4	5	6	7	8	9	10	11	12

We note that the material described further concerns the use of a usual office model (Section 1.4.1). The overwhelming majority of SCS are created in this configuration in our country. If it is necessary to design a cable layout for an open office, all the situations described here can be easily adapted for its features based on its technical parameters.

The design process diagram is given in Figure 10.19.

According to the accepted practice, the selection of type and the calculation of quantity of all signal transmission path elements is carried out during this stage, without equipment installed in the cross-rooms and the equipment rooms. This equipment is an administrative subsystem.

Connection of separate workstations to the cross-rooms

The process of horizontal subsystem design begins with connection of separate workstations to the cross-rooms. The number of cross-rooms and their locations are set according to the decisions made in the architectural design stage. For the small and average SCS, when only one cross-room is provided on a floor, the connection process is a formality and consists of carrying the total number of data outlets from the appropriate columns of Table 10.9 into Table 10.10. In cases when there are several cross-rooms on a floor, it is necessary to observe the following restrictions:

- the maximum lengths of horizontal cables should not exceed 90 m;
- the number of cable routes with lengths more than 70 m should be minimized;

- each cross-room should, whenever possible, serve an approximately identical number of workstations;
- the distribution of separate workstations among cross-rooms should be made by the principle of minimization of channels with average cable length

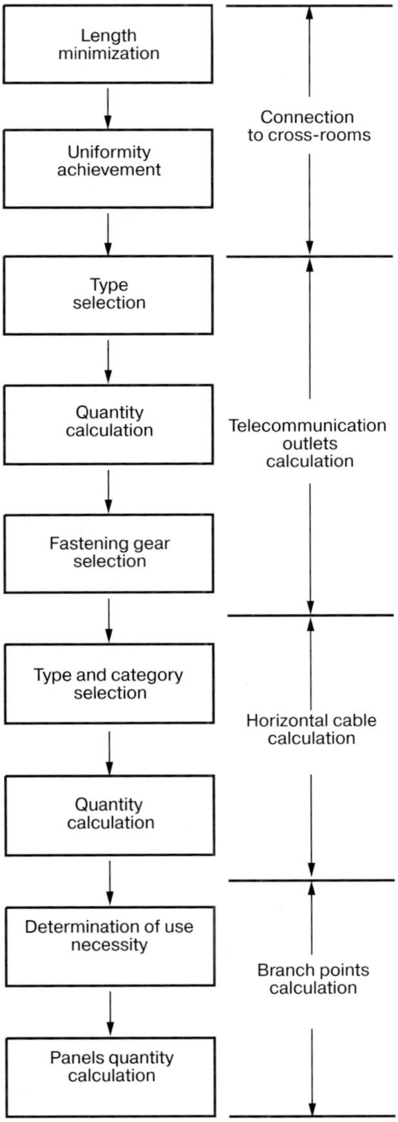

Fig. 10.19. Horizontal subsystem design process diagram

The observance of these requirements is also useful in those situations when, by virtue of architectural or organizational features, the restriction of the presence of even one cross-room on each floor is broken.

Selection of telecommunications outlet type

The selection of a type and category of telecommunications outlets (TO) is unequivocally assigned by the parameters created during the development process and the subsequent protection of the outline design and determination of the type of signal transmission medium. The basic data for outlets is systematized during workstation design. During horizontal subsystem design a concrete definition of the following is made:

- number of outlet modules on the workstations;
- principle of their fastening.

The design and use of various fastening methods at the installation points are essentially influenced by the telecommunications outlet selection.

Telecommunications outlets with one or two (less often, three) outlet modules can be used for a workstation (Section 3.3.3). Most TO are realized with two outlet modules and are formally intended for telephone and computer operation. Boxes with four to twelve outlet modules are more effective for operation of workstation groups. This approach is frequently applied when the users are located in halls of large spaces or when the SCS is created in spaces that are in open offices (Section 1.4.1).

The method of telecommunications outlet fastening is selected in view of the method of horizontal subsystem cable layout and is reflected in column 10 of Table 10.9.

Horizontal cable calculation

Selection of cable type and cable category. The selection of the cable type and category of horizontal subsystems depends on the decisions made during the development of the outline design and the development of the type of medium over which a signal is to be transmitted. In order to organize the SCS horizontal subsystems, according to the ISO/IEC 11801 standard, symmetrical electrical and optical cables can be used.

The category of symmetrical cables from twisted pairs is defined in Table 1.6 and depends on the maximum frequency of the transmitted signal. During the early stages of SCS equipment development, the formal observance of the minimum requirements of the standards (as of 1995) and the installation of one category-5 cable and one category-3 cable to the workstation was frequently practiced in Russia. The first is intended for connection to a computer, second to a telephone set. The use of this version of the horizontal subsystem construction slightly lowers the total SCS cost because of the lower price of cables and outlets of category 3. Nevertheless, such a scheme is not recommended, as it breaks the principle of versatility and limits the functional flexibility. In practice, in the overwhelming majority of cases, the leading system integrators install two cables of category 5 and appropriate outlet modules to the each workstation.

In the case of two-port workstations, some economy is achieved in the creation of horizontal subsystems by using dual cables, which allows the installation of two 4-pair elements to the workstation on one cycle of stretching. Mass implementation of this solution is restrained by the inconvenience of such cable installation due to its asymmetrical form, and the absence of dual constructions in the production programs of many manufacturers of cable products.

Multipair cables are installed directly to workstations only when six or twelve outlet modules are used. In all other cases, it is necessary to design branch points. The use of such cables on smaller capacity outlets is inadmissible, as the standards require the obligatory connection of all pairs to the outlet modules (conductors should not be "suspended"). It is impossible to distribute the twisted pairs of multipair cables to several modules without branching points, since it is required that the twisted pair not be outside the telecommunications outlet box without its sheath. This prohibition becomes void when multielement cables are used, however, they are in the structure of a very small number of SCS.

The standards forbid forming a parallel connection of electrical cable pairs or using a coupling to splice them. If it is necessary to use the SCS cabling to provide the operation of the network equipment, which is connected by a principle of "multipoint" and requires load resistors for its performance, the appropriate adapters should be used (see Section 3.4.2.4).

Calculation of quantity. During calculation of the horizontal cable length, the following obvious situations are taken into account. Each module of the telecommunications outlet connects to switching equipment in the cross-rooms of a floor by one cable. It is necessary to remember that its length, according to the ISO/IEC 11801 standard, should not exceed 90 m. The cables are installed on the channels without formation of any bays and loops. The descents, rises, and turns of these channels are also considered.

There are two methods to calculate cable quantity for horizontal subsystems: the *summation* method and the *empirical* method. The first consists of the calculation of the length of each horizontal cable route with the subsequent addition of these lengths. The technological reserve value of no more than 10% and the reserve to carry out cable preparation in the outlets and patch panels are added to the previous result. The advantage of considered method is its high accuracy. However, in the absence of a means of automation and in the case of SCS design with many ports, this approach appears excessively labor-intensive, and practically excludes the calculation of several versions of cable system organization. This method can be recommended when specialized programs for automatic design (for example, CADDY) are used, whereby the accounting for all descents and turns, as well as the calculation of the general length of all cables, is shifted onto computer technology.

The empirical method realizes, in practice, a rule known as the central limiting theorem of probability theory, and as experience shows, gives good results for cable systems with more than 30 workstations. In essence, it consists of using generalized empirical formulas in order to calculate the general horizontal cable length necessary for a certain cable system realization.

The unique and essential limitation of this method is the assumption that the workstations are allocated uniformly in the served area. If not, for instance, when the workstations are placed in groups, the principle of uniform distribution is carried out less accuracy. For each such group, the calculation is carried out separately. This technique allows developers to bring a design task to the previous case. It is easy to see that by further splitting of groups down to single cables, the empirical method transfers to the method of summation.

On the basis of the assumptions made, the average length L_{av} of cable routes is equal to:

$$L_{av} = \frac{(L_{\max} + L_{\min})^{\cdot}}{2} K_s + X,$$

where L_{min} and L_{max} – route length from cable channels inlet point in the cross-room to the outlet module of the telecommunication outlet of closest and farthest workstation accordingly, designed in view of cable layout features; K_s – technological reserve coefficient – 1,1 (10%); $X=X_1+X_2$ - reserve for carrying out cables preparation. The value X_1 – reserve from the worksatation is 30 cm. Parameter X_2 designates the reserve from the cross-room. The X_2 parameter depends on the cross-room dimensions and is numerically equal to distance from horizontal cables inlet point of cross-room to the most remote switching element, again in view of all descents, rises and turns.

Then the value of all cable routes is calculated – N_{cr}, taking into account the use of one cable drum:

$$N_{cr} = \frac{L_{cb}}{L_{av}}$$

where L_{cb} – cable drum length (standard value are 305 m, 500 m and 1000 m), and the result is approximated down to nearest whole.

In the last step we gain total cable length L_c, necessary for cable system creation:

$$L_c = L_{cb} \frac{N_{to}}{N_{cr}}$$

where N_{to} – outlet modules quantity of the SCS telecommunication outlets.

This algorithm can be used in Excel software. The used formula for the most widespread in practice 305-meters (1000-foot) cable packings, can be expressed in the following:

$$=APPROXIM.FROM\ ABOVE(N_{to}/(APPROXIM.DOWNTO(L_{cb}/(L_{av} \times \times 1,1+X);0));0 \times 305 \tag{10.1}$$

where N_{to}, L_{cb}, L_{av}, X – numerical values or the references to cells, in which contain digital values of the appropriate parameters.

We note that the approximation in different directions used in the formula 10.1 is applied in order to reduce the probability of calculation mistakes.

In the case of various categories of cable layout realizations, the calculation is carried out on each of them separately. It is natural that during the use of the same telecommunications outlets with different outlet modules in one building, the number of cables of various categories appears equal and the calculation is carried out once.

Branch point design

A branch point is that place in the horizontal subsystem in which the cable type is changed without alteration of the transmission quality characteristics. According to the ISO/IEC11801 standard, in a branch point the plain cable incorporating a usual round or pair cable is split into several four-pair cables (in practice, this version is carried out more frequently in Russia). The branch point can be used for splicing two identical cables in extreme cases, for example, if it is necessary to increase the cable length or in order to restore communications in case of failure during SCS operation. This solution is not certified by the SCS manufacturer, and at the first opportunity such a cable should be replaced by an uninterrupted cable.

Switching equipment is installed in branch points, but they are not intended for cable system administration or connection of active network devices of various designations. Experience shows that it is very convenient to use 110-type panels of different construction variants as switching equipment. Panels with parallel connections of outlet contacts (Section 3.3.2.1) are most preferable. When other types of panels are used, cables coming from the FCR side are split on the unseparated sides of contacts. For cable layouts that connect the branch points and the telecommunication outlets in the workstations, demountable sides of contacts with obligatory additional mechanical fixing are used. Thus, switching equipment capacity for the branch point is calculated using only the number of cable pairs, which comes from the FCR. The general information for horizontal subsystems is summarized in Table 10.10.

10.3.4 SCS Backbone Subsystems

During the backbone cable system design stage, the following fundamental tasks are solved:

- selection of cable type and category;
- calculation of backbone cable capacity and quantity.

Selection of cable type and category

The selection of cable type and category for SCS backbones is defined by the decisions made during the development of the outline design and the development of the type of medium for signal transmission.

According to the ISO/IEC 11801 standard, backbone subsystems can be constructed on symmetrical electrical and/or fiber-optic cables.

The symmetrical cable category is determined according to Table 1.6, depending on the maximum frequency of the transmitted signal. The fiber-optic cable type (single-mode or multimode) depends on the type of network equipment used and the backbone length (Table 1.10).

LAN network equipment with transmission rates not higher than 100 Mbps allows the use of multimode optical cables with maximum length not more than 2000 m, although in practice, this length is exceeded. However, at the existing price level for the separate components that are necessary for realization of fiber-optic communication lines (cables with accessories plus active network equipment), it is economically expedient and technically more feasible to use single-mode facilities when the cable route length is more than 1500 m [123].

Another picture can be seen in case of LAN Gigabit Ethernet use. According to the 802.3z standard, the maximum length of multimode optical cable cannot exceed 550 m. Taking into account this circumstance and reasons discussed above, it is possible to conclude that the optical subsystem of internal backbones should be constructed mainly of multimode optical cables, whereas the basis of external backbones with lengths exceeding 500 m should predominantly be single-mode cables. In situations when signals of other applications (for example, PABX) are transmitted by optical cables, it is possible to use combined constructions, which simultaneously contain single-mode and multimode fibers.

During selection of symmetrical multipair cables, except for verification of compliance of its characteristics upon the application class, it is necessary to carry out the additional verification of signal compatibility with these applications. If application incompatibility is found, the following methods are used. If 25-pair cables are used for the backbone subsystem construction, then the application signals are transmitted on different cables. If the backbone subsystem is constructed using cables of greater capacity, then it is possible to take advantage of the fact that the core of 50-pair cables, and also cables of greater capacity, is made of separate 25-pair bundles, each of which has the electrical characteristics of the 25-pair cable of the same category. In this case, the signals of the incompatible applications are transmitted on different bundles of one cable.

The use of two internal backbones of different categories, for example, categories 3 and 5, is permitted in a building. Usually it is used when the telephone systems do not require high-quality cables for operation over long distances. The backbones of different categories can originate in one or in different cross-rooms of a building.

The described solution allows the creation of a relatively inexpensive system, which will be quite adequate for the present situation. However, during final selection of the possible versions, it is necessary to take into account the following circumstances:

- need for use of backbone cables to support the operation of equipment that requires greater transmission capacity;

- separate cable extraction for signal transmission of various network equipment reduces the flexibility of the cable system to a certain degree.

Calculation of backbone cable capacity and quantity

The calculation begins with compiling the internal backbone cable list, which is carried out on the basis of the outline design. It is desirable to explain the supposed designation of each cable in the technical design documentation. The capacity of backbone cables is calculated in view of the approved workstation configuration and the selected type of transmission signal medium for the internal and external backbones. The following meanings are used for the calculation of quantity of pairs and fibers.

Configurations with a low level of integration, which have one module in the telecommunications outlet and accordingly one horizontal cable for the workstation (it is necessary a minimum 2 pairs for the workstation in the internal backbone cables).

Configurations with an average level of integration, which contain two or more outlet modules for the telecommunications outlet with the appropriate number of horizontal cables for the workstations (typical solution in mid-1999: a minimum of three pairs for one workstation in the internal backbone cables).

Configurations with a high level of integration, which include two or more outlet modules for the telecommunications outlet with the appropriate number of horizontal cables for the workstations. Thus, it is possible to use fiber-optic cables in these configurations for the organization of the internal and external backbones as well as the horizontal subsystem. Configurations with a high level of integration assume the use of a minimum of three pairs and 0, 2 fibers for the workstation1 in the internal backbone cables and a minimum of two pairs and 0,2 fibers for the workstations in the external backbone cables.

The following reasons are the basis for the selection of the stated meanings. First, one or two twisted pairs are used for connection of subscriber's equipment in modern telephone communication systems. In addition, modern telephone communication systems, in overwhelming majority of cases, are based on centralized systems without the use of external devices, substations, and other similar equipment that carries out the function of hubs. Therefore configurations with low levels of integration, the minimum of two pairs per workstation are provided in the backbone of a building. Second, during LAN construction in the cross-rooms of floors, the active equipment is installed: repeaters, switches, multiplexers, and other similar equipment. This equipment has, as a rule, one or two ports for connection to central equipment in the cross-room of a building and some ports (4–24) for connection to the users' workstations by the horizontal subsystem cables. The dependence of the delivery volume of Allied Telesyn LAN hubs on the number of ports (based on IT company data) is shown in Figure 10.20. The large number of samples (in total, 2207 pieces of equipment) allows us to consider it as representative and to form further arguments on its basis. From this data, we note that the average number of ports served by one hub is 10.96. The practice of LAN construction shows approximately an equal probability of LAN up-link port connection of workgroups to local servers or switches, and to the backbone subsystem cable. Therefore we can state that the average number of ports needed serve the workstations in this equipment is equal to ten with a reserve pf approximately 10%. That is, the operation of ten workstations is provided for with one path of the internal backbone. One path is formed by four pairs of electrical backbone cables, hence, one pair for the workstation should be added to the two pairs of the internal backbone during configuration with a low level of integration. Also, one pair is added (4 pairs/10 workstations=0.4 pairs for the workstation →1 pair for the workstation) for SCS internal backbone configurations with average and high levels of integration. Similarly, we shall use 0.2 fibers for the workstation for fiber-optic cables of a building backbone (2 fibers/10 workstations=0.2 fibers for the workstation), as the data transmission path through optical cables is formed by two fibers.

The stated capacity values of internal backbone subsystem cables are the lower allowable margin. As agreed with the customer, its total value can be increased. The need to increase the capacity of backbone cables follows from the analysis of Figure 10.20, which shows the high popularity of the use of 8-port hubs during LAN construction. The additional twisted pairs and lightguide inclusion in the backbone cables provides significant improvement of the cable system flexibility, provides redundancy, and forms the prerequisites for system functionality expansion.

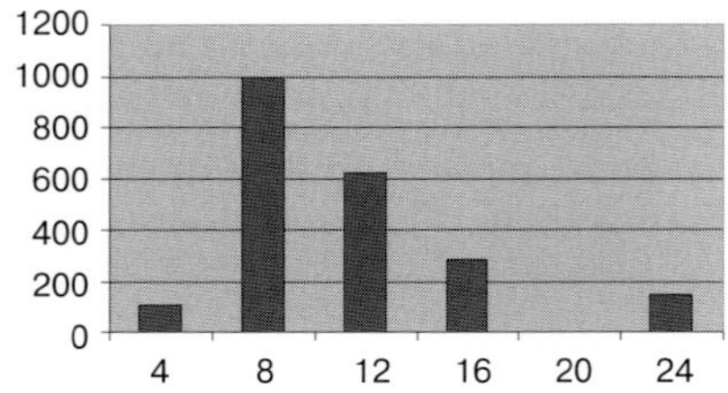

Fig. 10.20. Dependence between sales volume of LAN hubs and their number of ports

The required number of backbone cables is determined as follows. The defined minimum number of pairs/fibers for the workstation of each cross-rooms of a floor is multiplied by the number of workstations (column 3 of Table 10.10) that are served by this cross-room. The resulting value is approximated to the nearest whole number from the above number of pairs/fibers, which can be obtained using one or several cables of standard capacity (25, 50, 100, 200, etc. pairs, or 4, 6, 8, 12, 24, 48, etc. fibers). The resulting values for numbers of pairs/fibers and cables are brought to the appropriate columns (Table 10.11).

During creation of the distributed backbones the calculation of cable capacity is carried out by the same principles. If the optical cables are the basis of the internal backbone, then it is recommended, whenever possible, to provide a back-up of each backbone route by one or several 25-pair cables of category 5. This ensures the readiness for possible installation of equipment that functions with symmetrical cables, and also provides redundancy in the case of fiber-optic cable line failures or malfunctions of network equipment with optical interfaces.

The cable length is determined in view of all descents, turns, and other topological features of route, and includes reserves for cable preparation. The results of the calculations for backbone cables are inserted in Table 10.11.

Table 10.11. Bacbone connections table

Customer: ____________________

Object: ____________________

Building: ____________________

Item no.	Marking	Beginning	End	Cable type	Number of pairs/fibers	Cable quantity	Cable length, [m]	Purpose
1	2	3	4	5	6	7	8	9

Design features of external backbone subsystems

The design rules for external backbone subsystems coincide, in general, with design rules for internal backbones. Therefore, we note here only those features that were not mentioned earlier and that the designer must take account of by various means:

- During the design of the external backbone, cable routes are very frequently installed in the City Telephone Network (CTN) ducts and in ducts for various urban services. In this case, there is a problem of obtaining appropriate permits and technical terms in order to install and utilize only those cables that are included in the authorized cable list. It is also necessary to enter into lease contracts.
- Because of the relatively small capacity of cable routes of the external backbone subsystem, its calculation is carried out individually each time, and no universal recommendations can be given.
- In those cases when the cables of external backbone subsystems connect some buildings among themselves and are partially installed on one route, it is useful to consider using a splitting coupling on the route. Because of the existing prices in the Russian market (mid-1999), such a solution, as a rule, appears economically preferable for route lengths of more than 300–400 m. The cost advantage is achieved due savings n cable layout labor and in cable cost.
- The high cost and long duration of construction of the external backbones means increased capacity reserves must be provided. Thus, for example, it is necessary to use at least a double reserve of light guides for fiber-optic cables.
- Because of the complexities of restoring the physical integrity of cables quickly in case of errors the construction of the external backbones, it is recommended to widely use the redundancy principle (Section 10.3.4).

Redundancy of backbone cables

The redundancy of backbone cables is used in order to increase network reliability. This principle is realized by two approaches: increase the cable capacity or increase the cable layout use in spatially diverse routes. It is most expedient to use redundancy for the fiber-optic cables that:

- do not have rigid restrictions on the number of intermediate demountable and non-demountable connectors in a signal path;
- have approximately constant weight, irrespective of the number of light guides, in a wide range of cable capacities.

In the elementary case, which is applicable both to optical and to electrical solutions, redundancy is achieved by increasing the cable capacity by a factor of two, three, or more. The use of the several spatially diverse cable layout routes (for example, two different risers in a building), which ensures preservation of communications in case of physical damage to one of the cable routes, is more effective. In the most frequent practice, three cross-rooms K1–K3 are connected to the equipment room A without direct connection between them (Figure 10.21). The transit fiber junction is used in order to organize redundant links. This junction can be accommodate both the usual patching cords and the direct splicing of separate lightguides, which is carried out by welding or mechanical splices. The last variant is preferable both in terms of cost and in terms of losses.

Additionally, we note that redundancy of cable routes is urgently recommended in those common cases when the cable layout is carried out in the City Telephone Network (CTN) ducts. The one-time expense for construction of a redundancy link and the increased (though it is usually moderate) lease costs are quickly paid for in the case of the first serious failure on the backbone cable route. This savings is explained by the long time needed to restore both the cable duct and by the complexity of repair work organization, and also because of the need to obtain numerous sanctions and permissions.

Fig. 10.21. Connection of three cross-rooms with redundant links to the equipment room

10.3.5 Equipment Cable Subsystems

Equipment cables in this case are the terminated and installation cords and cables, with which SCS connects to the active network equipment, installed in the cross-rooms and equipment rooms. The calculation of switching cords to connect separate channels in the cross-field is not carried out during this stage. During design of this subsystem, the following selection is carried out:

- method of connecting network equipment to the cable system;
- type and category of equipment cables.

The calculation of the quantity of these elements is also carried out.

Selection of method to connect network equipment to the cable system

The basic function of the equipment cable subsystem is the connection of active network equipment (for example, switches and LAN repeaters) to the cable system. Such a connection can be carried out in any cross-room of the SCS. In cross-rooms of the upper levels (EBCR and BCR), the central network equipment (central switch, PABX, signaling system controllers, and other similar equipment) is connected to the SCS. In contrast, the FCR serves the active network equipment (for instance, common and switching hubs of workgroups and external blocks of telephone exchanges), which works only for a limited group of users.

The equipment cable subsystems as well as workstation cable subsystems are not included in the scope of the ISO/IEC 11801standard, since the particular applications have a strong influence over the component used in these subsystems. Therefore, design at this stage will be carried out in view of the manufacturers' recommendations for active equipment and standards for the given applications.

Nevertheless, the ISO/IEC 11801 standard contains a number of restrictions concerning the length and transmission capacity of these elements. Thus, the general length of terminated and switching cords (crossing wire) of horizontal subsystems, depending on the connection scheme chosen, should not exceed 9–10 m. When the network equipment is connected to the cable system in the BCR or in the EBCR, the length of connection cords should not be more than 30 m.

The active network equipment can be connected to the cable system in one of the following three basic ways (Figure 10.22):

- cross-connect connection;
- interconnect connection;
- connection with the help of a switch.

The "fixed" ports display of the active equipment and additional switching panel use (Figure 3.52), carried out with the help of so-called installation cords (Section 3.4.1.3) or with typical terminated cords used for special types of switching panels, are distinctive feature of *cross-connect connections* (Figure 10.22a). The similar solution is quite possible and is, in some cases, more convenient in the event that the output interfaces of the active equipment are based on Telco connectors. In this situation, it is connected to the so-called patch-panel, such as nectors. In this situation, it is connected to the so-called patch-panel, such as Telco. This approach requires approximately twice as many switching panels than the first approach. Its fundamental advantages are:

- reduced probability of an electrical port failure of expensive network equipment: practically down to zero since the number of switching operations are minimized;
- essential front panel "unloading" of switching fields from cords by a reduction in their length, and also since the majority of the wires can be "withdrawn" on the back sides of panels; hence, the aesthetic characteristics of the switching field and the convenience of reading labels are improved.

During *cross-connect junction,* the active network and switching equipment should be installed near each other. The information transmission channels are formed by direct connection between connectors on the network equipment housing and connectors of the switching equipment with the help of connection cords of the appropriate type (Figure 10.22b).

The principle of *connection with the help of a switch* can be considered as a development of the previous method, used in practice when switching and network equipment are installed in several cabinets. This method is widely used during construction of SCS with many ports. It also ensures independence from the connector types of the active network equipment. The connection is carried out by a symmetrical multipair cable, with one end prepared on the cross- or switching panel of the cable system, and the second on the output patch panel of the active equipment. The information transmission channels are formed by connection in each of these switches (Figure 10.22c). Because of the occurrence of an additional connector in the signal transmission path, the method is applicable only for low-level SCS.

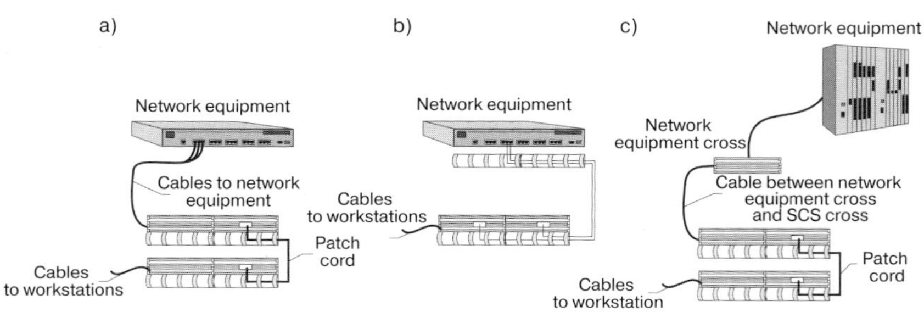

Fig. 10.22. Methods of network equipment connection: *a)* cross-connect connection; *b)* cross-connect junction; *c)* connection with the help of a switch.

Following two basic rules during selection of the method of network equipment connection is recommended:

- For the LAN network equipment, which differs by high rates of information exchange, the most preferable way is the interconnect connection, if it is permitted by the cable system scale. Interconnect connections are allowed; however they slightly reduce the reserve of interference immunity at the expense of the presence of a few additional cross-points;

- For other applications, which do not require so large a bandwidth of signal transmission path, connection by switches should be used.

Use of the cross-connect connection is highly attractive for average-scale networks, however it requires the use of special equipment (installation cords, panels with modular connector outlets, or Telco connectors on a back surface) and cannot be carried out in all SCS.

If the cable system is created for operation with specific active equipment, then it is necessary to provide the appropriate adapters and connectors, which carry out the coordination of network equipment interfaces and cable system ports in the cross-rooms and equipment rooms.

Selection of type and category of equipment cables, calculation of their quantity

The selection of the type and category of cables is based on the recommendations of the active equipment manufacturer and the standards for the application. In order to provide the maximum duration of cable system operation as well as expansion of its functionalities with a uniform element base, category-5 cable hardware products are preferable for use in construction of horizontal and backbone subsystems.

Two fundamental approaches can be used to calculate the number of equipment cables:

- calculation based on the number of workstations served;
- calculation based on the capacity of the active network equipment.

In the first case, the quantity is numerically equal to the sum of workstations served by given cross-room. For calculations based on capacity, the number of equipment cables coincides with the number of network device ports that are installed in the cross-room. Calculations based on the number of workstations served provide a cable reserve in the case of installation of additional active network devices.

The selection of a particular calculation strategy is defined by the specific conditions of the project. It is obviously not possible to give any fixed recommendations for this question in the absence of a priori information.

The equipment cable lengths are defined during the development of the network device location plans in technical apartments. The calculation is carried out separately for each cross-room. The results are found in Table 10.12.

Table 10.12. Equipment cables table

Customer: ____________________

Object: ____________________

Building: ____________________

Item no.	Cable and cord type	Number of pairs/fibers	Cable quantity	Total number of pairs	Length, m	Purpose
1	2	3	4	5	6	7

10.3.6 Administrative Subsystems

The development of an administrative subsystem is the most difficult SCS design stage. During this work the following tasks are solved:

- Determination of the functional sections of the cross-rooms and equipment rooms;
- Calculation of the capacity of the information transmission channels;
- Determination of switching equipment type;
- Development of equipment location plans in the cross-rooms and equipment rooms;
- Calculation of the number of switching equipment constructive units;
- Determination of switching cord types and quantity.

Determination of switching panels functional sections

The location, configuration, and type of switching equipment used for the organization cross-rooms, directly influences the method in which the administration and management of the cable system is carried out, and probably even dictates it. The careful design development of an administrative subsystem lowers the cost of SCS management and simplifies its operation with time.

In order to decrease costs for SCS administration, the signal transmission media and cable system physical interfaces are standardized. In addition, the design should include the ability for manual signal transmission channel switching by users without specialized instruments (rib-joint pliers, screwdrivers, soldering irons), and also without the attracting of highly skilled technicians of other companies.

The management of signal transmission channels in each SCS administration point is carried out by the use of cable connections of various subsystems with switching cords. The linear cables of any subsystem are installed at one time on a number of switching equipment connectors, which form the various functional sections. For simplification of identification, the standard TIA/EIA-606 gives the color-coding of these sections (Table 10.13). Use of this principle also increases the self-descriptiveness and presentation of the SCS structural diagrams.

Table 10.13. Section color designation of switching equipment

Color	Color according to Pantone system	Purpose
Green	353C	External network interface cables, external telephone communication lines
Purple	246C	Equipment cables for collective use (PABX, network hubs, multiplexers, etc.)
Yellow	101C	Special-purpose PABX cables (ISDN lines, etc.)
White	White	Internal backbone cables
Blue	291C	Horizontal subsystem cables (workstations), served directly from telecommunications cabinet or cross-room
Orange	150C	Data transmission system equipment cables (modems)
Gray	422C	Ancillary backbone lines between technical rooms
Brown	465C	External backbone cables
Red	184C	Special-purpose equipment cables

The data for the separate functional sections of the switching equipment of each cross-room are inserted in Table 10.14. During its formulation, the results from the design of the horizontal (Table 10.10) and backbone subsystems (Table 10.11), as well as calculations of equipment cables (Table 10.12) are used. The results in Table 10.14 generalize the initial data for calculation of the number of switching equipment components for each cross-room and equipment room of the cable system.

Table 10.14. Cross-room equipment table

Customer: ____________________

Object: ____________________

Building: ____________________

Cross-room: ____________________

Item no.	Color code	Purpose	Cable quantity	Pairs/ fibers in cable	Total number of pairs/ fibers	Pairs/ fibers in channel	Channels	Switching equipment type	Device quantity
1	2	3	4	5	6	7	8	9	10

Determination of capacity of information transmission channels

The capacity of the information transmission channel of each functional section depends on its function. For horizontal subsystems ("light-blue" section), the capacity of the channel is equal to number of cable pairs that serve one module of the telecommunication outlet, which is always four pairs or two fibers in the case of *fiber-to-the-desk* projects.

For the equipment cables, the "violet," "yellow," "orange," and "red" sections, the channel capacity is equal to the number of pairs that are used for transmission and reception of information by one port of the device, and depends accordingly on the type of network equipment for which service is intended in the SCS. For example, for the analog modules of private telephone exchanges connected by cables with Telco-type connectors, as a rule the channel capacity is equal to one pair. For small-capacity telephone exchanges, whose ports are frequently realized by 6-contact modular connector outlets, channel capacity is equal to three pairs. In the case of digital telephones, two-pair channels are used. For LAN Ethernet repeaters with ports based on 8-item modular connector outlets, channel capacity can be equal to two or four pairs (it is recommended to consider its equal to four pairs). For the same repeaters, which have output ports based on Telco connectors, channel capacity is equal to two pairs. In case of the absence of prior information on network equipment used, channel capacity is always selected equal to four pairs.

For the backbone cable subsystems ("white," "gray," and "brown" sections), only if they are not distributed to applications, it is difficult to determine the channel capacity without additional information. In this situation it is necessary to consider this SCS element as a means of organizing a great number channels whose capacities are one pair. When the types of applications served by the backbone subsystems are known beforehand, determination of channel capacity and the calculations for its separate parts are carried out separately. Thus, for data transmission backbones it is necessary to consider channel capacity equal to four pairs, for other applications, two pairs.

The channel capacity and the number of served channels for each functional section are inserted in Table 10.14 to calculation the equipment structure of the cross-room.

Selection of switching equipment type

The functionalities, fields of application, advantages, and disadvantages of the basic types of the switching equipment are described more fully in Section 3.3.2. Here we shall give only general provisions for the selection of this equipment type, which are only recommendations.

Switching panels, such as the 66-type, usually provide application signal transmission only for class C and lower. They are inconvenient for the performance of frequent overswitching and are now considered out-of-date. As a basic type of switching equipment, they can be used in very small SCS with especially rigid customer requirements for cost. In all other cases, 66-type panels are used only as elements of local cable layouts, for example, as input junctions for PABX.

The 110-type and S210-type switching panels are installed in "light-blue" section in those situations when known it is beforehand that the cable system will operate large numbers of telephones (at least commensurate with the number of computers). The second condition of their use is in the absence of combined one- and two-pair switching cords in the SCS state component structure. In this situation, their fundamental advantage is allowing easy administration by separate couples.

Switching panels with modular connectors are the most effective in the cable systems that are used mainly for ensuring LAN operation. This equipment is remarkable for its high aesthetic characteristics, ease of use, and effective spacing of mounting cabinets due to its high density of ports. At the same time, switching panels increase the SCS backbone subsystem cost and compel the use of four pairs for application signal transmission. Sometimes, the solution offered in practice, which is based on the connection of 1–2 pairs to modular connectors is ineffective, as it actually ties the cable system to certain equipment and brings to naught all the advantages of SCS versatility. If necessary, in order to provide the maximum load of horizontal and backbone cables, separate pairs with various signal transmission and the appropriate adapters should be used .

The selection of the type of fiber-optic switching equipment depends, first of all, on the approved scheme of network equipment with optical port allocation. If similar equipment is installed in 19-in mounting constructions, then it is preferable to install optical shelves. It is sometimes expedient to use wall-mounted coupling in small-capacity networks and during the realization of external subsystems based exclusively on fiber-optic elements that use a small number of low-capacity cables.

Development of equipment location plans in cross-rooms

An obligatory condition of administrative subsystem design is the development of equipment location plans in each cross-room.

The network equipment can be installed by three basic methods:

- on room walls;
- in 19-in mounting constructions, which function most frequently as mounting cabinets;
- by a mixed variant of wiring.

The number of workstations served from the cross-room influences the selection of the equipment location method. The general characteristics of location methods are presented in Table 10.15.

Table 10.15. A general characteristic of ways of arrangement of equipment

Location method	Number of workstations served	Switching method
On cross-room wall	≤24	Interconnection connection Cross-connect connection
In 19-in construction		Interconnection connection
• One cabinet	≤120	Cross-connect connection
• Two cabinets	100–300	
Combined version	>30	Cross-connect connection Connection by switches

Equipment location on walls of cross-rooms is most expedient for fewer than 24 served workstations. In this case switching panels and their accessories are installed on a wall by state or additional fixing elements, and the network devices are installed on tables, wall-mounted shelves, or special holders. This is the most economical and space-efficient method, and it is widely used in small offices. In this method, the active equipment can be connected to cable system as a switching junction, if the distance between the switching and active equipment allows the use of switching cords and switching connections.

The use of mounting cabinets provides compact location of practically any kind of equipment, and it also provides protection against nonauthorized access, as well as convenience of operational service. Past SCS realizations show that, in one mounting cabinet of height 42*U*, it is possible to place the connectors for a maximum of 120 workstations of switching and network equipment without loss of operational convenience. If it is necessary to increase this value, two mounting cabinets with removed adjoining sidewalls, which are adjacent to each other, are installed. In this case, the maximum number of workstations served increases to 300. The use of mounting cabinets allows designers to realize connection for a part of the network equipment to the cable system by switching junctions.

It is possible to use *the compound variant of installation* for any number of served workstations, and when the quantity of workstations is more than 250, the *compound variant* is preferable. In this case, 110 patch panels (cross-towers or cross-blocks with legs) and panels with modular connectors (with the use of mounting clamps or frames) are installed, as a rule, on a wall of the room, and the active equipment is placed in 19-in mounting constructions (cabinets or open racks). The free space around the switching equipment makes it convenient to work with a number of switching cords and gives compact location and convenient servicing of various devices. Locating active network equipment in closed mounting cabinets provides reliable protection against nonauthorized access to the most valuable equipment.

For compound installation variants, use of the cross-connect connection scheme is recommended in order to connect the network equipment to the cable system. If the number of workstations is large, then connection with the help of switches is possible.

The final selection a method of equipment location substantially depends not only on the number of workstations served, but also on the following factors:

- room height;
- location of cable channel inlet points into the cross-room;
- locations of power supply outlets for the network equipment;
- location of ventilating and air conditioning system inlets;
- convenience of equipment service and personnel movement inside room;
- local features of the illumination system;
- possible construction restrictions (beams, main supporting structures, and load limitations on certain points of interfloor overlap);

- prospective changes in room plans, say, in connection with expansion of its area.

The approved solutions for the method of equipment location in each cross-room is necessarily graphically fixed in the technical design documentation. The first drawing shows a top view of the room, the second drawing represents the equipment location scheme in 19-in mounting cabinets or on a wall.

The prepared plans should be included in the technical design structure. During the calculation of switching equipment quantity or during development of the technical design documentation, these schemes can be corrected as needed.

Calculation of number of switching equipment constructive units

All calculations are carried out separately for each cross-room functional section.

The use of one constructive unit of the switching equipment (cross-block or patch panel) for cable connections of different functional sections is not recommended. It complicates the section identification during operation, and it also fails to create a reserve for future expansion. Breaches of this rule are allowed only in small networks and only when all cables that enter the cross-room can be connected to a single constructive unit of the cross-equipment.

In view of this rule, each functional section of the cross-room is formed by one or several constructive units of the switching equipment, which are filled with cables to a greater or lesser extent. The calculation task consists of the determination of the required number of switching equipment units of the specified capacity for each functional section.

Horizontal cable sections. The dimension of an information transmission channel are always equal to four pairs in a "light-blue" section.

110 Cross-panels. One 25-pair contact ruler serves 6 channels, that is, six 4-pair cables can be connected to it, as can one bundle from 25 pairs of multipair cables (for connection to branch points of 6- or 12-port outlets). Hence, 24 channels for the operation of 24 outlet modules of telecommunication outlets will be placed on one 100-pair 110 cross-block. In order to calculate the total number of 110cross-blocks, the number of information transmission channels of the "light-blue" sections from the column 8 of Table 10.14 are divided by 24 and rounded up to the nearest whole number. The resulting value is placed in column 10 of Table 10.14.

Patch panels with modular connectors outlets. One 4-pair horizontal cable is prepared on the IDC contacts of the outlets of one 4-pair connector, and a 25-pair bundle of multipair cables is prepared on IDC contacts of six 4-pair outlets of patch panel connectors. In order to calculate the number of patch panels, the number of information transmission channels of the "light-blue" section from the column 8 of Table 10.14 is divided by the capacity of the selected patch panels and rounded up to the nearest whole number.

Backbone cable sections. The principle of calculation of backbone cable sections ("white," "gray," and "brown") and horizontal sections ("light-blue") basically coincide. Differences mainly arise since in the backbone subsystems section the multipair cables, which contain one or several bundles of 25 pairs in each, are usually entered. Therefore we shall stop only on calculation features.

110 Cross-panels. On one 100-pair 110 cross-block, 100 pairs from one or several multipair cables can be prepared. In order to calculate the total quantity of 110 cross-blocks, it is necessary to divide total number of section pairs from the column 6 of Table 10.14 by 100 and to round this result up to the nearest whole number. The resulting value is placed in column 10 of Table 10.14.

Patch panels with modular connectors outlets. One 25-pair bundle of multipair cables is prepared on IDC contacts of six 4-pair connectors of a patch panel. Thus, 100, 150, 200, or 300 couples of one or several pair cables are prepared on 24-, 32-, 48-, or 64-port patch panels, respectively. In order to calculate the number of patch panels, it is necessary to divide the total number of section pairs from column 6 of Table 10.14 by the number of pairs prepared on patch panels of the selected capacity, and to round this result up to the nearest whole number. The resulting value is entered in column 10 of Table 10.14.

In addition, we note that twisted pairs must be split on all IDC contacts of patch panel connectors. Only this solution provides complete functional flexibility of cable systems, while the preparation of less than four pairs of multipair cables on each connector appreciably limits the functionality of the cable system backbone.

In situations when the backbone is created with the use of 4-pair cables, the calculations of switching equipment quantities are carried out in accordance with the procedures described in Section 10.3.6.

Fiber-optic patch panels. The calculation of the number of switching equipment units during the creation of backbone subsystems using optical cables is carried out by the same principles as the case of the "electrical" realization. During calculation it is additionally necessary to take into account that, in one unit of switching equipment (shelf or wall-mounted coupling), the limited number of cables is equal to the number of cable inlets. For example, no more than two cables can usually be put into most common wall-mounted couplings or shelves of height 1*U*.

Equipment cables section

110 cross-panels. The number of equipment cables (installation cords) that can be prepared on the 100-pair 110 cross-blocks, in the case of individual connection of ports on the basis of outlets of modular connectors, is determined by dividing 25 by the number of pairs in one equipment cable, rounding this result down to the nearest whole number, and multiplying by 4. Then, the total number of equipment cables in a section from column 4 of Table 10.14 is divided by the number from the previous step, and rounded up to the nearest whole number to determine the total number of 110 cross-blocks. The result is entered in column 10 of Table 10.14.

This algorithm is easily used if the number of pairs in the equipment cables equals 25 (the Telco connector) or even exceeds this value. In order to calculate the total number of 110 cross-blocks, it is necessary to divide total number of pairs of a section from column 6 of Table 10.14 by 100 and to round up to the nearest whole number. The result is entered in column 10 of Table 10.14.

Patch panels with modular connectors. If the number of pairs in an equipment cable is less than 4, one such cable is an installation cord and is connected to the IDC contacts of one 4-pair connector. Therefore, in order to calculate the number of patch panels, it is necessary to divide the number of equipment cables from column 4 of Table 10.14 by the selected capacity of patch panels and to round up to the nearest whole number. The resulting value is entered in column 10 of Table 10.14.

If the number of pairs in an equipment cable is more than 4, then the pairs of one information transmission channel (column 7 of Table 10.14) are prepared on IDC contacts of one 4-pair connector. The number of information transmission channels that can be supplied by one equipment cable is determined by dividing the cable capacity (column 5 of Table 10.14) by the rate of information transmission channel (column 7 of Table 10.14), and rounding this result down to nearest whole number. The number of equipment cables that are prepared on patch panels of a selected capacity is determined by dividing the number of patch panel ports by the value gained in the previous step, and rounding this result up to the nearest whole number. The total number of patch panels is determined by dividing the number of equipment cables (column 4 of Table 10.14) by the value gained in the previous step. The result is entered in column 10 of Table 10.14.

Calculation of the number of switching equipment units

Location of 110 cross-panels on room walls. Location of 110 cross-panels on room walls in large-volume SCS (capacity over 100 ports) is usually carried out in case of the use of cross-towers (Section 3.3.2). In the case of small SCS, mounting blocks with legs are used.

First of all, it is necessary to determine the total number of 100-pair 110 cross-blocks. For this purpose it is necessary to sum up the values from the column 10 of Table 10.14. Then we shall find the quantity of cross-towers by dividing the total number of cross-blocks by their quantity in towers of selected dimensions and rounding this result up to the nearest whole number.

The cross-panels can be placed on cross-room walls in one or more vertical columns and in one or more horizontal lines. During design development it is recommended to consider some variants. One of the most important selection criteria is the minimization of the lengths of switching cord, which connect various sections of switching equipment. If any of these variants fails to satisfy the designer, it is necessary to choose other sizes of cross-towers and to repeat the calculation anew.

During final stage, the distribution of cross-blocks of different functional sections of cross-rooms on installed cross-towers is made. During selection of a location plan it is necessary to be guided by the following principles:

- Functional sections should be formed by continuous sets of cross-blocks. It is necessary to locate the cross-blocks of functional sections under one another (in one or several columns).
- Location of cross-panels is carried out in view of placement of cable channel inlets into cross-rooms and is made in order to minimize the lengths of cables installed in the cross-room.

- If the cross-blocks of two different functional sections are placed on one column of cross-towers, then the cable inlets are made from different column sides in opposite directions (from the column block top downwards for one section, and from the column block bottom upwards for other section). If the cross-blocks of more than two different functional sections are placed on one column of cross-towers, then the connection of cables on cross-blocks is carried out from above downwards for each functional section. The direction of cable inlets in columns is not important for this and is usually made so that the number of cables put into a column from above and from below are be approximately identical.
- It is necessary to take into account prospective future functional section expansions.

The vertical cable organizers are installed between columns of cross-towers. They must be placed between columns in which there are cross-blocks of different functional sections. The cable organizers are installed in three columns in large-volume functional sections. The height of the cable organizers is chosen to be equal to the height of the cross-towers; the installation of these elements is carried out at the same level.

The chosen types and numbers of switching equipment are entered in the appropriate columns of Table 10.16. Based on the calculation results, corrections are made to the equipment location plans in the cross-rooms (Section 10.3.6), if necessary. An example of a 110 switching equipment location plan on a room wall is given in Figure 10.23.

Table 10.16. Swithing equipment table

Customer: ____________________

Object: ____________________

Building: ____________________

Item no.	Cross-room	Switching equipment		Organizers		Ancillary equipment	
		Type	Quantity	Type	Quantity	Type	Quantity
1	2	3	4	5	6	7	8

Arrangement of 110 cross-panels in 19-in mounting constructions. Specialized structural elements (Section 3.3.2.1) are used in 19-in mounting constructions for locating 110 cross-panels, which do not have internal fastening elements.

In case of panel locations on cross-room walls and in 19-in constructions, the quantity calculation for patch blocks, such as 110 patch blocks, is carried out by identical rules. The total height of panels, such as 110-types, is determined on the basis of these results. The resulting value is the basis for selection of cabinet heights or open rack heights, in which the installation of switching equipment is made. It is also necessary to take into account that equipment should occupy no more than 60% of the general height of mounting cabinets. If it is necessary to use two cabinets because one is too small, these two cabinets should be installed close to one another and fastened by lag bolts with the adjacent sidewalls removed. If capacity of two mounting cabinets of maximum height fails to suffice, it is necessary to use the combined method of equipment location and to carry out the design of the administrative subsystem anew.

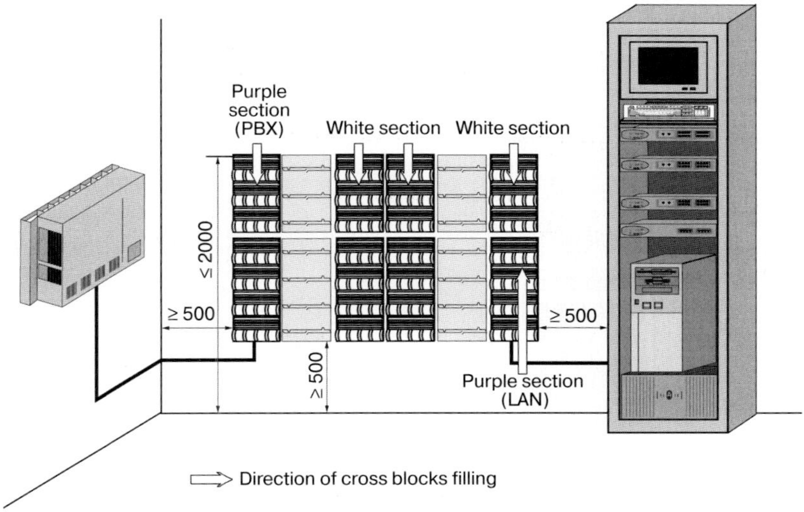

Fig. 10.23. Example of a 110 switching equipment location plan

As soon as the selection of cabinet height is completed, the location plan of the cross-blocks of different functional sections is determined. Thus it is necessary to be guided by the following principles:

- Functional sections are placed on continuous sets of cross-blocks, which are installed from left to right and from above to below for one functional section.
- For the use of two 19-in mounting cabinets, the distribution of switching equipment between them is carried out based on the customer requirements.
- It is necessary to take into account future expansion of any functional section.
- In the case of the use of two mounting cabinets, a vertical cable organizer is necessarily installed in one of them.

The results of these calculations are entered in the appropriate columns of Table 10.16 and can be used to correct the equipment location plans in the cross-room (Section 10.3.6).

An example of a 110 switching equipment location plan in this case is given in Figure 10.24.

Location of patch panels in 19-in mounting cabinets. The procedures and the common principles of calculating the number of switching equipment elements in the case of the use of patch panels with modular connectors and 110-type panels coincide. A unique feature is that the majority of patch panels with modular-type connectors have no state organizers of connection cords. Cable organizers are installed as separate elements between the functional sections in the necessary order, and inside large-volume functional sections are installed in every 72 ports. In practice, frequently one follows an empirical method of obligatory installation of horizontal organizers under patch panels of any capacity.

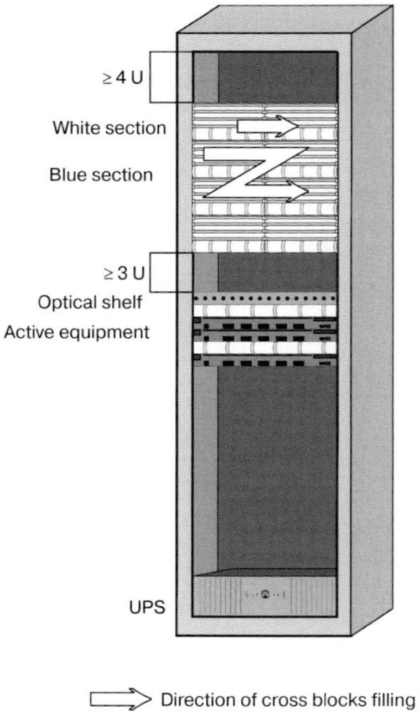

Fig. 10.24. Example of 110 switching equipment location plan in a mounting cabinet

The results of the calculations are reflected in the appropriate columns of Table 10.16 and can be used to correct the equipment location plans in the cross-room (Section 10.3.6).

An example of a 110 patch panel location plan in a mounting cabinet is given in Figure 10.25.

Location of patch panels on cross-room walls. The procedures and the common principles of calculations in this variant coincide with their installation in 19-in constructions. The basic criterion during design is the minimization of the lengths of patch cords that will be used for information transmission channel switching between various sections of the switching equipment.

During selection of the location plan for the patch panels of different functional sections in the cross-room, it is necessary to be guided by the following principles:

- The patch panels are placed in one or several vertical columns.

- Their installation locations are selected in view of the location of cable channel inlets into the cross-room and, whenever possible, should minimize cable lengths in this technical room;
- Functional sections should be formed by continuous sets of patch panels. The patch panels of one functional section are placed vertically in one or several columns.
- If the patch panels of two different functional sections are placed in one column, then the cable inlets of these sections are carried out from different column ends (from the column block top downwards for one section, and from the column block bottom upwards for another section). The cable connections to the panels are ordinarily carried out from left to right and from above to below. If the patch panels of three or more functional sections are placed on one column of cross-towers, then location of the patch panels and the connection of cables is carried out from above to below for each of them.
- It is necessary to take into account prospective expansion of any functional section.

The cable organizers are installed between patch panels. The various functional sections, in the necessary order, are separated from each other by the use of horizontal and vertical organizers. Moreover, horizontal organizers are installed in every 72 ports, and vertical organizers are installed in every three columns in large-dimension functional sections.

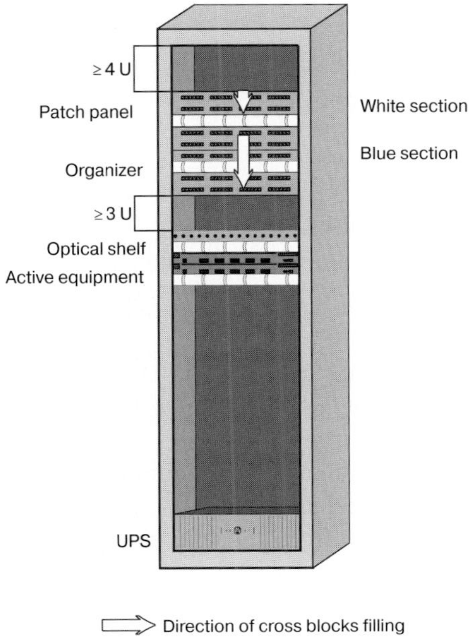

Fig. 10.25. Example of a 110 patch panel location plan in a mounting cabinet

The final stage is the height calculation of each column. In order to make this calculation, the heights of the patch panels and the horizontal organizers are added up. It is necessary to provide the appropriate number of mounting clamps and frames for fastening panels and organizers.

The chosen types and quantities of switching equipment are entered in the appropriate columns of Table 10.16. If necessary, corrections should be made to the outline equipment plans in the cross-room (Section 10.3.6).

The example of the plan of allocation patch panels on apartment wall is given in Figure 10.26.

On the basis of the carried out calculations results the list of cross-blocks, cross-towers, patch panels, cable organizers etc. is made for each cross-room and equipment room in view of the chosen method of equipment allocation. The calculation results are brought in the appropriate columns (Table 10.16).

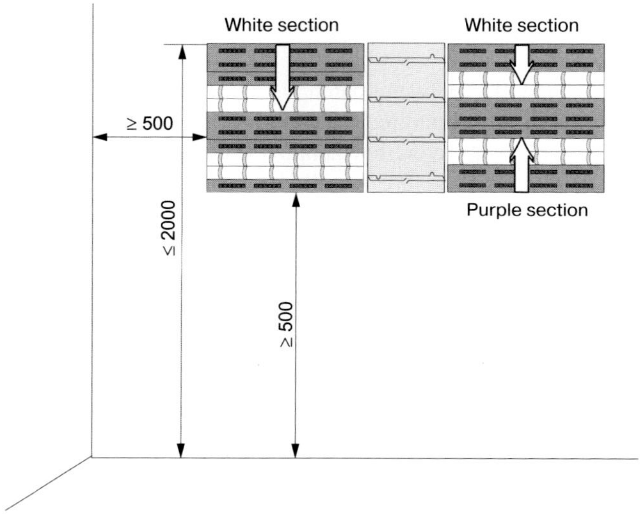

Fig. 10.26. Example of the plan of patch panels allocation on apartment wall

Determination of patch cord types and quantities. The patch cords serve for information transmission channel switching and are installed between connectors of the switching equipment.

It is recommended to use patch cords made in industrial conditions with potentially higher operational reliability and electric characteristics.

The calculation for the each cross-room begins from the determination of functional section pairs between which switching will be carried out. For each pair of sections the following are set:

- types of patch cord plugs, which make the connection to the switching equipment;

- number, lengths, and patch cord types, which are required to connect the two nearest and two most remote connectors of the switching equipment of these two sections.

In the case of a large number of sections that can be served by any particular cross-room, the enumeration of the pairs of these sections during the calculations should be carried out by a matrix (Table 10.17). In case of the absence of any section in the switching equipment structure, the appropriate line and column of the table should be deleted. It is very convenient to make such a matrix using Excel software (Microsoft). Any absent functional section is deleted by closing the appropriate lines and columns of the table; sections without connections are marked by filling the appropriate cells.

The results are entered in the appropriate columns (Table 10.18). For SCS of small and average capacity, the data for termination cords of network equipment is also frequently entered in this table. In this case, the table refers to terminated and patch cords.

Table 10.17. Connection matrix of various functional sections

	Green	Yellow	White	Blue	Orange	Gray	Brown	Red
Green	+							
Yellow	+	+						
White	+	+	+					
Blue	+	+	+	+				
Orange	+	+	+	+	+			
Gray	+	+	+	+	+	+		
Brown	+	+	+	+	+	+	+	
Red	+	+	+	+	+	+	+	+

Table 10.18. Patch cords table

Customer: ______________________

Object: ______________________

Building: ______________________

Technical room: ______________________

Item no.	Functional section		Channel quantity	Plug type		Cord length		Patch cord		
	1	2		1	2	Min	Max	Type	Length	Quantity
1	2	3	4	5	6	7	8	9	10	11

Despite the wide use of two-pair schemes of information exchange in LAN, for cable system versatility it is *always* recommended to use 4-pair patch cords for data transmission systems. For low-speed applications, the use of one- and two-pair cords is optimum.

In order to provide normal operation of administrative subsystem, it is recommended to include patch cords with different lengths into the SCS structure. Since the patch cords made in industrial conditions are delivered with discrete lengths, it is necessary to sort out some delivered lengths from the minimum up to the necessary maximum (columns 5 and 6 of Table 10.18). It is possible to provide for equal numbers of cords with different lengths, but, as practice shows, the distribution of their number on the graph in Figure 10.27 is optimum. Thus the length of the maximum number of cords ordered is $L_{middle}=(L_{max}+L_{min})/2$.

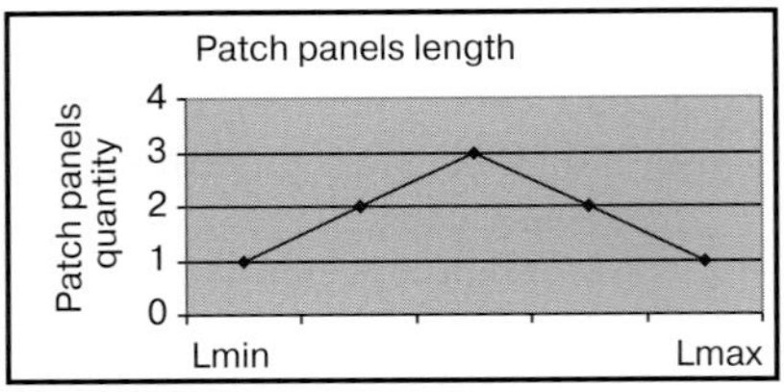

Fig. 10.27. Distribution of patch cord number by their length

Specification formulation

The main normative document that regulates the formulation rules of specifications is the All-Union State Standard (GOST) 21.110-95 [125], included in the system technical design documentation for construction. According to this intergovernmental standard of the CIS countries, the specification, with reference to the considered area, is the textual technical design document that determines the equipment structure, hardware products, and materials necessary for SCS realization. All equipment, hardware products, and materials stipulated by the technical design documentation are included in the specification. It is recommended to draw up this document in sections; the name of each section should be written on a separate line as a header and underlined. The informal part of the specification is given in a table. In column 1 of this table, the equipment designation stipulated by the working drawings is given. Column 2 contains the name of the equipment with its characteristics in brief. In column 3 are the type, model of equipment, and its technical terms. The equipment codes are entered in column 4. The manufacturers' information, country, and company (for imported equipment) are given in column 5. Columns 6 and 7 contain units of measurements and numbers of equipment units. Column 8, according to the All-Union State Standard (GOST) is allocated for the indication of equipment unit weight. Additional information is given in column 9. The standard does not fill out some columns. The first sheet of the specification is the title page, filled out on a special form.

The specification is usually a separate appendix to the contract for equipment delivery and cable system realization and, consequently, is approved by the directors of the customer and contractor, whose signatures are officially stamped.

10.4 SCS Design Example

We shall consider an example of the use of the above material for cable system design in a hypothetical project.

10.4.1 Initial Data

A SCS is created in a four-floor building (Figure 10.28) with the dimensions 50×15 m in a plan. The height of the floor in light between overlaps is 3.5 m, the general thickness of the overlaps is equal to 50 cm. The same type of corridor plan of working rooms is used, which has the dimensions 6×3 m on all floors. The 3-m wide corridor passes the longitudinal axis of building, Figure 10.29.

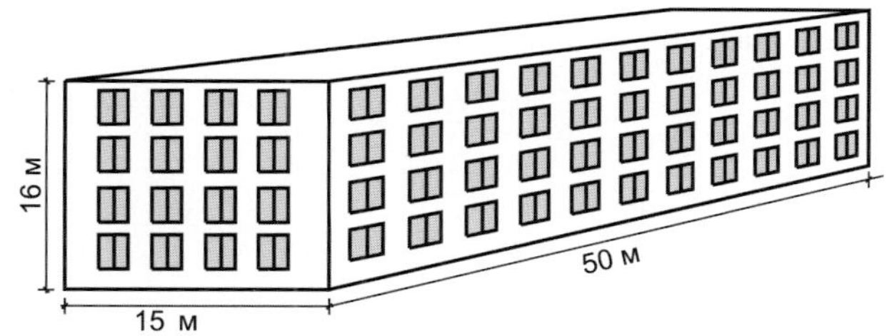

Fig. 10.28. Building dimensions

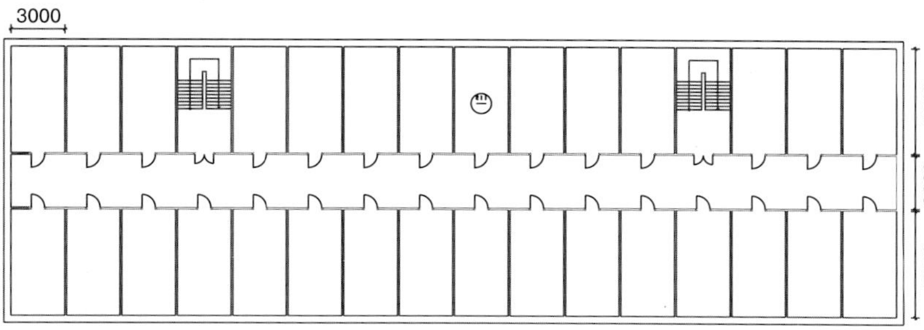

Fig. 10.29. Floor plan

Table 10.19. List of cross-rooms and equipment rooms

Room no.	Purpose	Space	
		Actual, [m^2]	According to norm, [m^2]
111	Equipment room	36.60	11.3 (14)
211	Cross-room	9.10	8.4
311	Cross-room	9.10	8.4
411	Cross-room	9.10	8.4

There is a false ceiling with free-space height of 35 cm in the corridors, and there is no false ceiling in the rooms of the building. The walls of the rooms are made of normal bricks and are covered with 1-cm thick plaster. No additional channels in floor and walls, which can be used for cable layout, are stipulated by the building design. The list of technical room allocated to cross-rooms and equipment rooms is given in Table 10.19.

A created SCS should provide operation of LAN equipment and telephone network of a building, that is, telecommunications outlets with two outlet modules on each workstation are installed. Connection of private PBX with a 100-pair input cross of public switch telephone network is also provided.

Besides the telecommunications outlets, two power supply outlets, connected to a guaranteed power supply network and one power supply outlet, connected to household power supply network, are installed on the workstation. The power supply cable layout and the installation of power supply-distributing equipment is carried out by an allied subcontractor company.

According to the customer requirements, outlet blocks are installed at 1-m height above the floor level. The cable layout in working rooms is carried out in decorative conduits.

10.4.2 Architectural Design Stage

In accordance with the plan, there are 30 working rooms, each with 18 m^2 of workspace, on each building floor; that is, the general working space of a floor is equal to 540 m^2. According to Construction Norms and Rules (SNiP 2.09.04-87), Section 3.2, 135 outlet blocks are necessary to serve this space, and there are 540 outlet blocks total in the building. The calculation on the basis of the separate working rooms gives a different value. In concordance with the Construction Norms and Rules, four outlet blocks are installed in each room, that is, there will be 120 outlets on each floor. The difference in the calculations by various criteria turns out that the space of the separate working rooms does not give 16 or 18 m^2. Usually, a large value (figure) is given in the design, which derives from the general floor space calculation. The additional outlets are installed in corridors, technical rooms, and in some working rooms, and are used, for example, for connection of active network equipment of collective use, such as printers and faxes, guard post telephones, and servers.

The outlets that were not used during SCS construction remain in the Spare Parts Accessories (SPA) and are used to repair the cable system during current operation, and for future expansions. When the customer requires an inexpensive SCS, the purchase of unused outlets can be provided after cable system construction completion by the system integrator at nominal cost, as given in the contract.

In order to carry out cable layout of horizontal subsystems on floors along corridors, cable ladders are installed behind false ceilings. The distance from the top edge of the cable ladder to the main ceiling is equal to 25 cm. The cross-room is located at the center of floor and consequently, the cables installed on each half of the ladder serve 270 m^2 of working space. The cross-sectional space of the ladder (on the basis of Section 10.2.6.2) with two outlets on each workstation gives

650×(270/10)×2/3=11,700 mm^2. The standard cable ladder with dimensions 200×60 mm has such space. Ladders with smaller areas can be used far from the cross-room.

According to the customer requirements, the cable layout in working rooms is carried out in decorative conduits. In order to make a link between cable ladders and conduits, holes are drilled in the walls of corridors, in which the ducts are installed. On the basis of Table 10.8, the internal diameter of the duct should be about 26 mm. According to the above calculations, four outlet blocks — two for each side — are installed in each room. Therefore, the decorative conduit capacity is selected to contain four horizontal information cables and two power supply cables (one for the guaranteed power supply system for the computer equipment (UPS), another for outlets of household power supply). The calculated diameter of the horizontal cable is equal to 5.5 mm (Table 3.5), which is 10% larger than its nominal external diameter. It allows designers to take into account the increase in the necessary space of a section at the expense of the roughness of its layout. The general cross-sectional space of four information cables is approximately 100 mm^2. For 50% filling of the conduit section in which these cables will be placed, the cross-sectional area should be 200 mm^2. Moreover, at least one separate section should be provided in the decorative conduit where the power supply cable layout is carried out. For these reasons, in this particular case a three-section conduit such as Legrand L/30026 is used. Its dimensions are 60×16 mm, the cross-sectional area of its central section is 200 mm^2, and the overall dimensions of the side sections are sufficient for location of power supply cables. In accordance with the Table 11.3, nylon dowels or jet plugs can be used as fixing elements for the conduits and outlet modules.

According to this plan (Figure 10.29 and Table 10.19), the cross-rooms and equipment rooms are located directly above one another. Therefore, the slots and sleeves with appropriate fixing accessories can be used as vertical risers for the backbone cables. The space of interfloor apertures for the installation of the sleeves or slots is selected in view of the provisions of the Section 10.2.5 and the location of the equipment room in apartment 111 (that is, on the first floor). We shall also take into account that the aperture from the first floor to the second floor serves three floors, whereas through the aperture from the second floor to the third floor pass cables that serve two floors. The calculation results of the apertures space are shown in Table 10.20. On this basis, the slot in the overlap between the first and second floors has overall dimensions 300×50 mm. The overall dimensions of other slots are selected as the same for reasons of uniformity or are accordingly smaller in view of Table 10.20.

Table 10.20. Aperture spaces and slot spaces in interfloor overlaps

Aperture no.	Served		Aperture space, [mm^2]	Dimensions, [mm]
	Floors	Space, [m^2]		
1	2–4	1620	5200	300×50
2	3, 4	1080	3500	200×50
3	4	540	1730	100×50

The spaces of technical spaces for cross-rooms and equipment rooms correspond to typical norms (Table 10.19). The environmental conditions described in Sections 10.2.2 and 10.2.3 are additionally controlled in these apartments. Sometimes it is necessary to give allied subcontractors the individual technical tasks for elaborating the rooms.

PABX, servers, and central LAN equipment will be placed in the equipment room, which is the principle of multipoint administration (Section 1.2.5).

10.4.3 Telecommunications Design Stage

The information on the number of information and power supply outlets in each room is given in Table 10.9. We note that one telecommunications outlet with two outlet modules, forming SCS ports, and three power supply outlets, which are integrated in the uniform block, are provided on each workstation.

The type of outlet module is determined by the requirements for transmission capacity, workstation configuration, and the chosen method of fastening. In this concrete case for telecommunications outlet construction, it is convenient to use two-port outlet modules. Their total quantity is also the same as the number of power supply outlets and should be calculated by summation of the values in columns 5–9 (Table 10.21). Additionally, if necessary the fastening elements for power supplies and telecommunications outlets in or near decorative conduits are included in the specification.

Table 10.21. Distribution of workstations

Customer: ______________________

Object: ________________________

Building: _______________________

Item no.	Floor	Room no.	Work-station quantity	Outlet modules			Power-supply outlets		Fastening method
				Cat. 3	Cat. 5	Fiber optic	UPS	Household	
1	2	3	4	5	6	7	8	9	10
1	1	101	4	–	8	–	8	4	On wall surface in frame near box
…									
120	4	429	4	–	8	–	8	4	On wall surface in frame near box
Total			-	–	1080	–	1080	540	

Horizontal subsystem design

Two category 5 telecommunications outlet modules are installed on each workstation. The number of outlets on the workstation was determined in the architectural design stage, the use of two category 5 outlet modules is defined by reasons of versatility.

In this building, there are no large halls or compact solitary groups of users. Therefore cable layouts under carpets will not be applied and the realization of separate sections and some routes of the horizontal subsystem using multipair cables is inexpedient. In turn, branch points are not required in the SCS.

The SCS horizontal subsystem is created on the basis of unshielded 4-pair category 5 cables, laid two to each outlet block. The required number of cables is calculated by the empirical method, as there are more than 30 telecommunications outlets on each floor, and the requirement of uniform outlet distribution on served area is executed.

The rise from the output hole of the mounting cabinet to the cable ladder trays in the corridors and the descent to the decorative conduit in the rooms makes 3.25+2.25=5.5 m. The cable route length, in accordance with the plan, from the cross-room to the nearest and to the most remote outlet block is 5 m and 35 m, respectively. Therefore the minimum and maximum length of cables is respectively 5.5+5=10.5 m and 5.5+35=40.5 m, and the average length of cable routes will be equal to (10.5+40.5)/2×1.1+3=31 m. There will be enough one-cable drums to lay out 9 average cable routes (305/31=9). Therefore, for horizontal subsystem creation it is necessary to use 120 cable drums (540×2/9=120) or 36,600 m of cable (120×305=36,600).

Design of internal backbone subsystems

The cables of internal backbone subsystems link cross-rooms and equipment rooms. The informational flows of the LAN network equipment and the telephone signals of private automatic telephone exchanges are basically transmitted on these cables. According to the principle of the use of two-port telecommunications outlets in workstations included in the system, and in the absence of external private automatic telephone exchanges, signal transmission of a significant number of phone calls should be expected on backbone cables. Based on this and according to the principle of multipoint administration, the following ideology of internal backbone subsystems construction is adopted:

- the part of internal backbone subsystems intended for telephone network operation is constructed on multipair category 3 electrical cables ;
- in order to organize part of the internal backbone subsystems served by LAN operation, fiber-optic cables with 100% back-up by category 5 electric cables is used.

We shall calculate capacity of cables in fibers and pairs. The SCS designed has a high scale of integration: two telecommunications outlets with the appropriate number of horizontal cables per workstation. Therefore, for each workstation in the internal backbone of the building, it is necessary to provide two pairs of category 3, 0.4 pairs of category 5, and 0.2 fibers, and accordingly for each floor: 270 pairs of category 3, 54 pairs of category 5, and 27 optical fibers. Using known values of floors height (4 m) and reserve for cable preparation (3 m from each end), taking into account that the vertical riser passes directly through the cross-rooms, we shall calculate the lengths of the backbones cable routes. The results of the calculations are entered in the appropriate columns of Table 10.22.

Summarizing the resulting values, we see that for a technological reserve of 10%, 95 m of 25-pair category 5 cables, 143 m of 100-pair category 3 cables, and 95 m of 16-fiber optical cables are necessary. Not all manufacturers of SCS equipment produce optical cables for internal layouts with capacities of 16 fibers. Therefore, in this case, we shall use two 8-fiber cables to build the system in accordance with the norms of fire-prevention safety cable layouts in vertical risers.

Table 10.22. Backbone cables table

Object: ______________________

Building: ______________________

Item no.	Marking	Beginning	End	Cable type	Number of pairs/fibers	Number of cables	Route length, [m]	Customer
1	2	3	4	5	6	7	8	9
1	KM021	111	211	Cat. 5	25	2	10	LAN (reserve)
2	KM022	111	211	Cat. 3	100	3	10	Telephony
3	KM023	111	211	Fiber optic	16	2	10	LAN
4	KM031	111	311	Cat. 5	25	2	14	LAN (reserve)
5	KM032	111	311	Cat. 3	100	3	14	Telephony
6	KM033	111	311	Fiber optic	16	2	14	LAN
7	KM041	111	411	Cat. 5	25	2	18	LAN (reserve)
8	KM042	111	411	Cat. 3	100	3	18	Telephony
9	KM043	111	411	Fiber optic	16	2	18	LAN

Administrative subsystem design

In SCS we have designed, the number of telecommunications outlet modules that are to be used for telephone system operation coincides with number of outlet modules for LAN connection. Therefore, we shall use 110-type panels as switching equipment.

Because there are many served cables in the equipment room, we shall use the combined method of equipment location—on a wall and in a cabinet. In the cross-room, all equipment can be placed in a cabinet. Another possible variant, which is not used in this particular project, is the installation of an open 19-in rack.

In order to connect network equipment, the interconnect connection method will be used in all technical rooms. The calculated results for switching equipment are shown in Table 10.22. The calculation of the number of separate functional elements for the switching equipment is made with the assumption that 24 horizontal cables or 4-pair cables are connected to one 100-pair 110 cross-block .

The optical cables of internal backbone subsystems are split in 19-in shelves from 16 SC outlets and height 1*U*. Welding is used for the installation of optical connector plugs; the layout of protective bushing sleeves and technological reserves of fiber lengths are installed in a common organizer, such as a splice-plate.

The space of cross-rooms exceeds the recommended value, therefore equipment can be placed in closed mounting cabinets. Because there are many served workstations, it is necessary to use mounting cabinets of maximum height 42*U* or 45*U*. The reserve for cable preparation is selected to be equal to 3 m, therefore the cabinets are installed near the risers. One version of switching equipment location in the mounting cabinet is shown in Table 10.24. In the left part of this sketch, units of height are marked, which will be assigned for installed equipment. We note that the assembled 19-in 200-pair 110 cross-panels contain the horizontal cross-cord organizers. In additional, it is necessary to provide the 19-in horizontal organizers of height 1*U* for optical shelves. It is recommended that the LAN network equipment be placed below the organizers of optical shelves. If the capacity of one concentrator is 12 ports, then the LAN network equipment will occupy a height of 12*U* in a cabinet, that is, 40*U* will be occupied from 42U in a cabinet.

In accordance with the approved location scheme, the switching equipment of all functional sections, except for backbone "white" category 5 and "white" optical, will be installed on a wall of the equipment room. This choice is explained since the backbone cables of the "white" section are used only for communication within the network equipment, without connection to the workstations. Cross-towers with not fewer than 40 cross-blocks (Table 10.22) will be necessary for the organization of the electrical part of this section. Providing about 10% reserve, we shall use 5 columns of cross-towers, each of which consist of a 400-pair and a 500-pair tower. For category 5 switching equipment of the "white" section, which is installed in the mounting cabinet, we shall choose a 200-pair 110 cross panel. Shelves with 16 SC outlets and height 1*U* are used to organize the optical part of the "white" section. Six shelves are used, as shown in Table 10.22. A patch cord organizer is also installed under each shelf. Table 10.23 shows an example of cross-equipment location in the equipment room.

Table 10.23. Cross-room equipment

Customer: ______________________

Object: ______________________

Building: ______________________

Item no.	Color code	Purpose	Cable quantity	Pairs/ fibers in cable	Total number of pairs/ fibers	Pairs/ fibers in channel	Channels	Switching equipment type	Device quantity
1	2	3	4	5	6	7	8	9	10
				Cross-room: 211 (311, 411)					
1	Blue	Horizontal cables	270	4	1080	4	270	100 paired blocks, 110-type	12
2	White (Cat. 5)	Internal backbone	2	25	50	4	12	100 paired blocks, 110-type	1
3	White (Cat. 3)	Internal backbone	3	100	300	4	72	100 paired blocks, 110-type	3
4	White (fiber optic)	Internal backbone	2	16	32	2	16	Optical shelves (16 ports)	2
				Equipment room: 111					
1	Blue	Horizontal cables	270	4	1080	4	270	100 paired blocks, 110-type	12
2	White (Cat. 5)	Internal backbone	6	25	150	4	36	100 paired blocks, 110-type	2
3	White (Cat. 3)	Internal backbone	9	100	900	4	216	100 paired blocks, 110-type	9
4	Purple (PABX)	PABX cables	46	25	1150	1	1150	100 paired blocks, 110-type	12
5	Purple (LAN)	LAN cables	135	4	540	4	135	100 paired blocks, 110-type	6
6	Green	PSTN cables	1	100	100	1	100	100 paired blocks, 110-type	1
7	White (fiber optic)	Internal backbone	6	16	96	2	48	Optical shelves (16 ports)	6

Table 10.24. Example of cross-equipment location in the cross-room

	Mounting cabinet	
28 U	White (Cat. 5)	White (Cat. 3)
27 U		
26 U		
25 U	White (Cat. 3)	White (Cat. 3)
24 U		
23 U		
22 U	Blue	Blue
21 U		
20 U		
19 U	Blue	Blue
18 U		
17 U		
16 U	Blue	Blue
15 U		
14 U		
13 U	Blue	Blue
12 U		
11 U		
10 U	Blue	Blue
09 U		
08 U		
07 U	Blue	Blue
06 U		
05 U		
04 U	White (Fiber optic)	
03 U	19" organizer	
02 U	White (Fiber optic)	
01 U	19" organizer	
	Network equipment	

Calculation of quantity and length of terminated cords and patch cords

Terminated cords on workstations. Terminated cords with modular connector plugs are used to connect PCs to telecommunications outlets on the workstations. In total in the SCS, in order to connect the workstations to the LAN, it is supposed that 540 outlet modules are used. We shall assume that at the first stage of network development, one-third of the user workstations will be connected to LAN, that is, 180 cords plus 10% for development reserves are required. We shall use 2.1-m long cords. Some of the cords (about 10%) should be quite long, for example, 5 m, in order to provide operational flexibility. Thus, 200 2.1-m long cords and 20 4.57-m long cords will be required.

Terminated cords for telephone set connection are usually delivered with telephone sets and therefore they are not included in the final specification.

Cross-rooms. The following types of patch cords are provided for the cross-rooms:

- One-pair cords with 110-type plugs for connection of floor outlets to PABX, that is, for connection of horizontal cables to multipair vertical cables—only 135 cords on a floor;
- 4-pair combined cords to connect horizontal cable to ports of LAN hubs— only 135 cords on a floor;
- Optical cords to connect optical up-link-ports of floor hubs to the vertical backbone—only 12 cords on a floor;
- Backup combined 4-pair cords to connect electrical up-link-ports of floor hubs to category 5 vertical cables—only 12 cords on a floor.

Table 10.18 contains the patch cord information necessary for the realization of switching in the equipment room and in the cross-room. In order to determine the minimum and maximum length, it is necessary to use the equipment location scheme (an example is given in Table 10.24). The distribution of patch cord lengths is made according to Figure 10.27.

Equipment room. Terminated cords to connect the PABX and to connect the first-floor workstations to the LAN network equipment are required in the equipment room.

Table 10.25. Example of cross-equipment location in the equipment room

Mounting cabinet

Unit		
15 U	White (Cat. 5)	White (Cat. 5)
12 U	White (Fiber optic)	
11 U	Organizer	
10 U	White (Fiber optic)	
09 U	Organizer	
08 U	White (Fiber optic)	
07 U	Organizer	
06 U	White (Fiber optic)	
05 U	Organizer	
04 U	White (Fiber optic)	
03 U	Organizer	
02 U	White (Fiber optic)	
01 U	Organizer	
	Network equipment	

Cross field on equipment room wall

Purple (LAN)	Organizer	Blue	Organizer	Purple (PBX)	Purple (PBX)	Organizer	White (Cat. 3)
Purple (LAN)		Blue		Purple (PBX)	Purple (PBX)		White (Cat. 3)
Purple (LAN)		Blue		Purple (PBX)	Purple (PBX)		White (Cat. 3)
Purple (LAN)		Blue			Purple (PBX)		White (Cat. 3)
Purple (LAN)	Organizer	Blue	Organizer	Green	Purple (PBX)	Organizer	White (Cat. 3)
Purple (LAN)		Blue			Purple (PBX)		White (Cat. 3)
		Blue		Blue	Purple (PBX)		White (Cat. 3)
		Blue		Blue	Purple (PBX)		White (Cat. 3)
		Blue		Blue	Purple (PBX)		White (Cat. 3)

Installation cords made as 25-pair cables with fixed Telco connectors on one end are used to connect to the PABX. Their length depends on the distance between the PABX and the cross-field. Cords up to 30-m long can be ordered. In this case, in view of the overall dimensions of the equipment room and the location of the PABX cross, we shall define the length of these cords equal to 7 m.

The 25-pair category 3 cables with fixed Telco connectors on one end will be used for PABX connection to SCS in the equipment room. The free end of these cords will be prepared on the violet section of the switching equipment of the equipment room. Each cable is intended for connection to the cable system of one 16-port card of internal PABX numbers or one 8-port card of external PABX numbers. Therefore, 46 cables are required to connect 540 workstations and 96 city telephone numbers.

Table 10.26. Terminated cords and patch cords

Customer: ____________________

Object: ____________________

Building: ____________________

Item no.	Functional sections 1	Functional sections 2	Channels quantity	Plug type 1	Plug type 2	Cord length, [m] Min	Cord length, [m] Max	Patch cord Type	Patch cord Length, [m]	Patch cord Quantity
1	2	3	4	5	6	7	8	9	10	11
				Cross-room: 211, 311, 411						
1	Blue	White (Cat. 3)	135	1 pair 110	1 pair 110	0.5	2.0	110–110, 1 pair	0.5	22
									1.0	45
									1.5	46
									2.0	22
2	Blue	Network equipment	135	4 pairs 110	Modular	0.5	1.5	110-modular, 4 pairs	0.5	33
									1.0	68
									1.5	34
3	White (Cat. 5)	Network equipment	12	4 pairs 110	Modular	1.0	1.5	110-modular, 4 pairs	1.5	12
4	White (fiber optic)	Network equipment	12	2×SC	2×SC	0.5	1.0		1.0	12
				Equipment room: 111						
5	Blue	Purple (PABX)	135	1 pair 110	1 pair 110	0.5	2.5	110–110, 1 pair	0.5	15
									1.0	30
									1.5	45
									2.0	30
									2.5	15
6	Blue	Purple (PABX)	135	4 pairs 110	Modular	0.5	2.5	110–110, 4 pairs	0.5	15
									1.0	30
									1.5	45
									2.0	30
									2.5	15
7	White (Cat. 3)	Purple (PABX)	405	1 pair 110	1 pair 110	0.5	2.5	110–110, 1 pair	0.5	45
									1.0	90
									1.5	135
									2.0	90
									2.5	45
8	White (Cat. 5)	Network equipment	36	4 pairs 110	Modular	1.0	1.5	110-modular, 4 pairs	1.0	12
									1.5	24

Table 10.26. (cont.)

9	White (fiber optic)	Network equipment	36	2×SC	2×SC	0.5	1.0		0.5 1.0	12 24
10	Green	White[a] (Cat. 3)	7	1 pair 110	1 pair 110	0.5	1.5	110–110, 1 pair	1.0 1.5	2 5
11	Green	Blue[a]	3	1 pair 110	1 pair 110	0.5	1.5	110–110, 1 pair	1.0 1.5	0 3
12	Purple (LAN)	Network equipment	135	Modular	Modular	7.6	7.6	Modular, 4 pairs	7.6	135
13	–	–	10	Modular	Modular	1.5	1.5	Modular, 4 pairs	1.5	10

[a]Direct city telephone numbers.

Installation cords with modular connector plugs are needed to connect floor workstations to the LAN. Their length depends on the distance between the mounting cabinet and the cross-field. The maximum length of installation cords is 10 m. Some manufacturers of SCS equipment do not produce installation cords, therefore in this situation, the usual terminated cords with modular connector plugs on both ends are ordered and are then cut in half or one of plugs is cut. In this case, we shall use the last solution, and, based on the assumption of 68 workstations on the first floor (135/2=68), we shall provide 70 standard 7.6-m long cords.

The connection of LAN hubs, located in the equipment room and installed in the mounting cabinet, to the central switchboard will most likely be carried out through electrical ports, because of the short distance covered. It is possible to assume that there will be no more than ten usual or lower-level switching hubs (typical capacity: 12–16 ports) in LAN, even in the foreseeable future. In order to connect them to the central switchboard, we shall include ten terminated 1.5-m long cords with electrical modular connector plugs in the equipment list.

In order to connect the optical ports of the central switchboard to the vertical backbone, optical cords are provided. Their quantity is equal to the total number of optical up-link ports of the LAN network equipment in the cross-room, which is commonly the number of optical cords in cross-room: 36 (12 per floor).

All patch panels and the active network equipment installed in cross is mounted in 19-in cabinets of height 42*U*. The depth of the cabinets is selected depending on the depth of the housing of the active network devices. A ventilating shelf is provided in order to improve cooling conditions. The analysis of Table 10.23 shows that 32 various devices are installed in the cabinet, each of which is fixed in four points. Three packages of M6 bolts and square nuts are used for their fastening. The additional shelf per cabinet is provided for the installation of equipment that does not have fastening elements in the 19-inch construction. The power supply of network devices is carried out from two vertical splitters.

The complete final variant of the excerpt from the specifications in columns 2, 6, and 7 is shown in Table 10.27. The filling of the other columns strongly depends on the particular SCS manufacturer and, consequently, is not given here. The technological equipment is also stipulated in the specification. Single-wire and 5-pair impact tools are used for cable connection to the outlet modules of telecommunications outlets and patch panels. Pentascanner and Certifiber devices are used for testing of cable lines. Both the impact tools and the testing equipment are handed over to the customer after completion of installation, and then this equipment is used during operation of the system, during various tests, during minor repairs, additional cable line layout, and in other similar situations.

Additionally, the nylon hoops of various lengths, which help form the cable bundles and which are consumable materials, are included in the specification. With the help of these hoops, cable bundles and single wires are fastened to the cable ladders during cable layout. The marking of separate cables, cords, and outlets is carried out by the use of adhesive labels. Four tokens are used for each cable (2 for technological and 2 for finishing marking), two tokens are used for the cord. The outlets are marked once. Marking issues are described in more detail in Section 13.1.4.

The general network diagram designed here is shown in Figure 10.30. In order to simplify the visual perception, the patch cords are not shown on the drawing.

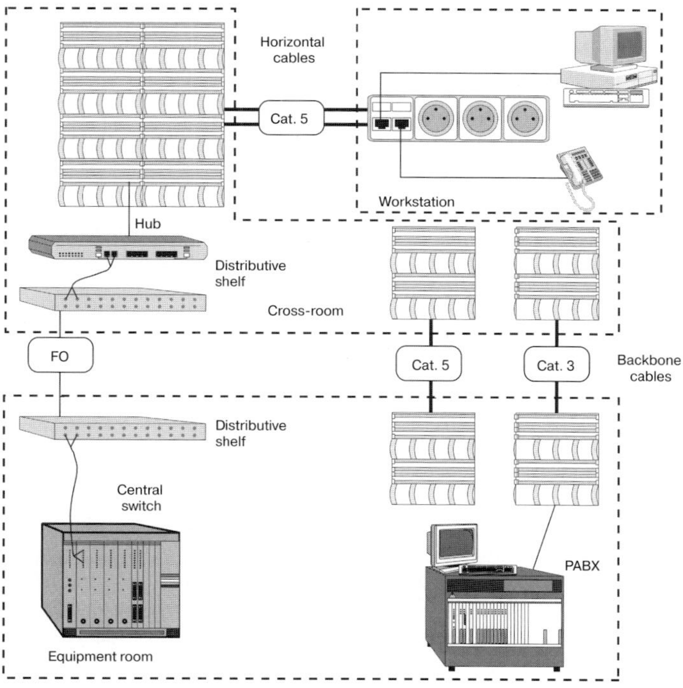

Fig. 10.30. General network diagram

Table 10.27. Excerpt from equipment specification

Denomination and technical specifications	Unit of measure	Quantity
2	6	7
Horizontal subsystem		
Outlet module of category 5, 2-ports, T568B, inclined	Pieces	540
Adapter 45 mm×45 mm (for installation of CT-type outlets in Mozaic frames by Legrand)	Pieces	540
Cable twisted pair of category 5, 4 pairs	m	36,600
Internal backbone subsystem		
Cables of category 5, 25 pairs	m	95
Cables of category 3, 100 pairs	m	143
Optical cables 8×62.5/125 fibers in PVC-sheaths	m	190
Terminated cords in equipment room		
25-paired patch cords, one-side “male”, 7.62 m	Pieces	46
Patch cords of category 5 with RJ45–RJ45 connector, 7.62 m	Pieces	68
Patch cords of category 5 with 110–RJ45 connector, 2.5 m	Pieces	15
Patch cords of category 5 with 110–RJ45 connector, 1.5 m	Pieces	10
Switching equipment		
Cross-panel, type-110, 19-in category 5, 200-paired with organizer	Pieces	8
Cross-tower, type-110, category 5, 400-paired	Pieces	5
Vertical organizer for 400-paired cross-towers	Pieces	3
Cross-tower, type-110, category, 500-paired	Pieces	5
Vertical organizer for 500-paired cross-towers	Pieces	3
Distributive shelf 19-in 1U-SC-MM-16	Pieces	12
Horizontal organizer, 1U	Pieces	12
Installation cord, multimode, 1 m (62.5/125)	Pieces	144
Protective bush sleeve	Pieces	144
Organizer for 16 fibers for patch shelves 19-in (splice-plate)	Pieces	12
Patch cords in technical apartments		
Patch cord, category 5 with 110–110 connectors, 1 pair, 0.5 m	Pieces	126
Patch cord, category 5 with 110–110 connectors, 1 pair, 1.0 m	Pieces	257
Patch cord, category 5 with 110–110 connectors, 1 pair, 1.5 m	Pieces	326
Patch cord, category 5 with 110–110 connectors, 1 pair, 2.0 m	Pieces	186
Patch cord, category 5 with 110–110 connectors, 1 pair, 2.5 m	Pieces	60
Patch cord, category 5 with 110–RJ45 connectors, 0.5 m	Pieces	114
Patch cord, category 5 with 110–RJ45 connectors, 1.0 m	Pieces	246

Table 10.27. (cont.)

Patch cord, category 5 with 110–RJ45 connectors, 1.5 m	Pieces	171
Patch cord, category 5 with 110–RJ45 connectors, 2.0 m	Pieces	66
Patch cord, category 5 with 110–RJ45 connectors, 2.5 m	Pieces	15
Optical cord with SC connectors, 62.5/125, duplex, 0.5 m	Pieces	12
Optical cord with SC connectors, 62.5/125, duplex, 1.0 m	Pieces	60
Terminated cords on workstations		
Patch cord, category 5 with 110–RJ45 connectors, 2.13 m	Pieces	140
Patch cord, category 5 with 110–RJ45 connectors, 4.57 m	Pieces	20
Installation equipment		
Floor cabinet 42U, 2033×600×600 mm	Pieces	4
Ventilating module 600 Series (installation from above) 2 high performance fans	Pieces	4
Grounding kit	Pieces	4
Legs for mounting cabinet 19-in (4 pieces kit)	Pieces	4
Perforated shelf for 19-in equipment L=454 mm	Pieces	4
Power-supply outlets for cabinets, vertical, 8 outlets	Pieces	8
Screw with washer and nut for 19-in equipment, 50 pieces	Pieces	12
Working tools for SCS		
Impact tool for 1 conductor (without blade)	Piece	1
Detachable blade for impact tool for 1 conductor, 110-type	Piece	1
Impact tool for 5 pairs, 110 type	Piece	1
Pentascanner+ with 2-way injector+ appliance	Piece	1
CERTIFIBER appliance	Piece	1
Power supply cable system		
Outlet 2P+T “ Mozaic” French style (for uninterruptible power supply)	Pieces	1080
Outlet 2P+T “Mozaic” German style (for household power supply network)	Pieces	540
Ancillary materials and equipment		
Nylon hoop for cable, 250 mm, package 100 pieces	Pieces	7
Nylon hoop for cable, 340 mm, package 100 pieces	Pieces	4
Nylon hoop for cable, 550 mm, package 100 pieces	Pieces	5
Tokens for marking on workstations, sheet 290 tokens	Pieces	5
Tokens for horizontal cables marking, sheet 64 tokens	Pieces	68
Tokens for patch cords marking, sheet 49 tokens	Pieces	69
Tokens for multipaired cables marking, sheet 21 tokens	Pieces	5

10.5 Conclusions

The SCS design procedure is a difficult multistage process and consists of two fundamental stages: architectural design and telecommunications design.

The main task of the first stage is the preparation of technical rooms and also of cable routes for horizontal and backbone subsystems to works on the SCS installation.

The requirements set by the standards for the cross-rooms and equipment rooms unequivocally allow designers to determine both their space and the environmental conditions inside and outside them, which, in turn, enables designers to formulate the requirements for the building engineering systems. The process of technical room design is facilitated, in many respects, by the unity of the requirements for the basic parameters of cross-rooms and equipment rooms, with slightly more rigid requirements for some characteristics of equipment rooms.

Depending on the architectural features of a building, various versions of in-floor and above-ceiling cable channels and vertical risers can be used, and separate versions of cable routes can be combined in a wide range.

The quantity calculation of separate components of particular SCS realization is carried out at the telecommunications design stage. It is expedient to carry out the calculation procedure using “from the particular to the general” principle, beginning from workstations, according to the model of hierarchical star structure of cable systems. The factor that determines the quantity of separate SCS components is the selected workstation configuration and the predetermined administration principle (centralized or multipoint). The equipment structure of technical apartments strongly depends on the chosen method of patch panel location (on a wall, in 19-in construction, or in accordance with the combined scheme).

The process of component calculation at the telecommunications stage can be of an iterative character. In order to simplify the transition from one stage to another and the preparation of the final equipment specifications, it is recommended that the calculated results of separate SCS subsystems be presented in tabular form by the use of computer technology.

11 SCS Installation

11.1 Organization

11.1.1 Structure and Provision of Installer Crews

The work of SCS installation should be performed by qualified specialists under the direction of supervisor who was trained at an authorized educational center. The installers are integrated into teams; some teams form a separate crew.

The crew can specialize in separate types of work or carry out all cycles of activities on a project. The latter is usually used during the installation of small SCS. During large projects, it is recommended to use some specialized crews. This provides more effective organization of the installation process, testing, and entering into operation.

The installation team, in most cases, consists of two people and can carry out cable layout, set up of mounting cabinets, and equipment installation, as well as almost all types of electrical and optical cable route testing. Usually, the individual installers carry out the layout of electrical cross-equipment layout outlet modules of telecommunications outlets.

Each installer crew should have a complete set of process equipment. The approved and coordinated technical design documentation is given to the supervisor before the beginning of the work.

The work clothes of the installer should be lightweight and durable, and fit closely to body to prevent difficulties during activity. The semicoverall or boilersuit, to which a vest is sometimes added, is often used as the work clothes. In order to hold instruments, documents, pens, markers, and other objects, it is necessary to have pockets with buttons, zip fasteners, or fastening tapes on the work clothes. For advertising purposes, a logo or the name of the company that is performing the SCS installation is sometimes put on the work clothes. It is convenient to hold some types of instruments, devices, and production materials in a special bag on a belt with some small pockets during the installation.

11.1.2 Technical Design Documentation

The installers should be acquainted and supplied with the technical design of the cable system as well as the technical design documentation before the beginning the installation. The technical design documentation includes the information necessary for the installers to perform and organize the installation process. In particular, the technical design documentation should contain the following:

- SCS structural scheme;
- Plans of cable conduit runs and/or other external backbone cable routes and cable channels (risers, ladders, etc.) inside a building;
- Plans of cable layouts with the indication of TO location points;
- Plans and schemes of location of equipment in telecommunications closets and equipment rooms;
- Equipment location scheme in mounting cabinets and racks;
- Tables of junctions and connections of backbones and horizontal cables;
- Cable register for documenting changes and fixing aberrations from the technical design documentation;
- Schedule for the project.

All identification labels and references that are necessary to bind the scheme components to other drawing sets and to objects are put on the structural scheme. The information on the characteristics of all cable system elements, including type, number, and capacity of backbone and horizontal cables, as well as their identification labels should be in the tables of junctions and connections and/or in the structural scheme.

11.1.3 Stages and Duration of Works Fulfillment

The basic criteria during the scheduling and the beginning of the performance of various stages are the building readiness conditions given in Table 11.1.

Table 11.1. Building readiness conditions for installation

	Task	Terms for completion
1.	Install cable channels and outlet boxes in floors	Before clear floor filling
2.	Make feed-through holes in corridor walls	Before the beginning of installation of ventilation system airways
3.	Install suspended consoles of backbone cable system ladders in false ceilings	Before the beginning of installation of ventilation system airways
4.	Install shelves and cable ladders on suspended consoles	Upon the completion of installation of ventilation system airways
5.	Install decorative conduits in building rooms	Upon completion of work
6.	Install backbone and horizontal cables	Upon completion of installation of cable channels, before installation of false ceilings
7.	Connect horizontal cables connection to the outlet modules of the telecommunications outlets	Upon completion of cable layout

Table 11.1. (cont.)

8.	Install mounting cabinet	Upon completion of work in telecommunications closets and equipment rooms
9.	Connect backbone and horizontal cables to the switching equipment	Upon completion of cable layout and assembly of mounting cabinets
10.	Test SCS lines and paths	Upon completion of installation
11.	Commission tests	Upon completion of SCS line testing

It is possible to use the typical parameters of work productivity subject to the absence of furniture, the presence of correct technological marking, proper and complete technical design documentation, standard organization of production as a reference point during the calculation of timing of the fulfillment of separate work stages and during schedule preparation (Table 11.2).

Table 11.2. Approximate productivity of SCS installers for some frequently performed operations

Type of work	Performed by	Performance in 8-h workday
Installation of raceways in heights up to 1.5 m	Team of installers	75 m (with corners)
Installation of raceways in heights over 1.5 m	Team of installers	40 m (with corners)
Installation of horizontal cables in heights up to 1.5 m	Team of installers	1000 m
Installation of horizontal cables in heights over 1.5 m	Team of installers	700 m
Installation of backbone cables	Team of installers	200 m
Installation of outlet modules and connection of horizontal cables to the telecommunications outlets	Installer	50 outlets
Connection of horizontal cables to telecommunications equipment	Installer	600 pairs
Connection of backbone cables to telecommunications equipment	Installer	900 pairs
Fiber welding, optical shelves, and enclosures	Team of installers	30 fibers
SCS line testing	Team of installers	100 electrical cable lines 30–50 fiber-optic cables

During the estimation of general duration of work to realize a large project (some hundreds of outlets or more), it is possible to be guided by the average empirical integrated productivity of the work of one installer of 1–2 SCS ports per working day.

11.1.4 Other Conditions for Installation

The customer representative gives instructions for safety before the beginning of SCS installation. The documentation of this instruction realization is given by registration of a report that is signed by the instructor and the installers who listened to these instructions and are consequently responsible for maintaining the safety techniques.

Almost all types of work will be performed at temperatures above 10°C. The temperature conditions for the tasks can strongly depend on the characteristics of the installed equipment, and the temperature can change to a greater or lesser extent. If the necessary temperature conditions cannot be supplied by means of central heating, then it is possible to use mobile and portable heat generators under observance of the appropriate fire safety requirements.

11.2 Inspection Test of SCS Components

11.2.1 Inspection of Electrical Cables and Other Electrical Components

The inspection of SCS electrical components is performed basically by the use of visual surveys; measuring is seldom used.

During inspection and testing of cables, it is necessary to:

1. Be convinced that the cable length on a drum corresponds to the value given on its packing. Markers of lengths on cable ends are used for this purpose. If the values of cable lengths fail to coincide, then it is necessary to correct the value on the cable drum. It is not recommended to spoil the order of the coils of a cable in a box while searching for the cables ends, since this makes its subsequent unreeling complicated or even impossible.
2. Check the conformity of the color coding of the cable pairs according to the IEC 708 standard (Section 3.1.5).
3. Check for the absence of failures in the packing and external cable sheath.

During inspection of other components it is necessary to be convinced of their completeness and integrity, of the absence of mechanical damage, and also absent of evidence of high temperatures and corrosive chemical substances.

Selective instrumental checking of cable electrical parameters is supposed. This will be performed in the factory packing, on coils or in boxes. If the results of the measurements of any parameter fail to correspond to the values required by the standards, then the cable should be unpacked, spread out on a floor by free coils, and the measurements repeated. If the measured electrical parameters still do not conform to the norms, then the cable is not suitable for use in SCS and should be rejected as defective.

11.2.2 Inspection of Fiber-optic Cables and Other Optical Components

The inspection of optical cables and fiber-optic components is performed by the visual supervision method and by tool testing. The visual survey is performed by rules similar to those for electric elements described above.

During the performance of tool testing, the absence of fiber breaks is determined and the attenuation is measured. At presence of If the installers use an optical reflectometer, the optical cable length can be determined and the absence of internal defects can be additionally controlled. The results of the measurements are necessarily compared to ratings data.

The elementary test for the absence of fiber breaks in short multimode cords and cables (1–1.5 kilometers) is performed by the usual lighting of separate fibers. A bright halogen lamp is used as a light source (it is possible to use pocket flashlight); it is also possible to use devices with lasers of red light (laser pointers or so-called visualizers of fiber defects, considered further in Section 12.3.4). During work performed in open air, the use of solar light use is allowed. In order to provide reliable input and output of light flux with small losses, the preparation of fibers face surfaces with the help of a cleaver and their termination by adapters on an exposed fiber can be necessary. The use of so-called operative connection devices is also possible [7].

The lighting method can be also used for single-mode cables and cords with the certain reservations. However, because of the small diameter of the fiber core, the supervision of the light flux on the output end should be made with the help of a control microscope. During checking of cables, a precision cleaver must be used for the preparation of fiber end faces.

The procedure to determine the common attenuation of the optical cable and other fiber-optic components as well as using a reflectometer to probe the optical cable is part of the tool testing methods at the inspection test. These procedures are described in Section 12.3 in more detail.

11.3 Construction of SCS Backbone Subsystems

In this section the general organizational and technological issues of the construction of SCS backbone subsystems are considered. Some additional information on this subject can also be found in the review [126].

11.3.1 Cables Layout in Conduit Runs

The layout of electrical and optical cables in conduit runs is performed by identical rules. The differences are mainly the allowable stretching and twisting. Optical cables, in comparison with electrical cables, have a slightly smaller diameter and lower weight, however, they require larger bend radii during their layout and operation.

The optical cable layout is usually performed in the free channel. Sometimes, some smaller diameter polyethylene pipes are installed in each free channel beforehand; these pipes form a subchannel system. One optical cable can be installed in each pipe. During layout of multipair electrical cables, it is necessary to control the value of the channel pipe filling (Table 10.7), which keeps stretching within allowable limits.

During construction the wire for stretching is left in the channels. If it is absent, it is most convenient to carry out the installation with the help of a channel preparation device. It is a round cassette by with diameter of about 1 m on which an elastic glass fiber-reinforced plastic stick with diameter of about 10 mm and length up to 150 m is reeled. The cassette should be placed at the well and the stick should then be pushed in the channel so long as its end appears in the adjacent well. A wire or rope is attached on the stick tip or sometimes directly on the cable end, then this stick is stretched back. For stick fastening on cable ends, it is recommended to use a special tip, which is fixed on the cable by its load-bearing element and shielding cover, and which can be supplied with a twisting equalizer. In the absence of a special tip, fastening should be performed by adhesive tape in two points located 15–20 cm from each other. In order to minimize the efforts of stretching, the cable should be necessarily drawn to the stick or rope at the forward point of fastening.

During cable stretching into the conduit run, the cable drum can be placed on a device for unreeling. These devices can be realized under two various schemes (Figure 11.1).

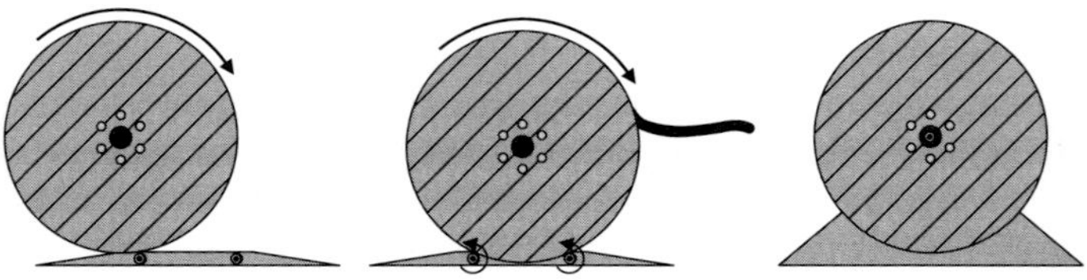

Fig. 11.1. Different types of unreeling devices

According to the first, the cable drum is moved with each side on two support rollers, and during unreeling the cable drum spins on them. The most popular approach is the second, in which the cable drum rises above a surface by some cm by means of a mechanical or hydraulic lifting jack, which allows it to rotate freely on its axis. During cable installation the drum rotates by hand. The installation (stretching) itself should be performed smoothly and without jerks. In the absence of this device for stretching, the cable is spread near the well on a plain surface in the form of loops or figure eights. The cable layout can be made both from the initial point, and from the middle of the route. The last method allows installation of routes of greater length without the use of intermediate couplings, however, it is a little bit more difficult to realize and results in a great deal of cable waste.

In those cases when the cable layout route has sharp turns, a slewing roller is installed in the well. In its absence, the cable is stretched from this well by loops and the further layout is performed the same as from initial point of the route. In order to decrease the construction time and in the presence of appropriate conditions, cable is frequently installed by hand directly in the well, by directing it into the conduit run.

11.3.2. Splicing of Construction Cable Lengths of External Layouts

Separate cable segments of external layouts are spliced in the following situations:

- cable layout by one construction length between the initial and final points of a route is impossible or is inexpedient because of local features (long distance, presence of numerous turns, etc.);
- local features of the installed system allow the use of splitting of one large-capacity cable into two or more cables of smaller capacities in order to minimize the cost and duration of construction;
- repair of out-of-order cables or scheduled replacement of one segment by another is performed.

The coupling of two or more (in the case of branching) construction lengths in the middle of a cable route is protected by an enclosure. This element provides fusion splicing of the constructions (in majority of cases, one-piece), while observing the technological limits on their bend radii, stretching parameters, etc. It also restores the integrity of the cables and provides a protective cover for the cables, as well as tight splicing.

The process of installation of the enclosure includes the following basic technological operations:

- separate thread or cable fiber splicing with the subsequent protection of the splice;
- layout of the threads or fibers in the coupling housing or its organizer (in the case of optical cables, the calculation of technological reserves of fiber length with observance of given bend radius and fixing protective bush sleeve or mechanical splice housing are additionally made);
- assembly of the enclosure;
- restoration of the cable integrity and strengthening of the cable elements with the subsequent or simultaneous sealing of the coupling housing;
- installation of the enclosure in a well, collector, or similar place with observance of the given cable bend radius during installation and operation.

The process of enclosure installation strongly depends on the selected technology of wire or fiber splicing, the coupling type (through-passage, splitting, terminal, or straight, see [7]), the construction of its housing, and the method of its sealing (cold, hot, with sealing tapes, etc.). Therefore, a detailed description of installation methods is obviously not possible, and it is necessary to use the companies' installation manuals in each specific case.

11.3.3 Installation of Optical Shelves and Wall Coupling

We shall consider process of installation of optical shelves (wall couplings are installed similarly). This procedure includes a number of separate technological operations, which are performed in the following order:

1. place the shelf housing on the work table, remove the top cover, and, in case of welding or mechanical splices, mount the organizer into the housing;
2. prepare an optical cable, having removed the external protective and strengthening covering. The length of the prepared cable is approximately 1 m in for installation cords and 1.5 m for the use of glue technology or for direct termination by other types of plugs. The label of cable technological marking is transferred to the rest of sheath or is replaced with label-finishing marking.
3. to install a cable into the housing of the shelf and to fix in the state holder, the fibers should be accurately put aside. Because they are more rigid, external layout cables should also be fixed on the cover of the work table by clamp or a plastic hoop;
4. reinforce the fibers by optical connector plugs; use a reflectometer to check for the absence of close breaks;
5. starting with the fiber with largest number, consecutively install the fibers in the self housing. In order to observe the bend radius, fiber organizers should be used and protective bushing sleeves of welded splices or mechanical splice housings should be installed in the organizer. The plug is connected to the outlet;
6. close the cover of the shelf and mount its housing into the 19-in construction;
7. the cable length reserve with observance of the given bend radius should be rolled in a coil, then attached to it by means of hoops and placed in the side cavity of the housing, under the false floor, or in the false ceiling. If necessary, the cable coil and cable are also attached inside with the help of plastic hoops or other fixing elements;
8. light all fibers with a reflectometer and check them with an optical tester, results of are entered in the protocol;
9. carry out the required connections by patch cords according to the technical design documentation;
10. unused optical connector outlets should be closed with protective caps;

The wall enclosure is installed similarly, except that the cable reserve is usually fixed on a wall with the help of dowels and plastic hoops, and the cable at the inlet of the enclosure is supported by special cable holders (Figure 11.2).

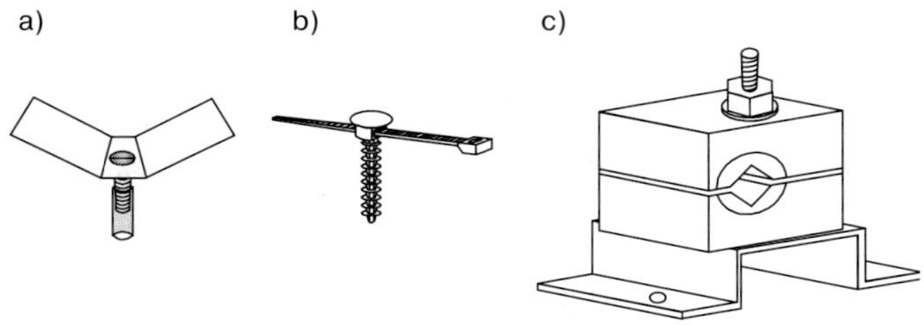

Fig. 11.2. Different types of cable holders: *a)* tin strip; *b)* dowel; *c)* cable holder.

11.4 Layout of Symmetrical or Fiber-optic Cables Inside Buildings

Symmetrical 4-pair cables, which are used for horizontal subsystem organization, are delivered in boxes or on drums. In the latter case an unreeling device is required for cable layout. The drum is suspended on the device axis or moved on its rollers and is unreeled according to length of the stretched cable.

The cable is more often packed into a cardboard box, in which the usual drum or a self-bearing coil is located (Section 3.1.2.9). In the process of pulling a cable from the box, the turns of the coils and the cable are unreeled uniformly and without loops. Use of this type of packing makes it impossible to jerk the cable while pulling, which would form a loop in front of the box inlet, making further work impossible.

It is possible to install the cables for several outlet modules simultaneously, using more than one box or drum. For this purpose the drums are suspended on special unreeling device, and the boxes are placed in a line or against each other. In this case, cable layout should begin on the workstation side. Therefore, the cables in the telecommunications closet should have the same termination lengths, and in the working rooms the cables are installed to the various outlets. It is possible to pull out the necessary number of cable from each packing for every workstation upon completion of cable layout to the telecommunications closet.

In most cases, it is expedient to begin performing the cable layout with the workstations that are most remote from the telecommunications closet. This allows equal and uniform cable installation in cable channels, without "puff-up" at branching points to the outlets. Furthermore, after laying long pieces of cable, the remaining cable on the drum is used for shorter sections.

Before the beginning of cable layout it is necessary to ensure that the remaining cable on each drum will suffice for cable installation from the outlet to the telecommunications equipment in the telecommunications closet with sufficient reserves for cable preparation.

It is necessary to take into account that the installation of the majority of cables that are designed for indoor operation cannot be performed at temperatures below 0°C. Therefore, in winter the cable must be warmed up to room temperature after delivery. Hence, it is expedient to deliver cables to a project in the evening, which allows installers to begin the cable layout in the morning.

In order to identify the cables at workstations and in technical rooms, they are labeled. These labels are frequently small strips of paper, fixed on the cable sheath by transparent adhesive tape. The application of so-called self-laminating labels is more convenient, in these labels the markings and the fixing tape are incorporated in a single unit (Section 13.1.4). In the latter case, marking information can be written not only manually on the label, but also during preparation by a laser printer. The marking information can be of any form, however, it should enable unequivocal determination of the second cable termination location. An example of markings is 2-202-04: second floor, room 202, and cable number 4. The markings form so-called identifiers, widely used during cable system administration (Section 13.1.1).

The cable terminations can be pulled together by adhesive tape and all cable bundles can be installed at once. In order to increase the convenience of stretching, it is possible to mount a wire loop on the cable bundle. The durability of loop fixing is increased if the loop ends of the adhesive tape are bent. In addition to this or similar improvised tools, it is possible to take advantage of stretching tips and factory-made caps with terminal loops or ring bolts. The last variant is preferable, as it eliminates cable twisting during layout.

Cable layout in ladders, raceways, and in open channels in floors is frequently performed by the usual cable installation methods in these channels. Stretching is used in order to carry out cable layout in ducts and in-floor channels.

Stretching can be done using the following:

- pieces of steel wire or plastic sticks with lengths up to 10 m;
- pieces of 10-pair or 20-pair telephone wires of the same length;
- special devices for channel preparation, DCP.

The DCP delivery set also consists of various types of tips and connectors. The cassette is usually installed in a body, supplied with rollers and handles for facilitation of moving. It differs from the DCP for use on external backbone subsystem routes, mainly by the stick diameter (approximately 5 mm), and by the diameter of the cassette (40 cm). Cassettes 30-m long can perform as drums with handles for carrying and are supplied with returnable springs, which pull the sticks into the barrel housings after the pawls are released.

During cable layout it is necessary to observe the following rules:

- The limits for maximum cable stretching must not be exceeded, see Technical Terms. If information on the stretching limit is absent, it is taken as equal to 110 N;
- There must not be more than 1.5 m between the fastening points for cable suspension and the cable should also be appreciably sagged (strong tension is not allowed). It is necessary to avoid squeezing the cable too tightly by the fixing clamps;

- The minimum bend radius of electrical cables should be not less than 16 external diameters of the cable during layout and 8 external diameters of cable after completion of layout. The optical cable minimum bend radius is not less 20 external diameters during layout and 10 external diameters after completion of layout;
- Cable bends can be no more than 90°;
- Do not overtwist a cable with regard to its longitudinal axis;
- In case of damage to the cable sheath, it is necessarily to replace it with a new one;
- The cable layout routes should be selected with regard to the location of strong electromagnetic field sources. For example, it is necessary to avoid installation near electrical motors, transformers, luminescent lighting instruments, and main power supply cables;
- The filling of cable ladders, conduits, inlets, and ducts should not exceed 60–70% (Table 10.7).

In case of absence of cable routes based on pipes, wells, and similar specialized elements, the routes are formed with the help of cable hoops, clamps, and staplers, and also by gluing separate cables on plain surfaces. These methods are considered in detail in [127]. A reserve for cable preparation from both terminations is provided. For electrical cables on the outlet side this reserve is equal to 20–30 cm. On the telecommunications closet side, the length of all cables should be sufficient for cable layout to the telecommunications equipment block, the most remote place from the cable inlet point, plus 30–40 cm for their preparation.

Upon completion of cable layout it is necessary to put markings on both terminations of each cable according to the tables of connections (junctions). The marking elements are attached about 10–15 cm from the cable termination. The markings should be identical on both terminations.

The fiber-optic cables of the internal layout are delivered on drums and their weights are close to those of 4-pair symmetrical horizontal cables. Therefore, the methods of their layout are rather similar. The basic differences are the following:

- During optical cable layout it is necessary to observe the minimum bend radius more carefully;
- The cable reserve for performance of cable preparation procedures in the telecommunications closets should be a minimum of 1.5 m (in practice, this value is frequently increased to 3–5 m). The type of optical connector plugs used in the installation process also influences the value of the cable reserve. Thus, for example, for cables of the internal layout, a reserve length of 1 m is quite sufficient for welding, whereas for the use of gluing, the cable reserve is increased at minimum by 0.5 m.

When the cable layout is complete and the cable is cut, it is necessary to note the remaining cable length on the boxes and the drums. For this purpose it is necessary to use length markings on the cable sheaths.

11.5 Installation of Raceways in Working Rooms and Outlets in User Workstations

The installation of raceways and telecommunications outlets of various types is performed with fixing elements that are selected on the basis of the wall material (Table 11.3 and Figure 11.3).

Table 11.3. Types of fixing elements

Wall material	Fastening element type
Concrete	Nylon dowel, screw for concrete, anchor wedge
Red and lime-sand brick	Nylon dowel
Cavity brick	Dowel for cavity brick
Gypsum concrete and plaster	Jet-plug
Gypsum carton, gypsum lath	Multifunctional dowel
Gypsum	Nylon dowel
Nature stone	Nylon dowel
Aerated concrete	Dowel for cavity brick

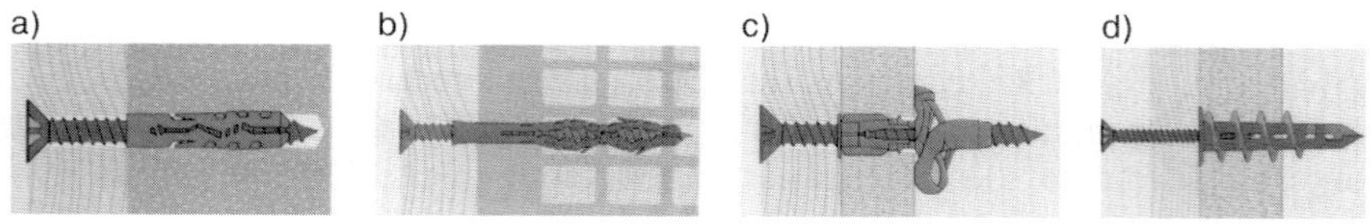

Fig. 11.3. Types of fixing elements: *a)* nylon dowel; *b)* dowel for cavity brick; *c)* multifunctional dowel; *d)* jet-plug.

The installation of raceways is performed in the following order:

1. Select fixing elements of a suitable type and dimension, depending on the surface material.
2. Draw a line for raceway installation according to the installation plan. Raceways are installed parallel to the floors or other building features by visual inspection or by the use of a bubble tube (level).
3. Note fastening points: it is necessary to provide 8 or more points in 2 m of length for raceways whose areas are not less than 100×50 mm; fastening is made in not less than 5 points in 2 m of length for smaller dimension raceways; the fastening points are located in staggered rows, only if the raceway construction does not provide integral fixing holes.
4. To apply the raceway on a surface, draw points of fastening and drill the holes for installation of fastening elements. The diameter and depth of the holes are selected based on the surface material and the type of fixing element.
5. Mount the fastening elements in the drilled holes (dowel, jet-plug, etc.).
6. Tighten screws into the holes in the fastening points, having applied the raceway to the surface according to the drawn line.

The installation of wall-mounted telecommunications outlet housings and other elements that are installed by the "along a profile" principle is performed similarly.

The standard length of a raceway section is 2 m. If it is necessary to work with shorter pieces, sections of the appropriate lengths are cut. The use of special scissors is strongly recommended, which, in contrast to a hacksaw, allow installers to achieve clean and exact cuts at practically any angle to the raceway axis.

11.6 Connection of Electrical and Optical Cables to Telecommunications Outlets

The connection of electrical and optical cables to the outlet modules of telecommunication outlets and patch panels is performed by similar rules and includes a number of stages (Figure 11.4). The differences arise mainly because of horizontal and multipair design features and also because of design features of electrical and optical cables, and are not of a fundamental nature. The general requirements for the performance of these procedures are accuracy and careful visual and instrument checking of results.

The final stage is always the installation of markings according to the technical design documentation.

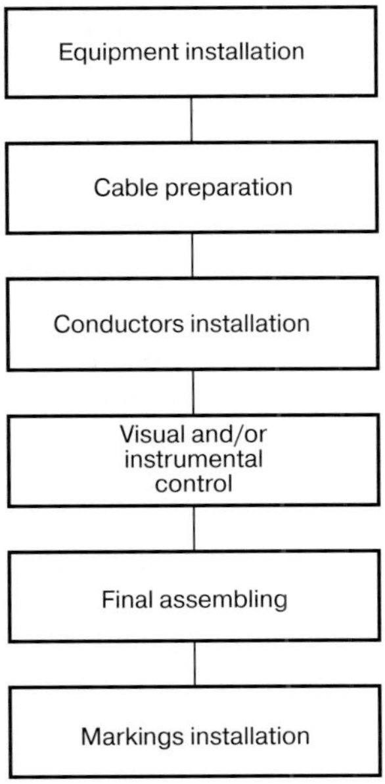

Fig. 11.4. Stages of the connection of electrical and optical cables to the outlet modules

11.6.1 Connection of Twisted Pairs to Outlets

The connection of horizontal cables to telecommunications outlet modules strongly depends on the housing construction and the installation principles of the separate conductors in the IDC contacts of the modules. In each specific case connection is made according to the instructions and requirements of the company. Therefore, we shall discuss here only installation by the single-wire impact tool, which is the most widely used in practice and is performed as follows:

1. Stretch cables through an aperture for the outlet module within the framework of the fixing mechanism.
2. Fasten the framework of the fixing mechanism.
3. Cut the cable surplus and remove the external sheath from it for about 4 cm. Check the insulation condition of separate conductors. In case of damage, it is necessary to perform the cable preparation again using the cable reserve, having first cut the conductors ahead of the damaged place.
4. If the construction of the outlet module provides cable attachment, then it must be fastened by a plastic clamp or other element, so that the end of the sheath is as close as possible to the IDC contacts of the outlet module. If the cable attachment elements are absent, then during connection the cable should be held in the appropriate position. In order to preserve the cable core structure, the sheath should be removed for a length such that its edge is located directly at the cut of the outlet module housing. This rule can be broken only in the case of cable preparation in small-dimension raceways, when the sheath impedes bending a cable for installation to the IDC contacts.
5. Install cable conductors in notches of IDC contacts, observing conformity between the color coding of cable conductors and the contacts. Thus, it is necessary to ensure that the twisted-pair conductors should be kept at the contacts. For this purpose it is possible to expand a clearance between them in one of the turns without unreeling the twisted pair conductors and thus install divided conductors in the notches.
6. Make sure that the installation order of the conductors in notches is correct.
7. Install the conductors between the IDC contacts using an impact tool for one conductor. The tool simultaneously cuts the extra length of the conductor, which extends beyond the contacts. Therefore, it is necessary to keep the orientation of cutting edge up during performance of this operation.
8. Perform the above-described procedures on all cable connections in the case of multiport outlets.
9. Install outlet modules in apertures of the fixing mechanism framework and mark the outlet modules of the telecommunications outlets.

If necessary, this process can be adapted to other designs without problems.

11.6.2 Connection of Fiber-optic Cables to Outlets

The connection of fiber-optic cables to the outlets is performed as follows:

1. Stretch the cable through a hole of the fixing box or the outlet housing.
2. Install the fixing box or outlet on the workstation.
3. Attach a cable on the housing inlet by integral fixing elements or plastic hoops. The length of the idle cable piece should be approximately 1 m.
4. Remove the external sheath of the cable to the fastening point, cut the strengthening Kevlar strings at the sheath edge. Sometimes they are cut with 3–4 cm reserve and bent back under the fixing hoop. In this case, it is first necessary to carry out operation 4, and then 3.
5. Reinforce the fibers by demountable optical connector plugs.
6. Check for the absence of close interruptions with a reflectometer.
7. Lay a technological fiber reserve in the form of rings in the housing with observance of the minimum bend radius. For this purpose, it is possible to use an internal organizer (if one is present).
8. Connect the plugs to the demountable optical connector outlets and close the housing.
9. Check serviceability of the fiber by a reflectometer. If the second cable termination is already prepared in the optical shelf or wall coupling, it is necessary to measure the overall line attenuation on the working wave-lengths with an optical tester and the appropriate protocol.

11.7 Equipment Installation in Technical Rooms

11.7.1 Organization of Installation of Equipment and SCS Elements

During this stage, it is necessary to place equipment in the technical rooms according to the plans developed during preparation of the technical design documentation. Beforehand, it is necessary to prepare cable channels for cable layout, to power up the mounting cabinets and racks, and also to perform other preparatory operations.

The installation of SCS equipment in the mounting construction is performed according to the plans developed during preparation of the technical design documentation. The equipment is fastened with the help of square nuts installed in holes on mounting rails, and screws. The installed equipment is marked according to numbers of components on the location scheme.

11.7.2 Connection of Symmetrical Cables to Telecommunications Equipment Components

In process of connecting cables to the telecommunications equipment, it is necessary to compare the table of connections prepared at the technical design documentation development stage and actual cable locations. If differences are found, it is necessary at once to amend the tables of connections. They are part of the operational documentation.

Connection of horizontal cables to 110-type cross-blocks

The connection of symmetrical cables to 110-type cross-blocks is performed as follows:

1. Install the 110-type cross-blocks on a base.
2. Divide the cables into groups, 6 pieces per group, according to the table of connections of the technical design documentation.
3. Lay the generated groups of cables and attach them to the base by the use of plastic hoops. The cables are installed to the cross-blocks from the back: from the side at the base of the cross-towers or from the mounting clamp at the installation of the cross-blocks in the 19-in mounting cabinets. It is first necessary to install the groups of cables that will be cut on the blocks that are the most remote from the inlet.
4. There are several holes on the cross-blocks between contact strips. The group of six cables can be easily put through one such hole, and three cables from it are prepared on contacts of the bottom contact strip, whereas the other three are prepared on contacts of the top contact strip. Starting with the cross-block that is the most remote from the inlet, move the cables to the appropriate holes. To move the cable markings so that it should be ahead of hole. To accurately remove an external sheath from cables with the help of cutting tool, so that its cut should be at the same level as a hole edge.
5. The cable preparation begins with the top contact strip and the most remote notches from the inlet hole.
6. For each cable, the pairs are installed in notches from left to right in the following order: blue, orange, green, brown, thus the white conductor is installed in the left contact notch. One of necessary conditions of organization of category 5 characteristics is the preservation of factory twisting of pair conductors whenever possible down to notches. For this purpose it is possible to expand a clearance between them in one of the turns and to install conductors in notches without unreeling the conductors of a pair.
7. Visually check correctness of installation of conductors in the notches.
8. After filling all notch routes with the conductors, press them to an operating position using the impact tool for five pairs with the cap and the knives in the cutting position. The tool simultaneously cuts excess conductors that flare over the notches. During performance of this operation it is necessary to keep the direction of the cutting edges up. It is recommended to incline the impact tool 10–15° from vertical for more precise cutting of conductors.
9. Install the connector blocks on the contact strips, remembering that their color markings must coincide with the color markings of the pairs prepared in the notches. Six connector blocks are installed on each contact strip: to the left— five 4-pair to the right — a 5-pair. The installation is made with the use of an

impact tool for 5-pairs. It is first necessary to turn the working cap of the tool into position for installation of connector blocks (knives inwards).

10. Insert the marking strips with cable identification numbers in the holders and install the holders in the cross-blocks.
11. Carry put other operations stipulated by the company's instructions for this specific type of telecommunications equipment.

Connection of multipair cables to 110-type cross-blocks

The connection of multipair cables to the 110-type cross-blocks is, in many respects, the same as the connection of 4-pair cables, with minor differences caused by differences in their construction. The process of connection is performed as follows:

1. Install the 110-type cross-blocks on the base.
2. Lay cables and attach them to the base by plastic hoops. The cables are brought to the cross-blocks from the back: on the side of the base at the cross-towers; from the mounting clamp at the installation of the cross -blocks in the 19-in mounting cabinet. The cables that will be cut on the most remote blocks from the inlet are installed first.
3. The connection procedure begins with the cross-block that is farther from point of inlet.
4. By the use of special cutting instrument, remove external cable sheath at the height of the cross-block to which the connection is made. Move cable markings on the rest of the external sheath. Separate 25-pair cable bundles from each other and fasten pairs of bundles near the end of external sheath by adhesive tape. Keep about 3 cm of identification tape, having cut the rest.
5. Stretch the paired bundles of cables through the feed-through cavities (at the front of the cross-block):

First bundle (blue and white marking tape)	top left cavity
Second bundle (orange and white tape)	top right cavity
Third bundle (green and white tape)	bottom left cavity
First bundle (brown and white tape)	bottom right cavity

6. It is necessary to connection the paired bundles from the top contact strip. The pairs of bundles from the left cavity are connected to the top contact strip; those from the right cavity are connected to the bottom contact strip.
7. For each bundle, it is necessary to begin the connection with the pair farthest from the inlet. The pairs are installed in notches from left to right, according to the recorded pair numbers. Hence, the connection begins with 25-pair (violet/gray) on the top contact strip, and on the bottom contact strip with the first pair (white/blue). In each pair, the conductor with the group color is installed on the left notch, and the conductor with the color of the pair in the group (Section 3.1.3)is installed on the right. It is recommended to incline the impact tool 10–15° from vertical for more precise cutting of conductors.
8. During installation of conductor pairs in the notches, it is necessary to ensure preservation of their factory twisting up to the notches. For this purpose, it is

possible to expand a clearance between them in one of the turns and to install conductors in notches, without unreeling the conductors of a pair.
9. Make sure of the correctness of the installation of conductors in the notches.
10. Press conductors into notches of the contact strip, using the impact tool for 5-pairs to press. The tool simultaneously cuts the excess conductor that flares over the notches. During performance of this operation it is necessary to keep the direction of cutting edges up.
11. Install the connector blocks on the contact strips, remembering that the blue markings at the block end are to be at the left side, and according to the chosen modularity of information channel transmissions, the last pair should be necessarily closed. The most used variants, in practice, are five 4-pair blocks and one 5-pair block, or seven 3-pair blocks and one 4-pair block.The impact tool for 5-pairs is used.
12. Insert the marking strips with identification cable numbers in the holders and install the holders in the cross-blocks.

Connection of horizontal cables to patch panels

The connection of 4-pair horizontal cables to the patch panels is performed as follows:

1. Install the patch panels in the 19-in mounting cabinet or wall-mounting frame.
2. Lay cables according to the table of connections in the technical design documentation. The cables are brought to the patch panels from the back: use the back mounting rails at the installation of the mounting cabinet or inside the mounting frame housing. It is first necessary to lay the groups of cables connected to the patch panels that are farthest from the cable inlets. Attach the cables in the operating position by plastic clamps.
3. Bring cables to a near end of the patch panel, starting with the most remote from the inlet. Attach cables to the forward mounting rail by plastic clamps.
4. Starting with the panel connector nearest to the point of cable fastening, bring an appropriate cable to the IDC contacts of the connector. Accurately remove the external sheath from the cables, trying not to damage the conductor covering, with the help of a cutting tool, so that its end is at the nearest point of IDC contact fastening. Install pairs of conductors in the IDC contact notches according to color markings of the notches, starting with the pair that is most remote from the cable fastening point. If such markings are absent, it is necessary to install pairs in the notches from left to right in the following order: blue, orange, green, brown. The white conductor is installed on the left notch in each pair, the conductor of the pair color is installed on the right notch. It is necessary to ensure preservation of their factory twisting up to the notches. For this purpose, it is possible to expand a clearance between them in one of the turns and to install divided conductors in notches, without unreeling the conductors of a pair.
5. Execute the previous operation for all cables connected to the patch panel. During connection, it is necessary to fix cables in the organizer or elements similar to it, if these are stipulated by the construction of the panel.

6. Make sure that the order of the installation of the conductors in the notches is correct.
7. Press conductors between knives of the IDC contacts, using the impact tool for one conductor, having ensured electrical contact of the cable conductors and the connector contacts. The tool simultaneously cuts the excess that flares over the contacts. During performance of this operation, it is necessary to keep the direction of the cutting edge of the tool up in order to prevent cutting of the conductors on the side of contact that are directed to the cable.
8. Attach markings with identification numbers of the telecommunication outlets and/or their ports on the face side of the panel.

Connection of multipair cables to patch panels

The connection of multipair symmetrical cables to the patch panels is performed in the following order:

1. Install the patch panels in the 19-in mounting cabinet or wall-mounting frame.
2. Lay cables according to the table of connections in the technical design documentation. The cables are brought to the patch panels on the back mounting rails at the installation in the 19-in mounting cabinet or inside the 19-in wall mounting frame. It is first necessary to install those groups of cables that are the most remote from the point of inlet, to the patch panels. Attach the cables to the mounting rails, wall clamps, or organizer (if present) by plastic clamps.
3. Starting with the panel that is located the farthest from the point of inlet, bring cables from the fastening point to the nearest end of the patch panel. Attach cables by plastic clamps to the front mounting rail.
4. Accurately remove the external sheath from the cables, trying not to damage the covering of the conductor, with the help of a cutting tool to the cable fastening point on the front mounting rail. Move cable markings on the rest of the external sheath. Separate 25-pair cable bundles from each other and fasten pairs of each bundle near the end of the external sheath by adhesive tape. Keep about 3 cm of identification tape, having cut the rest.
5. Bring the appropriate pair bundle to the IDC contacts, starting with the nearest point of the cable fastening of the panel connector.
6. Each 25-pair bundle is connected to six 4-pair connectors of the patch panel with IDC contacts, and the appropriate six modular 8-contact connectors on the face side of the panel. The connection should begin from the first pair. The final or 25th pair is not connected due to absence of a contact and is simply cut close to the sheath.
7. Install the pair conductors in the notches of the IDC contacts for each pair: install the white conductor of the pair on the left notch, install the conductor of the pair color on the right. It is necessary to ensure preservation of their factory twisting up to the notches. For this purpose, it is possible to expand a clearance between them in one of the turns and to install divided conductors in the notches without unreeling the conductors of a pair.
8. Execute the previous operations for all cables connected to this patch panel.

9. Ensure that the order of the installation of the conductors in the notches is correct.
10. Press conductors between knives of the IDC contacts using the impact tool for one conductor, having ensured electrical contact of the cable conductors and the knives, using the impact tool for one conductor. The tool simultaneously cuts the excess that flares over the contacts. During performance of this operation, it is necessary to keep the direction of the tool cutting edge up in order to prevent cutting the conductors on the side of contact that are directed to the cable.
11. Attach markings with identification numbers of the telecommunications outlets of the network equipment ports and other similar connectors on the face of the panel.

11.8 Channel Switching of Information Transmission and Network Equipment Connection

11.8.1 Channel Switching of Information Transmission on Telecommunications Equipment

The switching process is performed according to the tables of connections of the technical design documentation. The connection is performed by the patch cords, which are included in equipment delivery set. The length of the cords is selected depending on the distance between the connected outlets. After connection, the surplus cord is installed in horizontal and vertical organizers.

The cross-connect cords can be labeled by unique identification numbers on both terminations for further simplification of their search and identification. For the purposes of identification, patch cords from cables of various colors, colored collars installed on tangs of modular connector plugs, and icons and markers of various colors and constructions (Section 13.1.5) can be also used.

During patch cord installation, it is necessary to fill out a cable register, which is part of the operational documentation of the structured cable system. Table 11.4 contains an example sheet of a cable register.

Table 11.4. Example sheet of a cable register

Customer: ______________________

Object: ________________________

Building: ______________________

Cross-room: ____________________

	Operation					Patch cord		
No.	Date	Time	Dismantled/installed	Performed by	Signature	Identifier	Type	Purpose
1	2	3	4	5	6	7	8	9

Table 11.4. (cont.)

Connector 1			Connector 2		
Block number	Line number	Connector number	Block number	Line number	Connector number
10	11	12	13	14	15

11.8.2 Connection of Network Equipment to Workstations

At this stage it is necessary to choose suitable cord lengths for each specific workstation from the delivered SCS components and to connect the network equipment to the appropriate module of the telecommunications outlet. The connection is performed according to the tables of connections of the technical design documentation set.

If necessary, the adapters and connectors used for operation of some kinds of network devices are additionally connected or installed.

11.9 Conclusions

SCS installation is a crucial procedure, which in many respects determines the level of technical parameters of a cable system and their long-term conformity to norms. The installation can be performed either by one crew of the trained installers, or by specialized crews. The last variant is preferable for the realization of large projects.

The performance of timely and high-quality installations depends on the installers' special equipment and on the presence of the technical design documentation set. It is desirable that the crew should be divided into teams, as general technological operations are performed by two installers.

The list of the basic tasks that are performed during SCS installation includes: the inspection of separate components, backbone and horizontal subsystems cable layout, the installation of raceways and 19-in constructions, and connection of the cables to the outlets and telecommunications panels. The order of performance of those or other tasks is substantially determined by the conditions of construction readiness of a building.

The final stages of SCS installation are testing, network equipment connection, channel switching of information transmission, and filling out the cable register.

12 Testing of SCS Links and Channels

12.1 General Issues of SCS Testing

12.1.1 Measurement Purposes and Kinds

The complex of measurements of parameters of individual electrical and optical components of a structured cabling system as well as those of lines constructed from these components is aimed at the monitoring of the SCS state, fault prevention, and accumulation of statistical data to be used when developing measures to improve the communication reliability. The measurements are performed:

- during the inspection testing of individual components prior to starting work on their installation;
- when adjusting SCS;
- during the compliance testing of the SCS;
- during the acceptance testing;
- for maintenance, troubleshooting, and control tests during the operation of the cabling system.

The measurements during the *inspection testing* are to check the quality of the manufacturing of individual components and their compliance with the standard requirements. *Acceptance tests* are carried out by acceptance committees to check the quality of the work performed and the correspondence of the parameters of links and channels to standards and other normative documents. The *operational measurements* are carried out by technical personnel during SCS operation (Fig. 12.1).

Maintenance measurements are aimed at the detection of deviations of the parameters from the values defined by standards and the removal of these deviations. The plan and techniques of these measurements are basically similar to those of the acceptance tests. The plan of the maintenance measurements is specified by the respective standards and instructions. The interval between these measurements depends on the operating conditions, the state of the components to be tested, and the requirements for maintaining operational readiness.

Troubleshooting measurements identify the character and location of cabling line defects. It is of primary importance for these measurements to identify the nature of the defects with a high rate and to find their locations with high accuracy.

Control measurements are carried out only after completing repair and emergency work. These measurements are aimed at evaluating the quality of repair work and damage control. Usually, they involve the entire complex of testing link parameters and are performed in accordance with plans and techniques adopted for acceptance tests.

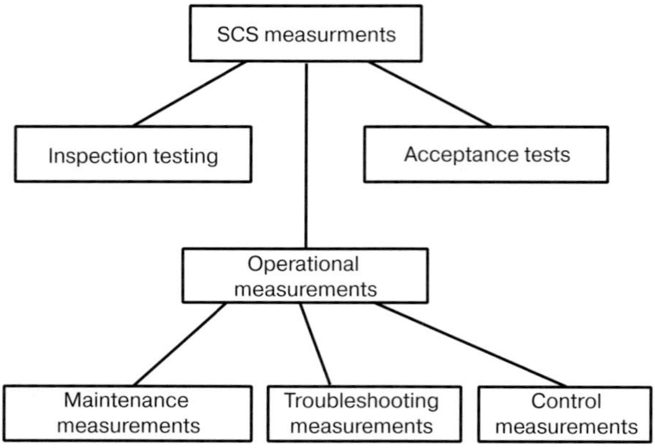

Fig. 12.1. Measurement kinds

Most of the methods and devices used for measurements in SCS coincide with those used for the inspection of electrical and optical public communication lines and their components. However, specific structural features of some of the SCS subsystems mean that both the reference models and measuring instruments used for testing must be modified. This is especially true for measuring instruments for testing electrical SCS channels.

The most important aspects of testing SCS electrical and optical subsystems are considered in Sections 12.2 and 12.3.

12.1.2 Documentation of Measurement Results

The basic forms of the reports about various measurements that have been conducted are the *protocol* and the *certification of cabling runs*. These documents are given to the customer.

The protocol contains information about (i) the date, time, and place of the measurements; (ii) the objects that have been tested with their individual identification; (iii) the methods of measurements and the devices used (including the model and the serial number of the devices as well as the software version), and (iv) the testing results. This document must be signed by specialists who carried out the measurements (with their positions indicated) and approved by the customer and the executive manager.

The passport of the run identifies:

- the number or designation letters of the run;
- the starting and terminal points of the run;

- the length of the run (measured by a reflectometer in optical subsystems and by a cable scanner in electrical subsystems);
- the measured parameters of the cabling lines.

The passport of a cabling run based on electrical balanced cables necessarily contains information concerning the attenuation and the value of near-end cross talk (NEXT). It may also indicate other parameters, according to the agreement with the customer. The lists of files with the testing results obtained by means of a cable scanner may be attached to the passport of an electrical communication line.

In addition, the passport of an optical communication line indicates:

- the total attenuation of the signal for operating wavelengths of 850, 1300, and (possibly) 1500 nm (measured by an optical tester or automatic meter);
- the presence of inhomogeneities, their location, and signal attenuation induced by these inhomogeneities (determined by a reflectometer).

The passport of an optical communication line may be supplemented with reflectograms of individual fibers, as well as lists of testing results obtained by means of an automatic meter (Section 12.3.2).

The passport of a campus backbone cabling system is supplemented with the scheme of pathways in the cabling runs and collectors, as well as the intermediate splicing manhole ties (when cabling in pipes of an enterprise or city telephone network).

12.2 Testing of SCS Electrical Subsystems

The basic normative document that most fully specifies various aspects of the testing of the electrical subsystem at the beginning of 2000 is the technical bulletin TSB-67 (Telecommunications Systems Bulletin). This document was issued by the Telecommunications Industry Association/Electronics Industry Association (TIA/EIA) in October 1995. It contains accuracy requirements for measuring devices to be used for determining basic parameters of SCS horizontal subsystems constructed on 4-pair cables whose electrical parameters meet the requirements of the standard TIA/EIA-568-A. Some information on the testing is also contained in the standard ISO/IEC 11801.

The cited documents apply both to unshielded and shielded cables. However, the influence of shields on the communications quality and the conditions of their grounding require further investigation, and the related issues are not considered in these documents.

12.2.1 Objects of Testing

As applied to SCS, one can divide the objects of testing into two basic groups – *SCS components* and *twisted-pair communication lines*.

The SCS electrical components to be tested are cables and connectors. The requirements of the standards for the electrical characteristics of these components are considered in Sections 3.1.2.5 and 3.2, respectively. The necessity of testing electrical cables as separate components arises when inspecting cables prior to confinement. This testing can be performed by means of portable instruments right at an object where the cabling system will be installed. Connectors can be tested only in laboratory conditions by means of stationary equipment.

The major objects of testing in an installed SCS are the twisted-pair communication lines. They are defined in the standard ISO/IEC 11801 and the bulletin TSB-67.

The communication line as an object of testing may be of two types that will be referred to as the basic link (or simply link) and the channel. The channel is defined as the total path of data transmission from the connector to the connector of the network equipment. The length of a channel in the horizontal cabling should not exceed 100 m. The latter type of the object to be tested is referred to as the permanent link in the standard ISO/IEC 11801 (2000 edition) and as the basic link in the standard TIA/EIA-568-A. The basic link of TIA/EIA-568-A is a channel without terminal cords for connecting network equipment. The permanent link of ISO/IEC 11801 is defined as a link between two SCS interfaces in one subsystem. Thus, a permanent link does not contain a cord or any other element playing its role. It is implied that during the measurement of parameters of a link, the terminal cords are replaced by the cords of the testing equipment. Note also that the term permanent link [128] was fairly widely used long before it was introduced by the standard ISO/IEC 11801.

Links according to ISO/IEC 11801

Figure 12.2 shows the scheme of a channel and a permanent link according to the standard ISO/IEC 11801 (2000 edition). The maximum length of a channel of the horizontal subsystem may reach 100 m. As follows from Fig. 12.2, the length of a channel going beyond the horizontal subsystem substantially depends on its configuration and, with reference to the data of Fig. 1.4 and Table 1.11, may exceed 2000–3000 m.

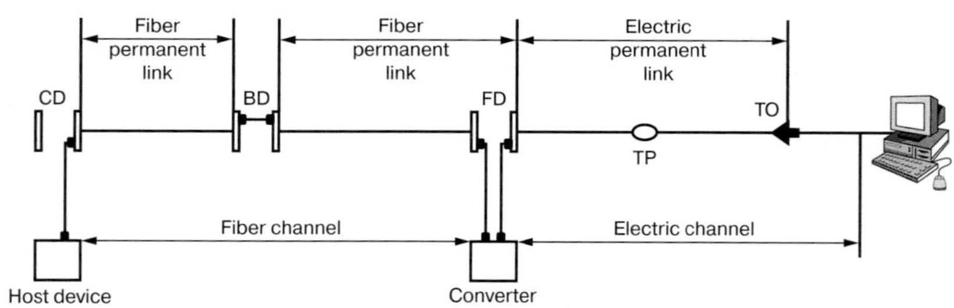

Fig. 12.2. Scheme of a channel and a permanent link according to the standard ISO/IEC 11801

The maximum length of a permanent link of a horizontal subsystem is 90 m. The permanent link consists of the horizontal cable and two connectors, one being located in a technical room and the other at the workplace.

The links may be subjected to the following kinds of testing:

- *Acceptance testing*: This is the simplest test. It is performed for the links designed in accordance with the standard requirements and assembled from the components that also meet these requirements.
- *Troubleshooting testing*: This test is performed in the case of the loss or interruption of communications.
- *Compliance testing (testing for compliance with the requirements of ISO/IEC 11801)*: This test is designed to check that the electrical characteristics of the link meet the requirements of class C or D.

Table 12.1 presents the lists of tests for all three types of testing.

Table 12.1. Table 12.1. The list of cable link parameters to be tested in accordance with ISO/IEC 11801 (2000 edition)

Type of testing	Acceptance	Troubleshooting	Compliance
Attenuation	–	+	+
NEXT	–	+	+
Impedance	–	–	+
Propagation delay	–	–	+
DC resistance	–	–	+
DC resistance of the shield	–	+	–
Impulse response	–	+	–
Continuity of conductors and shields	–	–	–
Short and open circuits	+	+	–
Return loss	–	+	+

The requirements of the standard ISO/IEC 11801 for the electrical characteristics of lines of class C and D have been worked out proceeding from the most rigid conditions for the totality of the electrical characteristics of the data transmission medium of maximum length for the applications to be supported by the lines. For lines of class C, it is 10 Mbit/s Ethernet (10Base-T), and for lines of class D, it is the twisted-pair specification of the network FDDI (TP-PMD).

In accordance with ISO/IEC 11801 (2000 edition), the maximum attenuation is determined for the permanent link and the channel and is specified for lines of all four classes, from A to D (Table 12.2).

The NEXT is specified in a similar manner for all classes of lines in Table 12.3. The power-sum NEXT (PS-NEXT) loss is specified only for lines of class D (Table 12.4). This is also the case for attenuation-to-cross talk ratio (ASR) and power-sum attenuation-to-cross talk ratio (PS-ACR).

The return loss (RL) is specified for channels and basic links for various classes of applications (Table 12.5). The experimental determination of this characteristic should be performed in accordance with the method of the international standard ISO/ICE 61935-1 at both ends of the cable link being tested.

Table 12.2. Maximum attenuation for the permanent link/channel for different classes of lines according to ISO/IEC 11801

Frequency, [MHz]	Class A	Class B	Class C	Class D
0.1	16.0/16.0	5.5/5.5	–	–
1.0	–	5.8/5.8	3.1/4.2	2.1/2.5
4.0	–	–	5.8/7.3	4.1/4.5
10.0	–	–	9.6/11.5	6.1/7.0
16.0	–	–	12.6/4.9	7.8/9.2
20.0	–	–	–	8.7/10.3
31.25	–	–	–	11.0/12.8
62.5	–	–	–	16.0/18.5
100	–	–	–	20.6/24.0

The impedance of the link must lie within the limits of 100±15 Ω for the frequency range 1–16 MHz for class C or to 100 MHz for class D. The DC resistance of a twisted pair is measured for conductors short-circuited at the far end. This parameter is important for Token Ring networks and is specified for applications of all four classes in accordance with Table 12.6.

Table 12.3. Minimum NEXT for the permanent link/channel for different classes of lines according to ISO/IEC 11801

Frequency, [MHz]	Class A	Class B	Class C	Class D
0.1	27.0/27.0	40.0/40.0	–	–
1.0	–	25.0/25.0	40.1/39.1	61.2/60.3
4.0	–	–	30.7/29.3	51.8/50.6
10.0	–	–	24.3/22.7	45.5/44.0
16.0	–	–	21.0/19.3	42.3/40.6
20.0	–	–	–	40.7/39.0
31.25	–	–	–	37.6/35.7
62.5	–	–	–	32.7/30.6
100	–	–	–	29.3/27.1

Table 12.4. Minimum power-sum cross talk, attenuation-to-cross talk ratio, and power-sum attenuation-to-cross talk ratio for the permanent link/channel for class D lines according to ISO/ICE 11801

Frequency, [MHz]	PS-NEXT	EL-FEXT[a]	PS-ELFEXT[b]	ACR	PS-ACR
1.0	58.2/57.3	59.6/57.0	57.0/54.4	59.1/57.8	56.1/54.8
4.0	48.8/47.6	47.6/45.0	45.0/42.4	47.7/46.1	44.7/43.1
10.0	42.5/41.0	39.6/37.0	37.0/34.4	39.4/37.0	36.4/34.0
16.0	39.3/37.6	35.5/32.9	32.9/30.3	34.5/31.4	31.5/28.4
20.0	37.7/36.0	33.6/31.0	31.0/28.4	32.0/28.7	29.0/25.7
31.25	34.6/32.7	29.7/27.1	27.1/24.5	26.6/22.9	23.6/19.9
62.5	29.7/27.6	23.7/21.1	21.1/18.5	16.7/12.1	13.7/9.1
100	26.3/24.1	19.6/17.0	17.0/14.4	8.7/3.1	5.7/0.1

[a]Equal-level far-end cross talk.
[b]Power-sum equal-level far-end cross talk.

Table 12.5. Return loss values for permanent links and channels for different classes of applications

Frequency f, [MHz]	Minimum return loss, [dB]	
	Class C	Class D
1–16	15	17
16–20	–	17
20–100	–	$17 - K \times \log(f/20)$[a]

[a]K=7 for the permanent link and 10 for the channel.

The propagation delay is also specified for the basic link and the channel for various classes of applications in accordance with Table 12.7.

In addition, the standard recommends that the conversion loss at frequencies up to 10 MHz (Table 2.6) and the shield transfer impedance be checked. Of course, the latter parameters should be checked only for shielded cables.

Table 12.6. Maximum DC-loop resistance for various classes of applications

Application class	A	B	C	D
Resistance, Ω	560	170	40	40

Table 12.7. Maximum signal propagation delay for permanent links and channels of different classes

Frequency f, [MHz]	Class	Delay, ns[a]
1–16	C	$K+36/f^{1/2}$
1–100	D	$K+36/f^{1/2}$

[a]K=486 ns for the permanent link and 544 ns for the channel.

The channel and the basic link according to TSB-67

The bulletin TSB-67 also distinguishes between two models of twisted-pair communication lines, the channel and the basic link. The schematics of these models are presented in Figs. 12.3 and 12.4, respectively.

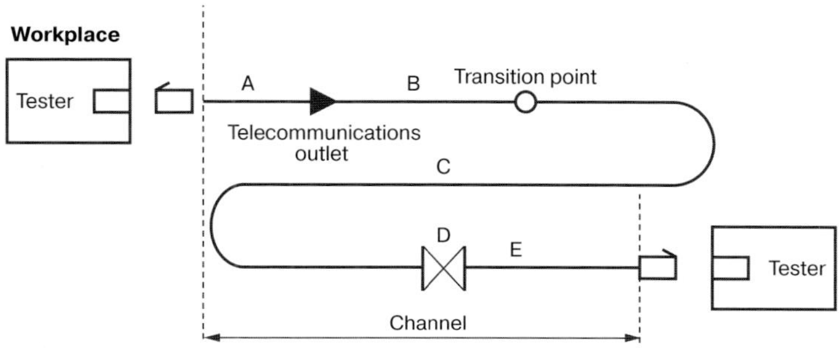

Fig. 12.3. Model of channel according TSB-67: A, E – terminal cords; B, C – horizontal cables; D – patch cord. *Max* B+C = 90m, *Max* A+D+E = 10m

The channel consists of a horizontal cable of length not exceeding 90 m, two terminal cords (at the workplace and in a telecommunications closet for connecting network equipment), the equipment for the transition point (if any) near the workplace, and a patch cord in a telecommunications closet. The total length of two terminal cords and one patch cord must not exceed 10 m. The 8-pin plugs of modular connectors of terminal cords for connecting network equipment, shown in Fig. 12.3, do not belong to the channel Thus, the channel contains at least four connectors (plug-type or permanent). These are:

- the connector of the telecommunications outlet at the workplace;
- the connector of two segments of the horizontal cable at the transition point;
- the first connector in the telecommunications closet connecting the cable of the horizontal subsystem to the patch cord;
- the second connector in the telecommunications closet connecting the patch cord to the terminal cord.

All measurements in the channel should be organized so as to exclude the influence of the plugs of the terminal cords connected to testing equipment.

The basic link (Fig. 12.4) consists of the horizontal cable of length not exceeding 90 m and two cords for connecting the measurement equipment, each cord having a length of 2 m. As was the case for the channel, the terminal connectors do not belong to the basic link. Thus, the maximum length of a basic link is equal to 94 m and it has only two connectors:

- the connector of the telecommunications outlet at the workplace connecting the patch cord to the cable of the horizontal subsystem;
- the connector in the telecommunications closet connecting the cable of the horizontal subsystem to the patch cord.

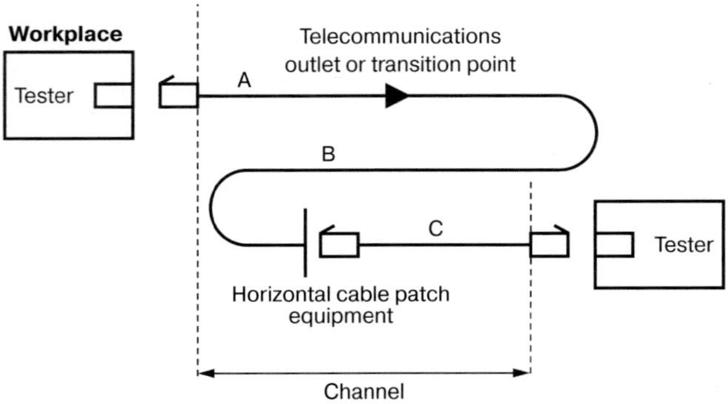

Fig. 12.4. Model of base link according TSB-67.; A, C – terminal cords (maximal length 2 m); B – horizontal cable.

One can readily verify that the model of the basic link allows for the evaluation of characteristics only for the horizontal cable that has been installed in the run and connected to the terminal connectors of the equipment in telecommunications closets and to the telecommunications outlet at the workplace. As regards the attenuation parameters and NEXT, the values of these quantities yielded by this model are somewhat better than those corresponding to the channel model.

Applicability of the models of lines to be tested

The introduction of two fairly close models of communication lines can be justified as follows. The electrical parameters of the channel are of major importance for the user, since these parameters fully determine the quality of the communication line. On the other hand, it may happen that during acceptance tests of the SCS, the terminal cords were not connected to the telecommunications outlets. It may also happen that by the time of finishing the SCS installation, the furniture has not been arranged in the working rooms and/or the network equipment is absent. In addition, it is frequently the case that terminal cords are manufactured by the customer or purchased from a company other than that delivering and installing the other SCS components.

The conclusion that the characteristics of the channel meet the requirements of the given category is valid only for the terminal and patch cords and only for the connectors in the telecommunications closet that have been used during the measurements. Meanwhile, the operation of an SCS implies the possibility of switching cords and that the user is allowed to change cords at the workplace without informing the administrator. The change of any element in the channel changes its electrical characteristics and, hence, for the repeated certification it is necessary to carry out a new cycle of measurements.

In contrast to this, all parameters of the basic link remain unchanged during operation. For this reason, the time during which the testing results remain valid is longer for basic links, as compared with that for channels. It is also important that by testing basic links one can achieve higher accuracy, compared with that achievable by testing channels. This will be demonstrated later.

The modern editions of both basic standards do not involve rigid requirements for the choice of testing models to be applied in various cases during the installation and operation of SCS. The analysis of the principles of construction of the models discussed above and the major properties of these models shows that during the construction and commissioning of the SCS, it is better to use the basic link or permanent link model. Sampled or global measurement of parameters of the channels during the acceptance testing can be performed in especially important cases or at the request of the customer. The channel model is convenient to deal with during the current operation of an SCS for detecting and removing faults. These observations have been commonly recognized in the engineering practice.

12.2.2 Parameters to be Measured

Channels and basic links are divided into categories 3, 4, and 5 in accordance with the definitions given in the standard TIA/EIA-568-A. The bulletin TSB-67 defines the following parameters to be necessary and sufficient for judging the correspondence to a given category:

- attenuation;
- NEXT;
- length;
- wire map.

The first two parameters should be measured in the frequency range corresponding to working frequencies of categories 3 to 5.

Apart from the basic parameters, modern SCS testers can measure also a number of additional characteristics.

Requirements for the attenuation of electrical transmission channels

The attenuation of a channel or a basic link is the sum of attenuations due to all individual components, that is, the horizontal cable, terminal and patch cords, and connectors. The maximum admissible attenuation can be expressed as follows:

$$A=\Sigma A_{\text{connector}}+A_{\text{cable}}\text{ per }100\text{ m}\times(L_{\text{cable}}+1.2\times\Sigma L_{\text{cord}})\ /\ 100\text{ m}, \qquad (12.1)$$

where $\Sigma A_{\text{connector}}$ is the sum of maximum admissible attenuations due to all connectors (Table 3.19). A channel may have up to four connectors, whereas a basic link always has two connectors.

The A_{cable} per 100 m is the maximum admissible attenuation of the horizontal cable per 100 m of its length (Table 3.6). The value of L_{cable} is the actual length of the horizontal cable of the channel or the basic link. The value of ΣL_{cord} is the actual total length of all cords of the channel or the basic link. The coefficient 1.2 takes into account the fact that the maximum attenuation allowed by the standard TIA/EIA-568-A for terminal and patch cords is 20% higher if conductors in the horizontal cable are twisted, as compared with the case of the horizontal cable with monolithic conductors.

In some publications, the quantity $L_{\text{cable}}+1.2\times\Sigma L_{\text{cord}}$ is referred to as the electrical length of the channel. Table 12.8 presents the maximum attenuation admissible for a channel and a basic link at a temperature of 20°C. The attenuations have been calculated by relation (12.1) for L_{cable}=90 m, ΣL_{cord}=10 and 4 m for the channel and the basic link, respectively.

If the temperature of the environment during testing differs from 20°C, one should correct the maximum attenuation from Table 12.8 by the expression:

$$A(t^\circ)=A_{\max 20^\circ}\,(1+k\,(t^\circ-20^\circ)),$$

where t° is the temperature of the environment during testing, °C; $A_{\max 20^\circ}$ is the maximum attenuation from Table 12.8; and k is the attenuation temperature coefficient (equal to 0.015 for cables of category 3, and 0.004 for cables of categories 4 and 5).

According to TSB-67, the measurement of the attenuation should be performed in the range from 1 to 16, 20, and 100 MHz for cables of categories 3, 4, and 5, respectively. The frequency change step size should not exceed 1 MHz. All four pairs are to be tested. The worst of the four values obtained is compared with the norms given in Table 12.8. Because of inevitable measurement errors (which are discussed in Section 12.2.3), the values of attenuation lower than 3 dB are not compared with the norms.

Table 12.8. Maximum attenuation in channels and basic links in accordance with TSB-67

Frequency, [MHz]	Attenuation, [dB]					
	Category 3		Category 4		Category 5	
	Channel	Basic link	Channel	Basic link	Channel	Basic link
1.00	4.2	3.2	2.6	22	2.5	2.1
4.00	7.3	6.1	4.8	4.3	4.5	4.0
10.00	11.5	10.0	7.5	6.8	7.0	6.3
16.00	14.9	13.2	9.9	8.8	9.2	8.2
20.00	–	–	11.0	9.9	10.3	9.2
31.25	–	–	–	–	12.8	11.5
62.50	–	–	–	–	18.5	16.7
100.00	–	–	–	–	24.0	21.6

To increase the accuracy and reliability of the tests, TSB-67 recommends measuring the attenuation for each pair and at both ends. The testing results should not differ from one another by more than the admissible error of the measuring instrument in the entire frequency range.

Requirements for NEXT in electrical transmission channels

The minimum value of *NEXT* admissible for a channel or a basic link is calculated by the expression:

$$NEXT = -20\log\left(10^{-NEXTcable/20} + n \times 10^{-NEXTconnector/20}\right), \qquad (12.2)$$

where $NEXT_{cable}$ is the minimum cross-talk loss in the horizontal cable per a length of 100 m; see Table 3.7; $NEXT_{connector}$ is the minimum cross-talk loss of the connector; see Table 3.17; and n is the number of connectors at the near end (n=2 for a channel and 1 for a basic link).

Analysis of relation (12.2) indicates two features. First, the compilers of TSB-67 used the model in which the powers of interference components generated by individual elements of an electrical transmission channel are added. Second, when testing a channel, one should expect more powerful cross-talk interference to come from the telecommunications closet, since in this case n=2, whereas for testing at the telecommunications outlet the channel and basic link models coincide.

Table 12.9 presents the minimum values of NEXT calculated by relation (12.2) for channels and basic links.

Table 12.9. Minimum values of NEXT for channels and basic links according to TSB-67

Frequency, [MHz]	Category 3, [dB]		Category 4, [dB]		Category 5, [dB]	
	Channel	Basic link	Channel	Basic link	Channel	Basic link
1.00	39.1	40.1	53.3	54.7	60.0	60.0
4.00	29.3	30.7	43.3	45.1	50.6	51.8
10.00	22.7	24.3	36.6	38.6	44.0	45.5
16.00	19.3	21.0	33.1	35.3	40.6	42.3
20.00	–	31.4	–	–	33.7	39.0
31.25	–	–	–	–	35.7	37.6
62.50	–	–	–	–	30.6	32.7
100.00	–	–	–	–	27.1	29.3

The value of *NEXT* for a channel and a basic link should be measured in the range from 1 to 16, 20, and 100 MHz for cables of categories 3, 4, and 5, respectively. The step size of the frequency change should not exceed 0.15 MHz for the range 1–31.25 MHz and 0.25 MHz for the range 31.25–100 MHz. The testing should be performed at both ends for all six possible combinations of pairs.

The worst of the values measured is compared with the data of Table 12.9, and by this comparison one judges the correspondence of the channel or the basic link to the normative requirements.

Measurement length

The length of a channel or a basic link is defined as the sum of lengths of the horizontal cables and all terminal and patch cords. The length is measured by reflectometry. This method involves launching a short electrical pulse into a twisted pair and measuring the delay time of the reflected signal. Then the length of the cable can be evaluated by:

$$L_{\text{cable}}=NVP\times C\times T_{\text{meas}}/2, \tag{12.3}$$

where NVP is the ratio of the velocity of wave propagation in the given cable to the velocity of light in vacuum (Section 2.1.3.5); C is the velocity of light in vacuum; and T_{meas} is the time interval between the instant of launching of the testing pulse and the time of arrival of the reflected signal.

The quantity evaluated by relation (12.3) is sometimes called the electrical length of the cable. By the polarity of the reflected pulse one can additionally determine the degree of matching of the twisted pair at the far end of the cable. If the load resistance R_L exceeds the absolute value of the cable impedance Z_w (for example, when the ends of the cable are open), then the reflected pulse has a positive amplitude. If $R_L < |Z_w|$ (for example, if the cable is short circuited), the amplitude of this pulse is negative. If the load is matched at the far end, the reflected pulse is absent and, hence, the length of the cable cannot be measured.

TSB-67 requires measuring T_{meas} for all four pairs. The relation of (12.3) is then applied to the least of the four values T_{meas}. The tester must have sufficient dynamic range for working with cables of lengths no less than 310 m. This enables one to inspect cables delivered in the most widespread packing (305 m, or 1000 ft).

The measured electrical length is compared with the maximum length allowed for a channel (100 m) or a basic link (94 m).

Checking the wiring of conductors of the pairs among the contacts of modular connectors

In a correctly configured channel or basic line, the conductors of twisted pairs should be connected to contacts of terminal modular connectors with the same numbers, as is the case for usual terminal and patch cords (Section 3.4.1). The correct connection of the conductors of pairs is shown in Fig. 12.5a. Figures 12.5b–d show typical errors committed by fitters when installing modular connectors.

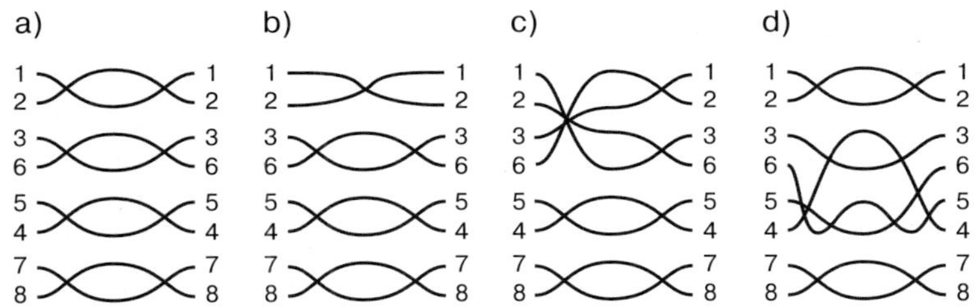

Fig. 12.5. Typical errors made installing modular connectors

All errors in the wiring of modular connectors can be divided into the following basic groups [129]:

- *Reversed pairs* (Fig 12.5b) occur in the cases where the numbers of contacts of the modular connector to which the conductors of the pair are attached mutually change place at both ends of the cable link. This leads to a change in the signal polarity at the receiving end.
- *Transposed pairs* (Fig. 12.5c) imply the connection of any pair to contacts of one of three other pairs at the other end of the cable link. Transposed pairs almost always lead to loss of communications. For example, it may happen that the transmitters at different sides are connected to one another. For that reason, this error is easy to detect even at early stages of adjustment or operation testing of the installed SCS.
- *Split pairs* (Fig. 12.5d), in this case, conductors physically belonging to different pairs (not twisted with each other) are attached to the contacts of the modular connectors designed for the connection of one pair. Split pairs lead to a considerable deterioration of the electrical characteristics of the cable because of the increase in attenuation and decrease in the values of cross-talk loss.
- The errors that are not involved in the list of wiring errors are *short and open circuits* of conductors.

All other errors represent combinations of the basic errors listed above.

Modern devices for testing SCS are equipped with graphic displays. After performing the test, the actual wiring of the twisted pairs among the terminal connectors of the channel appears on the screen of the display. By looking at the picture on the screen, an experienced fitter will readily catch the error and identify its type.

Performance grading

At the present time, the quality of 4-pair cables and other components of category 5 SCS of many manufacturers substantially surpasses the formal requirements of the standards ISO/IEC 11801 and TIA/EIA-568-A. Cables with conductors having a diameter of 0.5 mm possess more or less the same attenuation of 22±1 dB for a length of 100 m and a frequency of 100 MHz. On the other hand, the values of NEXT meeting the requirements of category 5 may differ by more than 10 to 15 dB. Apparently, SCS channels constructed on such cables will have substantially different quality.

To objectively compare operating characteristics of cabling systems constructed from components of different manufacturers one can use the attenuation-to-cross talk ratio (ACR) at the near end. However, this parameter was only clear to specialists and was not adopted by most users. Therefore, another parameter was needed that was comprehensive and easy to understand. Some manufacturers of SCS components, when advertising their products, guaranteed a substantial increase in the length of lines constructed on their cables. However, such an increase is prohibited by the working standards. In addition, when testing lines having the maximum length but different quality, the process of recalculation of the interference immunity margins to the limiting length is not fully clear.

The concept of *performance grading* suggested by specialists of Microtest (USA), producers of testing equipment for SCS, was more successful. According to this concept, the measured values of NEXT are successively compared with a set of threshold values, the first of which corresponds to the standard requirements for category 5. The threshold values are defined by the expression:

$$NEXT_{\mathrm{thr}}(N_{\mathrm{thr}}, f) = NEXT_{\mathrm{min}}(f) + N_{\mathrm{thr}} \times [1 + \log(f)], \tag{12.4}$$

where $NEXT_{\mathrm{thr}}$ is the threshold value of $NEXT$, dB; N_{thr} is the threshold number; f is the frequency, MHz; and $NEXT_{\mathrm{min}}$ is the minimum value of $NEXT$ for cables of category 5, links of class D according to ISO/IEC 11801, and channels or basic links of category 5 according to TIA/EIA-568-A.

Relation (12.4) implies that for $N_{\mathrm{thr}}=0$, the threshold value of $NEXT$ coincides with that required by standards for systems of category 5. For $f=100$ MHz, the threshold values of $NEXT$ are separated by 3 dB or differ in power by a factor of 2. The thresholds of $NEXT$ are shown by graphs in Fig. 12.6.

The regions between the thresholds are referred to as *quality bands*. The tester refers a cable for which the nearest threshold not exceeded by the measured values of $NEXT$ has the number N_{thr}, to the quality band number $N_{\mathrm{thr}}+1$. The higher the number of the quality band, the better the electrical characteristics of the cable and the higher the signal-to-noise ratio.

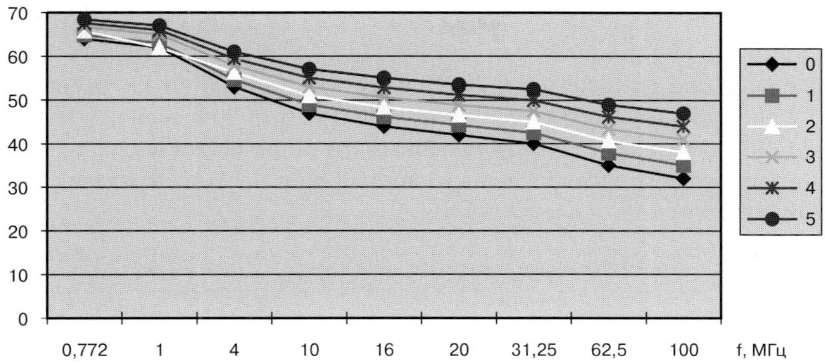

Fig. 12.6. Thresholds of *NEXT*

The possibility of utilizing performance grading instead of ACR follows from the definition of these concepts and the fact that horizontal cables of different manufacturers have approximately the same attenuation values in the entire working frequency range.

To be just, note that this simple and quite clear concept also has not become widely practiced and is rarely encountered in catalogs and other technical documents.

12.2.3 Measurement Errors of the Equipment for Field Testing of SCS

The basic devices for field testing the electrical parameters of an installed SCS are specialized testers. The first instruments of this type for cabling systems of category 5 appeared in 1993. The measurement accuracy of these devices was substantially lower than that of network analyzers designed for similar measurements in laboratory conditions. According to TSB-67, this was accounted for by the following reasons:

- The testers had low accuracy of balancing of intrinsic elements, in particular 8-pin modular connectors. This led to larger values of the measured NEXT as compared with the actual values.
- The size of linear or logarithmic frequency step set up for measuring the attenuation and cross-talk loss was too large. This increased the probability of missing the worst values on excursions of these parameters during the testing.
- To simplify the design of the devices, the level of testing signals was lower than that of real signals, and the final estimate was based on extrapolating the measurement results obtained at the limiting sensitivity of the tester.
- To connect the tester to the link to be inspected, an external patch cord was used, which increased the attenuation and decreased the value of cross-talk loss.

At the same time, network analyzers were connected to the cabling segment to be tested without any intermediate means.

TSB-67 defines two accuracy levels (I or II) for SCS testers. Level I corresponds to the devices produced before 1994 [130]. Table 12.10 presents requirements for operating characteristics exerting the most impact on the accuracy of measuring the attenuation and cross-talk loss for testers of both accuracy levels. In addition, TSB-67 requires the measurement results of the testers to coincide fairly well with the results of measurements performed with network analyzers; this is especially true for testers of accuracy level II.

Table 12.10. Requirements of TSB-67 for operating characteristics of SCS testers of accuracy levels I and II

Parameter	Notation	Value in the range 1–100 MHz	
		For testers of of accuracy level I	For testers of accuracy level II
Random noise floor at the receiver input	Δrnf	≥50 – 15 log (f/100) dB	≥65 – 15 log (f/100) dB
Residual NEXT	Δr_next	≥40 – 15 log (f/100) dB	≥55 – 15 log (f/100) dB
Output signal balance relative to ground	Δosb	≥27 – 15 log (f/100) dB	≥37 – 15 log (f/100) dB
Common-mode rejection	Δcmr	≥27 – 15 log (f/100) dB	≥37 – 15 log (f/100) dB
Dynamic accuracy	Δda	≤ ±1 dB	≤ ±0.75 dB
Length accuracy	Δla	≤ ±1 m ±4%	≤ ±1 m ±4%
Return loss	Δrl	≥15 dB	≥15 dB

The residual *NEXT* of the tester is affected primarily by self-interference of internal data-processing circuits and the patch cord modular connector. Figure 12.7 shows the graphs of the minimum *NEXT* allowed for connectors of category 5 in accordance with TIA/EIA 568-A and the minimum residual NEXT allowed for testers of accuracy level II in accordance with TSB-67 (Table 12.10).

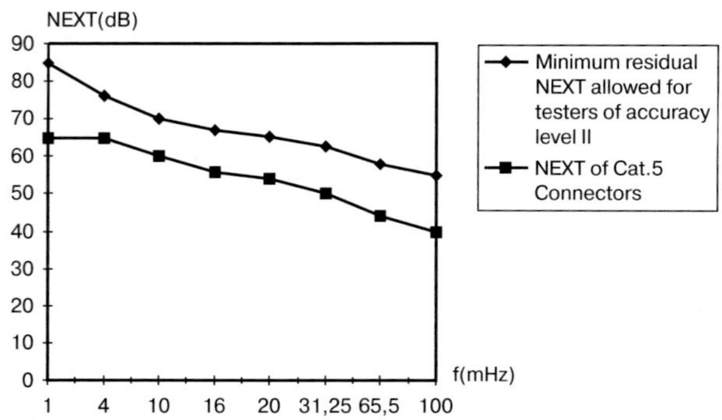

Fig. 12.7. Minimum NEXT allowed for connectors of category 5 and minimum residual NEXT allowed for testers of accuracy level II

It is apparent that testers of accuracy level II are subjected to more rigid requirements for residual *NEXT*. We must note that the terminal 8-pin modular connectors of patch cords are not involved in the model of a channel, in accordance with TSB-67 (Fig. 12.3). All measurements when testing a channel should not take into account the influence of these connectors on the electrical characteristics. If the terminal cord plug is connected to the tester jack, the *NEXT* of the modular connector will not meet the requirements for SCS testers of accuracy level II, whatever the proper parameters of the tester were. Accordingly, the accuracy of testing channel parameters always corresponds to level I.

The terminal cords connecting the tester and the injector to a basic link for measurements are parts of the measuring equipment. When manufacturing connectors of these cords, special technologies are used to provide very high values of NEXT (70 dB or higher for a frequency of 100 MHz) when connecting to the basic block or injector. To provide such high performance characteristics, one can either use a special design of the modular connector jack or reject the jack for high-frequency connectors of alternative designs possessing higher quality. As a result, the residual NEXT of SCS testers when testing basic links may correspond to the requirements for measuring devices of accuracy level II.

Errors in attenuation and NEXT measurements

Relying on the data presented in Table 12.10, TSB-67 defines the mathematical model for evaluating the measurement accuracy of testers. According to this model, the error in measuring the attenuation and *NEXT* can be calculated by the relations:

$$\Delta attenuation=|\Delta da|+20\log[1+10-\Delta rl / 10], \tag{12.5}$$

$$\Delta NEXT=|\Delta da|+20\log[1+10-\Delta rl / 10+(10(NEXT_{meas}-\Delta r\_next) / 10+ 10(NEXT_{meas}-\Delta rnf) / 10+10(10-\Delta osb) / 10+10(5-\Delta cmr) / 10)\ 0.5], \tag{12.6}$$

where $NEXT_{meas}$ is the measured value of *NEXT*; the notation of the other quantities is defined in Table.12.10.

The maximum errors of measuring the attenuation and *NEXT* by testers of accuracy levels I and II in the range 1–100 MHz are given in Table 12.11. These errors have been calculated by relations (12.5) and (12.6) with reference to the dependence of Δrnf, $\Delta rnext$, Δosb, and Δcmr on the frequency according to Table 12.10 for $NEXT_{meas}$, which is equal to the minimum value allowed for cables of category 5 according to Table 12.10.

Table 12.11. Maximum errors admissible for SCS testers of accuracy classes I or II

Parameter	Accuracy class of the tester	
	I	II
Δattenuation	≤ 1.3 dB	≤ 1 dB
ΔNEXT	≤ 3.4 dB	≤ 1.6 dB

Errors in length measurements

The requirements for the accuracy of measuring the length of a cabling segment for SCS testers of accuracy classes I and II are presented in Table 12.10. However, these requirements do not take into account additional errors from inaccuracies in the specification of the velocity of signal propagation in a given specific cable. The accuracy of the evaluation of the distance by relation (12.3) is influenced by the error of the specification of NVP and the accuracy of measuring T_{meas}. For cables of category 3, the spread of values of NVP among different cables, as well as among different pairs of the same cable, may reach 5%. For cables of categories 4 and 5, this spread is somewhat smaller. The modern testers contain an electronic table with the values of NVP for all basic types of cables in their internal memories and allow one to adjust this parameter. To that end, one should first calibrate the tester on a reference segment of a cable with known length. On this segment, one should measure the time between the launched and reflected pulses with the tester and then use relation (12.3) to obtain:

$$NVP=2\ L_{cable}\ /\ (C\times T_{meas}), \tag{12.7}$$

where L_{cable} is the length of the reference cable segment; C is the velocity of light in vacuum; and T_{meas} is the measured time between the launched and reflected pulses.

The calibration is strongly recommended for reducing the error of measuring length. To achieve higher accuracy, one should perform the calibration on a cable segment of length no less than 25 m.

The parameter NVP tends to increase with increasing frequency. The values of NVP for the same cable can vary within the limits of 5% in the frequency range 1–100 MHz. This leads to a distortion of the testing pulse and, as a result, to an error in determining the arrival time of the reflected pulse, T_{meas}. One more factor influencing the accuracy of the measurement of T_{meas} is finiteness of the frequency of the internal clock generator of the tester. This frequency defines the minimum time interval distinguished by the tester. The higher this frequency, the more accurate is the measurement of T_{meas}.

TSB-67 requires T_{meas} to be measured for all four pairs. The calculations by relation (12.3) should be carried out for the minimum of the four measured values of T_{meas}. The maximum length of the cable that can be measured by the tester must be no less than 310 m.

Apart from the listed factors that directly affect the accuracy of measuring the length of a cable, there is one more factor that can lead to quite unexpected (and inaccurate) measurement results. It is the variations in the impedance of the cable segment. As has been mentioned previously, impedance inhomogeneity causes reflection of the signal. It may happen that for a high-sensitivity receiver, the pulse reflected by the place of such an inhomogeneity will be interpreted as the pulse reflected by the end of the cable segment. In this case, the evaluated length of the link will be substantially less than the actual length. To avoid such errors, the SCS tester should enable one to adjust the threshold of sensitivity to the minimum value sufficient for the reception of only the basic reflected signals.

12.2.4 Measuring and Testing Equipment

Types of equipment for testing SCS electrical channels

Measuring and testing equipment for SCS on twisted pairs can be divided into three basic groups (Fig. 12.8). These groups include, respectively,

- network analyzers;
- SCS testers (field-test equipment, FTE);
- electrical testers (multimeters, continuity testers).

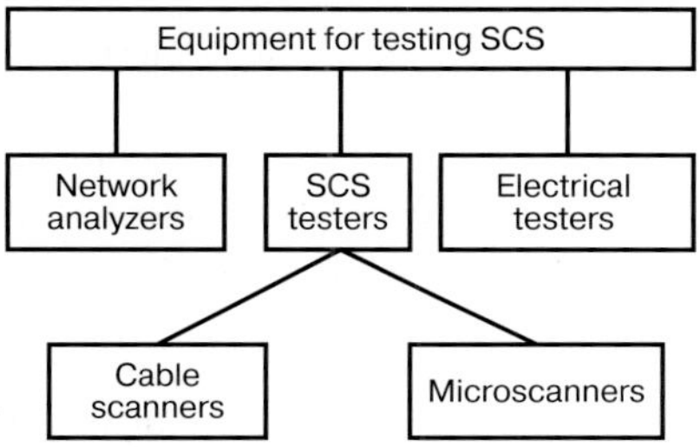

Fig. 12.8. Testing equipment for SCS

Network analyzers (not to be confused with protocol analyzers) are reference measuring instruments for inspection and standardization of cables and cabling systems. These devices have high precision and large overall dimensions and are designed for operation in laboratory conditions. Network analyzers are rather expensive; their cost exceeds $20,000. Typical network analyzers are, for example, Hewlett Packard HP4195A and HP 8510C. The equipment of this class is used at standardization centers and research laboratories. The consideration of basic design principles and structural features of this equipment is beyond the scope of this book.

SCS testers (*field testers*) have been designed for the inspection and testing of SCS at the place of their installation (in other words, for field testing). Often, these devices are called cable scanners. These are the basic equipment to be used for operational measurements in subsystems of SCS implemented on twisted pairs. The field testers allow one to carry out complex inspection of 4-pair cables, links of ISO/IEC 11801 classes C and D, and TSB-67 basic links for compliance with the requirements of categories 3, 4, or 5.

Electrical testers (multimeters) are simple, inexpensive, devices in widespread use that allow one to measure AC and DC current and voltage, as well as DC resistance. Most advanced instruments of this class can also measure frequency, capacitance, temperature, parameters of field and bipolar transistors, and a number of other characteristics. Earlier, such devices mostly used analog pointer indicators, whereas most modern multimeters have a digital LCD with backlighting. They are used for simple tests, mostly for continuity tests to identify short and open circuits, in the absence of an SCS tester. For a fairly detailed review of the functional capabilities of modern digital multimeters, see [131].

To test the SCS cable wiring one can use not only the devices listed above but also some models of protocol analyzers, which allow the corresponding measurements as an additional option, for example, the Fluke 68X Enterprise LANMeter .

SCS testers

General information and functional capabilities. An SCS tester is the basic testing and inspection equipment widely used at all stages of installation and operation of SCS. It is light (typical mass is less than 1 kg), portable (typical overall dimensions are 20×10×5 cm), and powered by batteries or an electrical network. The battery capacity is designed so as to provide measurements during one working day. Among the manufacturers of SCS testers, the companies Datacom Technologies, Fluke, Microtest, Scope Communication, and Wavetek are most familiar on the Russian market. General information about the products of these companies is given in Table 12.12.

Table 12.12. Basic parameters of SCS testers of level II produced by various companies [132]

Model	Measured parameters[a]	Maximum frequency measured, [MHz]	Time of Autotest	Two-channel testing	Number and type of batteries	Operating time with completely charged batteries, [h]
Datacom Textron						
LANcat System 6	A, AC, FC, I, L, N, NC, PA, PF, PN, R, RL, V, W	250	20	Yes	NiMg	12
LANcat System 5	A, AC, I, L, N, NC, PN, R, V, W	100	20	Yes	NiMg	12
Fluke Corp.						
DSP-4000	A, AC, C, FC, I, L, N, NC, PA, PF, PN, R, RL, V, W	350	10	Yes	NiCd	10–12
DSP-2000	A, AC, C, I, L, N, NC, PN, R, RL, V, W	155	20	Yes	NiCd	10–12

Table 12.12. (cont.)

Microtest Inc.						
OMNI Scanner 8222-XX	A, AC, FC, L, LR, NC, PA, PF, PN, RL, V, W	300	27	Yes	NiCd	10
Penta Scanner-350 8180-XX	A, AC, C, L, L, LR, 100. LT, N, NC, RL, W	21	Yes	NiCd	8	
Hewlett Packard/Scope Communication						
WireScope 155	A, AC, AR, FC, I, L, N, NC, PA, PF, PN, RL, V, W		160	13	Yes	NiMg
Wavetek/Wandel & Goltermann						
LT-8000	A, AC, C, FC, N, I, 100, L, NC, PA, PF, PN, R, RL, V, W	100	10	Yes	NiCd	12
LT-8155	A, AC, C, FC, N, I, 155 L, NC, PA, PF, PN, R, RL, V, W	155	10	Yes	NiCd	12

[a]A, attenuation; AC, ACR; C, capacitance; CN, continuity; FC, ELFEXT; I, impedance; N, noise level; LR, DC-loop resistance; L, cable length; NC, NEXT; PA, PS-ACR; PF, PS-ELFEXT; PN, PS-NEXT; R, DC resistance; RL, return loss; V, NVP; W, wiring.

The modern level of development of hardware for microelectronics allows one to achieve a high degree of automation of measurements and interpretation of the results. During the general test in the Autotest mode, a number of relevant parameters are measured consecutively within several seconds without interference by the operator, then the measurement results are compared with the requirements of a standard or a network protocol (if the latter is indicated explicitly), and a PASS/FAIL report is displayed (Fig. 12.9).

The test is considered passed only if all requirements of the standards have been satisfied. If at least one parameter has not satisfied standard norms, the test is considered failed. In case the result of the testing is negative, the parameters that failed to satisfy the standard requirements are displayed.

Apart from general testing with the overall result PASS/FAIL, SCS testers allow one to measure an isolated specific parameter or a selected set of such parameters that are explicitly indicated by the operator.

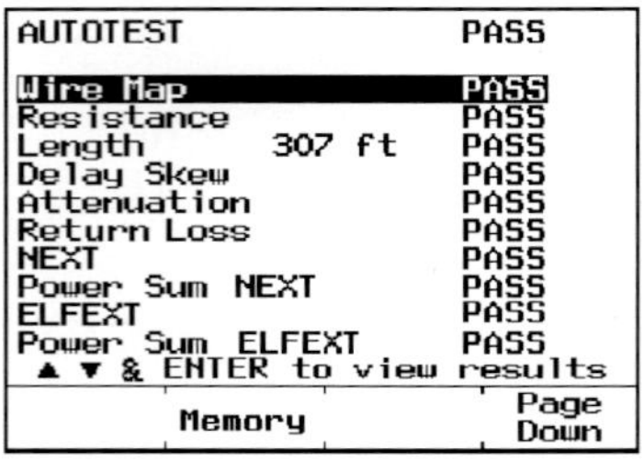

Fig. 12.9. SCS tester screen

SCS testers always consist of two units that are referred to as the basic block and the injector. When measuring, the injector is connected to the far end of the link to be tested. This is accounted for by the fact that most measurements and tests (for example, those checking the attenuation, *NEXT*, or wiring among the contacts of a modular connector) require the performance of certain operations at the far end. When measuring the attenuation, the injector emits a sinusoidal signal of prescribed amplitude and frequency into the cable, and the basic block receives this signal, passes it through a narrow-band filter to suppress noise, and measures the amplitude. When measuring *NEXT*, the injector at the far end provides a matched load. TSB-67 requires that *NEXT* be measured at two ends of the cable. Some modern testers do not permit one to change places of the basic block and the injector when measuring NEXT at the second end of the cable, which substantially reduces the time and effort of the testing. In such devices, the basic block and the injector change places functionally rather than physically.

The injectors of the first tester models were intended primarily for matching the load of the link to be tested to provide the measurement accuracy required by TSB-67. In this case, the injector was a purely passive unit. Second-generation testers use an active injector that performs part of measurement operations. This solution substantially increases the functional capabilities of the tester, in particular, it reduces the measurement time. The active injector is powered by a nickel-cadmium battery, and the device is supplied with a second charging unit.

The major components of the basic block are (i) the controller, (ii) the storage for the operational system and measurement programs, (iii) the control keyboard, and (iv) the graphical liquid-crystal display (LCD) with a typical resolution of 128×64 pixels The display shows the measurement results in a digital or graphical format and also presents a context-dependent menu and help prompts. The overwhelming majority of testers have black-and-white displays. Recently, the first models with color displays have appeared. To facilitate the work in insufficient visibility conditions, the displayed working files are backlit.

Control commands are entered from the keyboard, which has 12 alphanumeric and several additional functional keys. The keyboard may use various types of keys and/or control manipulators. The data are entered in accordance with a menu by means of a cursor or by setting certain parameters in an explicit form; if necessary, one can use the help information.

In some models of testers, the basic block and the injector have facilities for connecting a handset. This makes it possible to organize service communication between operators at different ends of the line subjected to testing, which accelerates the measurement process.

There are three basic schemes for connecting the object to be tested to the basic block and the injector (Fig. 12.10). The first scheme implies a fixed configuration and requires several types of connectors (usually, a modular jack and a BNC) to be located on the device, as is the case for Combo network adapters. This scheme is outdated and is no longer used in new devices.

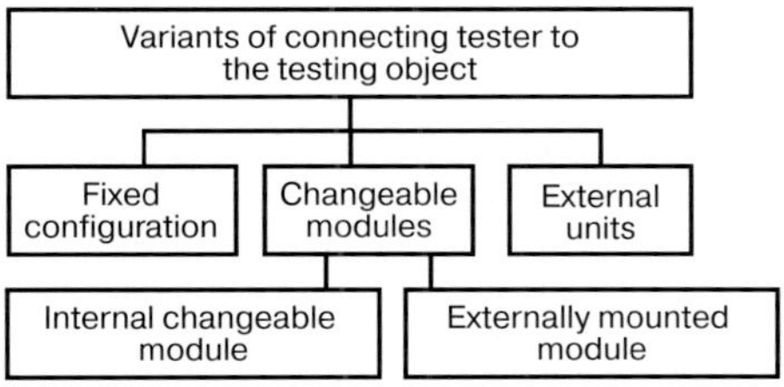

Fig. 12.10. Basic schemes for connecting the object to be tested to the tester

The second scheme uses a modular structure in which a changeable module is established if it is necessary to pass to a new type of connector. The device controller automatically recognizes the connection of the respective module and performs necessary switches and adjustments. This solution has two modifications. The first modification uses an internal changeable module entirely covered by the casing of the device when in the "on" mode. The second modification uses an externally mounted module. The external module may have a modular connector jack or a short cord with a modular connector plug. The popularity of the latter solution has considerably grown in connection with a rapid increase in the occurrence of level III testers. One cannot use universal cords when testing systems of category 6, because these systems produced by different manufacturers have unique circuits to compensate for cross talk and therefore are incompatible with each other. The device with a mounted module is somewhat less convenient to work with, but has smaller overall dimensions when being transported in a bag in the "off" mode. The solutions in question are easy to adapt to a specific physical medium, and the sensitivity of the tester that is lost from wear on the modular connector contacts during long-term operation can be easily restored by merely replacing the failed module.

Finally, the third modification involves external units that are attached to the basic block and the injector by means of short patch cords. Most frequently, this design solution is used in the testers that permit one to test fiber-optic SCS channels along with electrical channels.

Note that testing of all types of media is carried out in accordance with the same rules. This substantially facilitates the work and training of operators and reduces the probability of faults.

Testing results (500 or more lines) are stored in the device internal memory and can be read through a serial port as a file to be stored in a computer, for printing out as a protocol of measurements, or for recording into a special data base. Apart from the results of the measurements, this file necessarily contains the date and time of the testing, symbolic name of the channel, and other additional data facilitating the analysis and interpretation of information contained in the file.

To facilitate the adjustment of the device, its internal memory contains a built-in library of parameters of basic types of cables. In addition, the user can supplement this library with data for cables that are not involved in the original list.

Test parameters and operations .The standard characteristics to be tested in accordance with TSB-67 comprise:

- wire map;
- electrical length;
- attenuation;
- NEXT.

The measurement errors of modern SCS testers produced by leading manufacturers usually satisfy the TSB-67 requirements with a substantial margin (Table 12.13).

Table 12.13. Measurement errors for some models of modern level II SCS testers

Model	Manufacturer	Measurement error, [dB]	
		Attenuation	NEXT
TSB-67 norm		1.0	1.7
PentaScanner+	Microtest	0.3	0.5
WireScope 155	Scope Communication	0.8	0.5
LT-8100	Wavetek	0.6	1.6

Apart from the basic parameters, one should also test

- DC-loop resistance: measured for a pair short circuited at the far end and allows one to reliably detect open circuits and check the quality of contact termination;
- Capacitance of conductor pairs: a considerable difference from the nominal value indicates low quality of termination of the conductors at the connector plugs and jacks or an inadmissibly large extension of the cable during cabling;
- Impedance;
- Structural return loss (SRL);
- Propagation delay;
- Electrical noise level: permits one to assess the level of noise due to external sources.

Some devices may have a NEXT locator option. This option, when engaged, displays the value of NEXT at a specific point versus the distance. This option is rather helpful when searching for various kinds of defects.

Users can choose between the full and abridged testing programs. In the latter case, they should explicitly indicate the parameters to be tested.

Most types of SCS testers operate in a range of frequencies not exceeding 100 MHz. However, systems of category 6, which are expected to be widely used after 2001, operate at higher frequencies. This gave rise to a rapid increase in demand for instruments capable of testing in the range with an upper limit 250–350 mHz. For the same reason, the list of the parameters to be measured is substantially extended by characteristics important for creating high-rate channels of categories 5e and 6. Probably, such devices will be referred to as testers of levels IIe and III, by analogy with the available testers.

SCS standards specify parameters of cable channels for the worst case. Due to the use of pairs with the maximum values of NEXT, some types of network equipment permit have considerably better characteristics than can be achieved on category 5 cable channels according to the standards. When checking the possibility of a specific channel to be used for transmitting signals of the network equipment, testers measure the attenuation and ACR for the frequency ranges and pair combinations that support operation of the given application. During such measurements, the tester performs all adjustments automatically, and the operator should only indicate the name of the protocol in the menu.

To facilitate the creation of an SCS and its operation, some types of testers have tracing and office-locator options. When in the tracing mode, the tester launches a special signal into the cable, which is detected by the locator when it is above the cable. This allows one to identify and trace enclosed cabling pathways. The office-locator option allows one to substantially accelerate the process of restoration of the cabling log when it is lost or outdated. To that end, special plugs are inserted into the telecommunications outlets and the basic block is connected in the telecommunications closet. When a plug has been identified at the other end of the link, its number is displayed on the screen of the device.

Microscanners

A usual cable scanner is a complex and fairly expensive device, costing several thousand dollars (US). Typical prices on the Russian market in early-2002 were U.S.$3000–8000, depending on the models and the set of appliances attached. This device should be used for standardization tests of cabling channels, as well as for the identification and localization of defects. For operational inspections of SCS channels, one can use a simpler pocket-sized device that is frequently referred to as the microscanner. As compared with usual cable scanners, the functional capabilities of microscanners are substantially lower. However, they are much cheaper (U.S.$300–500) and have smaller mass and overall dimensions. Microscanners can test continuity, wiring, and length of cable channels. They may also have the office-locator option. As a measuring device, microscanners allow one to provide normal operation of SCS, and to extend an SCS by organizing several additional cable channels, without their certification and receipt of the manufacturer's warranty. The Microtest microscanner is the most widely used in Russia.

12.2.5 Alternative Devices for Testing SCS Electrical Subsystems

Reflectometers for electrical cables

The action of time-domain reflectometers (TDR) is based on the analysis of the signal reflected by various kinds of inhomogeneities in the link during probing by powerful short-term current pulses.

The electrical wave excited by the reflectometer pulse generator in the tested link is reflected backward by all inhomogeneities in its path. The amplitude of the reflected wave is proportional to the change in the impedance at the point of the inhomogeneity. The receiver's analyzer controls both the time of arrival of the reflected signal and the change of its shape in time. The analyzer results can be displayed as a graph (reflectogram) or a table. Additionally, the device calculates the distance to the inhomogeneity by the time delay between the launch of the probing signal and the arrival of the reflected signal and displays the result.

The reflection is absent if the load impedance is equal or very close to the cable impedance. In this case, the reflectogram shows only the probing pulse. If the impedance of the inhomogeneity exceeds (is exceeded by) that of the cable, the reflected pulse has positive (negative) polarity. The amplitude of the reflected pulse also allows one to judge the degree of matching between the cable and the load. If the cable is loaded by an active unit or an antenna, the reflected pulse may have a more complex S-shaped profile (Fig. 12.11). SCS installers frequently encounter the phenomenon of distortion of the shape of the pulse reflected from an active load. When testing, they look to see that the user's workstation is not attached to the telecommunications outlet. If this condition is violated, the scanner frequently shows lengths that may be several times different from the actual values.

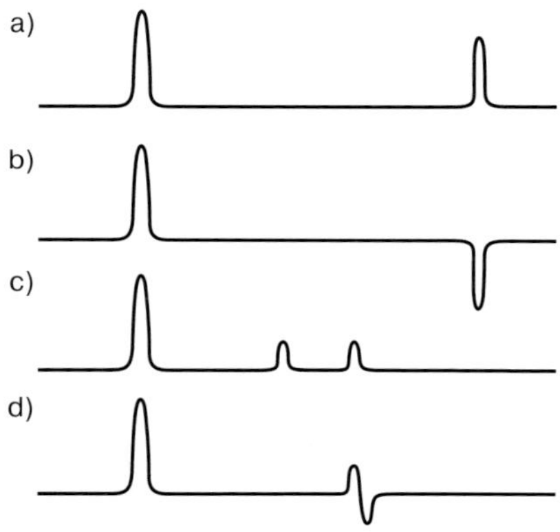

Fig. 12.11. Electrical cable reflectograms: *a)* cable breakdown; *b)* short circuit; *c)* defects; *d)* resistive load.

A graphic display permits one to carry out a complex analysis of all inhomogeneities in the cable. The devices with a numerical display are frequently used as length meters or probes for detecting simple defects.

Reflectometers for electrical cables have been widely used in city and long-distance communication networks. Because of the difficulties associated with the analysis of the starting portion of the reflectogram, they are effective only for testing cables in backbone subsystems and have not been widely used in SCS engineering. When testing horizontal subsystems, cable scanners successfully play the role of reflectometers. All types of cable scanners implement functions of reflectometers, to a certain degree. Operating characteristics of reflectometers for electrical cables available on the Russian market can be found in [133, 134].

Wiring testers

These devices, sometimes referred to as cable testers [135], are designed for simple tests to check the correctness of wiring of isolated twisted-pair conductors, as well as the absence of short and open circuits. They can be used for testing separate patch and terminal cords, as well as entire installed lines.

This device is designed as a pocket-sized unit with a battery supply. On its case, there are several jacks of 8-pin modular connectors, one of which is the reference connector and the others are working connectors, each corresponding to a specific wiring scheme. The most widespread design contains three working jacks for wire maps USOC, T568A or T568B. Sometimes, the testers are specialized for only one of these schemes. The testing results are displayed on an LED indicator. Some models, in addition to the optical indicator, also have an aural indicator.

Most of the devices of this type are designed for working with systems on 4-pair UTP cables. The only difference between these devices and those designed for testing S/UTP cables is that the latter units additionally test the continuity of the shield and the absence of short circuits between the shield and cable conductors. Typical examples are the Molex SLT-3 (33.D0020) and the HCI HC-45 devices.

There are also devices of this class that possess extended functional capabilities. They allow one to measure larger number of parameters and to interpret the results with a higher degree of reliability. A typical example is the device KARPi of the German company BTR. The additional options provided by this device include monitoring extraneous DC voltage on each of the conductors with the indication of its polarity, and an office-locator function.

The indication is provided by ten LEDs (nine red: one for each conductor and the shield, and one green: to signal the implementation and completion of measurements), and a seven-segment digital indicator. The digital indicator combined with the switch key allow one, first, to obtain on-line information about the character of the defect detected and, second, to analyze complex errors. Some devices designed for wiring tests are equipped with a BNC jack, which also enables testing of coaxial cables.

12.3 Testing of Fiber-optic Subsystems

12.3.1 Components and Parameters to Be Tested

The list of fiber-optic components to be tested involves

- *Cables:* The standard requirements for the optical characteristics of the fiber cables are presented in Section 4.1.2. The cables are tested on reels or drums at the inspection stage before starting cabling. Depending on the cabling complexity, testing of cables and their fibers can be sampled or global. For cables to be wired inside buildings, sampled tests can be applied, whereas for cables to be installed outside, global acceptance testing of all fibers is strongly recommended.
- *Connectors*: The standard requirements for the attenuation of optical connectors are presented in Section 4.2.1. One may not perform acceptance tests before installing a connector; however, it is necessary to test the attenuation level caused by this connector when it has been installed.
- *Permanent welded or mechanical optical connectors*: Inside buildings, it is not necessary to test the attenuation level due to each connector of optical fibers. The attenuation of optical connectors to be installed in outside straight joints and branch sleeves should be tested immediately upon their installation, before sealing the joints and sleeves.
- *Links*: A fiber-optic link is an installed structure comprising a cable and all relevant cable fittings (joints, sleeves, electrical and optical connectors, etc.), except for terminal cords. The optical links are to be tested at the stage of acceptance testing of the cabling system and for troubleshooting during operation.
- *Channels*: A channel is a fiber-optic link with terminal cords. The optical parameters of the channel characterize its quality from connector to connector. Channels are tested for troubleshooting during operation.

According to the standard ISO/IEC 11801, network equipment that uses fiber-optic cables as the transmission medium form the optical class of applications. Applications of this class have high and superhigh data exchange rates; clock frequencies of these applications are no lower than 10 MHz. With reference to recent engineering achievements, the transmission bandwidth of optical cables is not considered the limiting factor for ranges characteristic of SCS. Fiber-optic cables can be used to construct all three SCS subsystems.

The scheme of a channel and a permanent link of the horizontal subsystem for applications of the optical class is shown in Fig. 12.12. This scheme virtually coincides with the schemes of the respective objects for the electrical subsystem (Fig. 12.2), and differs only in that two additional permanent connectors (PC) have been introduced in the transmission channel. These permanent connectors can be implemented by either welded joints or mechanical splices [7].

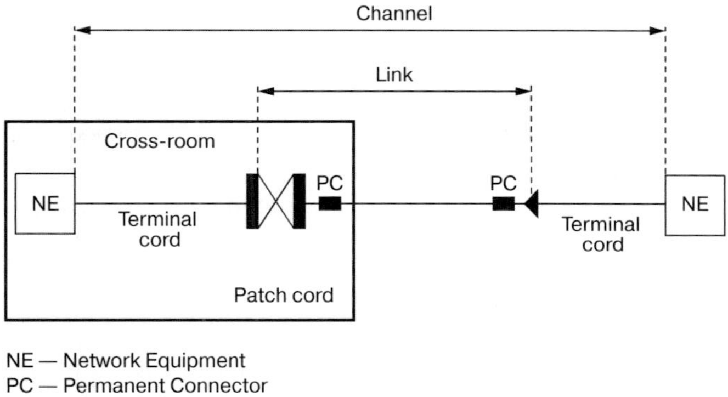

Fig. 12.12. Scheme of a channel and a permanent link of the horizontal subsystem for applications of the optical class

The maximum length of the lines of the backbone subsystem is 2000 m when constructed from multimode cable and 3000 m when constructed from single-mode cable. Single-mode cables, in principle, allow one to construct lines of larger length, however, such lines are not covered by SCS standards.

Building backbone subsystems are usually constructed on optical cables for internal wiring. This cable connects a floor distributor (FD) with the building distributor (BD) and has a length not exceeding 500 m. Campus backbone subsystems are constructed on the cable for external wiring connecting the campus distributor (CD) and the building distributor.

Channels of the backbone subsystems are formed by adding two terminal cords, each 30-m long, to connect network equipment. The terminal cords may be longer, however, in this case, the length of the backbone cable must be reduced. If the system involves cross-connections, the length of patch cords at each cross-connection must not exceed 20 m. Table 12.14 presents data on the signal attenuation in cables of different types for operating wavelengths. The analysis of these data shows that by constructing SCS optical channels on usual commercially available components, one can readily satisfy all attenuation requirements.

Table 12.14. Maximum channel lengths and signal attenuation allowed for SCS optical class subsystems

Subsystem	Channel length, [m]	Attenuation, [dB]			
		Single-mode		Multimode	
		1310 nm	1550 nm	850 nm	1300 nm
Horizontal	100	2.2	2.2	2.5	2.5
Building backbone	500	2.7	2.7	3.9	2.6
Campus backbone	1500 (3000)	3.6	3.6	7.4	3.6

The basic measuring instruments applied during construction and operation of SCS fiber-optic subsystems are the optical reflectometer and the optical tester. For optical class applications, the bandwidth of the line whose characteristics satisfy standard requirements is not the limiting factor and, therefore, devices of both types can test physical continuity and the attenuation of signal transmission channels at operating wavelengths of the network equipment. The reflectometer is more frequently used during the construction of the channel and for troubleshooting measurement to identify open faults and internal defects of isolated fibers. The optical tester is used for measurements after the construction has been completed and check tests with the reflectometer have been performed, as well as during maintenance. The measurement data obtained by means of the tester show the actual value of the attenuation "from connector to connector," thereby adding to the results of measurements by the reflectometer. In addition, the optical tester allows one to test the correctness of connection of separate plugs to the jacks of plug-type connectors in a switch unit. The parameters to be measured during the construction and operation of the lines are given in Table 12.15.

Table 12.15. Parameters to be measured during the construction and operation of optical communication lines

Parameter	Inspection test	Construction	Acceptance test	Operation
Attenuation	+	+	+	-
Attenuation of				
• Optical joints	-	+	+	-
• Fiber-optic channels	-	+	+	+
Distance to the defect and/or inhomogeneity	+	+	+	-
Optical power level				
• At the transmitter output	-	+	-	+
• At the receiver input	-	+	-	+

12.3.2 Optical Testers

Methods for measuring attenuation

Most frequently, Method 6 of norm IEC 874 is used to measure attenuation. It implements the *insertion attenuation* technique, which is well known in optical measurement technology. By measuring insertion attenuation one understands the difference between the levels of optical power at the input of the receiver when connected to the source directly and through the object to be tested. This method belongs to the group of "point-to-point" methods requiring the power source and the meter to be placed at different ends of the object to be tested. The advantage of this method is that it takes into account the power loss at the input and the output and eliminates it from the measurement results. However, it is necessary that the power loss during the calibration be approximately equal to that in the operating mode, which is a drawback.

The measurements are performed in two stages. The first stage involves the calibration of the workplace and is performed according to the upper diagram of

Fig. 12.13; the reference value is stored in the receiver's memory. After finishing this stage, the receiver automatically switches to the mode of measuring relative power.

At the second stage (lower diagram in Fig. 12.13), the attenuation is measured and the results can be read on the receiver's indicator. For these measurements, it is necessary to use high-quality jacks to connect the cord of the meter (testing cords) to the tested cable. These jacks are included in the appliance set of the tester. In addition, the testing cords provide effective suppression of parasitic radiating modes and leaking modes, thereby increasing the measurement accuracy. To increase the accuracy, the measurements should be performed in both directions and then averaged.

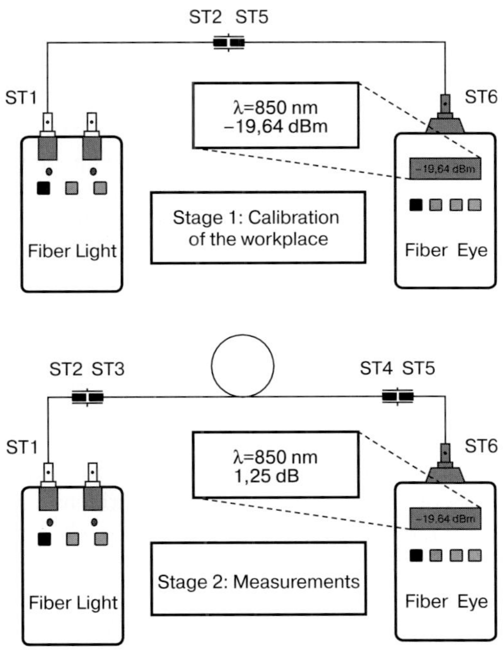

Fig. 12.13. Measuring attenuation using insertion attenuation technique

Note that the diameter of the core and the numerical aperture must be the same for the testing cords and the tested cable. Since optical cables with 50/125 fibers are widely used in Russia, two single-mode 50/125 cords are usually added to the standard appliance set of the tester.

As is apparent from Fig. 12.13, the measurement process requires the introduction of two additional plug-type connectors into the optical signal propagation channel. Since the repeatability of the power loss in the connector is finite (usually it lies within the limits of ±0.1 to 0.15 dB), one cannot guarantee the reliability of the measured attenuation value if the attenuation level is low. For this reason, if the device shows 1 dB or lower, one can make the record <1 dB in the protocol. This case is typical of measuring the attenuation in multimode optical patch cords, and the passports of these cords frequently have such a record in the "Loss" column.

Sometimes, the *direct measurement method* is used. According to this method, the absolute level of the optical signal is measured at the output of the signal source and at the output of the element (link) to be tested. The difference between the measured values gives the attenuation value. This method requires preliminary calibration of the devices and patch cords. The method provides good results if the value of the attenuation to be measured is significant. It can be applied to long channels when the preliminary calibration of the receiver, necessary for the insertion attenuation method, is either impossible or highly complicated.

The cut-off method is a destructive testing method. It is frequently used during the inspection test of an optical cable. To implement this method one should equip the fiber with an end piece, connect it to the signal source, and measure (by a meter) the level of the output signal at the other end of the cable. Then one cuts off a piece of the fiber of length 1–1.5 m at the transmitting end, splits off its face, and measures the signal strength by means of an adapter connected to the bare fiber. This strength is regarded as the input level. The attenuation is defined as the difference between the values obtained. To increase the measurement accuracy, it is recommended to repeat splitting off the fiber several times and to take the average of the measured values as the level of the input signal.

Design features of optical testers.

Optical testers (optical loss meters) are designed to measure the average level of the optical radiation power at operating wavelengths of fiber-optic communication lines (850, 1300, and 1550 nm) and to determine the signal attenuation in cables and isolated components of the line. The testers are applied during the construction, fitting, and maintenance of fiber-optic communication lines, as well as during prechecks and tuning of the network equipment with fiber-optic ports. The testers may operate both with multimode and single-mode fibers. They are supplied with several changeable adapters for various types of connector plugs.

The optical tester consists of two basic units – the optical power meter (OPM) and the light source. The optical power meter is used to measure the power of optical signals, as well as the attenuation of these signals in links, channels, and isolated components of an optical cabling system. The meter design involves a photodiode (usually germanium or InGaAs), a photocurrent amplifier, a signal processor, and a digital display. The photodiode converts the light flux incident on its window into electrical current that is analyzed by the signal processor. As a rule, the result of the processing is shown on the digital display (the Pocket Fiber Tester device by Darkstar Technologies uses a bar indicator with a resolution of 2 dB for that purpose). The signal processor also (i) compensates for the nonlinearity of the amplitude characteristic and the nonuniformity of the spectral characteristic of the photodiode, (ii) converts the input analog signal into the digital output signal, (iii) suppresses fluctuations by means of accumulating several measurements, and (iv) displays the control signal. Upon switching on, the meter tests the operational integrity of separate electronic units. The meters are usually calibrated for several fixed wavelengths whose values are displayed.

Stabilized light sources (SLS) are designed for launching an optical signal of a prescribed power and wavelength into the fiber-optic component to be tested. The constancy of the output power of such a source is maintained by feedback control of the forward current of the source by the difference between the reference voltage and the voltage of the photodetector in the feedback circuit. The light sources can be divided into light-emitting diode (LED) and laser diode (LD) ones. Laser multiwave sources are inferior to LED ones in terms of the cost and stability of the characteristics. However, these sources are substantially more convenient in operation, since they have only one connector. Signals of different lasers are fed to it by means of an internal optical splitter. This unit is rarely used in LED sources to reduce the cost of the device.

Many of the sources of high-accuracy optical testers may operate in the mode of modulation of the output signal intensity with a frequency close to 2 kHz. Such signals allow one to eliminate the influence of extraneous light in the optical range and low-frequency noise in the electrical channel, thereby improving the measurement accuracy. To increase the sensitivity of the receiver, one can use resonance amplifiers and synchronous detectors. Some types of optical power meters display the modulation frequency of the received signal or indicate the reception of such a signal by an aural indicator. An optical signal modulated with a certain frequency is also used during the operation of active fiber identifiers (Section 12.3.4).

Some types of testers combine the light source and the light detector with matched optical characteristics in one case. Such testers are sometimes referred to as integrated meters or optical cable attenuation analyzers. Such a design reduces the measurement time if two operators working at different ends of the cable channel have identical devices.

Multimode optical testers are usually powered by NiCd batteries or galvanic cells. Laser light sources with high energy consumption are frequently supplied with adapters for network power supply.

Table 12.16. Basic operating characteristics of optical testers

Characteristic	OTU-94 Optronic, Russia	AQ4250(085) +AQ2752 ANDO, Japan	Fiber Solution Kit Microtest, USA	FOT-02 EXFO, Canada
Operating wavelengths, μm	0.85, 1.3, 1.55	0.85	0.85, 1.3	0.85, 1.3, 1.55
Dynamic range for measuring power level, dB	–50...+3	–80…+10	–55...+3	–50...+6
Light modulation frequency, Hz	270	–	0, 2000	–
Measurement error	10%	5%	0.02 dB	6%
Working temperature, °C	–10...+50	0...+50	0...+50	–10...+50

Table 12.16 presents basic operating characteristics for some models of optical testers. For more detailed information about characteristics of optical testers, including those produced by other companies, see [136] and [137].

To increase flexibility and reduce the user's nonrecurring costs, some companies offer specialized models of light sources and meters designed for a single operating wavelength. Depending on the specific need, the user can purchase a certain laser or LED source and a detector with appropriate accuracy characteristics and the dynamic range for measuring the optical power.

Adapters for cable scanners and automatic meters

Some manufacturers of testing equipment for SCS produce adapters for their cable scanners enabling these to test fiber-optic cables. The controller of the scanner automatically identifies the adapter, when connected, and runs the corresponding program. The results of measuring the optical signal level are displayed on the standard indicator of the scanner and, if necessary, can be stored in the scanner memory for subsequent documentation.

A typical adapter of this kind is, for example, the Wavetek Fiberkit. This device is designed to connect to the cable scanner Lantek Pro. The standard version of the adapter is designed for testing at a wavelength of 850 nm; there is a modification for testing at a wavelength of 1300 nm. The internal memory capacity of the scanner suffices for recording the results of 500 tests.

The adapter Fiber Smartprobe can be used with the meter Wirescope 155 of the company Scope Communications. This meter is recommended as a standard device for testing the Molex structured cabling system. The functional capabilities of this device have been extended by the length measurement function. The results of this measurement are displayed on the screen of the basic block and compared with the prescribed standard value. Because of the growth of the popularity of single-mode meters, the first models of single-mode adapters permitting one to test at wavelengths of 1310 and 1550 nm have appeared on the market. Usually, such adapters are designed as a separate device or block, which is inconvenient during testing. This drawback is removed in the Omnifiber adapter for the Omniscanner2 scanner, permitting one to test both single-mode and multimode cables.

In the DataCom LANCat V tester, the photodetector is designed as a changeable block that is plugged into the device jack and fixed by means of screws. The light source is implemented as an additional module powered by a galvanic cell or a network adapter.

Adapters for cable scanners somewhat reduce the costs of the testing equipment, since in this case the basic block contains part of the functions of the meter. Even less expensive is the Wavetek FPC850 device. This device is also designed as an adapter and provides a value of 1 mW/mV for the transfer characteristic slope. Therefore this device can be connected to a standard electric multimeter in the voltmeter mode to utilize its indicator when measuring power.

When testing and certifying cabling systems based on multifiber cables, it is convenient to use automatic meters. As compared with traditional testers and optical reflectometers (which will be considered later), these devices determine not only the length and the total attenuation but also the degree to which the channel subject to testing satisfies the requirements of a specific standard chosen from an available list. The Canadian company EXFO produces the FOT-910 meter, which allows one to measure the attenuation for two wavelengths and store the measurement results (up to 600 values) in the internal memory. Two identical devices, when established at different ends of the channel, automatically perform the preliminary adjustment of the parameters and calibration, which substantially increases the accuracy of measurements. Using the RS-232 port integrated in this device, one can enter the measurement results in a computer or print them on a printer.

The Microtest CertiFiber measurement toolkit is more versatile. It contains the basic and remote blocks to be connected to a pair of fibers at both ends of the tested channel so that one does not have to switch the fibers. Testing is allowed at two wavelengths (850 and 1300 nm). The results of up to 1000 measurements, together with the dates and times of these measurements and (if necessary) the alphanumeric ID of the channel, can be stored in the internal memory of the basic block. This information can be read from the memory for analysis or to print the standard test record sheet.

Apart from the attenuation, this toolkit also measures the total length of the optical line and the signal propagation delay. All measurements are performed automatically; the only parameters that are entered manually are the number of plug-type connectors and splices. The toolkit has an option (Symmetry) to compare the attenuation values in the fibers of the pair being tested. If the difference between these values is significant, the device generates a warning message. In the opinion of the developers of this toolkit, it reduces the testing time by 75% over the usual testers.

12.3.3 Optical Reflectometers and Locators

Optical time-domain reflectometers (OTDR), frequently referred to as reflectometers, are one of the most powerful means for testing fiber-optic cables. They are applied during construction of cabling systems, their certification, maintenance, inspection, repair, and other sorts of work. Such a wide application is explained by versatile functional capabilities of reflectometers. The reflectometer allows one to determine during one measurement cycle a large number of basic parameters of an optical cable, including the length, the attenuation per unit length, the location of inhomogeneities and defects (if any), the nature of these defects, the loss in connectors and splices, as well as other characteristics without complex preparations. It also permits one to perform all measurements at one end of the optical cable (in contrast to optical testers). The major drawbacks of the reflectometer involve its relatively narrow dynamic range (not exceeding 45 dB for available devices), accounted for by low power of the backward scatter signal; its high requirements for the quality of the input of light into the fiber to be tested; and its high cost. In addition, the reflectometer does not allow real-time measurements (at least 30 s dition, the reflectometer does not allow real-time measurements (at least 30 s are required to obtain a reflectogram of acceptable quality).

Operation principles of reflectometers

Measurements performed with reflectometers implement the backward scatter method. This method is based on the phenomenon of Rayleigh's backward scatter (Section 2.2.4). During measurements, the tested fiber is probed by short-term powerful optical pulses launched through the splitter. These pulses are reflected by distributed or local inhomogeneities, creating the backward scatter flux. By measuring this flux, one determines the attenuation in the cable as a function of its current length. The analysis of this function allows one to determine the location of inhomogeneities, their nature, as well as local and distributed losses due to these inhomogeneities. The results are represented in the form of a diagram (reflectogram), permitting one to determine the characteristics of the inhomogeneities and find out the cause of their appearance with more accuracy and reliability.

Figure 12.14 shows a simplified block diagram of a reflectometer. The controller processor coordinates the operation of the semiconductor laser and the electronic oscilloscope. The directed splitter with the optical connector socket serves for launching pulses into the fiber. The backward scatter flux is fed through the splitter to the photodetector. In the photodetector, this flux is converted into the electrical voltage that is fed to the oscilloscope Y-plates. The oscilloscope displays the backward scattering curve.

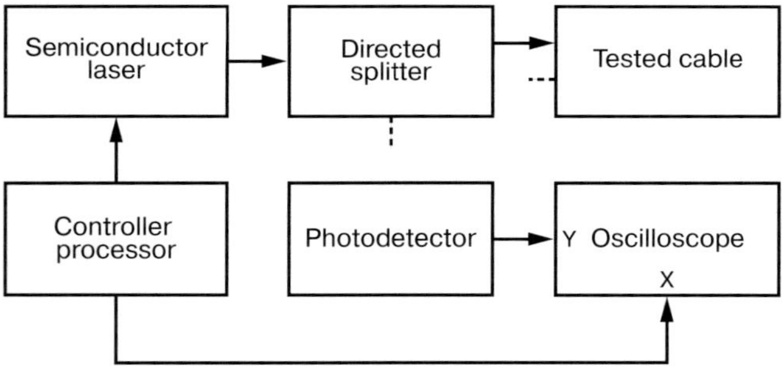

Fig. 12.14. Block diagram of a reflectometer

Many reflectometers, especially portable ones, construct point-to-point reflectograms using the data stored in the internal memory in digital format and are equipped with a liquid crystal display. This design reduces the mass and overall dimensions of the device and extends the functional capabilities for representing the measurement results and their storage for the further analysis.

Figure 12.15 shows a typical reflectogram displayed on the screen of a reflectometer. Together with smooth changes in the flux power, the reflectogram also shows local excursions due to inhomogeneities of various kinds. The initial excursion 1 is accounted for by Fresnel reflection in the optical connector between the device and the tested cable. The shape of the right front of this excursion is determined by the process of formation of the mode content of light at the initial stage. Excursion 3 indicates the plug-type connection and excursion 4 appears at the end of the cable. Splices (welded or mechanical) of fibers, at which reflection usually does not occur, correspond to steps 2 on the reflectogram, where the height of the step is proportional to the loss in the respective splice. The broken cable or the end of the cable are identified by the peak 4 due to the Fresnel reflection, followed by sharp noise-like changes 5 in the level of the detected signal.

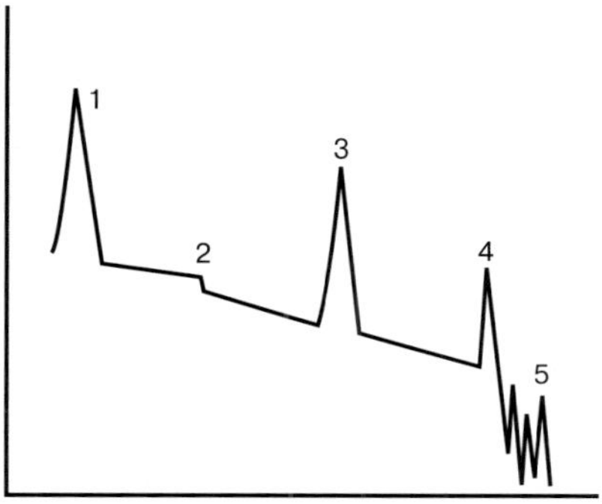

Fig. 12.15. Reflectogram example

In some cases, the end surface of the cable is not planar or perpendicular to the cable axis. In these cases, the reflectogram shows a sharp decrease in the strength of the backward scatter signal, with zero or low levels of the noise-like change in the signal strength beyond the cut-off point.

When testing short channels with low loss, high-quality sensitive devices frequently fix an inhomogeneity at the distance exceeding the channel length by a factor of 2. This effect is accounted for by double reflection of the probing pulses from the far and near ends of the fiber.

One can calculate the specific loss by the slope of rectilinear portions of the reflectograms and the total loss by the difference between the signal strengths at the ends of the channel. All calculations are performed automatically. For measurements on local segments, the operator should use special markers and cursors.

Design features of reflectometers

At the present time, both single-mode and multimode reflectometers that operate in all basic spectral windows of optical fibers are used in practice. These reflectometers can be designed as fixed devices whose size is comparable with that of a professional oscilloscope (this is the case for long-range reflectometers), or as portable minireflectometers. Minireflectometers combine low mass and small overall dimensions with good operating characteristics when working on cable channels several dozens of kilometers long. This combination makes minireflectometers popular among system integrators creating SCS fiber-optic links. Operating characteristics of some minireflectometers are given in Table 12.17.

Prior to measurements, the operator can set the duration of the probing pulse, the time of accumulation of results, the threshold for detecting inhomogeneities, as well as other parameters affecting such characteristics as the accuracy, the duration of measurements, and the stability of the reflectogram.

Table 12.17. Operating characteristics of minireflectometers

	FTB-250-202	FTB-250-303	FTB-250-303	OFT-50	OFT-30	AQ7225A	AQ7229A
Manufacturer		EXFO		Wandel & Goltermann			Ando
Parameter							
Fiber type	SM	SM	MM	SM	MM	SM	MM
Wavelength, nm	1310	1550	850	1310	850	1310	850
Dynamic range, dB	28	26	24	30	22	31	31
Near dead-zone, m	3	5	2	–	–	25	25
Resolution, m	15	25	7	14	8	5	5
Mass, kg	6,52	6,52	6,52	2,5	2,5	4,6	4,6

The losses from various defects and inhomogeneities, whose values are characterized by the heights of the respective steps on the reflectogram, are calculated by the two-point or four-point method. To increase the accuracy of calculation of the total and specific loss, the operator can manually set appropriate markers; all other operations are performed automatically by the reflectometer. To facilitate reading the reflectogram, the X-axis of the indicator is scaled in meters and the Y-axis in decibels.

Reflectometers have a time-magnifier option, permitting one to analyze certain portions of the reflectogram with higher resolution. In this case, the reflectogram of the entire link, with the analyzed segment with higher resolution indicated, is also displayed but in a smaller scale.

To facilitate the operation, modern reflectometers allow automatic adjustment of the range of measured lengths and of the duration of the probing pulse, as well as the option of displaying the reflectogram in real time. These devices have a resolution of 0.01 dB with respect to the strength of the optical signal and about 10 cm with respect to the length. This permits one to analyze reflectograms, both in graphical and in table forms, with a high degree of accuracy. To facilitate the analysis of complex channels with a large number of inhomogeneities, the Wandel & Goltermann FiberFox reflectometer has a certifying option. When this option is engaged, the device automatically compares the parameters of all inhomogeneities with standard allowances and forms the integral result "CORRESPONDS/DOES NOT CORRESPOND."

The measurement data can be stored in the internal random-access memory of the device or saved on a magnetic card, internal hard disk, or 3.5-in floppy disk. If necessary, one can read the reflectogram on an external unit by means of an RS-232 or Centronics interface.

To make the measurement results more graphical and informative, modern minireflectometers are equipped with color LCDs, typically with a diagonal measurement 7–8 in, although previous models had monochrome screens to reduce the device cost. Most devices use button switches to control the measurements, data processing, and reading the results. The Canadian company EXFO uses a touch screen for that purpose. To facilitate he operation, a comprehensive system of help information that can be requested by the user and output on the screen is organized.

Reflectometers of the companies EXFO (Fig. 12.16) and Laser Precision are equipped with a built-in alphanumeric keyboard. This allows one to supplement the reflectograms with text comments without attaching an external keyboard.

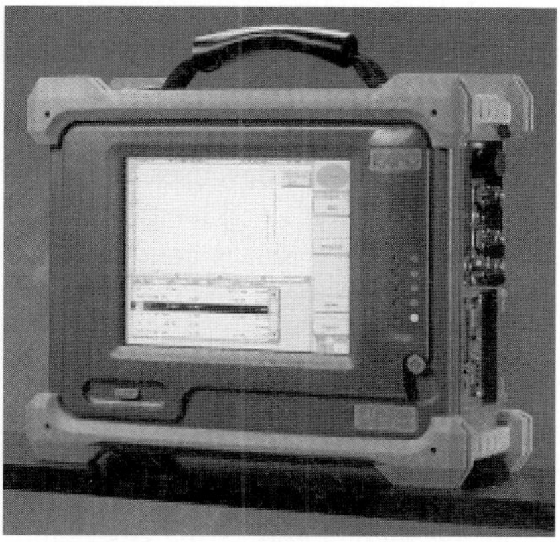

Fig. 12.16. Reflectometer FTB-300 of the company EXFO

Usually, reflectometers have a dead-zone, since it is impossible to measure the backward scatter flux until the probing pulse stops acting. To avoid this drawback, the Wandel & Goltermann OFT-30 and OFT-50 reflectometers have an internal extending fiber reel whose end is taken as zero of the scale.

Minireflectometers are powered by either network sources or NiCd batteries. When completely charged, the battery enables some models of minireflectometers to operate during 8 hours. The devices are supplied with two batteries and a charging unit, which provides the possibility of continuous measurements. Many models of such devices have an analog or digital indicator of the battery charge level on the display.

First-generation minireflectometers were usually designed as complete devices. In modern models, a tendency toward modular solutions has been clearly manifested. The devices are designed as a base with several slots for different changeable units, enabling one to test single-mode and multimode fibers at wavelengths of 850, 1300, and 1550 nm. Such reflectometers frequently have a built-in thermal printer, a driver for 3.5-in floppy disks (for storing the measurement results), and a slot for a magnetic card. The floppy disk or magnetic card can be also used to update the software of the device. The changeable modules also include an optical tester, a defect visualizer, an optical phone, an optical spectrum analyzer, and other units. Usually, there are sockets on the case to connect a keyboard and a mouse-type manipulator. The Wavetek MTS-5100 reflectometer has 22 changeable modules.

Reflectometers of virtually all leading manufacturers are available on the Russian market. The most widely used are devices of the companies Wavetek, Ando, Hewlett Packard, EXFO, and Anritsu.

To reduce the cost of reflectometers, designs in the form of cards to be used on PC-compatible computers has been suggested (AOC cards of the company Antel and FCS-300 and FCS-400 cards of the company EXFO). The card carries electronic components for forming the probing pulse, receiving the reflected pulse, and converting the latter into an electric signal. It should be inserted into a slot of a fixed or portable computer. The computer processes the measurement data and constructs the reflectogram, provided that it has appropriate software. If necessary, several cards can be used on one computer.

When testing by means of reflectometers, one sometimes uses normalizing reels, which are reels of fiber that has 0.25-mm buffer coats and lengths of no less than 1 km. One of the ends of the fiber has an end-piece that is connected to the reflectometer. The other end of the fiber is connected to the tested cable by means of a weld, a mechanical splice, or a plug-type connector. The major aim of the reel is to suppress leaking modes, thereby providing steady-state modes. Thus, the reel works as a mode filter. This permits one to substantially increase the accuracy of the analysis of the initial segments of optical cables that do not have end-pieces. The reel can also remove the dead-zone of the reflectometer. To that end, an optical connector plug should also be attached to the second end of the fiber, and the reel itself should be placed in a case that has sockets on its faceplate. This design is the case for the FSM-1000 reel of the American company FiberPlus.

Optical locators

The traditional reflectometer, even when designed as a minireflectometer, is a complex and expensive device. For this reason, many medium and, especially, small businesses dealing with the construction of structured cabling systems cannot purchase it. To satisfy the demands of this segment of the market, the designers of the fiber-optic technological equipment have developed simplified models of optical probes—optical locators and optical link length meters. The action of the locators and meters is based on backward scatter, as is the case for traditional reflectometers. The simplification has been achieved primarily by simplifying the software that constructs reflectograms and by the removal of the graphical display. This has reduced the cost of the locator 10–15%, as compared with conventional reflectometers.

In the more complex Photodyne (ser. 5200) meter of the American company 3M, the graphical display has been replaced by a simple alphanumeric display that can read either the distance to the defect or the attenuation of the signal on this defect. The range of this meter can reach 82 km.

Fiber Ranger optical link length meters of the American company Rifoc have substantially less mass and smaller overall dimensions, but their range does not exceed 5 km. These meters have single-mode and multimode models. They consecutively output the distance to inhomogeneities on the screen. If users have PCs on which appropriate software has been installed, they can reproduce the complete reflectogram of the tested link on the monitor.

12.3.4 Active Fiber Identifiers and Defect Visualizers

Active fiber identifiers are used during the maintenance of terminal switching and distribution systems. They provide quick and precise identification of the fibers without interrupting communications and show the direction of propagation of the signal. These devices have a bending splitter and a photodetector, which records the light flux from the splitter. For identification, one should put the head of the device on the fiber. After the clamp has been placed, the fiber is subjected to a calibrated bending force. At the place of bending, the normal conditions of wave propagation are violated, which leads to the appearance of a large number of leaking modes. These leaking modes cause intensive light emission, which is recorded by the photodetector. The identifier not only identifies the active fiber but also shows the direction of propagation of the optical signal through this fiber. The sensitivity of the photodetector used in the identifier suffices for operation with fibers in a colored 0.9-mm buffer coat. The measurement result is fixed by an LED and by aural indicators. For additional increases in accuracy and functional flexibility when working with background, one can tune the identifier to a signal with a prescribed modulation frequency generated by the optical tester. The well-known identifiers AFD-100 (NHC), LFD-100 (EXFO), and OFI (Wilcom) are designed as hand-sized devices with battery power supplies (Fig. 12.17).

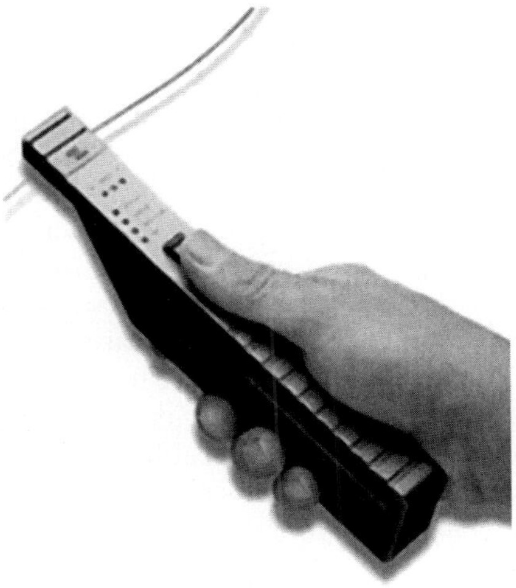

Fig. 12.17. Active fiber identifier AFD-100 of company NHC

Defect visualizers are designed for the detection of open faults and other defects of optical fibers by means of transillumination. The range of visualizers does not exceed 5 km and, hence, they can only be used to detect faults that are relatively close to the ends of the cable. The basic unit of the defect vizualizer is a powerful laser that generates red light. To facilitate observation, the laser light flux can be modulated with a frequency of 1 Hz. When the visualizer is connected to the fiber, the emission of red light is observed at the place of a defect.

Some models of optical reflectometers have a modular design, which allows users to plug the defect vizualizer modules into the basic blocks. This is the case, for example, for the Hewlett Packard E6000A reflectometer.

12.4 Conclusions

Structured cabling systems are subjected to measurements at all stages of their constructions and operation. These measurements are necessary to maintain normal functioning of the system and quickly restore the operability of channels and links case of emergency.

The measurements are carried out during the inspection testing, acceptance testing, and operation of an SCS. The operational measurements are divided into those for maintenance, troubleshooting, and control. All these measurements are based on the same principles and require the same equipment for their implementation. The content and technique of a specific measurement (test) depend on the purpose of the test as well as on the requirements for accuracy and speed.

The basic methods of measurements in SCS coincide with those used for testing common cable links. However, the specific features of structured cabling systems means that both the measurement techniques and the designs of meters must be seriously modified.

The models of the objects to be tested and measurement techniques standardized by normative documents allow one to reduce errors in determination of the basic parameters to a level of 1.0 to 1.5 dB. This accuracy suffices for both operational measurements and for certification tests to receive a warranty from the SCS manufacturer.

The general conclusion on the correspondence of an electric link subject to testing to the standard requirements is based on the measurement results for four characteristics: attenuation, cross-talk loss (both in a range of frequencies not exceeding 100 MHz), length, and wire map. Modern SCS testers measure also a number of additional parameters. After approving new editions of the standards, one should anticipate the appearance of new testing equipment with extended frequency ranges and lists of parameters to be measured. For optical links, it is necessary to measure only the length and the total attenuation at various wavelengths. Devices with microprocessor control and an internal memory for 500 or more testing results (SCS testers and automatic meters) substantially facilitate the testing of a large number of objects and the measurement of a large number of parameters in electrical and optical SCS links. These devices carry out the complete testing of one link over several dozen seconds. Testing is performed in the automatic mode without preliminary adjustments. However, these devices are rather expensive. For this reason, for current measurements, cabling system engineers widely utilize simplified non-automatic meters.

The main direction in the development of equipment for testing SCS optical channels involves the creation of specialized monoblock automatic meters or optical adapters for electric SCS testers. Such devices can simultaneously test a pair of fibers for all operating wavelengths of the network equipment. When testing, they implement some functions of the reflectometer and carry out primary analysis of the measurement results.

13 SCS operation

13.1 Administration

A medium-scale structured cable system (SCS) includes several thousand or even tens of thousands of components that interact according to a certain scheme. The scheme for the linkage of the elements may be changed during its operation, often substantially. New pathways may be added, or some lines may be dismantled. It is obvious that the efficient operation of such a complicated system as an SCS is impossible without conformity to a certain range of strict requirements that are called "administration."

It is obvious that the efficiency of the administration depends on the initial stages of the cable system development, that is, the design stage. Therefore, the administration principles must be observed when preparing working documentation.

13.1.1 Administration Concepts

The principal normative documents, which specify various administration issues for cable systems, are the standards TIA/EIA-606 [138] and ISO/IEC 14763-1 [17]. The recommendations included are reviewed below. The purpose of these documents is to create a uniform cable system administration scheme, regardless of application type. The standards additionally include documentation rules.

The administration is based on the creation and support of a database containing the verified information on cable system characteristics, its components, and their linkages. The database enables quick access to the information required for reconnection during troubleshooting, communications recovery in an emergency, system reconfiguration when relocating personnel, and creation of new work areas as well as other similar arrangements.

The database shall contain information on the current structure of a specific SCS, including its permanent components:

- cable pathways;
- cables;
- telecommunications outlets of work areas;
- cable termination on termination hardware in telecommunications outlets and equipment rooms;
- telecommunications closets and equipment rooms;

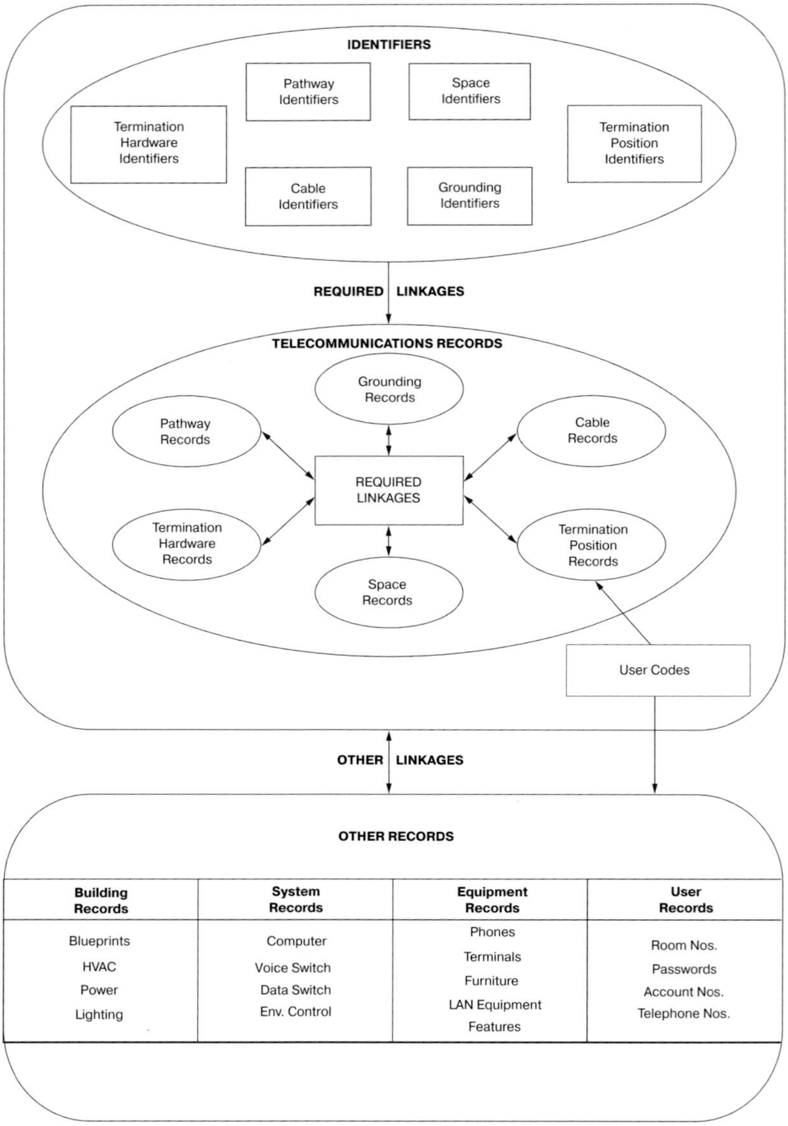

Fig. 13.1. Interaction between identifiers, records, and linkages according to TIA/EIA-606

The database must include information on:

- patching and terminations;
- cable system component failures.

The database can be accomplished by either paper-based or computer-based systems. The paper-based system is only preferable for small SCS. The presence of structured information on permanent SCS elements in the form of a relation database enables:

- real information on the current cable system state;
- easy planning and performing of necessary reconnections;
- emergency troubleshooting and troubleshooting service.

The administration concept is based on using identifiers, records, linkages between the records, and supplementary information for each of the permanent elements of cable system. Figure 13.1 shows the interaction between the concepts.

Identifiers

An identifier is assigned to an element of the telecommunications infrastructure linking it to its corresponding record. Identifiers shall be marked at the elements to be administered.

The identifier is a number of user-friendly alphanumeric symbols in compliance with TIA/EIA-606. It is recommended that the identification scheme of cable system elements be planned so that single-type elements use unique identifiers. Such uniqueness is achieved by means of prefixes. Selecting prefixes, principal conditions are the same as for DOS file names: the created identifier should be clear to the SCS administrator. The length of identifier and the list of restricted symbols when using an electronic database are limited only by the software used for its support (Table 13.1).

Table 13.1. Examples of unique prefixes for cable system elements

BCxxx	Bonding conductor
BCDxxx	Backbone conduit
Cxxx	Cable
CBxxx	Backbone cable
CDxxx	Conduit
CTxxx	Cable tray
ECxxx	Equipment (bonding) conductor
EFxxx	Entrance facility
ERxxx	Equipment Room
Fxxxx	Fiber
GBxxx	Grounding busbar
HHxxx	Handhole
Jxxx	Jack
MCxxx	Main cross-connect
MHxxx	Manhole or maintenance hole
PBxxx	Pull box
Sxxx	Splice
SExxx	Service entrance
SLxxx	Sleeve
TCxxx	Telecommunications closet
TGBxxx	Telecommunications grounding busbar
TMGB	Telecommunications main grounding busbar
WAxxx	Work area

When creating the identification scheme, it is possible to include specific information on the identified element directly into the identifier. For example, the identifier P701-1-1 may be assigned to telecommunications outlet 1, in work area 1, in room 701 (Fig. 13.2).

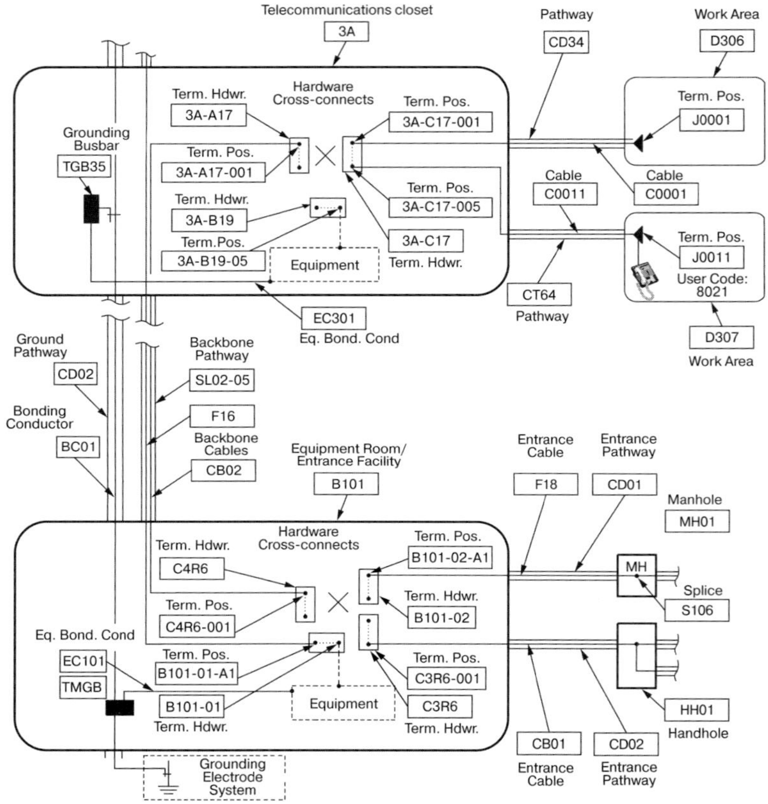

Fig. 13.2. Sample plan for identification of cable system elements

Unlike its American analogue, the international standard ISO/IEC 14763-1 gives specific recommendations on identifier construction principles and their field structures. According to this document, the element of the administration system has a format specified in Table 13.2. Field 1 includes general information on the building or its location. The minimum length of the field shall comprise 5 alphanumeric symbols. Field 2 defines room locations or any other premises and has a minimum length of 7 alphanumeric symbols. Field 3 is left for the identification of a specific SCS component and includes at least one character cell to record the identifier of the component, and three characters to record its number. Field 4 numbers the active equipment port, field 5 contains supplementary information on the component.

Labeling is the marking of an element of the telecommunications infrastructure with an identifier and other relevant information. Labeling shall be accomplished in either of two ways: labels with identifiers may be securely affixed to the administered element, or the element itself may be marked. The mark should be durable, clear, and easy readable. The issue is reviewed in detail in Section 13.1.4.

Table 13.2. SCS administration system identifier format according to ISO/IEC 14763-1

Field 1	Field 2	Field 3	Field 4	Field 5
Information on building	Information on room	Component identifier	Port number	Physical data

Records

A record is a collection of information about or related to a specific element of the telecommunications infrastructure. It is required that identifiers of separate elements clearly point to their corresponding records.

Each record allows fields of four types:

- required information;
- required linkages;
- optional information;
- optional linkages.

The required fields define the minimum information that is necessary for efficient system administration. For example, according to ISO/IEC 14763-1 the required information on any electrical cable includes information on its end locations, type, and number of pairs. Information in the optional fields makes the administration of SCS convenient.

The spaces fit records with numeric or textual characteristics of cable system elements. The records shall be mandatory updated when any changes are made in the cable system.

Linkages

Linkages are the logical connections between records and enable linkage between records. Linkages are placed in the dedicated fields of records linked with other records. For example, one linkage field of a cable record may include an identifier of the termination position or any other similar element of termination hardware. This identifier points to the record with information on this termination hardware.

The rule is applicable not only for information, but for other engineering systems of the building as well, such as power supply, ventilation, and air conditioning.

Information on neighboring systems

In addition to information on the cable system itself, the database may contain actual and reference information on neighboring systems. For example, it is useful to supply a panel jack that accommodates a PSTN cable with corresponding telephone number.

Alternatives for information presentation

Besides identifiers, records, and linkages, the standard TIA/EIA-606 recommends the use of *summary tables, drawings*, and *work orders* when administrating a cable system.

The summary tables and other similar documents are summaries of administration information, and include information on several records of the same type or relative records of different types. If the administration system is filed in electronic form, the tables can easily be created by the standard tools of database control systems, such as reports, diagrams, or forms of any kind.

Drawings serve for presentation of cable system elements in graphical form at various stages of planning and implementation. Commonly, drawings show the location schemes of cable system elements inside any object (for example, the disposition of termination hardware inside the mounting rack), pathways, and outlets locations on a building floor.

Depending on the type and kind of information presentation, the drawings are divided into structural schemes, working drafts (used by installers during mounting work) and maintenance documentation, which contains information on cable system configuration once the mounting is completed and takes into consideration all corrections made in the working drawings during SCS installation.

Work orders document the operations needed to implement changes affecting the telecommunications infrastructure. Typical work orders document operations such as moving a patch cord, installing a conduit, or relocating an outlet box. A work order should list both the personnel responsible for the physical operations and those responsible for updating various portions of the documentation to assure its accuracy. After completion of any step that changes the configuration of the telecommunications infrastructure, all records must be updated.

Records and their content

Table 13.3 presents the list of mandatory fields for records of any type of cable system elements according to TIA/EIA-606.

The standard ISO/IEC 14763-1 recommends a somewhat different structure and mandatory content of the fields. In particular, cables, telecommunications outlets and termination hardware in telecommunications closets, cable pathways and spaces, and also active network equipment as well shall be described. It is recommended to include measurement outputs of parameters for lines and pathways.

Table 13.3. Mandatory record fields according to TIA/EIA-606

	Record	Required information	Required linkages
Pathways and spaces	Pathway	Pathway identifier	Cable records
		Pathway type	Space records
		Pathway filling	Pathway records
		Pathway load	Grounding records
	Space	Space identifier	Pathway records
		Space type	Cable records
			Grounding records
	Cable	Cable identifier	Termination position records
		Cable type	Splice records
		Unterm. pair/cond. nos.	Pathway records
		Damaged pair/cond. nos.	Grounding records
		Available pairs/cond. nos.	
Wiring	Termination hardware	Term. hardware identifier	Outlet connection modules
		Term. hardware type	Spaces
		Damaged position nos.	Groundings
	Termination position	Term. position identifier	Cable records
		Term. position type	Other term.position records
		User code	Termination hardware records
		Cable pair/cond. nos.	Space records
	Splice	Splice identifier	Cable records
		Splice type	Space records
	TMGB	TMGB identifier	Bonding conductor records
		Busbar type	Space records
		Grounding conductor identifier	
		Resistance to earth	
		Date measurement taken	
Grounding bonding	Bonding conductor	Bonding conductor identifier	Grounding busbar records
		Conductor type	Pathway records
		Busbar identifier	
	TGB	Busbar identifier	Bonding conductor records
		Busbar type	Space records

13.1.2 Administration of Telecommunications Elements

Administration of cable pathways and spaces

The cable system spaces imply the following premises, areas, and constructions that are suitable for location of network equipment, patch panels, and cables:

- equipment rooms;
- telecommunications closets;
- work areas;
- cable entrances;
- manholes;
- access points.

All spaces must be labeled. It is recommended to place the labeling at a space access point.

Cable pathways are intended for cables that link telecommunications together. If a cable pathway is made by two or more separate segments of different types or sizes, then each such segment is administrated separately.

Any cable pathway gets a unique identifier indicating the corresponding record. In a cable pathway that consists of several sections, a unique identifier must mark each section.

Cable pathways must be labeled at all endpoints. Additional labeling may be desirable at intermediate locations, or regularly spaced throughout the length of the pathway at equal intervals. Intermediate points with three or more pathway endpoints shall have labels on each endpoint.

If is impossible or not advisable to mark cable pathways by common means (for example, pathways in cellular floors or under-floor distributor structures), they must be marked on drawings.

Table 13.4 shows an example of a cable pathway record; Table 13.5 gives an example of a space record.

Table 13.4. Cable pathway record

Parameter	Required information
Cable Identifier	
Type	Description of physical parameters of cable pathway: type, size, other characteristics
Fill	Current fill
Load	Current load
Required linkages	Record identifiers
Cables	Cables in pathway
Space (end 1)	Space where pathway begins
Space (end 2)	Space where pathway ends
Spaces (access points)	Intermediate spaces where cable access is available
Cable pathways (others)	Neighboring cable pathways (extension of the given one)
Grounding busbar	If pathway needs grounding
Parameter	**Supplementary information**
Cable length	
Maximum fill	
Maximum load	
State	If any mechanical damage to cable is present
Assignment	
Number of bends	
Drawing number	Reference to drawing with designated pathway
Other	
Other linked records	
Reference 1	Building identifier
Reference 2	

It is recommended that summary tables include a list of corresponding identifiers indicating, at minimum:

- for cable pathways – their types, current fills, and current loads;
- for spaces – their types and locations.

It is necessary to show top views and sectional views of all telecommunications closets, equipment rooms, and cable entrances on space drawings. Drawings of cable pathways should include raceways, bend radii, pull boxes, wall ducts, and fire-retardant components.

Sections of cable pathways in work orders shall include at least the identifiers of cable pathways, their types, and their corresponding space records, but space sections must include space identifiers and their types.

Table 13.5. Space record sample

Parameter	Required information
Identifier	
Type	Equipment room, telecommunications closet, work area, accesspoint, etc.
Required linkages	Record identifiers
Cable pathways	Cable pathways ended in this space
Cables	Pathways ended within the space
Grounding busbar	Grounding busbar within the space
Parameter	**Supplementary Information**
Size	Space size (length, width, height)
Location point	Reference to building axis
Service zone	Name of building part
Key	Number of door key or code
Other	
Other linkages	
Reference 1	Power supply panel identifier
Reference 2	Air conditioning identifier
Reference 3	Active equipment record identifier
Reference 4	

Administration of cable lines

Horizontal and backbone cabling is marked at both ends. It is sometimes desirable to mark intermediate cable points, such as pathway endpoints, access points, and pull boxes, and to also have cable markings with indication of outlets identifiers.

The most preferable marking method is to fix markings of various types on the external cable jackets. The cable jacket can be temporarily marked with a soft-tip pen or special marker; the marking should be replaced with a permanent marking at the first opportunity.

When splicing two identical cables, the total construction is considered as one cable. If hybrid cables are used in the cable system, markings are installed on each functional element of such cables.

Termination hardware includes several termination positions or functionally identical elements. Therefore unique identifiers are assigned to each element of the termination hardware, and to each of its connectors.

Tables 13.6, 13.7, 13.8, and 13.9 show how to record elements of a cable line.

Table 13.6. Cable record sample

Parameter	Required information	
Identifier		
Type	Cable physical description, manufacturer code, category, etc.	
Disable	Number of disabled pairs/conductors	
Damaged	Number of damaged pairs/conductors	
Free	Number of free (unused) pairs/conductors	
Required linkages	Record identifiers	
Connection points	End 1	End 2
	Connector point identifier	Connector point identifier
List of pair numbers		
Splicing	Identifiers of splice record	
Cable pathways	Cable pathways with routed cable	
Grounding busbar	If the cable jacket requires grounding	
Parameter	**Supplementary information**	
Length		
Owner		
Other		
Other linkages		
Reference 1		

Table 13.7. Termination hardware record

Parameter	Required information
Identifier	
Type	Physical description, manufacturer code, type
Damaged	Number of damaged positions
Required linkages	Record identifiers
Connection point 1	Identifier of connected cable
Connection point N	Identifier of connected cable
Space	Equipment room space
Grounding busbar	If equipment requires grounding
Parameter	**Supplementary information**
Protection	Information on line protective device: type, protected pairs, current, voltage, cut-off time
Other	
Other linkages	
Reference 1	

The following types of summary tables are recommended:

- Cable log – includes at least a list of cable identifiers, their types, and connection points. Other useful information from cable or other records, for example, cable length records, is also available.
- Cable pathways table – shows port-to-port connections in a cable system. It contains, at least, a list of users, pathway cables, and appropriate plug&jack connectors of various patch panels, which are interconnected.

- Table of connections – describes all patching made within the given space. The minimum required information is a list of connectors for termination hardware within the given space that are connected with patch cords, with an indication of the cord type.

It is mandatory to draw all spaces that fit cable ends and backbone pathways on drawings. The indication of horizontal raceways is desirable. It is also necessary to include in documentation the building layout profile and show raceways for backbone cabling through cable pathways, telecommunications closets, equipment rooms, and cable entrances.

Table 13.8. Connection point record

Parameter	Required information
Identifier	
Type	Physical description, manufacturer code, category, etc.
User	Only on work areas
Number of pairs/fibers	
Required linkages	Record identifiers
Cable	Identifier of connected cable
Connected at other end	Second cable end connection point
Patching	Connection point that connects the given connection point with patch cord
Connecting equipment	(The connection point is part)
Space	Equipment room space
Parameter	**Supplementary information**
Assignment	Pathway application and identifier
Patching type	Type of patch cord
Attenuation	Patch cord attenuation
Other	
Other linkages	
Reference 1	

Table 13.9. Splicing point record sample

Parameter	Required information
Identifier	
Type	Splicing technique
Required linkages	Record identifiers
Cable	Identifiers of cables to be spliced
Space	Space with located splicing
Parameters	**Supplementary information**
Used equipment	
Other	
Other linkages	
Reference 1	

It is necessary to show mounting locations of telecommunications outlets on drawings. The disposition of all intermediate boxes and other devices for cable splicing must be shown on drawings.

Grounding administration

As there is only one grounding circuit of telecommunications equipment in a building, it is given a unique identifier. The conductor, which bonds the main telecommunications equipment grounding circuit with the building protective grounding circuit, is marked at both ends by mounting a table (Figure 13.3). All grounding conductors are marked by unique identifiers at both ends.

Fig. 13.3. Grounding conductor marking

Tables 13.10 and 13.11 give examples of records for main grounding circuits and grounding conductors.

Table 13.10. Example of record for telecommunications main grounding conductor

Parameter	Required information
Identifier	
Type	Type and physical characteristics of telecommunications grounding busbar
Grounding conductor label	Is the label installed on the grounding conductor (Fig. 13.3)?
Grounding resistance	Resistance of grounding conductor to grounding electrode
Date of last measurement	
Required linkages	Records identifiers
Grounding conductor	Record of grounding conductor to be connected with building grounding circuit
Space	Space to house telecommunications main grounding busbar
Parameter	Supplementary information
Other	
Other linkages	
Reference 1	

The grounding system summary table contains, at minimum, a list of buses for separate grounding circuits with indications of grounding conductors for connection to a high-level circuit.

The drawings of the grounding system must show the locations of the grounding electrodes, pathways of the conductors that connect the grounding electrodes with the telecommunications main grounding circuit, all grounding circuits of the building and, if possible, the raceways of grounding conductors from these circuits to the telecommunications main grounding circuit. The documentation must also include the building profile with raceways of grounding conductors through cable pathways, telecommunications closets, equipment rooms, and cable entrances.

Table 13.11. Example of record for grounding conductor

Parameter	Required information for grounding conductor
Identifier	
Type	Type and physical characteristics of conductor
Grounding busbar	Which grounding circuit connects conductor?
Required linkages	Records identifiers
Grounding busbar	Which grounding circuit is connected by this conductor?
Cable pathways	Cable pathway of conductor
Parameter	**Supplementary information**
Length	
Other	
Other linkages	
Reference 1	
Parameter	**Required information for grounding busbar**
Identifier	
Type	Type and physical characteristics of grounding circuit busbar
Required linkages	Records identifiers
Grounding conductor	Record of grounding conductor to be connected with telecommunications main grounding circuit
Space	Space housing grounding circuit
Parameter	**Supplementary information**
Other	
Other linkages	
Reference 1	Reference to records of grounding conductors for active equipment
Reference 2	

13.1.3 Interactive Control Systems for SCS

PatchView System

System view. The PatchView system was developed by RiT Technologies (Israel) in the mid-1990s, and represents a successfully advanced approach of interactive control for telecommunications infrastructure by using elements to monitor the states of separate ports across SCS. The system automates a number of routine administration procedures, thereby easing searches for misconnections and speeding up the SCS configuration process by the introduction of automation facilities to control the state of patch panel ports.

The computing reserve of controllers enables the system to perform a number of auxiliary functions by polling various sensors and issuing instructions to low-speed execution elements. The PatchView system, like the SCS, has a hierarchic tree structure and can be construed by actual overlapping of its equipment on the SCS without causing considerable change.

The central element of the PatchView system is a control station, which functions as the interface device for the operator, supports the database, and creates instructions, warnings, and other messages. The second level consists of one or several scanners. The scanners process the information from separate panels, provide communications support with the control station, and send information to panel LED displays. The third and lowest level is formed by intellectual (in the terminology of RIT Technologies) panels. Besides connections of ports with patch cords or switches, these panels create messages about the occupation of some outlets and transmit them to the scanner, indicating the network address.

Element base. *Patch panels* of the series SMART and SMART-Giga are produced in 16-, 24-, and 32-port options. They differ from products of other manufacturers by sensors that are placed on each port to define the moment of connection or disconnection of the patch cord plug.

Patching of ports is carried out by means of special 9-wire cords, terminated into 10-wire plugs (the neutral contact is left unterminated). If required, it is possible to use appropriate standard 8-wire cords with 80-position modular plugs, however, the system monitoring the state of the ports becomes inefficient. The necessity to use a special tailor-made 9-wire cables in the cord is stipulated by the fact that the valid SCS standards strictly prohibit forming any parallel connections to the conductors of raceways.

Each jack of the panel is equipped with a red LED, which controls the interaction of an executive program with the user when working with the panel. The patch panel is connected to a cable scanner by a ribbon cable via a control port, which is located on the rear panel aside. All types of patch panels by RiT Technologies are available in shielded and unshielded options with or without PatchView sensors.

By the statistics from mid-2000, the optical intellectual panels consist of a single model, SMART F/O 96. The device looks like a shelf 5*U* high, with 48 duplex SC-outlets placed in four rows. Each outlet is equipped with an LED. Like the electrical panels, the control port is paced on the rear surface of the shelf at the bottom and contains special 1–5-m long 2-wire optical cords (4 models). Unlike the common ones, they have an auxiliary copper conductor with diameter of 26 AWG in the cable and an electrical contact in the optical plug. Standard patch cords can be used, the same as for electrical modular panels, however, it is impossible to monitor the state of ports.

A cable scanner consists of a special electronic device, which is equipped with fastening elements in 19-inch hardware, and located in the same rack as the panels. There are master and satellite scanners, which form a general system by means of a local bus over the standard RS-485 interface. A port with a 10Base-T interface serves for messaging to the wiring management station. One cable scanner may connect up to 5 or 10 patch cables, depending on their modifications. Patch cable connectors are located on the rear of the scanner, so when the cable connected, it is not visible to the user, which complies with the general approach of RiT Technologies. The device is basically intended to permanently control the state of patch panel ports without affecting the data transmission process. Processing of messages and issuing instructions to the scanner are carried out by a remote management station. To work directly on site, a technician at the wiring center can use a PatchView hand-held Control Pad, which is attached to the PatchView scanner via a dedicated port on the front panel.

As an option, a so-called controller, which tracks the states of various devices in the rack (for example, locks and humidity sensors), can be connected. The controller is made in two versions: a full version and an auxiliary adapter. One controller supports the work of a maximum of six different adapters and issue instructions to four executive elements by means of relay contacts. The cascading of two controllers is allowed. The lowest cost and most compact auxiliary adapter serves two controllers. The controller and auxiliary adapter have no fastening elements on mounting rails in 19-inch hardware.

A wiring management station is a PC with preinstalled Windows-friendly control software. The station is usually located in telecommunications closets or equipment rooms so that the system administrator may control the entire network up to its physical level. The graphical display of racks, patch panels, and scanner is very realistic, and fully corresponds to the familiar graphical Windows interface.

PatchView system operation. The scanner functions automatically and starts periodic scanning of the status of ports as soon as the power is on. The scanner collects the patching information and reports it in standard SNMP protocol to the PatchView management station via LAN. Alternatively, the data can be sent via modems or RS-232 connections to control remote devices.

The collected information is recorded in appropriate fields in a database. The remaining fields contain textual information and are filled by the designer on preparation of the design documentation and by the system administrator during the current maintenance. The database is created in strict compliance with the requirements of the standard TIA/EIA-606 and enables the creation of all the documents foreseen by this standard.

The selection of pathway is made by the database inquiry indicating user's name, room number, port, or any other field. All changes may be stored as a file, which is activated when reconfiguration starts. The control software analyzes changes in the current configuration, as compared to the given one, and then sends instructions about changes made in links to a properly rack-installed scanner. Luminous or blinking LEDs help the technician control the patching. The following ideology is accepted: the communications channel has always only two ends, therefore, LEDs are activated in pairs. First, all unnecessary cords are disconnected. The ends of a patch cord indicate the disconnection status by blinking LED indicators. Then those ports that must be connected are indicated with constantly lit LEDs. At a wrong connection, the indicators start blinking, but the LEDs of the next pair of ports being connected stay switched off. Once the interconnection procedure is completed, the database is updated automatically. It guarantees an immediate and full record of all changes made in the architecture of SCS.

The scanner additionally processes and sends all messages from the sensor and execution elements controller to the management station. The management station immediately displays the received messages on its monitor.

The main advantage of the VisiPatch system is that it enables the interactive control of wiring. These advantages become essentially apparent for large-scale networks with numerous ports, as the system helps to dramatically reduce the paperwork and enables automatic updating of the connection database without the aid of a system administrator.

We must note the main drawbacks of this undoubtedly successful solution, which hamper its commercial introduction:

- somewhat increased cost of SCS as the inevitable result of its functional extension;
- the need to employ special 9-wire cords terminated into tailor-made 10-position modular plugs;
- the solution is supported only by one company;
- the lack of direct permission to use the interactive control elements in patch panels in the valid versions of the standards.

The reviewed system only monitors the configuration of the SCS elements that are covered by the valid standards (Fig. 13.4). Thus, it does not control the state of the equipment cords, whose breaks and incorrect connections cause the majority of failures in office information systems.

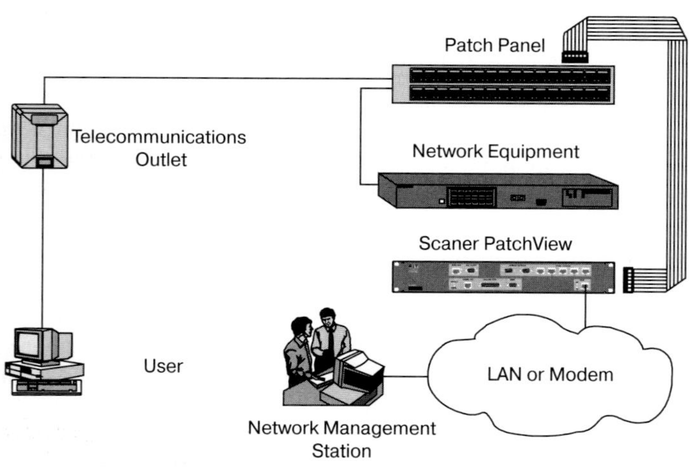

Fig. 13.4. PatchView structure

The information, which occasionally appears in the Russian technical literature, on the introduction of PatchView principles into official drafts of normative documentation has not been yet been officially confirmed.

Enterprise 1 system

The Enterprise 1 system, which sometimes is called PatchView for Enterprise and was developed by RiT Technologies in 1999, consists of further development of the PatchView system with a focus on functioning in large-scale corporate networks. The Enterprise 1 system is a program complex (Table 13.12), which employs the PatchView hardware. It differs from its primary standard by the following [139]:

- the system is based on the client–server architecture and relational database;
- the product is aimed at simultaneous support of several users, whose maximum number is limited by the software license;
- the software ideology is based on the "port-to-system" principle, which is more convenient and familiar to maintenance personnel;
- the Enterprise 1 system integrates powerful tools to support day-to-day operations of a system administrator, which far exceed the functional possibilities of the system PatchView.

As in the PatchView system, the network configuration is controlled by a work-task module, which arranges and automates planning and reconfiguration of tasks. The database of connections is updated only once notification is generated for the administrator and approved.

The built-in module for network graphical representation enables tracking of any line from port-to-port of active equipment. The analysis of state and network reconfiguration is essentially simplified and speeded up because of the visual representation of signal transmission channels. All elements and their identifiers are displayed by the administrator's management station.

Table 13.12. Software and hardware requirements of Enterprise 1 system

	Server	Work station
Operating system	Windows NT 3.51 and higher	Windows 95, 98
Minimum timing frequency of processor, MHz	266	166
Minimum RAM, Mb	64	32
Minimum HDD memory, Mb	1000	250

The open architecture of the product enables the very efficient importation of various data of other applications. As of mid-2000, such exchange is possible for network management systems by such leading network equipment manufacturers as Cisco and Cabletron, and also for the well-known HP Open View system by Hewlett Packard. Therefore, it is possible to control the entire signal transmission channel and to inventory physical resources of all active and passive network elements at their physical layer. The presence of a wide range of reporting facilities, which provide users with on-line information on the network at any layer, essentially contributes to the accomplishment of the latter function.

The maintenance and development of informational and computing systems for a large-scale enterprises is carried out by a large group of specialists, who create departments and sectors with assigned "areas of responsibility" for each department or specialist. The Enterprise 1 system, as a client–server class product, provides for simultaneous work with connection databases of several users, where each user gets an individual access level and the right to make changes.

The software has a Web-interface to access a module for report generation and network graphical representation. The interface does not allow users to make any changes, in order to protect the network against unauthorized access.

The Avaya iPatch system

The iPatch system (the prototype solution was known as the Smart Patching System) was shown for the first time by the Avaya company in 2001 at the CeBIT exhibition. The iPatch system implies a software and hardware complex, which includes special-purpose panels, scanners[2] in two various options, and software (Table 13.13).

Table 13.13. Comparison betwwen iPatch and PatchView

PatchView	iPatch
Smart Panel	Info-Panel
Satellite Scanner	Rack Manager
Master Scanner	Network Manager System
PatchView Software	Management Software

The panels of the iPatch system are manufactured in 24- or 48-port options, and depending on the type of termination positions used, may provide performance of category 5e or 6. The telecommunications panel includes a standard modular jack with three external and independent auxiliary sensors: a sensor with a plate-type sensing element, LED, and button. The sensor plate is located in the upper part of the modular jack and partly overlaps its opening. An appropriate plug presses the plate angle, rotating it up. The scanner registers activation of the contact.

Both of options are of height 1*U*. The depth of the device corresponds to the depth of the panel, which enables mounting it in shallow hardware such as wall frames. The distinctive main visual feature of an iPatch system scanner is a rather large, square liquid-crystal display, which serves for various data messages and other information. The scanner is administrated by means of a contextual on-

[2] Elements controlling the state of the iPatch system ports are further referred to as scanners in order to use uniform terminology in this section.

screen menu and six buttons. The rear side of the housing accommodates connecting leads.

A single Rack Manager supports the function of the interactive control system, which serves up to forty 24-port iPatch panels. The Rack Manager, which can be connected to the Network Manager, can manage a maximum of 99 scanners. The System Manager software for cable system control uses Windows 95/98 and has a state-of-the-art Windows Explorer-type user graphical interface.

Depending on the complexity and topology of the SCS being controlled and the presence of LAN as well as its structure, the iPatch system can be set up in three basic ways. A so-called standard configuration supposes integration of iPatch patch panels and their patching to the Rack Manager. Data exchange between Rack Manager scanners and the software System Manager runs in TCP/IP. A so-called alternative configuration differs by the integration of Network Manager scanners via RS-485 interfaces. One of the Network Manager scanners, which assumes a central position in the configuration, is connected to the workstation with activated software in TCP/IP. Both options for development of interactive control network monitor connections and notify administrators of failures in equipment premises, and enable the sending of planned tasks to the iPatch panels of all racks.

Finally, the *simple configuration* is based on the integration of only the Rack Manager scanners via RS-485 interfaces. Its provides connection tracking and notifies the administrator of trouble in the equipment premises.

The iPatch system logic implies message processing of sensor responses. The sensor response is recorded by a Rack Manager scanner, information about that response is recorded on the hard disk of the management station, and this information is used to create a database of connections and then is processed as usual. The use of such principle enables:

- making connection information nonvolatile (it remains on the hard disk when the power is off);
- use of standard cords with modular plugs;
- considerable decreases in any changes to the panels (for example, replacing modules of category 5e by those of category 6) on the introduction of the solution to mass production, because of the full electrical discontinuity of informational and control functions, .

The work tasks are created by the network manager through the System Manager software. The work tasks are then sent and displayed on iPatch panels in equipment rooms.

The major drawback of the iPatch system is, in our opinion, the strict control of the connection of cords: the scanner controls the operation order of scanner contacts, but not the physical connection of two ports.

13.1.4 Software Products for Interactive Control of Cable Systems

The have been stated standard TIA/EIA-606 recommends to file documentation on SCS operation in a computer-based form. There are many custom software products on the market for problem solving.

The first products appeared on the market in 1987. At present, the Russian market offers program packages Crimp for Windows (the company Cablesoft), System Manager (the company Unilogix Technologies) and some others. Each of the products is provided with individual records, tables and functions. Despite external distinctions, modern solutions within the field are specified by following common features:

- all systems are based on databases;
- graphic user-friendly interface is used to ease administration;
- cell library includes in standard delivery;
- an advanced checklist to check the correctness of various connections;
- support of various objects and images import from program packages CAD and various graphics editors;
- possibility to form different reports, work orders and similar documentation.

Hierarchic representation of both the whole cable system, and its objects make convenient the filing of maintenance documentation.

Separate elements of SCS being administrated are presented on plans as a simple image, vector icon-type image or compound graphic symbol.

There are two different administration strategies based on the reviewed systems. According to the first one, the system administrator plans changes to be carried out and proceeds them as reports. Once the required connections and other operations are made, the system automatically updates the database using the information recorded in the file-order. The alternative strategy is based on setting of special marks for certain objects and filling of required textual comments in appropriate fields. The database updates after the confirmation that the works are made. When realizing the both strategies, the database stores information about the date the tasks were created and executed, surnames of responsible persons and other service information.

Databases can make the user's work easier in a number of ways, including those in the brief list that follows:

- by partly automating administration by means of macro instructions;
- by searching for an optimal pathway to connect two given points;
- by checking for the presence of free lines and ports of patch panels.

More detailed information on the reviewed products, representation of their interfaces at standard operation modes, and other supplementary information is given in [140].

13.1.5 SCS Marking Elements[3]

Scope

An SCS is a complicated engineering product, which, in addition to the elements providing its creation and functioning, also includes thousands or even tens of thousands of separate components. It is obvious that the efficient maintenance and control (administration) of such a system is impossible without clear and accurate labeling of its separate components. According to the standard TIA/EIA-606, the list of SCS elements to be labeled includes:

- cables;
- patching equipment;
- work area and patch cords;
- outlets;
- various plug&jack connectors;
- trays and conduits;
- grounding elements.

The marking elements used during the introduction and maintenance of SCS must meet the following requirements:

- must comply with the requirements of the test UL969;
- must enable manual, printed, or field-made marking lettering of the required length;
- must enable color coding;
- must be easily installed and possess high mechanical strength and stability to environmental stresses;
- must enable labeling of devices having the same functions, but various gauges.

The standard TIA/EIA-606 gives no strict requirements for the labeling design. Both adhesive labels and markers with replaceable legends (marking inserts) can be used. It is recommended to use self-laminating markers to mark cable products (Section 13.1.5).

Marking elements that are used during SCS construction are called technological ones. They are easy to use and speed up the installation. Markers used during operation are called finishing ones. The presence of the finishing marking is a mandatory condition for SCS effective administration. There are also marking elements that are a part of many of the SCS products (panels and outlets), which are called the standard ones. Marking elements reviewed below, which are not a part of the equipment that is marked and delivered separately by the producers, are called auxiliary.

At present, one may state that the majority of SCS products covered by the standard TIA/EIA-606, include standard finishing marking elements. However, they do not completely meet the requirements. For example, the labels for patch

[3] This section partly includes materials by A. Nyshuller.

panels enable only manual legends in majority of cases, which does not give a pleasing appearance and leads to potential errors. Such elements often have only white backgrounds in the operating fields, which somewhat goes against the principle of color coding. To eliminate such drawbacks, a number of supplementary marking elements have been developed, which are reviewed below. The most well-known companies within the field are Brady, Legrand, Hellermann-Tyton, and Panduit.

The following factors affect the selection of the marker type:

- marking conductor or cable diameter, gauges of patch panel marking fields, etc.;
- space to set a label (end or middle of cable);
- information content;
- maintenance conditions.

The text may be written by hand or by means of various printers. Ballpoint pens or pens with indelible colored (including gray, red, and white) ink are used for manual markings. Sometimes such pens are included in the auxiliary equipment of some SCS (for example, markers in the PFX and PX series by Panduit).

The comparison of different printers is given in Table 13.14. To make it easier to prepare marking legends, many companies (AMP, Brady, Panduit, and others) offer special-purpose PC-compatible software (Table 13.15), which makes it possible:

- to operate in DOS or Windows;
- to support the use of laser and dot-matrix printers made by major equipment manufacturers;
- to use all type of Windows fonts, including TrueType;
- to scale separate signs within a broad range (6–300 points);
- to rotate and justify text, and to import various images and data, including .bmp and .wmf files;
- to provide linkage with ASCII, Excel, and other files;
- to form serial marks.

Table 13.14. Comparative characteristics of various printers for label marking

Printer type	Dot-matrix	Laser	Thermal-transfer	Inkjet
Operation speed	Low	Average	High	Average
Resolution	Low	High	High	Average
Roll carrier	Available	No	Available	No
Material choice	Wide	Low	Standard	Low
Cost	Low	High	High	Average
Dimensions	Good	Poor	Average	Good

The operation of such programs in Windows requires a processor not lower than 386-type and a minimum 4MB of operating storage (for DOS it requires at minimum a 286-processor and 2 MB of memory), that is, it may be used on the majority of modern workstations. Once the program is installed on the computer, it uses less than several megabytes on the hard disk.

Table 13.15. Custom software to create tags for SCS equipment

Title	Manufacturer	Required volume, [Mb]	
		RAM	HDD
PAN-MARK-Win2.0	Panduit	4	3
CodeSoft	Brady	4	3
Safety Easy	Panduit	8	5
TagPrint 6.0	Hellermann-Tyton	4	–
Logicab	Legrand	–	–

Experience shows that SCS marking elements are applicable for identification legends and programs for general use (text editors, database control systems, and electronic worksheets). They are inferior to the custom software in some kinds of services, which they lack. For example, creation of commercial marks is complicated. For these operations Excel electronic worksheets are most often used. They are recommended, in particular, for the creation of marking legends by such well-known manufacturers of SCS patching equipment as the company Siemon.

We must note that when delivering clip-on labels, hoops, and other similar nonplanar marking elements packed as blocks, the text can be machine-printed by means of special hand-held plotters and electronic typewriters (Section 13.1.5.3). The capacity of built-in accumulator of these devices is enough for three hours of uninterrupted work.

Adhesive labels

Adhesive labels have become widely used during construction and maintenance of SCS, and serve as technological and finishing marking elements. The labels are divided into:

- self-adhesive;
- self-laminating;
- flag-markers.

Self-adhesive labels are most efficient when the marking elements have a flat surface of a certain size. Therefore, SCS engineering uses them mainly for various patch panels and for large constructions such as conduits, trays, and termination hardware. The majority of self-adhesive labels are rectangular or square. Labels with warning legends, signs, icons, and inspection dates are often round or triangular and include tamper-evident coverings. For more resistance to various external exposures, these labels may be covered with an auxiliary laminated transparent film (for example, products of the LWS-11P series by Panduit).

A self-laminating label or marker is the main element to mark a cable end. The element may be considered as the evolution of a typical adhesive label and is made as a polymer strip of any size, with one side covered with a thin layer of glue. Its main distinction from other markers is that a self-adhesive label is always fixed overlapping the cable or wire. The strip has transparent and opaque parts. The latter serve for textual marking, and therefore their surfaces are nonporous. After installation the transparent part protects the legend against external mechanical and environmental exposures. The marker is steadily fixed along its length on the cable jacket by the layer of glue. The glue compound is selected to provide reliable self-fastening of the strip. Depending on the cable diameter, one may use gauged markers whose lengths are selected to make it possible to wrap the transparent part a minimum of 1.5 times around the cable.

Vinyl, polyester, and other materials may be used for self-laminating markers. The standard operating temperature range is 40–70°C, that is, it fully corresponds to the operating frequency range for external cabling. The operating range may be extended, if necessary, up to +135°C by the selection of an appropriate base material (Table 13.16). The latter property is important when routing external backbone cables in headers, since their marking must survive the leakage of hot-water supply pipes.

F*lag-markers* are used to mark thin conductors 1–2 mm in diameter. They often serve as elements of finishing marking for electrical pig-tails (in optical enclosures of various structures) and power cables (in electrical outlets of work areas).

All three types of adhesive labels are supplied on:

- standard sheets for laser or inkjet printing;
- fan-fold paper;
- in paper rolls for hand-held printers;
- pocket marker cards;
- leaflets in the form of a pocket book with a firm cardboard cover.

Table 13.16. Some materials for the production of labels for SCS elements

Material	Temperature range, [°C]	Application	Tolerance	Printing technology
Vinyl		Cable pathways	Oil, water, solvents, abrasives	Dot-matrix printer
Polycarbonate	–40…+120	Cable products	–	–
Paper	–40…+90	Termination hardware	–	Dot-matrix printer
Polyolefin	–46…+135	Termination hardware	Contaminants, moisture	Dot-matrix printer, thermal-transfer printer
Tedlar	–18…+135	Cable products	UV-radiation, contaminants, fats, abrasives	Dot-matrix printer
Polyester	–40…+145	Cable products	Oils, solvents, UV-radiation, moisture	Preprinted
Polyethylene	–40…+50	Cable products	Oils, solvents, moisture	Dot-matrix printer, thermal-transfer printer

The first two options enable the printing of labels by means of a laser or any other printer, which essentially reduces the effort for preparation of identification labels for major SCS. The latter two options are principally intended for manual marking. Book-type labels are supplied with pre-printed legends or icons (so-called preprinted labels).

The roll-type markers packed in a special cassette are used for manual label printing. The cassette has a window with a steady base, through which the markers move forward when the tape is pulled. This engineering solution is suggested, for example, by 3M, Hellermann-Tyton, and Panduit (Fig.13.5). Preprinted adhesive roll-packed labels are convenient for storage in the PMD-0-9 cassette (dispenser) by Panduit. The product has compartments for ten rolls, which can be replaced as needed. To make a high-quality label, the cassette is equipped with a cutting blade for even and accurate cutting.

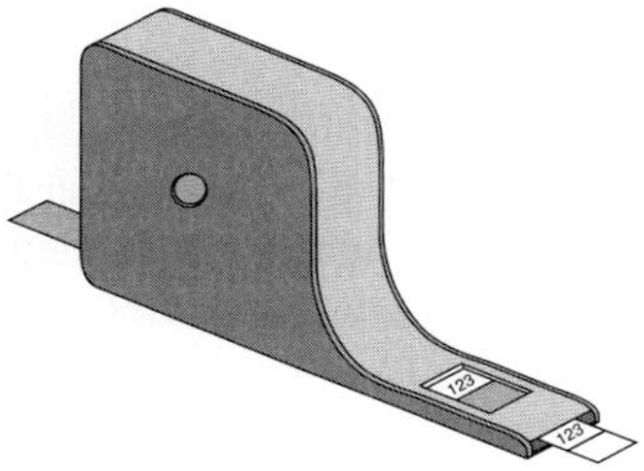

Fig. 13.5. Cassette for roll carrier of marking labels

The machine-printing is performed, for example, by means of hand-held dot-matrix printers IDPro and LS2000 by Brady (maximum line length: 16 and 20 characters), see Table 13.17.

Table 13.17. Specifications of hand-held printers

Type	Company	Printing method	Dimen-sions, [mm]	Weight, [kg]	Display
LS5E	Panduit	Dot-matrix, 9 pin	92×108×337	1.2	LCI, 2-lines, 16-positions
ID Pro Plus	Brady	Dot-matrix, 9 pin	280×105×85	0.73	LCI, 2-lines, 16-positions
LS2000	Brady	Dot-matrix	440×200×120	3.5	LCI, 4-lines, 20-positions
LS2200	Brady	Thermal-transfer	300×100×18	1.25	LCI, 2-lines, 16-positions
TT10SM	Heller-mann-Tyton	Thermal-transfer	200×135×226	1.81	–

At the end of 1998, Brady introduced a hand-held thermal transfer printer TLS2200, with a working field width of up to 50 mm. The device enables the creation of barcodes in addition to alphanumeric symbols, and differs from its dot-matrix analogues by a wide range of materials. The hand-held LS3E dot-matrix printer by Panduit also enables vertical printing, which may be required for manufacturing of some labels. All the printers reviewed possess a wide range of built-in service functions, such as automatic numeration, justification, and centering.

The majority of adhesive labels have white marking spaces. Many companies, for example, Brady, also deliver markers with variously colored spaces. It should be noted that several companies produce adhesive labels with enhanced bonding with the carrying surface. This is achieved by using glues that polymerize under pressure.

Custom cable product marking elements

Besides the labels reviewed in Section 13.1, other elements can be used for the marking of cable products.

Wire markers. A wide range of elements have been developed for marking wires that are reviewed below.

Shrink markers are available in two versions, the first of which are heat-shrink markers. There are in turn two variants of those. A marker of thermal transfer material is a tube 5–25 mm in diameter, with identification information printed or hand-written by a printer or a pen. For convenient printing, such markers are delivered in form of strips; the tubing is flat and fastened by adhesive tape. For marker shrinkage, a flame-free gas burner or small electric heat gun (for example, the HSG series of devices by Panduit [142]) are used. The fan is powered by the electrical mains and air temperature and volume can be adjusted, since it is equipped with a built-in controller. The shrink ratio is 1:2–1:3 on heating, the diameters of marked wires are 0.5–25 mm. The functional capabilities of the markers may be extended by making them of colored materials. These products are the only products of wide application that provide electrical insulation and therefore are very efficient for marking power leads. For example, the thermal transfer markers of the PAN-SHRINK series by Panduit ensure a voltage breakdown stress not less than 25.6 kV/mm.

There are also thermal transfer markers of low-insulation materials, which consist of two flat fused elements. They are somewhat less expensive than the classic shrink markers, however, they have poorer aesthetic properties. Finally, it is possible to combine common adhesive labels with transparent thermal transfer tubing (for example, products in the HSDL9 series by Panduit).

The second variant of thermal transfer markers can be called chemically shrunk markers. These labels are tubes impregnated with volatile chemical compounds, which shrink on volatilization. They may be stored only in hermetically closed packages, therefore they is not convenient and are rarely used.

A clip-on marker (Fig. 13.6) looks like a open element made of plastic that is specially designed with spring legs, whose marking space contains only one factory-made symbol (letter, digit, or special electrical signs). The clips can be differently colored to comply with international color codes[4] (Table 13.18). Such markers are delivered separately in blocks. It is convenient to install small-sized clips by means of an applicator. The other variant of a mounting device looks like a flared application wand. For convenience the wands can be sometimes delivered prefilled with clips and installed on a reel-type holder, which is equipped with a latched hook for belt fastening [143]. Once the clip is installed, it moves toward the wire by thumb. The diameter of cables that can be marked is 0.8–18 mm.

Table 13.18. International color code system

No.	Color	No.	Color
0	Black	5	Green
1	Brown	6	Dark blue
2	Red	7	Violet
3	Orange	8	Gray
4	Yellow	9	White

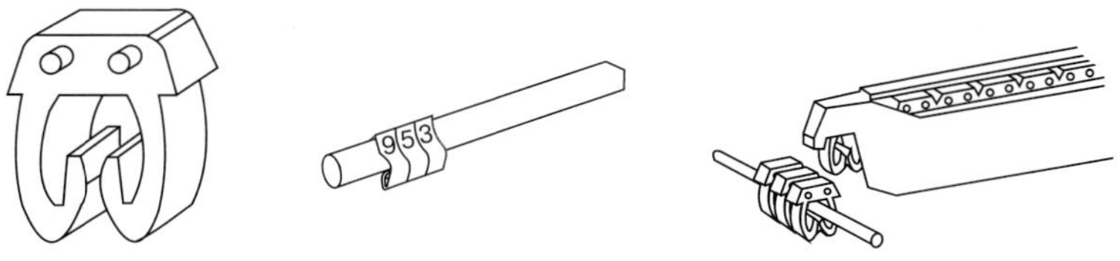

Fig. 13.6. Clip-on marker: *a)* U-shaped clips with fastening lugs; *b)* horn applicator to mount clips on wire.

The drawback of clips as marking elements is their free rotation relative to each other on a wire. Therefore, many manufacturers of the products recommend their use only for two-position marking. There are two main methods to eliminate this drawback. Clips by the company Unix are V-shaped, with a chevron cut to keep multiple clips aligned. Clips by Legrand have cylindrical lugs and grooves to protect them from turning relative to each other after their installation on the housing. The same idea is used by the company Weidmuller, however, the fastening element is a rectangular lug. This provides very efficient fastening of clips and enables the extension of the marking length up to five characters.

For large-sized cables, Legrand suggests the use of a holder in the form of an open ring that is put on the cable jacket. There is a round plastic rod on the top of the holder, which enables fastening of up to four standard small-sized clips. The clips are recommended when cables are prevented from unauthorized access and the identification requires fewer than four to five characters.

Plastic hoops also have one alphanumeric symbol. They are used for finishing marking in cases when a cable end is not terminated or equipped with a connector. They are free from accidental loss of the marking element, which makes them differ from clips, however, they are larger then clips.

---

[4] The German engineering literature calls this code the "color code for resistor coding" (Widerstandsfarbcode).

Thermal marking is done by means of special devices. The heated engraved symbols in their heads leave foiled marks on the plastic cable jackets. Such devices enable creation of differently colored legends of up to 12 characters per 2–3 s. The color depends on the color of the foil that is loaded in the instrument.

Removable labels are convenient when it is necessary to change the legend without replacing the marker. In this technique, a transparent closure in the form of a sleeve with a marking label holder is put on the cable. When it is necessary to make a change, a new label is placed in the holder. The labels are predominantly used to mark various patch cords.

A tag is made of thick cardboard, plastic, or thin sheet metal. It is a long rectangular or oval plate with a marking space. This relatively large tag is very efficient as a marking element for large cables: multipair electrical, optical, external, and power cables. The tag can be installed in different ways. Tags with multipoint fastening have two or more pairs of openings through which plastic ties are laced. The surface of the marking plate is close to the cable after installation. An advanced option is based on using two tags with one or several integrated two-sided ties. Tags with one-point fastening (or, by analogy with adhesive labels, flag markers) are fixed on cables at only one point, and therefore, do not sit flush with the cable. A strap, which is laced through anchoring openings or is a part of the construction, serves for its fastening (Fig. 13.7). An adhesive strap is rarely used for this purpose. The company Panduit has suggested the use of loop-type fastening. The products of series PDL-500 have tags equipped with long, narrow end hoops. During mounting, the loop with a tag is placed around the cable and tightened. The flag markers protrude above the cable surfaces and may snag other wires and elements; they are easily lost during maintenance. Therefore, the use of this type of marking has not become wide spread.

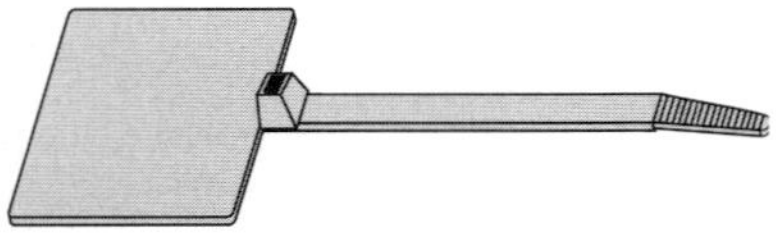

Fig. 13.7. Flag marker with integrated fastening tie

Identification tags possess two drawbacks. First, the tags are visible only from one side, which cannot be defined prior the complete routing of cables. The company Legrand has designed an element called “Duplex O”, which consists of two identical holders that are fixed from opposite cable sides with two plastic hoops, to eliminate this drawback The marking itself is made by means of special flat clips (up to seven characters), and by replaceable legends. The functions of the element may be extended by the use of five colors of holders. Second, tags fail to provide high durability of identification legends. Panduit suggests the use of so-called self-laminating tags of the SLCT series to eliminate this drawback. They have identification labels that are placed under transparent covers, which perform lamination after they are glued to the bases.

When marking individual wires, marker pens in various colors have been widely used. However, their use is not so advisable since it is rather difficult to immediately find such identification labels on the cable jackets.

One may put identification legends on tags made of plastic and cardboard by using techniques that are common for the rest of the identification elements. Hand pliers, compact presses, dies, and electric engraving instruments are used to mark metal tags [144].

Marking elements for bundled cables. Any identification elements used to mark single cables may be used as marking elements for bundled cables. This mechanically holds single cables in bundles. Some types of adhesive labels are used in practice. Moreover, various tags, which are reviewed in Section 13.1.5, are widely used. The following custom elements are also widely popular:

- hoops;
- adhesive straps;
- tapes;
- two-part tubular markers.

The easiest special element to mark bundles made of various wires is a *plastic hoop* with a designation space. The hoop strap has a perpendicular or angled end. The latter solution somewhat simplifies its entry into the fastening slot. The space, which usually has a rectangular form and is positioned along or across the strap axis, is marked by a pen or self-adhesive labels (Fig. 13.8).

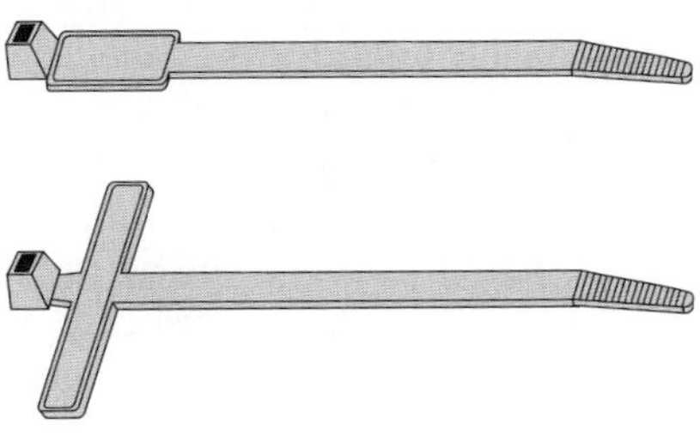

Fig. 13.8. Plastic hoops with designation fields

As a rule, hoops are used to hold large bundles. There are also hoops specially intended for bundles of wires. They differ by their straps, which are covered with numerous low cylindrical lugs along the part of the length that contacts the cables, in order to avoid sliding of the cables in the bundle.

Adhesive straps are 150–300 m in length. These elements serve to bundle horizontal cables and to form bundles of patch cords. They are more aesthetically pleasing than plastic hoops and enable color coding. The strap is quite often equipped by an end buckle, which may be made in two versions. The first has hooks and loops on different sides of the tape. The buckle serves as an auxiliary element when pulling, the end of the strap is laid on the upper surface of the tape without changing its position. The second option has loops on a greater part of one of the sides, the end part is equipped with hooks. After lacing through the buckle, the strap end is cinched back in the form of a loop. If required, hoops and adhesive straps can be laced through lugs of appropriate fastening platforms and fixed in the required position.

The tapes are made by the company Brady, which has suggested a polyolefin thermal-transfer shrink tape with a glued layer. A bundle is first formed by winding the tape around the cable. The end fastening is made after heating and shrinking the tape. A supplementary benefit of this solution is the possibility to make identification labels by means of hand-held printers when working on site.

A two-part tubular marker consists of a rolled tape of 10–15 cm width of hard plastic with a preprinted identification legend. It sits on the bundle because of spring power. During installation the tape draws itself up and lays on the wires. Products of the CMF series by Hellermann-Tyton serve as examples of this product type.

Additional information on identification elements for SCS products may be found in [145].

Identification of patch panels and outlets

The identification of patch panels and outlets is carried out by means of marker pens, adhesive labels, and replaceable legends. Adhesive labels and marker pens are used to mark patch panels with modular connectors. The adhesive labels quite often are used as standard elements of finishing marking, and therefore are included in the panel delivery kit. The length of label, the number of windows, and other identical geometries only depend on the faceplate design of the panel and the modular outlet grouping scheme. In the majority of cases, the labels have white backgrounds. Recently, several models of panels with field colors to be customized in any specific order have appeared. Marker pens are used to mark identification fields. Pens with colored inks are on the market, but black is the mostly used ink color. Identification by means of replaceable legends is a standard element of 100 panels. The same approach is used in a limited number of modular panels.

The general drawback of the standard marking for various patch panels is the difficulty in creating machine-printed identification legends. Therefore, it is recommended to use auxiliary identification elements delivered on standard sheets, which can be printed on printers.

Identification of outlets is often made by replaceable legends and adhesive labels. Their housings have appropriate designation fields and windows.

It should be noted that besides symbol identification of outlets and modular panels, it is also possible to use very efficient color and icon coding. The available elements (icons, colored modules, and covers) are described in their respective sections.

13.2 Troubleshooting

13.2.1 Faults in Cable Systems Based on Twisted Pairs

The following failures may occur in the electrical parts of various SCS subsystems during their implementation and day-to-day operation:

- cable breaks;
- breaks or short circuiting of single conductors;
- electrical contact loss between cable conductors and jacks of patch panels or termination positions;
- wrong wiring;
- disturbances of electrical performance in lines;
- increased noise levels from external high-electromagnetic sources.

Cable scanners are used for troubleshooting. The functional possibilities of cable scanners enable users to accurately find the locations and the reasons for failures in cable systems. In some cases, failures can be visually detected. Troubleshooting methods, elimination of some failures, and their reasons, are reviewed in detail below.

Cable breaks

When a cable is broken, it must be immediately replaced with a new one. Cable splicing is made in cases when:

- cable routing is complicated due to some architectural or other features;
- there is a lack of reserve transmission channels;
- interruptions of communications line result in the shutdown of important equipment.

Two plugs, plug&jack, and two jacks are installed on its ends. The plugs are connected by means of an I-adapter (Section 3.4.2), the plug&jack simply fit to each other, and the jacks are connected by a common patch cord. It is possible to use connection modules (Section 3.5.2), however, they have not become popular in Russia. Connection of cable conductors by screwing or soldering is not allowed, as it breaks the twisting regularity and causes a considerable deterioration of the electrical performance in the line. Finally, it is possible to route a temporary cable along passages and flights of stairs.

Any repairs are made taking into consideration the presence of appropriate accessory hardware as well as the length allowances and extension of damaged parts to be cut off prior to splicing. We must note that when a cable breaks from a lack of pulling strength, the axial damage is much greater than for impact and cutting breaks. This fact must be taken into consideration when selecting the damaged length to cut off. Spliced or temporary cables must be replaced at the first opportunity.

Breaks or short circuits of cable conductors

Such failures occur for the following reasons:

- at points where cable conductors connect the contacts of connectors – their insulation is most often damaged by tools due to inaccurate mounting;
- along the cable pathways – as a result of impact, too much pressure, perforation, or cuts in the external jackets, insulation, and conductors.

Failures are detected by means of testing devices; breaks or short circuits can often be visually revealed. If the failure is found at a point where conductors are terminated to the equipment contacts, it is necessary to strip the cable, once again using the length allowance. One usually may replace the cable when failures occur along the pathways.

If routing of a new cable is impossible due to some reasons or is a time-consuming process that may lead to shutdown of vital resources, then the damaged piece of the cable is cut off, and the maintenance personnel follow procedures specified for cable breaks.

Lack of electrical contact between cable conductors and outlet contacts

To eliminate the lack of electrical contact between cable conductors and outlet contacts, it is necessary to repeatedly impact conductors to IDC contacts by one-wire or five-pair impact tools. If the procedure does not help, it is necessary to repeatedly strip the cable using the cable allowance. If no allowance is available, a new cable must be routed.

Incorrect reversing, transposing, and splitting of pairs

Defects are often detected during testing by means of a cable scanner. To eliminate the problem, it is necessary to visually define the defective cable end and to repeatedly wire the conductors using the allowance. If the allowance is not available, a new cable must be routed.

If mounting of the outlet construction on a cable does not require use of an impact tool, then the whole outlet has to be replaced (the majority of such outlet constructions are not available for reinstallation).

Disturbances of electrical performance in lines

Troubleshooting the disturbances of electrical performance in lines is a very complicated procedure. Some defects that frequently occur in practice, leading to the deterioration of NEXT, are caused by improper installation (Table 13.19).

When the actual performance of a line or pathway does not correspond to the specified performance, it is first necessary to visually check the accuracy of the cable wiring to the equipment contacts. If required, it is necessary to wire the cable, once again using the allowance. If the procedure does not solve the problem, it is necessary to check the compliance of all the components of the line with the requirements of the standards, and, if necessary, to replace all misplaced elements. The checklist also includes the limits set by standards for the lengths of cables and cords.

Table 13.19. Effect of installation quality on channel performance [146]

Defect	Deterioration in NEXT, [dB]
Cable pairs untwisted up to 12 mm	1.5
Cable pairs untwisted up to 50 mm	3.8
Cable twisting with bending radius of 35 mm	1.9
Cable twisting with bending radius of 12 mm	2.1
Cable rupture	2.4

Devices with NEXT locator functions and other similar options may aid in troubleshooting.

Increased noise levels from external electromagnetic sources

When noise occurs from external sources, it is necessary to increase the efficiency of the screening and/or physically separate the noise sources and cable runs or equipment locations as much as possible. Such noises are usually specific to industrial buildings.

13.2.2. Failures in fiber-optic cable systems

The following failures may occur in the optical parts of various SCS subsystems during their implementation and day-to-day operation:

- cable faults or breaks;
- attenuation increases in connectors;
- faults in patch cords;
- improper connections of patch and equipment cords.

Troubleshooting failures and establishing reasons for their occurrence in SCS optical subsystems is done by means of the measuring devices described in Chapter 12, Section 12.3.

Cable faults or breaks

When mechanical damage appears as a result of overly high pulling, cutting, or pressing stresses, the easiest troubleshooting method is by optical reflectometer or locator. In cases when the cable run consists of several cable segments that are patched between each other at optical cross-connects, or when it is difficult to analyze the reflectograms due to numerous reflections, it is recommended to verify these measurements on each individual segment by means of the reflectometer. The same approach is advisable when the maintenance personnel do not have reflectometers and when optical testers are used to make measurements.

The so-called near-end fault, that is, a fault at a distance of 10–15 m from the beginning of the cable run, is not detected by the reflectometer because of the available dead area. In this case, it is most efficient to use visual troubleshooting.

Fiber-optic cables predominantly organize backbone SCS subsystems. Since the cable runs are quite long, the cables are not usually replaced in these subsystems. Instead, repairs are made using intermediate sleeves. If there is a cable allowance close to the location of the fault, then one sleeve is installed. For the remaining cases, the damaged areas are fitted in cable insertions and two intermediate sleeves are installed.

Fibers are spliced in enclosures by welding or mechanical splices. The type of enclosure (common box, hermetic construction, etc.) is selected depending on the specific conditions at the place of its installation.

Attenuation increases in connectors

The main reasons for the increased attenuation are the following:

- pollution of the ferrule end surfaces of spliced plugs;
- improper assembly of the connector;
- fiber near-end breaks or fault.

The end surface of the fiber ferrule should be cleaned by a smooth tissue moistened with solvent.

The improper assembly of the connector implies the improper or incomplete insertion in the outlet. Typical examples are: the plug of an SC-connector is not properly patched in the outlet; the nut of an ST-connector plug bayonet connector is not fully or partly covered; or the nut of plug for an FC-connector is not screwed. Such faults commonly appear during testing and pilot launching of cable systems.

The near fiber break or fault implies a fault that happened along the fiber entry section (at a distance not more than 15 m from its end). These faults cannot be detected by a reflectometer because of its dead area, which appears from the response arriving before the pilot pulse scattering. Cable breaks are easier to detect by means of defect browsers or testing using extension reels. If there is a near break, it is recommended that the optical plug be reinstalled.

Faults in patch cords

Patch cords are usually damaged during day-to-day operation of SCS mainly as a result of abrupt cable pulling, pinching, or kinking during rerouting and repatching, and from falling of tools, equipment, and other heavy objects. These faults are detected by lighting separate fibers and measuring the insertion attenuation by optical testers.

Improper patching of equipment and patch cords

The improper patching of equipment and patch cords implies the wrong order of patching the plugs into the outlets. This is specific to systems that are equipped with ST and FC connectors, whose constructions do not use duplex plugs with mechanical blocking elements to prevent wrong connections. In systems with SC duplex connectors, such faults may occur only when terminal patching devices are improperly assembled. This kind of fault is not a real fault as it can usually be tested and detected even during construction. To eliminate this kind of error one may control optical signal traffic via optical ports according to Figure 4.16.

13.3 Maintenance

13.3.1 Scope of Maintenance

The maintenance work that should be carried out on an installed structured cable system consists of procedures described in Table 13.20.

Table 13.20. Maintenance procedures

No.	Description	Purpose	Interval
1	Visual inspection	Control physical continuity of cable system components	Monthly
2	Dusting telecommunications outlets	To protect electrical characteristics from dust content	Every 6 months
3	Dusting telecommunications connectors of work areas	To protect electrical properties of cable system against dust content	Every 6 months for unterminated connectors; once every 2 years for active connectors
4	Rerun patch cords and cross-connect wires	To provide rerun of patch and equipment cords and cross-connect wires, and easy administration of cable system	Once per year
5	Verification of cable logs	To ensure correspondence between information and maintenance documentation and connections of cross-connect fields	Once per year
6	Random testing	Verification of performance of optical and electrical lines to ensure that electrical and optical components possess permanent stability	Once per year

13.3.2 Maintenance procedures

All kinds of maintenance procedures are carried out by personnel who are responsible for operation of the structured cable system.

Visual inspection

The following elements should be visually inspected:

- housings of telecommunications connectors (faceplates) in work areas. These elements shall not have mechanical damages and cracks. The telecommunications connectors shall be laced through openings in the housings and fixed in anchoring brackets;
- opposite sides of cross-connect blocks and patch panels in mounting racks and cabinets. It is prohibited for cables entering cross-connect blocks to have mechanical damage and breaks or be pressed by heavy parts of the equipment installed in the immediate vicinity. The patch panel connectors shall be free from mechanical damage; all cable conductors shall have electrical connectivity with the appropriate connectors of the patch panels;
- the front side of the cross-connect blocks in mounting cabinets and cross-connect panels in ATE premises. It is necessary that cable conductors shall be free from mechanical damage, but cable conductors and cross-connect wires shall have electrical connectivity with appropriate connectors of cross-connect blocks.

If any of these conditions has been violated and further testing has revealed that the violation negatively affects the quality of signal transmission along the channel, then the following maintenance procedures are required.

Dusting

Dust is removed by a household vacuum cleaner. It is desirable that accessory kit of the vacuum cleaner includes various nozzles for efficient cleansing of difficult areas. The procedure additionally requires the control of the state of the cable elements being treated.

Rerun of patch cords and jumpers

The procedure is aimed at accurate routing of cords and jumpers, which improves the electrical performance of cable pathways and enables good visibility of marking by removing kinking. During new cabling, it often works well to replace some long cords with short ones, and thereby reduce the total loss in signal transmission channel.

It is recommended that new cables be laid at night, on weekends, or holidays, when the valuable information content transmitted across SCS is considerably reduced. Prior the procedure it is advisable to additionally verify cable logs.

The cable rerun procedure is carried out stepwise:

1. Patch cords are removed from patch panels and classified according to their length;
2. The cross-connect wire is removed from 110 and 66 blocks, and depending on its state, is sorted for further utilization;
3. The connections are recovered according to cable log records. First, by the cross-connect wire, then by cross-connect cords of required lengths. It is simultaneously verified that identification labels of SCS elements correspond to identifiers in the cable log.

Information on the date and time the cable rerun procedures were made is registered in the cable log.

Verification of cable logs

The verification of cable logs implies determination of correspondence between current records and actual connections in structured cable system. When the verification is made, each signal transmission channel from a terminal device (for example, telephone set) to the network equipment is subsequently verified with the records of cable log. If some discrepancies are found, it is necessary to make changes in the records or to perform the required connection.

Information on date and time of verification is registered in the same log. It is recommended to carry out the verification of cable logs and rerun of patch cords and cross-connect wire at the same time.

13.3.3 Emergency Maintenance

The list of emergency states includes:

- breaks or faults of cables in horizontal or backbone SCS cabling;
- breaking of work area telecommunications connectors;
- faults of termination hardware elements in mounting racks or telecommunications closets;
- undetected errors or cable system failures.

Troubleshooting is required as soon as possible in emergencies. The troubleshooting can be carried out by instruments or visually. In cases when the maintenance personnel cannot perform troubleshooting by available instruments and tools, it is necessary to contact the company that installed the structured cable system, or its local authorized partner for consultation or to submit a request for service or warranty repair. For quick recovery it is necessary to make reconnections to a reserve transmission channel, or if such a channel is not available, to provide temporary communications.

13.4 Conclusions

The SCS is a complicated high-technology product, and its maintenance has to be carried out by personnel who are specially trained in compliance with the standards TIA/EIA-606 and ISO/IEC 14763-1. The maintenance personnel perform control (administration) of the cable system as well as troubleshooting and maintenance procedures.

The clear and accurate identification of separate components and elements, and the subsequent filing of maintenance documentation are the necessary conditions for efficient operation and maintenance of the cable system. For more convenient administration, electronic documentation in the form of a database is used.

The main task of the personnel during troubleshooting in emergencies is exactly locating failures and implementing diagnostics by instruments and visual controls. The appropriate control instrumentation helps the personnel to ensure quick troubleshooting. For efficient restoration of SCS, changeover to reserve channels that were created during the cable system design is widely practiced, in addition to the physical restoration of communications lines.

The high quality of separate SCS components and manufacturer's requirements for the qualifications of installers reduce the scope of maintenance procedures. The main procedures during maintenance are rerunning cords and jumpers, dusting, and verification of cable logs

14 Additional Information

14.1 Cabling Systems of Different Manufacturers

Here we present general technical information about some SCS of different manufacturers that are to a greater or lesser extent present in the Russian market and are known to the authors of this book. The first criterion for inclusion in the list is the presence of main and additional features, considered in the Introduction, that make it possible to refer to this product as an SCS. The second criterion is the presence of permanent representation or a partner in Russia to promote the system in the Russian market while providing a manufacturer's warranty. The list does not claim absolute completeness, since the number of these systems is growing rapidly.

Additionally, the products of several foreign SCS manufacturers are offered by their Russian partners at the component level. Among such manufacturers one could note FMT, Telegertner, Infra+, and Honaco.

14.1.1 3M Volition Fiber-optic SCS

In 1998, the American company 3M created the Volition fiber-optic cabling system. The system was developed with the goal of reaching the maintenance and price level that is equal to or exceeds that of its electrical cable functional analogue.

The basic system element is the VF-45-type optical connector, having original construction and no ceramic centering tip.

The optical linear cables that are a part of the system have a maximum of 72 fibers of 9/125, 62.5/125, and 50/125 types and are manufactured in versions for internal and external layouts. In the latter case, either dielectric fulfillment or corrugated steel-cord belts are used to increase the resistance to external actions. Cables that are oriented to operate in horizontal subsystems consist of two fibers (mini-breakout-type construction). Backbone cables with up to 24 fibers have one conduit construction. When there are a greater number of fibers, modular structures are used.

Optical patch cords with VF-45 plug connectors are factory built only; there are single-mode and multimode variations (with 50/125 fiber, and 62.5/125 fiber in the latter case) in lengths of 1 to 30 m. To increase the functional system flexibility, combined cords with ST- and SC-plugs at the second end are introduced. When using a two-layer version (glass with an external diameter of 100 µm plus a polymer coating with external diameter of 1256 μm), the operating characteristics are improved appreciably.

The switching equipment of traditional constructions consists of 19-in switching shelves and wall couplings. A shelf of height 1*U* has 24 standard optical ports. If the shelf height is increased to 2*U*, it is equipped with integral circular-type patch cord organizers. The wall couplings have 6, 12, or 24 ports, depending on the variant.

Two-port outlets are intended for installation at work areas.

Manufacturing equipment contains the original set of instruments for glueless VF-45 connector outlet mounting. It normally takes 2 min to mount one outlet. VOL-2290-type optical testers, which are additionally supplied with test cords for operation with VF-45 connectors, monitor the mounted information transmission routes.

To arrange the horizontal subsystem in its classical form, the limited nomenclature of category 3 electrical cables in the versions of plenum class and riser fire resistance are provided for in the Volition system. The patch cords and patch panels for keystone style outlets are included in the electrical equipment.

The significant expansion of the Volition SCS functional opportunities has been achieved by introducing the advanced nomenclature of LAN active network equipment (medium converters, hubs, Ethernet and Token Ring switches of different types). The devices have both optical and electrical interface and provide bit rates up to 1000 Mbps in older models.

Training for system designers and installers lasts three working days. When operating with active equipment, an additional one-day training course is taken.

The standard 20-year system guarantee is provided for certified systems. In order to obtain a system guarantee, the technical design must be performed using 3M rules, and the installation must be done by trained personnel. In especially critical cases, 3M directly inspects the installed SCS with its own inspectors. The duration of the guarantee for the electronic components is 1 year.

14.1.2 The Advanced Electronic Support Products SignaMax system

The SignaMax structured system is related to one of the newest SCS systems being promoted in Russian market. It was created in 1997, and the Moscow office of Advanced Electronic Support Products (AESP) had a hand in its development.

The SignaMax electrical subsystem is a part of the HiLink system. The HiLink element base makes it possible to create a cable system with a throughput up to category 7. The mains shielded cable is also included in the HiLink composition to provide voltage to the mains outlets, which allows additional expansion of functional SCS opportunities.

The horizontal subsystem can be created using unshielded cables of categories 3, 5, and 5c. When using shielded constructions with cutoff frequency ratings of 300 and 600 MHz, systems can be created with characteristics that correspond to the requirements of categories 5, 6, and 7 [147]. HiLink is one of the few SCS using 2-pair cables of category 5. If operation in a severe environment is planned, then it is better to use cables with Teflon insulation that can guarantee operation in the temperature range from –70°C to +200°C. All types of horizontal cable are delivered in boxes. Pairs of boxes are located in common cardboard packages that provide a great deal of safety during the long time of transportation.

Vertical subsystems are created using 25-pair cables of category 5 (UTP and S/UTP versions). Cables of category 3 are intended for telephone signal transmission and have a maximum of 100 pairs.

The switching equipment consists of panels with modular connectors using shielded or unshielded variants with a maximum of 120 ports; cable preparation is done on 110- or Krone-type contacts. Most of these products are intended for mounting into 19-in constructions. Also, there are versions for wall mounting. Modular construction panels can be used, if necessary. The filler plates are completed by the outlets of modular connectors, BNC outlets, and ST outlets to form optical ports.

Category 5 110-panels are available in variants of 100 and 200 pairs; they can be mounted on walls or in 19-in constructions. To specifically support telephone subsystem functioning, panels with 6-position outlets having a maximum of 120 ports are offered.

The unshielded and shielded one- or two-port outlets are intended for installation at the work areas. Patch cords and terminal cords are also available in shielded and unshielded versions with maximum lengths of 10 m.

When creating fiber-optic subsystems, the equipment manufactured by the company BCC is used.

The functional capabilities and the flexibility of SignaMax SCS are provided by including well-known components in the design: 19-in mounting equipment (floor and wall closets, open cabinets) and cable boxes (metallic or plastic) manufactured in England.

The standard system guarantee for SignaMax SCS is provided for 25 years.

One more feature of SignaMax systems is its inclusion of FOCUS network equipment. In this case, in contrast to other solutions, in cases of complex project realization (SCS and LAN), the customer gets a total 20-year guarantee for SCS and FOCUS active equipment.

14.1.3 Alcatel Cabling Systems

Alcatel cabling systems meet the requirements of the main International SCS standards. All the components of SCS under consideration are manufactured at Alcatel's own facilities.

The cable production for horizontal subsystem creation consists of 4-pair cables in UTP, S-UTP (in general, double film shields are used), and S-STP versions. The cable impedance is 100 and 120 Ω and the cable conductor diameters are 0.5 and 0.6 mm, respectively. Cable parameters of the Gigatek and Symtek series are certified up to a frequency range of 600 MHz. The telephone signals along the vertical parts of the routes inside the building are transmitted over multipair cables of category 3 with impedances of 100 and 120 Ω and with a number of pairs equal to 10, 25, 50, 100, and 150.

Optical cables can be used in all three SCS subsystems and are supplied by optical waveguides of three main types: 9/125, 50/125, and 62.6/125. Cables without helium filling with 900-μm tight buffer sheaths and 24 fibers are intended for the creation of internal backbone subsystems. Cables for external layouts are of both modular and single-tubular construction. The maximum number of fibers is 60. Hybrid cables of the Twintek series with two tubular elements serve to arrange two-port work areas. The first of these elements is a typical or shielded 4-pair electrical cable, the second is a fiber-optic element containing 2, 4, or 6 optical waveguides in a 0.9-mm buffer covering.

The electrical patch cords with lengths from 1.5 to 20 m are manufactured in usual and shielded variants. Three-meter connecting cords with Telco connectors are provided to connect telephone exchanges and apparatus for low-speed applications. Functions of splitting elements for realizing the cable-sharing principle are done by so-called adapted cords with lengths of 3, 5, and 8 m using 4- or 8-positional modular connector plugs of at the second end of the cord.

The standard lengths of optical patch cords are 2, 5, and 10 m. The cords are supplied with the plugs of ST, SC, SMA, and MIC connectors.

The great nomenclature of internal and external outlets with direct and angular installation of outlet modules of electrical and optical connectors has been developed to install at work areas. The optical outlets are manufactured both as single filler plates and as duplex product variants (including ST outlets). The built-in elements to allow marking with changeable inscriptions are provided for at the front panels of some outlet modules.

The electrical modules of Epsilon series are manufactured in three versions: unshielded, shielded, and so-called EMC variant. The last two versions differ from the EMC, which has a module case covered entirely with shielding casing, whereas the shielded outlet has a shielding casing that covers only the female connector for plug installation. Module preparation is done by the usual single-pair impact tool. The preparation procedure is simplified significantly by using manual support under the module being prepared.

The process of module installation into the external outlets (cases with 2, 3, 6, and 12 outlets of the LAMBDA 40 series) is made easier by using one-, two-, and three-outlet filler plates. The panels with modular outlets that are provided to cross- and equipment rooms are also installed by modular principles. The main construction element of such a panel is 12-port electric filler plate in shielded and unshielded variants. Besides the outlets for Ethernet systems, there are also outlets for Token Ring systems.

The optical panels are made by the same principle and typically have 36 or 60 optical ports. The optical connector plug installation is done using 3M Hot Melt technology.

The mounting process of optical and electrical panels of the OMEGA series is made easier by using the ability to move the shelf case forward into the front position and tilt it down by approximately 30°.

The Alcatel cabling system uses the usual telephone terminal blocs and the corresponding combined cords with modular connector plugs at one end and so-called IDC plugs at the other ends to provide convenient connection to telephone exchanges. The length of such cords is 1.5–8 m.

The equipment installation of SCS under consideration can be carried out using 19-in constructions of the AC and TC series with glass-front doors. There are also modular-type cabinets for wall mounting, and cases to allocate the telephone terminal blocs for intermediate interconnection inside it. Open frames can be installed in the technical premises, if necessary.

Connection of equipment that does not have standard interfaces based on modular connectors is done by baluns, adapters, and medium filters.

The Alcatel cabling system components also include Ethernet and Token Ring 4-port installation microhubs. The microhubs are manufactured either in horizontal variants (all four outlets are located in a single horizontal row) or in vertical variants (the outlets are located in pairs one above another).

The technical design process and current system maintenance is made easier by special Cabling Management Software. Functionally, this software is divided into modules of visual administrator and technical design and is offered in versions from 250 to 5600 ports, depending on the network capacity.

14.1.4 The AMP NETCONNECT System

The Netconnect system by the American corporation AMP (part of Tyco International Holdings) is created according to the TIA/EIA-568-A and ISO/IEC 11801 standards in current and future releases. The system contains the entire set of components that is needed to realize all three SCS subsystems on the basis of unshielded and shielded, twisted pairs and fiber-optic cables (Fig. 14.1, [47]).

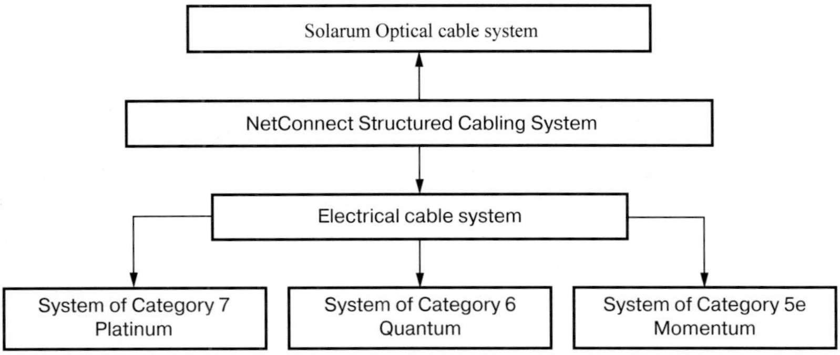

Fig. 14.1. Structured cabling systems of AMP Tyco Electronics Company

The electrical cabling system is built on the basis of multipair and multielement cables of category 3, usual and expanded cables of category 5 and of category 6 and 7, using usual and fire-safety versions (with sheaths made of fluoropolymer). The unshielded category 6 cables provide ACR=4.5 dB at the frequency 250 MHz.

To make the creation of 2-port work areas easier, a broad range of double cables is offered. In this case, the separate elements of such cables can be both similar and different, including those with different impedances.

The optical cables are delivered for the main types of multimode (50/125 and 62.5/125 μm) and single-mode optical waveguides for the variants of internal and external layout. The maximum number of fibers in cables for internal layout is equal to 72 optical waveguides (16 with LSZH sheath); cables for external layouts also have at maximum 72 fibers and can be made with sealing elements or entirely with dielectrics.

To create communication equipment, two different construction designs are offered. The bases of the so-called AMP communication system whose separate parts meet technical the design requirements of category 6 are the constructions that are installed into standard electrical boxes and that contain electrical connectors. The 4-pair horizontal cable is prepared for this connector. The functional system flexibility is provided by a wide set of shielded and unshielded single and double filler plates with different types of electrical interfaces (modular connectors, BNC, IBM connectors, etc.). Use of double filler plates makes it possible to very effectively realize the cabling sharing principle. Built-in baluns, load resistors, filters, and high-voltage protection elements are provided for, if necessary.

When there is need to improve the EMC (electromagnetic compatibility) characteristics, it is possible to use connectors of the EMT (electromagnetic technology) series. They differ from their analogues by shielded circular contacts as well as by low transient resistance of the plug and outlet [148].

Usual patch panels with category 5e modular connectors are part of the 110 connect system and have from 12 to 96 ports. This panel type is manufactured both in the usual and modular versions. Panels 110-type for wall mounting and in a cross-tower variant with 50, 100, and 300 pairs are merged into the XC cross-connect system. To install inside the work premises, a large number of internal and external outlets as well as floor boxes and multimedia outlets is provided for.

Fiber-optic switching equipment consists of numerous models of wall couplings with 6, 12, 24, and 48 of outlets of the ST, SC, MT-RJ, MIC, and ESCON types. Sometimes, coupling is supplied by additional covers that protect the plugs of optical connectors from accidental mechanical effects. The functional flexibility of coupling is provided by using modular filler plates with the standard number of connector outlets (six single or three double outlets). Both usual and combined outlets are allowed. To develop fiber-optic subsystems, shelves with a maximum of 128 outlets are used. The shelves with a greater number of outlets should be added to the shelves to store the protection mold tubes of welded connectors or the cases of the mechanical splices. The termination of the optical cables should be carried out by mounting cords and Corelink mechanical splices.

In 1998, the Solarum fiber-optic cable system was introduced into the Netconnect system. The basis of the system is a multimode cable with 50/125-type fiber and MT-RJ optical connectors. The use of nontraditional 50-μm fiber for SCS with improved frequency response makes it possible to transmit Gigabit Ethernet system signals over distances up to 550 m in both spectral windows. Use of the MT-

RJ optical connector makes it possible to get the same number of ports as electrical modular connectors. When mounting the optical connectors of different modifications, mechanical connectors of the LightCrimp series and epoxy-based glue technology are widely used.

The electrical patch cords in usual and shielded versions have lengths 2–20 ft (0.61–6.1 m). There are nine different color termination points for color cord marking. The optical patch cords in their usual and combined versions in standard lengths of 1, 2, 3, and 5 m are delivered in single and duplex versions.

The AMP Undercarpet cabling system contains the entire set of cable products, accessories, and technical tools that allows users to create the power cabling system laid under the carpet.

The specific expansion of cabling system functional capabilities is achieved by including lower-level LAN network equipment (medium converters, hubs, etc.) in SCS components.

Netconnect D/S software for cabling system technical design has been introduced into the system composition as have the a set of glue labels for final cables, cords boxes, and marking of other components. The machinery manufacturing of marking signs is made easier by using specialized software for Windows-compatible computers and printers from the major printer manufacturers.

For SCS systems mounted by AMP, two different types of 25-year guarantees are provided: the so-called enhanced component guarantee and the system guarantee, which mainly corresponds to component and system guarantees of the SCS manufacturer in its classical understanding (Section 1.6).

AMP has organized two courses to train their partners. The aim of the ACT1 course is to get basic and practical experience that is necessary during the installation and SCS warranty period. The ACT2 course has been introduced to get the knowledge and experience to perform diagnostics and SCS certification. The condition to get training at ACT2 course is preliminary training at the ACT1 course.

The authorized partners of AMP are divided into two categories: NI Contractor or Partner for the installation makes the mounting of SCS using the legally bought components and strictly adhering to the norms and rules of AMP, as well as providing guaranteed support of the installed system. If even 25% of the installers are trained at the ACT1 course in the AMP company training center, then the partner has the right to provide a 25-year enhanced component guarantee. The next grade of partnership is ND&I Contractor or System Partner. This company should have on its staff a minimum of one designer trained at the ACT2 course, and more than 50% of the team members should be certified. The system partner has the right to provide a 25-year system guarantee to the customer.

In early 2000, more than 50 AMP partners with TNT and ND&I [149] status were operating in Russia and the CIS.

14.1.5 Brand-Rex Millennium SCS

Millennium SCS was developed in 1990 by the English company Brand-Rex (in early 2000, this company is in the BICC [156] group). It is widely known by its communication and power cables of different assignments. This SCS includes a number of independent products that are distributed under its own trademark [151].

The elements of electrical SCS subsystem are included in the CAT5Plus system. The subsystems are based on 4-pair UTP, STP, and S-STP cables in the usual and fire-safety versions. In this case, the UTP GigaPlus cable that is a part of the system provides ACR=3 dB at 350 MHz, and at 100 MHz this parameter value is 25 dB, which is 15 dB higher than the standard value. Multipair category 5 cables have 25 pairs, category 3 cables contain 25, 50, and 100 pairs.

The switching equipment of the CAT5Plus system consists of patch panels with modular connectors with 16 and 24 ports (height 1*U*), as well as 32 and 48 ports (height 2*U*). The cable conductors are connected to the contacts of the 100 or LSA-Plus type. In addition, the shielded panels with 16 and 24 ports are part of the system as are modular constructions of the Combi series. Six-port shielded and unshielded modules are provided for installation in the latter units.

The 110-type patch panels are merged into the 5310 cross-connect system and include 50- and 100-pair modules for mounting on the base (heights of 17 *U*, 22 *U*, and 42*U*) as well as 50-, 100-, and 300-pair wall modules with mounting legs.

Work area telecommunication outlets are manufactured in single- and two-port versions with direct and angular installation of the outlet modules. There are style variants of the external appearance for shielded and unshielded outlets. Module and outlet coding is carried out by changeable inscriptions of different colors.

To switch and connect the network equipment in the cross-rooms and at the work area, there are many types of usual and combined cords with modular connector plugs and 110-type connectors, as well as 25-pair cords with Telco plugs and connector outlets. It should be noted that 110-type cords can be made under field conditions.

Baluns and medium filters consist of numerous elements providing the connection of Token Ring equipment, IBM, 3270 units, and TV devices of different assignment to SCS. Besides the hinged variant, several rack-type models of the devices designed \ as panels for 19-in constructions are offered.

Millenium SCS optical subsystems can be formed on the basis of a wide variety of cables with different capacities and protection levels from external action with single-mode and multimode fibers. In rigorous environmental conditions it is better to use cables with protection coating formed by armor, corrugated steel tape, and wires of different diameters.

The switching equipment for SCS optical subsystem creation is merged in the frame of the Helious system. In the cross-rooms, 19-in shelves with three places for modular filler plates are installed. The filler plates are provided with six or eight connector outlets of the main type. Coupling for wall mounting is carried out with steel casing and has sites for similar filler places. The multimedia outlets with maximum 8 outlets of optical connectors are mounted in the user premises. The termination of the optical cables in field conditions is made by gluing (two-component anaerobic glue or Hot Melt technology) or by welding.

To speed up the process of optical subsystem creation, the MTConnect system is provided. Its main elements are the predetermined assemblies with cables of lengths not more than 50 m. The size of the end protector is decreased at the expense of mounting of the MT-type group optical connector to the cable parts.

Another unique element used in Millennium subsystem is Bluelite technology. Its element base and features are described in detail in Section 4.7.

When Millennium SCS are designed and mounted by certified specialists, 25-year system guarantees and termless guarantees for applications are provided.

14.1.6 Elgadphon Cabling Systems

Elgadphon cabling systems by Elgadphon Communications (Israel) is promoted in Russia as a complex product since the end of 1999 (before this time, its separate components were known).

The basic system component is a proprietary outlet module that, as a separate element or assembly, is a part of telecommunications outlet and patch panel. The module is formed by a Steward female connector and divided termination (split termination) by Molex. In this case, the female and the terminal are mounted on the printed board from one side and oriented in the same direction. This configuration provides the ability to connect the cables to the panels and outlets from the front without dismounting the case or the module. Therefore, the mounting time is decreased by 50%, according to Elgadphon data.

The main type of switch equipment is a panel of the Alpha series with height 1.25*U* that is always mounted in pairs with 0.75*U*-high circular organizers. The number of panel ports is equal to 8, 16, or 24, depending on the number of assemblies installed in it. The unused locations are covered by plugs with modular connector outlet depicted at their faceplates, which gives a highly pleasing look to the product installed. Thirty-two-port panels of height 2*U* are supplied with a switch that connects two outlets situated one above another, which makes it possible do without patch cords in some cases.

The telecommunications outlets are manufactured for internal and external mounting and are gathered by one or two outlet modules for flat installation. All switch equipment is manufactured in UTP and STP versions, and its characteristics are normalized up to 200 MHz, taking into account the possibility of unconditional support of the ATM-155 system functioning with NRZI linear signal coding.

The length of patch cords is 0.3–7.5 m; the cord products are offered in UTP, STP, and S/STP versions.

Any special technical means providing PBX operation and other low-speed applications are not provided for in the system composition.

When organizing horizontal subsystems, the use of 4-pair UTP, STP, and S/STP cables that are, most probably, OEM-products is recommended. The normalized cutoff frequency for unshielded designs is 100 MHz, and for shielded cables this parameter reaches 600 MHz.

The optical switch equipment is presented with the Delta and Beta series of shelves with 12 and 24 ports of ST, SC, and SMA outlets. In this case, SMA outlets are recommended to support single-mode functioning of applications. Using the telescoping guide makes it convenient to work with these panels. Additionally, the shelves can be completed with splice plates, if necessary.

It is recommended to mount SCS equipment in the technical rooms in 19-in cabinets of the Intelligent series. The cabinets are manufactured in floor variants (the maximum height is 44U) as well as in wall variants, and have a complete set of accessories (shelves, ventilation equipment, blankets, etc.). The characteristic feature of the Elegant series of wall cabinets in MIDI, MINI, and MICRO versions is the original glass vee-type front doors in its cross section.

The Elgadphon SOHO systems based on 3-section 10-in cabinets are intended for the creation of small systems.

A15-year guarantee is provided for SCS mounted systems.

14.1.7 IBM Advanced Connectivity System

The Advanced Connectivity System (ACS) SCS of IBM [152] is the further development of the first cabling system created, in general, to support Token Ring LAN functioning. The following subsystems are the functional elements of ACS: external backbones (campus cabling subsystems), internal backbones (building backbone subsystems), horizontal subsystem (horizontal cabling subsystem), and work premises (work area cabling).

Structurally, the ACS cabling system is divided into two products designated as Type C and Type G in the IBM documentation. Type G products appeared at the end of 1990s and consist of several separate subsystems with different capacity levels for the routes created. It is presented by the developer as a high-quality and inexpensive solution primarily for mass users.

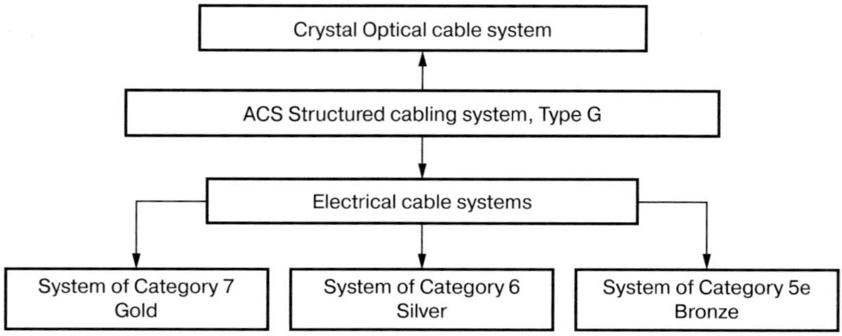

Fig. 14.2. ACS structured cabling systems of IBM Company, Type G variant

Type-C ACS contain all the main types of 4-pair UTP, STP, S-UTP, and S/STP cables of category 5. The cables can have 100-Ω and 120-Ω impedance and have both PVC and LSOH sheaths. The cables are delivered on rolls with cable lengths of 500 and 305 m. To make the arrangement of two-port work areas easier, double products are offered (Siamese by IBM classification). The broadband horizontal cables consist of S-STP versions and have a cutoff frequency of 600 MHz (the attenuation is not more than 50 dB and cross-talk attenuation is not worse than 60 dB).

The majority of multipair cables belong to category 3 and have 20, 25, 40, 50, 100, 200, and 300 pairs in UTP versions, and 25, 50, 100, 200, and 300 pairs in S-UTP variants. Category 5 constructions are manufactured in multielement 20-pair versions (five separate 4-pair elements inside the general sheath).

The patch panels in shielded and unshielded variants are created by modular principles. The main element is an assembly of four or five modular connector outlets, in this case, these assemblies may be installed horizontally (the full number of ports is 16 for a panel of height 1*U*) or vertically (48 or 60 ports for height 3*U*). To make the outlet mounting process easier, the easy lock system is proposed. This system makes it possible to carry out cable termination without using special impact tools. Panels of the 110 type are not used in the ACS system.

The patch cords are from 0.5 to 10-m long with a discrete metric step and are supplied with modular connector plugs. Detachable plastic termination tails of eight different colors are provided for individual coding. Special additional adapters are put on plugs and outlets, if necessary, which allows users to connect the plugs with various outlets.

The system composition under consideration provides for a large number of different-type outlets with a maximum of 6 ports for wall installation as well as for duct mounting. The outlets are manufactured with straight and angled female connectors to mount the outlet parts of modular connectors.

The ACS optical subsystem uses a predetermined assembly that is manufactured using different types of cables with 4–12 fibers and lengths up to 2000 m. Single-mode and multimode E2000 connector plugs are available for cable termination from Fiberdesk. A unified optical connector outlet fastening mechanism is used, which allows mounting it in the patching shelves, together with the electrical outlets of modular connectors. The length of optical patch cords is 1–5 m.

There is a developed set of adapters and baluns: the ACS Type C system that is primarily oriented to connect the IBM 3270, System/36, System/39, and RS/6000 equipment.

The solutions used in Type G products, in general, are the same as those for the Type C series. They do not differ in principle, but in some construction changes and fewer versions of the separate products. It should be noted that original MiniC electrical connectors with characteristics certified up to 1 GHz are used in Gold products. In Crystal optical products ST, SC, and MT-RJ connectors are used. Welding technology and Unicam mechanical splicing technology are used in their installation.

A large list of active and passive LAN network equipment is introduced into ACS system composition. Modular constructions are based on 2-, 4-, and 10-slot chassis that are installed into 19-in constructions or wall cabinets. Unpackaged electrical and optical modules of medium converters as well as Ethernet and Token Ring are included in the chassis. Modules are also available for network management, passive hubs, and repeaters of different types. The equipment in its fixed configuration consists of hubs and Ethernet, FDDI, Token Ring, and ATM switches that have electrical and optical interfaces and allow cascading to increase functional flexibility.

For the IBM ACS system mounted by authorized partners, a 15-year system guarantee is provided from the manufacturer.

Four-day long mounting and technical design training courses are organized at the company's training center.

14.1.8 IT SCS of the IT Company

The IT SCS of IT Company [153] is the first Russian structured cabling system. It was created in mid-1996 and by late 1998 moved to the lead in the Russian telecommunications technology market. The commercial success of the system is provided by good adaptation to Russian conditions together with high-quality Russian-made components.

The system has been built according to International ISO/IEC11801 Standard. The IT SCS is formed by external and internal backbones and by horizontal subsystems. Cross- and network equipment is mounted in the cross-rooms and equipment rooms. The electrical characteristics are entirely according to the requirements of category 5e. In August 1999, the IT GigaSystem SCS, providing technical design characteristics of category 6, was created. In June 2000, the third product in the composition of IT SCS appeared—the IT SCS-mini, which includes the SCS technical systems on the basis of 10-in mounting constructions, which are more convenient and economically beneficial in the cases when the number of cabling system ports is from 5 to 50.

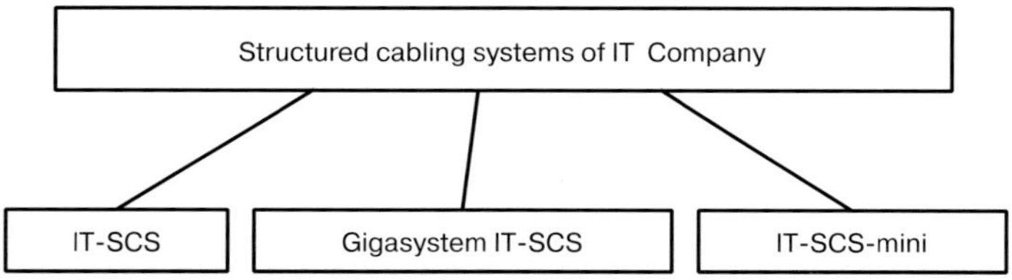

Fig. 14.3. Structured cabling systems of IT Company

The electrical IT SCS subsystems can be realized using shielded and unshielded cables. Four-pair cables with cutoff frequencies up to 550 MHz and category 3 and 5 multipair cables for internal and external layouts are included in the system composition. The backbone subsystems are built, in general, using fiber-optic techniques. To realize these systems, a wide range of multimode and single-mode optical cables with metallic armoring or dielectric versions has been introduced into the composition of IT SCS.

The 110-type panels in the cross-tower variant and for wall mounting variant are used as a switch equipment in the electrical subsystems. The panels with modular connectors giving 12–48 ports in shielded and unshielded versions are recommended for horizontal subsystems. The input telephone line layout is done at the 66-type patch panels. The modular patch panels of the CT-PNL series with 16–48 ports are presented, mainly, to realize small networks and make it possible to do easy mounting of outlets both electrical and optical connectors.

Functions of specialized optical switch equipment are carried out by wall couplings with up to 42 outlets and 19-in shelves with 16, 32, and 48 ports with single-mode or multimode outlets of the main types (ST, SC, and FC) that are already installed. When mounting the plugs of optical connectors at an optical cable waveguide, gluing (aerobic two-component glue is recommended) and distributed mechanical fixation technology are used. It is possible to order predetermined assemblies.

The electrical patch cords with modular connector plugs and 110-type connectors in usual and combined versions have lengths 0.91–7.6 m, cords with 110-type connectors are available in 1-, 2-, 3-, and 4-pair versions. To switch network equipment with Telco ports, 25-pair cords up to 3 m long are provided. A wide variety of the usual and combined multimode and single-mode cords with connector plugs of the main types and standard lengths of 3.5 and 10 m are used in optical subsystems.

For cable systems of the new-generation GigaSystem, equipment containing cables and modular connector outlets is provided. When using this equipment, the route characteristics correspond to the requirements of category 6 for technical design.

The developed set of 19-in mounting equipment and decorated cable channels as well as the elements of technological and final marking of cable and switching products are included in the design of IT SCS and IT SCS-mini. The system functional capabilities are enhanced by introducing power cable elements to power the computer hardware. When creating electrical and optical subsystems, different technological and measuring equipment including welding apparatus and optical reflectometers are used.

For IT SCS mounted equipment, a 15-year system guarantee is provided; when the GigaSystem is installed, the guarantee period is increased to 20 years.

Certified specialist training is done by the IT Academy at the authorized courses. The general duration of training is 5 working days and contains two parts: 2 days for electrical subsystems and 3 days for optical subsystems.

In addition to IT, 40 authorized partners install IT SCS in the Russian Federation.

14.1.9 ITT Network Systems & Services SCS

ITT (NS&S) offers four types of SCS that include ISCS GigaPath shielded, as well as LANConnect 5 and LANConnect 200 unshielded systems (Table 14.1 [154]).

The main original element of all unshielded systems is a small-sized outlet of a 808-type modular connector. ISCS and GigaPath systems use shielded variants of modular female connectors that are identified as modules of the 808 MK2 type. To provide identification, the front panels of the unshielded modules are made out of plastics of seven different colors.

Table 14.1. ITT Network Systems & Services SCS

	Category 5	Category 6
Unshielded design	LANConnect 5	LANConnect 200
Shielded design	ISCS	GigaPath

In the cross-rooms, these modules can be mounted inside the distribution panels of the 300 and 500 series with a maximum of 96 ports. The cases of wall outlets are manufactured in internal and external installation versions and have 1, 2, or 4 ports; the construction design of the internal outlets corresponds to those of Britain, France, and other styles.

The horizontal layout is done by 4-pair cables with usual sheathing and with LSZH sheathing. Connection is made to the outlets with 0.5–10-m long connection cords. The cable sheathing can be five different standard colors.

The distinctive feature of the ISCS SCS is that its channel can have seven plug&socket connectors per channel instead of four, which is allowed by the standard. Therefore the increased channel lengths, in comparison with the standard, presented in Table 14.2 for different applications [155] are guaranteed.

In practice, the lines with enhanced length are divided into parts. The horizontal parts of these line have maximum lengths of 90 m; the remaining parts are for vertical sections. Such an artificial approach makes it possible to formally retain the requirements of the standards and at the same time to give the SCS new features. So, in particular, it is possible to create SCS by centralized schemes due to increases in cable routes.

Table 14.2. Channel lengths for different applications

Application	Route length, [m]	Application	Route length, [m]
RS 232	>1200	100VG AnyLAN	160
DEC423	>1200	TP-PMD (CDDI)	140
IBM3270	820	ATM OC-3	145
Token Ring, 4 Mbps	480	Apple Talk	800
Token Ring, 16 Mbps	250	Telephone	1000
10Base-T	177	Videoconference	350
100Base-TX	140	AS400/System 3X800	800
100Base-T4	160	ISDN	500

Because of the high requirements for the quality of signal transmission on electrical routes, only factory-made cords may be uses in certified systems.

Mini-LAN systems include a complex of active and passive technical means by which cable layouts and simple telecommunications systems are created in the offices of small companies and in houses with minimum costs.

Specialist training is done in two stages, with a duration of 2 working days each. During the first course stage, the installation training is done; during the second stage, the trainees are taught rules and testing methods.

The installed system must be tested by the of WireScope cable tester; for this, software especially developed for ITT NS&S is used.

Six-year and 20-year system guarantees are provided for the mounted systems, depending on their type. So, for example, a LANConnect 5 system gets a 16-year guarantee when the conditions listed in Section 1.6 are carried out. Additionally, channel testing is required by measuring devices that are recommended by ITT NS&S and that are calibrated annually.

The software of the Crimp for Windows graphical database makes the work of technical designers, installers, and SCS administrators easier.

By the end of 1998, ITT NS&S has more than 33 first-hand partners (by their classification system). In addition, there are approximately 50 companies with partner status that differ from system partners only because they cannot install shielded systems and category 6 systems.

14.1.10 Krone Link and HighBand Components

Krone Concerns was founded in 1928. It is well known all over the world as a manufacturer of high-quality components for PSTN systems. Elements for SCS creation are manufactured in full accordance to EN 50173, ISO/IEC 11801, and TIA/EIA 568-A standards [156, 157].

The advantage of the designs of Krone is the ease of SCS interfacing with PSTN networks, which is done by the switching equipment being made by Krone.

The apparatus for SCS includes the so-called units for line termination APL, the distribution units for data transmission, floor distribution units, and connection outlets (Figure 14.4). It should be noted that the SCS structure being proposed coincides with the classic one by the accuracy of its designations. Equipment of distribution units and line termination units is mounted on the wall, in open telephone-type racks, and in 19-in wall and floor cabinets.

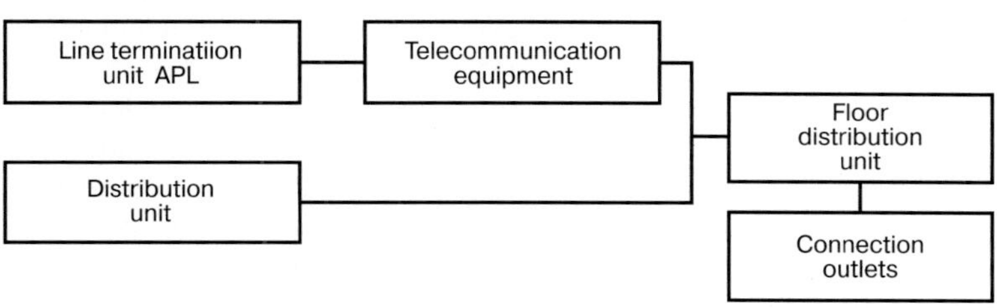

Fig. 14.4. SCS structure of Krone Company

There are two technical solutions of the system that are delivered in the market. The Krone Link system provides category 5 characteristics, whereas the characteristics guaranteed by the Krone Highband system exceed category 6 of the technical design requirements with an essential reserve All switching products are supplied only by firm IDC contacts of the LSA type.

The SCS under consideration allows the use of both electrical and optical cables. For this, the optical cable solutions are preferable for the backbone (primary and secondary) subsystem arrangement, whereas the designs based on balanced electrical cable are typical, in general, for horizontal or third-class subsystems. The SCS under consideration does not include cable products and any cables from the approved list are suitable for its creation (for example, those by Alcatel Company, by Belden Company, and others).

When realizing Krone Link SCS either open frames or 19-in constructions (cabinets and open racks) having maximum heights up to 2500 mm can be installed in the technical premises. When realizing SCS, it is possible to use 16- and 24-port patch panels for shielded and unshielded outlet modules. The switching is done by shielded and unshielded cords 2, 3, and 5 m long with modular connector plugs. To minimize SCS cost when ATM signals, TP-PMD signals, and telephone signals of the exchanges are transmitted, two-pair cord versions are available. It is convenient to code the cords by flexible termination ends of five different colors.

Single- and two-post outlets with straight and angular installation of 6- and 8-positional outlet modules are mounted at work areas. The outlets are designed in various styles (in German, English, and other) and, in case of straight module installation, sometimes are supplied with a protection blind.

For optical subsystems, the Krone FiberPlus switching products are offered as 19-in shelves and panels for mounting in open distribution frames. These products are completed by the outlets for the main types of optical connectors (ST, SC in usual and duplex versions, MIC, DIN FC, and E-2000). Patch cords in single and duplex versions have lengths 1–20 m with 1-meter discrete length steps. Additionally, when realizing the optical subsystem, optical splitters, switches, and spectrum multipliers are used, if necessary.

The Krone Highband system, from point of view of its creation, is the classic cabling system. It is realized in unshielded and shielded variants by the principals of structured cabling and is considered for the frequency range 100–600 MHz. If switching based on modular connectors is required, the original design of the RJ-K type by Krone is used. This is an entirely shielded unit that is functionally similar to those with modular connectors; it is certified to frequencies up to 300 MHz and essentially exceeds the norms of category 5 and the technical designs of category 6 [158]. In the technical premises, the outlets of such connectors are mounted in patch panels having 16 and 24 ports. Besides modular solutions, solutions based on the Highband-S terminal blocs are possible. For switching, special cords with connector plugs similar to 110-type plugs are used. The cords of this type have 1-, 2-, 3- and 4-pair versions and are available in shielded and unshielded versions.

For installed SCS, a standard 15-year system guarantee is provided.

14.1.11 Lucent Technologies Systimax

The Systimax SCS of Lucent Technologies in its electrical part is formed by two products: PowerSum and GigaSpeed. Both systems are formed in according to TIA/EIA-568-A Standard; GigaSpeed characteristics meet the requirements of technical designs of the category 6 standard. The technical solutions being used in these products provide backward compatibility.

The Systimax electrical subsystem is created only by unshielded cables of twisted pairs. Ninety percent of the entire volume of category 5 cables being delivered have the NEXT reserves relative to the standards of 6 dB. GigaSpeed-series cables are certified up to a frequency of 550 MHz and are intended for category 6 systems. To provide operation of telephone exchanges, one can use armored cables of the ANMW series with 25, 50, and 100 pairs to transmit telephone signals. The cable products are delivered both in cardboard boxes and on rolls.

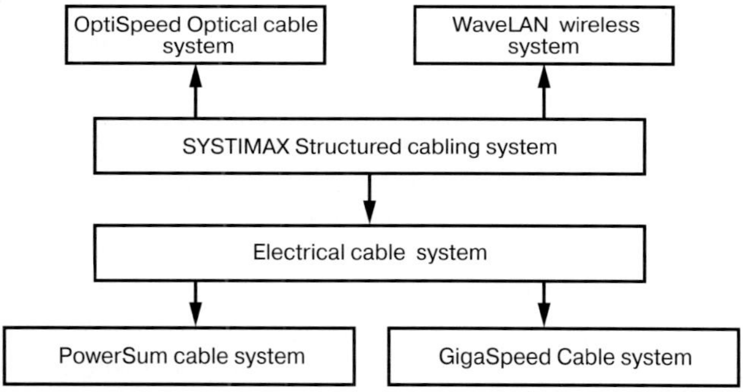

Fig. 14.5. SYSTYMAX SCS structure of Lucent Technologies Company

The electrical subsystem switching equipment is based on 110-type panels (versions of cross-tower for 300 and 900 pairs and wall versions and 100 and 300 pairs), as well as on panels with modular connectors having up to 64 ports. The VisiPatch new-generation panel gives significantly better conditions for maintenance and reading markings in comparison with its prototypes because of a new 110-type plug construction with backward layout of the cord cable. To switch the electrical ports, the developed set of cords with modular connector plugs and 110-type connectors as well as combined cords are provided. Cord lengths for all types are 0.6–4.6 m in increments of 1 ft.

The fiber-optic equipment used for creation of Systimax SCS optical subsystems is merged with the OptiSpeed system. The optical cables have single-mode fibers and 62.5/125-type multimode optical waveguides. Cables for internal layout of the Accumax series have a maximum of 36 fibers and are delivered in plenum variants; the construction of cables for external layouts is modular or is made on the basis of a central tube. Cables for external layouts are manufactured both with metallic armor and with entirely dielectric variants. The OptiSpeedPlus system differs from its prototype by using LazrSpeed cables that support 1000Base-SX signal transmission over distances up to 600 m with the optical carrier wavelength of 850 nm.

More often in practice, one can find fiber-optic switching equipment that consists of 110A3-type and 200A-type wall couplings with up to 24 outlets, and with 600-series and LST-series shelves, the number of ports increases to 72. Switching is made by cords with lengths 1.5–9 m and connectors of the ST, SC, LC, and MIC types, including the combined versions.

Multimax modular panels with 24 sites for electrical and optical outlets of different types are intended for small networks.

The internal and external outlets are manufactured in versions with straight and angled installation of electrical outlet modules (the outlets of optical connectors are always located at right angles) and are mounted at work areas. In networks created by the fiber-to-the-desk principle, 40A1-type multimedia outlets can be mounted at work areas.

The Systimax SCS includes the developed set of adapters and baluns that make it possible to use very flexible designs of the electrical cable layout to transmit the signals of different network equipment with interface connectors that are nonstandard for SCS.

When mounting the electrical routes, single-wire and five-pair impact tool are used. Plug fastening of optical connectors is done by gluing using a 1032 tools set. For fiber splicing, mechanical CSL splices are installed. They also serve as the latches of STP-series mounting cords with ST connector plugs.

The system under consideration differs from others by its lack of instrumental testing of the characteristics of the electrical routes after installation. The correctness of the layout is only controlled visually The achievement of network parameters that are required for normal functioning of the network apparatus is guaranteed by the equipment quality and by the high level of training of the technical designers and installers. To check the attenuation value of optical routes, use of a 938A1 optical tester is recommended.

The Wavelan wireless system is used in the Systimax system, which makes it possible to connect dedicated segments and separate Ethernet and Token Ring workstations to the local network.

The full training cycle for specialists is three one-week courses: technical designing, and installation of electrical and optical subsystems. The study is done in two stages. At the first stage, which lasts for two days as is the case for all courses, the basic information about SCS are given, the remaining three days are for special training.

For installed Systimax PowerSum SCS, a 15-year system guarantee is provided; if GigaSpeed is mounted, the guarantee is increased to 20 years. Simultaneously, Lucent Technologies also provides guarantees for its applications.

In mid-2001, according to the data of the Moscow office of Lucent Technologies, Systimax SCS installation in the Russian Federation is performed by over 55 authorized partners.

14.1.12 Molex Premise Networks

Molex Premise Networks system (until 1999 it was named Mod-Tap) is related to the classical SCS. It was created according to American Standard TIA/EIA 568-A and includes the backbone subsystems and the horizontal subsystems as well as a system of distribution points. Electrical and optical solutions are possible. In electrical solutions the use of cables of UTP and STP is allowable.

The basis of the electrical modules are contact of the KATT type, described in Section 3.2.1. The available designs make it possible to create channels for equipment of the following types: CDDI, FDDI, 10Base-T, Token Ring, IBM 3X/AS400, IBM3270, Apple Talk, and ATM. Network equipment in which standard interfaces based on modular connector outlets are lacking can be connected by means of systems of different assignment adapters.

The electrical cable products for the Molex Premise Networks SCS consist of 1- and 2-pair category 3 crossing wires, by horizontal cables of category 3 (2, 3, 25, 100, 200, and 400 pairs), by 6- and 8-conductors of flat cables to connect telephone sets and low-speed (up to 1 Mbps) network equipment, and by 1-, 2-, and 4-pair horizontal cables of category 5.

The internal telecommunications outlets at work areas are based on 1- and 2-post modules of Euromod series. The modules are delivered in versions with straight and angular installation of outlet modules and sometimes are supplied with protection blinds. One- and 2-post blocks are also intended for external installation, which, in case of fastening on the metallic furniture, can be equipped with magnetic latches. Modular mounting is made easier by a wide range of boxes, frames, and adapters of different styles and sizes.

The electrical patch panels consist of both demountable and undemountable versions. The demountable variant has 12 and 24 sites for modules of different types. The undemountable panels have 24 and 48 ports. A special panel with 24 6-position female outlets of category 3 with USOC layout is intended for telephone applications. To make multipair cable layout easier, so-called distribution MOD panels with Telco connectors are at the back.

Various 110 panels with 100 pairs are manufactured for wall mounting and for mounting on panels for installation into 19-in constructions. Convenient wall mounting without mounting legs of the 100-pair blocks is provided by the special KATT PDS wall-mount frame.

The construction elements for the creation of shielded cable layouts, in general, repeats the solutions for unshielded subsystem and differs only by the presence of shielding elements by providing the continuation of the shields in the switching products.

The electrical panel switching is done by cords of the 110–110, 110–RJ-45 (length 1-5 m) types, by cords with Telco connectors (lengths 1–30 m) and by the cords (lengths 0.5–15 m) with modular connectors. In the latter case, the cable sheath can be painted in five different colors.

The Molex Premise Networks system has all the main elements to create optical subsystems: telecommunication outlets and multimedia outlets, wall coupling, and shelves in 19-in constructions, including modular variants. The optical cables for internal layout with 62.5/125-type fiber have the optical from 4 to 24 waveguides; cables for external layouts are used for cable layouts in conduit runs and for air suspension. The standard lengths of patch cords are 1.5, 2.3, and 10 m; the cords are armored with plugs of ST, SC, and MIC connectors. The connectors are installed by the Hot Melt method and by using two-component glue or mounting cords. In the latter case, Ultrasplice-type mechanical splices are used. The number and types of optical components of the Molex Premise Networks system is very large after the merger of the developer (Mod-Tap Company) and Molex [159].

The process of optical cable layout creation is speeded up by the use of the ModLink series of preterminated assemblies.

To create cable layouts in small offices and for separate working groups, systems based on the USN-32 cabinet are used.

The Molex Premise Networks system installation and maintenance is made easier by the vast set of technological and sizing tools recommended for this purpose. The equipment mounting in the technical premises is done in 19-in floor and wall cabinets. Shelves, ventilation modules, vertical and horizontal outlets, modules of power supply system with built-in network filters, and other auxiliary elements are offered.

For the mounted systems, a 15-year system guarantee with a certificate number is provided. The authenticity of the certificate is confirmed by a hologram of the company logotype.

14.1.13 Ortronics GigaMo System

Ortronics GigaMo structured cabling systems were first available in Russia in 1999. The systems under consideration are known in two variants: GigaMo and GigaMo+. The first provides the technical design characteristics of category 5, whereas the second provides characteristics of category 6 and guarantees, in particular, the transmission of Gigabit Ethernet[160] network interface signals with given grade of service.

All types of products that are necessary for the entire channel creation, except the cables, which must be purchased from certified manufacturers, are included in the composition of the systems listed. Moreover, a wide set of equipment for use in the cable systems of category 5 is manufactured at the company's plants.

The electrical patch panels consist of patches with modular outlets in shielded and unshielded versions with a maximum of 48 ports. Modular solutions are available, if necessary. The 110-type panels are known in versions of cross-towers for 300 and 900 pairs as well as in 100- and 300- pair blocks with legs for wall mounting [161].

The subscriber outlets are offered in versions for internal and external mounting and have different style of embodiment. The electrical outlet modules are manufactured both for straight and angular with a projection installation

The electrical cords of different assignment are manufactured with modular connector plugs and 110 connectors as well as in combined variant. The length of the cords is 0.9–5.5 m.

The optical switching units consist of 19-in shelves and wall couplings. Shelves of the ORMMAC series with 36, 72 and 108 ports are manufactured in modular variants. The so-called low-profile high-density rack-mounted Fibre Cabinet is a shelf of height 1*U* with 24 ST or SC outlets. Six- and 12-port wall couplings are supplied with upper and lower compactors for cable and cord passing and have two sites for plates with six optical ST and SC outlets.

Convenient mounting of this equipment in the cross-rooms and equipment rooms is provided by using open racks of the Mighty Mo II series and both universal custom mounting cabinets. Different original models of cord and cable organizers make the service of the panels and mounted network equipment easier.

A wide list of different adapters and elements that make it possible to arrange MUTO multiuser outlets and consolidated points are in the range of the equipment that can be delivered.

The distinctive feature of Ortronics equipment is the presence of a system of different color identifiers of cords and port outlets of different kinds.

The original fiber-optic medium converters of Transoptix are included in GigaMo SCS. These products support Ethernet and Fast Ethernet standards and are made as unpackaged plates that are inserted into the idle slots of workstations or hub chassis.

A 25-year expanded system guarantee is provided for Ortronics SCS.

14.1.14 PAN-NET System of Panduit Company

The Panduit PAN-NET system is formed by merging separate independent subsystems that are each distributed under their own trademarks. In MOD-COM and MINI-COM systems, the solutions are merged based on modular connectors. The difference between them is that MINI-COM system equipment needs, on average, 20% less space for mounting. In addition, this equipment was initially developed for use in offices and had the corresponding characteristics and design. The PAN-PUNCH system includes 110-type patch panels and accessories and is used, in general, in technical premises. The OPTI-COM system contains elements that make it possible to create SCS fiber-optic subsystems.

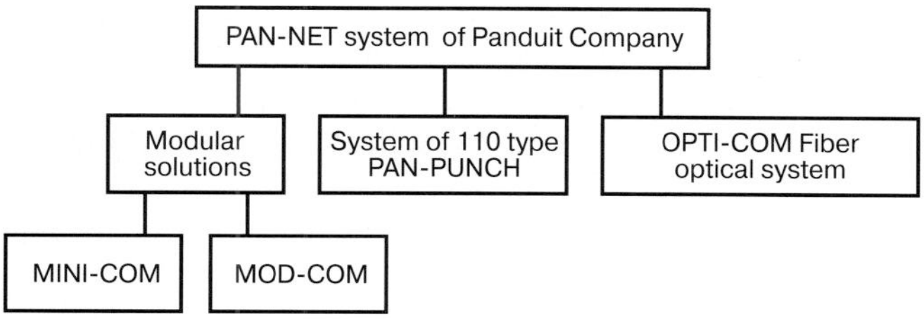

Fig. 14.6. PAN-NET system structure of Panduit Company

The MOD-COMand MINI-COM systems contain outlet modules of category 3 and 5 for external and internal (in profile and internal space) mounting. For this, modules of the internal mounting are delivered in variants of German, English, European, and other styles, and straight and angular outlet installation are allowed.

Modular connector outlets of the PAN-NET system are manufactured in usual and shielded forms in two some different versions—MINI-JACK and MOD-JACK. The panel front surface color of the outlet modules and the scheme of their layout can differ. Besides 8-position modules, the use of 6-position modules with six and four conductors is allowed. Cable connection to the outlet is done without using impact tools.

The 110-type equipment is a part of the PAN-PUNCH system and is manufactured as cross-towers for 300 and 900 pairs and as panels for wall mounting of 100 and 300 pairs. 3-, 4-, and 5-pair variants of rulers are available , the mounting is done by usual single- and 5-pair impact tools. In case of telephone service applications, it is very useful to use 100- and 300-pair prelaid wall panels with the output of the second ends of the multipair cables to Telco connectors. The 110-type connector plugs of the PAN-PUNCH system are adapted for direct mounting on the object.

The GIGA-PUNCH system equipment is the development of a 110-type system and differs from it by locating the rows of contacts in two staggered rows. It allows locating 36 pairs on one ruler instead of 25, as in traditional constructions. Simultaneously, the equipment of this variety provides category 6 electrical characteristics.

The fiber-optic subsystems that are realized by OPTICOM equipment are based on shelves with a maximum of 72outlets , wall couplings with 12 and 24 outlets, and multimedia outlets. It is possible to install up to 12 optical outlets into the case of the multimedia outlets. ST and SC connectors in the usual and duplex versions and MIC connectors of the FDDI system are used. Cords with lengths 1–3 m are intended for switching, including the combined variant with connector plugs of different types at different ends.

The FIBER-DUCT and PAN-DUCT systems are recommended to organize the layout of internal optical cables. They differ from usual decorative ducts mainly by the presence of a set of angles that makes it easier to pass turns with the given radius. The channels are mounted by special brackets.

Cable layout and outlet mounting in the office premises is carried out by system of decorative PAN-WAY ducts. The system has monolithic and composite ducts as well as a large number of component elements (including angles and plugs).

It should be noted that the additional accessories of the PAN-NET system include a vast set of self-laminated markers, labels, and glued labels that can be marked by using computer printing. Accurate cable beam layouts and their additional marking is provided by using Velcro adhesive loops, plastics belts with marking spaces of the PAN-TY series, and PAN-STEEL brace rods made of steel belts.

14.1.15 Reichle & De-Massari Freenet system

The Swiss company Reichle & De-Massari is known for its cable and switching equipment for telephone network creation. The Freenet cable system [162] is a relatively new product for this company and is not yet widely distributed.

The Freenet SCS is built according to International ISO/IEC 11801 Standard. Currently, its electrical parts correspond to the requirements of category 5 and technical design of Categories 5 and 6.

The horizontal subsystem is built mainly using 4-pair electrical cables in UTP, STP, and S/STP versions. The upper cutoff frequency of some constructions reaches 600 MHz. To make the layout easier in cases of two-port work areas, double cables ("Siamese") are used.

The distinctive feature of the electrical subsystem switching equipment is that it uses specially developed modular connectors. When mounting cables in the outlets of such connectors, separate crimping tools or impact tools are not required because their functions are fulfilled by two protection covers in form of levers that are integrated parts of the outlet.

The patch panels have 16–84 ports and are characterized by their seldom use of vertical ruler locations with modular connection outlets. For greater convenience of mounting, the IDC contacts of Reichle &De-Massari equipment are located on the front panels and, after the completion of mounting, are closed by flaps which allow marking of elements of separate ports. For this, one ruler makes it possible to locate from four to six such outlets. Besides vertical versions, horizontal 16- and 30-port panels are used. The panel capacity can be changed in wide range and is flexible at the expense of special 4- or 5-port modular insertions with shielded and unshielded outlets on it.

The characteristic feature of the Freenet system is its use of a system of electrical connector coding that increases the convenience of cable system service during maintenance. Coding done by means of adapters that are put on the plugs and modular connector outlets and that provide mechanical blocking to prevent wrong connections. Color marking is used for visual indication of coded connectors.

Freenet systems do not use 110-type patch panels. Use of telephone terminal blocs is recommended for service of low-speed applications that are rarely switched.

The fiber-optic equipment organizes the backbone SCS subsystems. One can see cables with rod wires when laying out the system inside the building, which is very rarely done in modern SCS. The optical port density is increased at the switching units by two custom connectors: SC Contact (for usual cables) and Mini-MPO (effective in cases of installation of ribbon cables).

This set of internal and external telecommunications outlets of different styles with straight and angular installation of the modular outlet connectors are intended for installation at work areas. The systems that are popular in German-speaking countries contain large numbers of elements to provide ISDN system operation. To increase the maintenance flexibility of the system the set of connector adapters, adapters, and baluns of different assignments is included in its composition.

For systems mounted by certified specialists, a 20-year guarantee is provided; the duration of the guarantee for applications has no time limit.

14.1.16 RiT Technologies SCS

RiT Technologies delivers four different types of cable systems: the SMART cabling system, the SMART-Giga cabling system, the STP 600 MHz, and the STM 400 MHz (Table 14.3). The first two systems are the most popular in Russia. The difference between them is that SMART SCS parameters correspond to category 5, whereas SMART Giga has characteristics certified up to 200 MHz and correspond to technical designs of category 6.

Table 14.3. RiT Technologies SCS

Unshielded solutions	SMART (category 5	SMART Giga (category 6)
Shielded solutions	STP 400 MHz	STP 600 MHz

RiT Technologies Company manufactures the whole spectrum of products for its own SCS.

The patch panels with modular connector outlets in fixed configurations have 24–96 ports (heights 1–4*U*) and are manufactured both in shielded and unshielded versions. Cable wire connection is done by 110-type contacts or Krone contacts. The modular solutions consist of panels of Double 5 with ten sites in each; 2-port modules with UTP and STP outlets can be installed. In addition to the panels with connectors, there is a wide range of SMART ICS 32-series panels provided for connecting 150-Ω cables. Patch panels of RiT Technologies are characterized by the range of products with switches, which makes it possible to connect to outlets located above one another. Thus, it is possible to decrease the number of patch cords by 80%, according to data from RiT, and by the right technical design fulfillment, at least, at the first system maintenance stage.

Panels of 110 and 66 types are not used in these SCS.

A large set of internal and external outlets is intended for installation at work areas. The maximum number of ports for external Giga-Flash outlets is 6. The internal outlets of German, American, and other styles are available both with straight and angular installation of outlet modules. In the latter case, there are constructions with a projection and with sunken installation. The outlet modules are used both in usual and shielded versions.

A range of connector adapters, baluns, and adapters of different assignment and special means to support Token Ring network apparatus operation is included in the composition of RiT Technologies SCS.

The use of horizontal cables, usual and shielded, is limited to the frequency range 100–600 MHz. Versions with twisted conductors and enhanced stability for bending are intended for the manufacture of patch cords. The backbone subsystem creation is based on 25-pair cables of category 5.

Panel switching with modular connectors is done by shielded and unshielded cords having lengths 0.3–7.5 m. Use of cords manufactured not only at the plant but also at the object is allowed.

Fiber-optic components consist of 19-in shelves with 96 and 12 outlets, by wall outlets, and outlet modules for installation into decorative ducts. Fiber-optic cables of the SMART system serve for creation of all three subsystems. Cables for external layouts are delivered in corrugated and uncorrugated versions as well as with integrated ropes for air suspension. The greatest number of fibers in the standard variants is 24. Cables for internal layouts also have a maximum of 24 fibers. The patch cords in simplex and duplex versions with lengths of 1–5 m are supplied with plugs of ST and SC optical connectors. When installing the plugs of optical connectors in field conditions, epoxy gluing technology is used.

For SMART SCS it is typical to use the PatchView monitoring system of cable system status. The system operation principles, technical means, and software are described in detail in Section 13.1.3.

RiT Technologies provides a system guarantee for 15 years (SMART) and for 20 years (SMART-Giga).

The usual training course for technical designers and installers lasts 2 days. So-called PatchView consulting 4-day course is also avaliable.

14.1.17 Corning ICCS and FutureLink Systems

Corning Integrated Communication Cabling Systems (ICCS) and FutureLink Systems were created by Siemens (Figure 14.7 [163]). In February 2000 after purchasing Siemens, the American company Corning became the owner of the trademarks. The system includes the entire set of passive components, accessories, and additional equipment that are necessary to form SCS. ICCS covers SCS electrical parts and uses mainly shielded cables and the SSL (Siemens standard link) and SML (Siemens modular link) connection techniques. FutureLink is used for SCS fiber-optic system creation. The switching products of ICCS and FutureLink are carried out with unified designs and can be combined in any fashion.

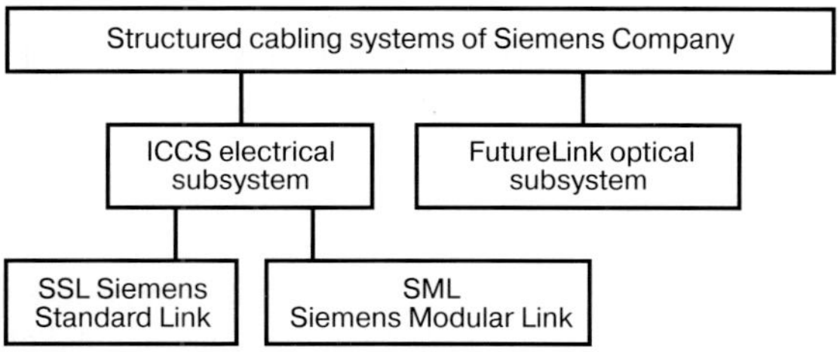

Fig. 14.7. SCS of Corning Company

All components of ICCS and FutureLink systems are built entirely corresponding to the norms and statements of the ISO/IEC Standard and German version of the EN50173 Standard, and their technical characteristics in great number of cases very often exceed the requirements of these normative documents by a large margin.

SCS using this equipment generally contain four subsystems. The primary subsystem, or campus subsystem, usually connects several buildings. The main task of the secondary subsystem is merging the floors of one building. The third or horizontal subsystem serves to connect the subscriber outlets to the cross of the floor. Work area subsystems allow connection of the network equipment to the outlets. It is allowable to lay electrical and optical cables on equal terms inside the building. The primary subsystem is built exclusively on the basis of fiber-optic cables.

ICCS contains both unshielded and shielded cables in usual and fire-safety variants. UTP cables of category 5 provide the operation of analog telephone communication systems as well as LAN Ethernet and Token Ring. S/UTP and S/STP cables are recommended in situations of complicated interference at the object of mounting as well as to support operation of applications that require broader frequency bands. Some cable samples have parameters that are normalized at frequencies up to 300 MHz or even 600 MHz. Cables for ICCS differ by the presence of a fourth twisting in their designs.

The patch cords are mainly manufactured in the shielded variants and have standard lengths of 2, 3, and 6 m. In addition to 4-pair cords, less-expensive 2-pair cords were also introduced in the design of ICCS, oriented to the traditions of the European market. Two-pair cords are available to support the functions ISDN, CDDI, and ATM systems.

Patch panels and telecommunications outlets with modular connector outlets are part of SSL and SML systems and are manufactured only in shielded variants. The SSL system panel has 16 ports and height 1*U*, and the SML system panel has 16 sites for changeable insertions and height 2*U*. In the latter case, the cable can be laid out not only to single but also to double insertion, therefore it is possible to realize the cable-sharing principle.

The broad range of designs of internal and external outlets is intended for use at work areas. The internal outlets are manufactured in one-, two- and three-post versions and contain adapters for mounting in ducts, floor boxes, and other similar products.

Fiber optical components are merged within the FutureLink system and are used both for creating SCS backbone subsystems and for horizontal layouts to realize fiber-to-the-desk solutions. In addition to cable and switching products, technological and measuring equipment is also included in the composition of FutureLink, including the welding apparatus and optical reflectometers.

Cables for external layout have up to 60 fibers and are sometimes supplied with strengthening coverage made out of steel ribbon to enhance their mechanical characteristics and to protect them from rodents. Cables for internal layout typically have up to 24 fibers. Several versions with different mechanical characteristics and fire stability can be selected.

The fiber-optic patch cords are manufactured with the main types of 9/125, 50/125, and 62.5/125-μm fibers and with maximum lengths of 30 m. Depending on the frequency range, the lengths of the standard cords increase by 0.5, 2, and 5 m. Cords with 62.5/125 fibers are armored by the following types of plugs: ST, SC in usual and duplex versions, E-2000, and ESCON.

The switching products consist of shelves of heights 1*U* and 2*U*; the component density reaches 48 fibers per each *U* of height. There is a wall coupling with the capability for installation of a maximum of 12 ST, SC, and SMA connector outlets and 4 fiber wall outlets of the FO4-W type.

For optical waveguide armoring in field conditions, so-called FuseLite mini-pigtails and UniCam plugs with mechanical distributed fixation are used. Two-fiber splicing is carried out by welding or by using mechanical Camsplice splices.

A 15-year guarantee is provided for a mounted system.

14.1.18 Siemon Cabling System

Siemon delivers the entire spectrum of SCS equipment, except cables. The system under consideration is built according to American Standard TIA/EIA-568-A. The design of the Siemon cabling system allows for the use of cables made by seven leading cable manufacturers.

The equipment that is installed directly in the work area consists of different types of outlets. The internal outlets of the CT series are manufactured in 1- and 2-port versions with straight and angular installation of outlet modules that can be supplied with flaps or slide lid. If a high density of ports is required, modular connector outlets of the MAX series are used. Wall mounting is provided by 1-,2-, 4- and 6-port external outlets of the SM series. Modular connectors for all types of outlets can be changed by optical and coaxial ones of different types, if necessary.

Modular patch panels with 16–96 ports, which are offered in demountable and nonseparated versions, are mounted in the CT-series modules of the cross-rooms and are suitable to be installed in demountable panels. The panel layout is carried out by 110-type rulers . There are panels of category 3 with Telco connectors. In addition to the usual panels, 24-port so-called modular patch blocks are offered with 1-, 2-, 3-, and 4-pair layouts and Telco connectors to connect multipair cables. Installation of panels and other equipment is made easier by the presence of several types of open racks and wall-mounting panels. The vast range of cable organizers of different kinds is also introduced into the Siemon rack system.

Patch panels of the 110 type in the cross-towers variant have 300, 400, and 500 pairs; in versions for wall mounting, the number of pairs is equal to 100, 200, and 300 pairs. In large networks it is possible to use wall frames of the XLBET series that are built as open cabinets with a maximum of 10,800 pairs.

To support telephone system operation and data transmission system operation with bit rates not more than 10 Mbps, 66-type patch panels with a maximum of 125 pairs are used. Switching processes of such panels are made a little easier by the presence of the vast range of prelaid panels.

Patch cords with modular connector plugs in shielded and unshielded versions have lengths 0.9–7.6 m and are supplied with identifying termination points, if necessary. Cords with 110 connectors have the same lengths and are manufactured in 1-, 2-, 3-, and 4-pair versions.

In mid-1998, System 6 cabling system electrical elements with enhanced characteristics became available. The totality of the elements makes it possible to create cable routes with upper cutoff frequencies of 200 MHz and with parameters corresponding to technical designs of category 6.

The fiber-optic equipment consists of 19-in shelves of height $4U$ with 72 and 144 ports. The functional flexibility of the shelves is increased by using adapter insertions under ST and SC outlets and shelves of height $3U$ for patch cord layout. The wall coupling of classical constructions with 12, 24, and 48 ports are also built by the principle of modular insertion installation and are supplied by flaps closing the external sides of the outlets with the cords connected to them. The optical patch cords in single and double versions with ST and SC optical connector plugs in single-mode and multimode versions have lengths 1–5 m. Plug installation to the cables in field conditions is carried out by gluing with two-component aerobic glue.

The creation process and the following SCS maintenance are made easier by the developed system of technological instrument for electrical and optical subsystems, diagnostic equipment for electrical subsystems, and a large number of adapters to connect the different monitoring devices and network equipment to SCS.

14.2 Decibel Unit, Signal Level, Gain, and Attenuation

When transmitting information over cables, the power of the electrical signal changes exponentially during propagation, therefore, the use of the widely used idea of “transmission gain” is not convenient in engineering calculations. To simplify the calculations, it is expedient to transfer to logarithmic units. In this case, the operations of multiplication and division are performed by addition and subtraction, and instead of the idea of signal and interference power, signal and noise levels are used. According to GOST 24204-80 (State Standard) [164], the signal level is logarithm of the ratio of the power value or voltage at the point under consideration to the point value (P_0) or voltage (V_0) at the circuit point selected for comparison. It is assumed in wire communication techniques that the zero level has a signal with power P_0=1 mW at the load resistance 600 Ω. The values shown are selected from the historical point of view: 1 mW is the power of a usual telephone signal and the resistance 600 Ω has the diaphragm case. The signal level measured relative to this level is called the absolute signal level.

Simple calculations show that the signal with zero level with a load of 600 Ω is characterized by U_0=0.7746 V and current 1.201 mA. The signal with any other power, by definition, has the absolute level:

$$p_S = 10\lg\frac{P_S}{1mWt} = 20\lg\frac{U_S}{0,775V} + 10\lg\frac{600}{R_l} \quad (14.1)$$

Analysis of (14.1) shows that:

- Signal power with positive level is always greater than 1 mW; when the power is less than 1 mW, the signal level is negative;
- Power increases or decreases by a factor of two correspond to its increasing or decreasing at 3 dB;
- Signals with a similar (equal) voltages can have different levels that are entirely determined by load resistance at the measured point.

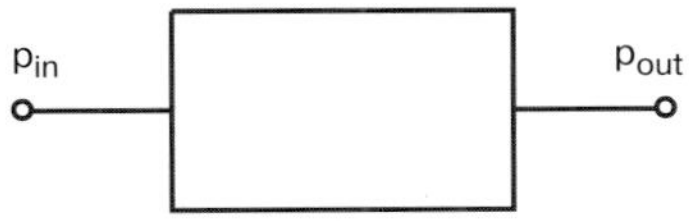

Fig. 14.8. To definition of gain and attenuation conception

Now, review Figure 14.8, which shows a circuit that is analyzed as 4-pole. According to GOST24204-80 this 4-pole has a gain

$$S = p_{out} - p_{in},$$

and attenuation

$$A = p_{in} - p_{out},$$

where p_{in}, p_{out} are the levels of input and output signals, respectively. Usually, the conception gain is used with signal component, and attenuation for the interference because, to provide normal quality of communications, the latter must be significantly lesser than signal power. Introduction of the gain and attenuation concepts makes it possible to always operate with positive values of transmission gain presented in decibels.

14.3 AWG and SWG Value Conversion and Copper-wire Line Resistance

The vendor cable conductor diameter is often indicated in AWG (American Wire Gauge) units. The latter designation is connected with technologies of wire manufacturing by a series of passes through the rollers. The distance between each following roller is less than the previous ones, thus the more rollers the wire passes, the thinner it will be. Two-digit numbers after the letters AWG show how many rollers a wire passed during manufacturing. Table 14.4 gives the AWG value conversion in millimeters. There one can see the corresponding square of the cross section and the running active resistance of copper wire to direct current.

Table 14.4. AWG values versus diameter, cross section, and running resistance of copper wire

AWG	Diameter, [mm]	Cross section, [mm^2]	Running resistance at 20°C, [Ω/km]
16	1.290	1.307	13.6
18	1.024	0.824	21.6
20	0.813	0.519	33.2
21	0.724	0.411	41.9
22	0.643	0.324	53.2
23	0.575	0.259	66.6
24	0.511	0.205	84.2
25	0.455	0.162	106
26	0.404	0.128	135
27	0.361	0.102	169
28	0.320	0.081	221
30	0.254	0.051	357
32	0.203	0.033	559
34	0.160	0.020	857
35	0.142	0.016	1080

In practice, the analog English unit, SWG (Standard Wire Gauge), whose values are shown in Table 14.5, is used less frequently. Comparing Table 14.4 and Table 14.5, one can see that for the most popular numbers in the range 22–26, the wire diameters given in AWG and SWG are very close to each other.

14.4 IP protection Levels

Protection levels providing safety equipment maintenance and the protection of equipment itself from the ingress inside dirt, dust and water is determined by DIN 40050 and IBC 529 Standards. In the Standards, the protection levels are designated by two symbols: IP (index of protection) and by two figures from 0 to 6, for example, IP-65. It is allowable to supplement the index by additional and auxiliary signs.

Table 14.5. SWG values versus wire diameter

SWG	Wire diameter, [mm]
14	2.1
16	1.6
18	1.2
20	0.92
22	0.71
24	0.56
26	0.46
28	0.38

The first digit shows the protection level of operators from touching the body parts, instruments, or by any other objects to the mechanical parts under voltage or moving mechanical parts of the equipment, as well as the protection level of the equipment itself from the ingress of foreign objects and dust. The requirements of each level of such protection are shown in Table 14.6.

Table 14.6. Protection levels from touch and ingress of foreign objects and dust inside equipment

First digit	Protection	Comments
0	No protection	Protection from accidental touching of parts under voltage or of moving parts of the equipment is not provided for. Protection of the equipment from ingress of foreign objects is not provided for.
1	Protection from penetration of large-size foreign objects	Protection from accidental touching by a body or body parts (for example, hand) to the parts of the equipment that are under voltage or to the moving parts of the equipment. Hence, there is no protection from premeditated touching of these parts. Equipment protection from penetration of foreign objects with diameters >50 mm.
2	Protection from penetration of medium-size foreign objects	Protection from finger touch to parts of the equipment under voltage or moving parts of the equipment. Equipment protection from the penetration of foreign objects with diameters >12 mm.
3	Protection from the penetration of small-size foreign objects	Protection from touch to parts of the equipment under voltage or moving parts of the equipment by any objects having diameters > 2.5 mm (for example, tools). Equipment protection from penetration of foreign objects with diameters > 2.5 mm.
4	Protection from penetration of grain-sized foreign objects	Protection from touch to parts of the equipment under voltage and moving parts of the equipment by any objects with diameter >1mm. Equipment protection from penetration of foreign objects with diameters > 1mm.
5	Protection from dust layers	Complete protection from touch of body parts or other objects with diameters >1mm. Complete protection from dust penetration is not provided for; hence, penetrated dust cannot form layers that can effect the equipment operation.
6	Protection from dust penetration	Complete protection from touch of body parts or any objects to parts of the equipment under voltage or moving parts of the equipment. Complete protection from dust penetration.

The second digit refers to the protection level of the installed equipment from water penetration. The requirements for water protection are shown in Table 14.7.

Any additional protection from the effect of aggressive fluids and vapors is not provided for.

Additional signs determine the protection level from access to the electrical and mechanical parts of the equipment that are dangerous to life and health. It is pointed out that only in cases when the protection, in fact, exceeds the level given by the first digit and where the digit means only protection from access to the dangerous details.

Table 14.7. Protection levels from water penetration in equipment

Second digit	Protection	Comments
o	No protection	Protection from water penetration is not provided for.
1	Protection from vertically falling drops of water	Drops of water falling vertically cannot cause dangerous effects on the equipment.
2	Protection from “slanting” rain drops	Drops of water falling at angles not more than 15° from the vertical cannot cause dangerous effects on the equipment.
3	Protection from small drops of water	Drops of water falling at any angle up to 60° to the vertical cannot cause dangerous effects on the equipment
4	Protection from fine drops of water	Fine drops of water falling at any angle cannot cause dangerous effects on the equipment.
5	Protections from sheets of water	Sheets of water directed at any angle (for example, from a hose) cannot cause dangerous effects on the equipment.
6	Protection from strong sheets of water	Sheets of high-pressure water directed at any angle cannot cause dangerous effects on the equipment.
7	Protection from short-term underwater exposures	Short immersion of the rack cannot cause dangerous effects on the equipment located in it.
8	Protection from long-term underwater exposures	Lengthy immersion of the rack cannot cause dangerous effects on the equipment located in it.

Additional signs give further information about the protection level, for example, they can refer to special weather conditions.

Summary

The state-of-the-art engineering and standardization of cable systems enables the following conclusions:

- At present, there is a vast variety of electrical and optical cables, connecting products, and all kinds of accessories in the market, which enable the creation of SCS with capacities from tens to tens of thousands of ports, by means of standard hardware and application of typical solutions. This also enables the creation and augmentation of SCS for small one-to-two room offices and several buildings with possible exits to PSTN.
- Since it is a uniform system, SCS provides a wide range of network equipment with various capacities and signal transmission media, thereby allowing the creation of so-called intelligent buildings.
- By means of SCS infrastructure and a large collection of various adapters, designers and maintenance personnel may establish data exchange between network hardware and coaxial, triaxial, and other kinds of interfaces that were not initially designed for operation along balanced and optical cables.
- The technical level of the element platform specified by the standards ensures efficient operation of installed cable systems and enables the support of existing and prospective applications for a minimum of 15-–20 years.
- Thanks to a number of manufacturers who offer a wide range of individual standard components and fully completed SCS, it is possible to create cost-effective systems for any specific need that possess superior technical performance.
- Since the standards specify the compatibility of category-5 and category-5e equipment produced by various manufacturers, and because many manufacturers ensure backward compatibility of their products with category-6 projects, customers are guaranteed reliable protection for their investments, which is independent of the economic position of a particular SCS manufacturer.

It is also very important to possess a wide scope of supplementary and auxiliary components, such as raceways, 19-in hardware, and marking elements, which are widely applied during SCS construction. These components provide:

- attractive appearance of office and equipment premises, and also ensure additional mechanical, and, if required, electromagnetic protection of cables and connecting products of SCS;
- compact and convenient disposition of network and termination equipment with control of unauthorized access;

- high-quality identification marking of all SCS components and auxiliary elements during construction and maintenance, as well as ensuring database and maintenance documentation support.

A large variety of measurement equipment for testing electrical and optical SCS subsystems is available. These measurement models and techniques enable:

- quick diagnosis of whether the actual state of the tested object (separate components and entire communications lines) complies with the requirements of the standards, and the use of SCS for signal transmission of various applications, where possible;
- quick and precise troubleshooting to detect the nature of faults and to control quality of installation and maintenance.

Today, the standards have experienced rapid development, which does not apply to platforms of SCS construction, but promises a considerable improvement of their parameters and functions in the future. For example, the next version of the standard ISO/IEC 11801, which is planned for adoption in 2001, suggests:

- to give some release of the present restrictions on cable system qualitative parameters – building square footage and number of outlets connected to one telecommunications closet;
- to officially permit use of the integrated optical architecture of SCS;
- to standardize use of consolidated points and multiuser MUTO outlets in open offices;
- to retain the existing restrictions on maximum length of a cable pathway to 3000 m, to remove the restrictions on the number of users and general area served.

Taking into consideration our own experience in SCS implementation, an analysis of proposals from world-leading equipment manufacturers, our knowledge of the state-of-the-art information technologies market, and information on prospective solutions, the authors expect that SCS engineering will develop in further directions.

In the areas of construction principles and applications we project:

- the launch of commercial network equipment with gigabit data exchange rates and absolutely new levels of automated control of various processes, which can be expected once new versions of standards are adopted. Their adoption is planned for 2000–2001;
- the extension of functional possibilities for cable systems in view of the increase in the number of equipment types for different systems and building services, which are connected to SCS with respect to the intelligent building concept;
- the commercial penetration of SCS engineering into the small office/home office (SOHO) and so-called home networks, including wide use of category-7 cables and higher (frequency limit: 850 MHz) and the cable sharing principle

(transmission of signals from several information sources across one cable, including air and cable TV, Internet, TV, and broadcasting);.
- standard SCS equipment may incorporate wireless facilities of various types of small preferential working groups for connection to LAN. Today, only Lucent Technologies and Krone have suggested the solutions. However, their benefits as possible ways to quickly deploy new segments of LAN and to connect remote groups of users in areas where the use of cable solutions is impossible or not cost effective positions them for penetration into the very large SCS market;
- the incorporation of individual kinds of active network equipment, as well as remote control and diagnostics elements, as needed for applications in SCS will make it easier to solve typical problems that may occur in practice, and also to simplify day-to-day operation of systems.

In the sphere of electrical SCS subsystems we forecast:

- rapid growth of category-5e solutions and repeated classification of previously installed SCS of category-5 to the requirements of category-5e, once standard 802.3ab is adopted (Gigabit Ethernet across balanced cable);
- commercial introduction of category-6 solutions after 2000;
- an increase in the number of high-performance systems that are based on shielded platforms, which will ensure potentially high signal transmission characteristics, superior electromagnetic compatibility with other kinds of electronic equipment, and cause less harmful environmental exposure;
- commercial launching of new type of electrical connector, which operates on frequencies up to 1 GGz and higher, and possesses dimensions and operation performance that are at least as good as a traditional modular connector;
- introduction of SCSs in routine engineering practice (if this is specified by standards and other normative documents), which operate up to 1 GHz and higher, taking into consideration cables with upper frequency limits up to 1–1.2 GHz.

The sphere of special-purpose fiber-optic engineering for SCS may include:

- increased volume of fiber-optic solutions and their wide use, predominantly because of an increase in the number of cable systems and their volume, including fiber-to-the-desk cable lines;
- increased numbers of single-mode solutions, as only the single-mode platform enables the transmission of signals of the latest and prospective ATM 1.2-Gbps and 2.4-Gbps applications, and also Gigabit Ethernet at a distances of 500 m and higher;
- considerably increased optical cabling volume with 50/125 and 62.5/125 fibers that possess superior frequency characteristics (for example, GIGAlite by Alcatel), due to the adoption of the standard 802.3z (Gigabit Ethernet). The platform creates prerequisities for use in building backbone cabling that are technically well thought-out and easily installed on site, without the use of complex

production equipment, as well as cost-effective multimode engineering without limitations;

- standardization and introduction of optical connectors for high-density mounting in wide engineering practice, since it is possible to increase the density of patch panel ports up to the level of electrical solutions or greater. At the same time, it is possible to replace the basic optical connector (at present, SC).

We must also note that existing prospective solutions for fibers enable us to forecast commercial use of this platform for horizontal subsystems of SCS. At present, there is a complete set of components and technology that has been developed for their installation; thus, it is quite possible to use them for SCS construction. The factor that hampers broad application of the equipment is the somewhat increased attenuation level of polymer fibers; manufacturers cannot guarantee steady operation of high-speed applications at distances further than 80 m.

The authors are satisfied with existing and rapidly growing interest in the national industrial manufacturing of individual components of SCS and accessories in Russian plants. Along with foreign manufacturing equipment, previous experience and various original solutions and ideas are widely applied. According to statistical data from mid-2000, the situation within the field is as follows:

- four-pair horizontal category-5 cables are commercially manufactured by two domestic plants, and pilot lots have already been produced by four cable plants;
- a number of integrators have implemented small-lot production of electrical patch and equipment cords, which are widely used as auxiliary cords and for installation in uncertified SCS;
- optical cables for external routing manufactured by a number of Russian plants possess the same quality as those of foreign production, and because of their considerably lower cost can successfully compete with them. The manufacturing problems of cables with 62.5/125-type fibers, which existed prior 1997, have been successfully solved. Commercial manufacturing of cable samples for inner routing with fibers in 0.9-mm buffers has been launched;
- optical patch products manufactured by Russian producers possess superior quality and are successfully exported;
- there are a number of solutions for optical measuring devices, predominantly optical testers, which are produced in small lots;
- starting in 2000, a number of Russian companies have broken into the market with commercially produced 19-in equipment and accessories for multipurpose use. Some chemical plants are interested in setting up production of raceways and their standard accessories;
- the first samples of raceways of most applicable gauges, which are compatible with products of international companies and produced in Russian plants, have broken into the market.

According to statistics, by the middle of 2000 there were many education centers in Russia to train qualified personnel in SCS design and installation. For example, the IT Academy has trained over 500 specialists on IT-SCS. Almost all leading manufacturers arrange short seminars from one to five workdays in Russian. Those specialists who successfully pass the qualification tests are certified by the company.

We would like to point out a certain interest in training of SCS specialists, which was expressed by higher engineering institutions and schools. According to the available information, since 1999 a number of Russian engineering schools and universities have included elective courses for students in advanced training departments.

In 1998 the Russian company IT attempted to elaborate a national standards for SCS. The international standard ISO/IEC 11801 has been set as the platform for the document. Once it is officially adopted by SCS, it can be "legalized" in Russia as an independent engineering sector.

Almost all leading periodicals for specialists in electrocommunications and information technologies publish materials devoted to actual issues in SCS engineering and interfacing issues. A national school of specialists has been established within the field in Russia.

Summing up all the above points, we would like to underline that the great variety of equipment by leading manufacturers, the well-developed design concepts, construction and testing techniques, and maintenance rules enable systems integrators to offer cost-effective solutions to their customers. These solutions possess superior technical performance, not only for the present, but also for the future.

Glossary

100VG-AnyLAN – LAN with bit rate of 100Mbps developed by Hewlett Packard. Operates over four twisted pairs of category 3 as well as over multimode fiber-optic cables. Not in widespread use.

Absorption losses - signal attenuation when propagating over an optical waveguide based on energy absorption in the core material.

ACR – attenuation-to-cross talk ratio.

Adapter –matching unit that is used in network equipment systems. According to the TIA/EIA-568-A standard, adapters make it possible to connect connector plugs of incompliant sizes or types to each other, to change wire layout schemes, to distribute a single multipair cable into several cables with fewer pairs, and to connect cables with each other.

Aperture angle – the greatest angle between the core optical axis of an optical waveguide and a beam in space at which the directed modes appear.

Application – any telecommunications system with corresponding network equipment in which SCS channels serve for information exchange.

Application class – classification introduced by ISO/IEC 11801. Depending on the rate of information exchange over the twisted pairs, the applications are divided into four classes: A (up to 100 kHz), B (up to 1 MHz), C (up to 16 mHz) and D (up to 100 mHz); the optical class is separate. Class A is considered as the lowest class and class D is the highest one.

ATM – universal transport network for voice, data, and video transmission, with transmission speeds of 25, 155, 622, and 2400 Mbps. The first two varieties can operate over two twisted pairs of category 5; apparatus operating at 155, 622, and 2400 Mbps uses optical cables for transmission medium

Attenuation – effect of power loss by the signal during its propagation along the communications line. It is equal to difference in signal levels at the output and input of the line.

Automatic cross – unit having several inputs and outputs. Switching between them is made by keying signal to a special input. Provides throughput not greater than category 3, is not the standard equipment of SCS and is not widely distributed.

AWG – American wire gauge, a unit for electrical cable wire diameter designation. It is equal to the number of drawing bench cylinders through which the wire was pulled. The most widely distributed SCS cable wires are characterized by AWG equal to 24, which corresponds to the diameter of 0.51 mm.

Backbone grounding bar – element of telecommunications grounding. It is an isolated grounding conductors connecting grounding plates in the cross-rooms with the main grounding plate.

Backward compatibility – principle of passive SCS component design. It supposes the entire mechanical and logical identity of functionally similar elements of different categories. By this principle, elements of higher category can be included in SCS of lower category without any limitations.

Balanced cable – cable containing one or several balanced cable elements made from twisted pairs or 4-pairs.

Balanced circuit – circuit whose conductors both have similar construction and electrical parameters. In the vast majority of cases it is created by using a matching transformer to provide additional galvanic isolation; it is more stable to external electromagnetic radiation than an unbalanced circuit and almost fails to radiate itself. It has greater broadband character in comparison with an unbalanced circuit, therefore it is widely used for SCS creation.

Balanced transmission – see Balanced circuit

Baluns – adapter intended to connect twisted pairs and coaxial or twin-axial cables. Provides matching of impedances and transition between balanced and unbalanced transmission schemes.

Bayonet connector – connector in which plug fixation in outlet is done with a spring bayonet nut. The principle of bayonet fixation is used in some types of connectors for coaxial cables and in optical ST connectors.

BloTwist – BloLite system combined element of the BICC Brand-Rex company. It contains a cable of category 5 and a tube where up to four single-mode or multimode optical waveguides can be simultaneously or sequentially blown in with compressed air under a general sheath. Used for horizontal SCS subsystem creation.

Branching scheme – scheme of one point administration with interconnection based on optical cable layout to work area at which backbone cable is laid out to floor cross-room; further layout is carried out with horizontal cables connected to backbone with nondetachable connectors. Maximum communication line length with interconnection is selected to be equal to 300 m.

Breaking thread – strong thread that is located under external protective cable sheath, which when pulled makes a longitudinal cut on the sheath, giving access to the elements of the cable core. It is used, in general, in backbone cables of external layout.

Breakout cable – variety of fiber-optic cable for internal layout with additional individual protection of each optical waveguide by a layer of Kevlar threads and protective hose with an external diameter of 2–3 mm. It is primarily used to create preterminated assemblies and multifiber connection cords.

Bridge adapter – unit containing several modular connectors with parallel contactors of the same name. It provides parallel connection to a single SCS route of network units operating by a "multipoint" scheme.

Broadband coefficient – parameter of optical waveguide broadband. It is equal to the pass bandwidth of an optical waveguide with length of 1 km. It depends on the operating wavelength. It is used for multimode optical waveguides.

Buffer – protection sheath that protects optical fibers from mechanical effects and moisture.

Bus topology- network topology with general physical transmission medium as a bus to which network equipment is connected.

Cable – unit for information transmission over long distances. It contains one ore several isolated electrical wires or optical waveguides located in general (usually hermetic) protective sheathing. Depending on the construction variant, one finds coaxial, triaxial, balanced, and optical cables.

Cable armor – cable construction element to protect the core from mechanical efforts and from damage from rodents. Metallic armoring of different densities, crimping steel tape, and circular steel wire of different diameters serves as an armor. Armored covers out of polymeric materials are often used in optical cables, and if increased protection is necessary, multilayer armor is used.

Cable clamp – construction element of switching equipment for backbone or horizontal cable fastening at the entrance into switching unit case or coupling. It can be carried out as clamp, latch, collet, or perforated bar for bracing wire.

Cable core – aggregate of cable elements providing signal transmission. Core construction can also contain supporting and strengthening structures, element for moisture protection, and so on.

Cable element- minimum construction element (for example, pair, four, or separate fiber) in the cable. Cable elements can have individual shielding and protective sheaths.

Cable for cords- intended for manufacturing of patch and terminal cords. The electrical cable for cords differs in that its conductors are manufactured out of seven thin twisted copper wires with diameters of about 0.2 mm each. Therefore, it is characterized by high attenuation in comparison with horizontal cables. Fiber-optic cables for cords have identical transfer parameters to those of backbone cables in spite of their high flexibility.

Cable for under-carpet layout – flat construction cable with twisted pairs for maximum pressure distribution over the product surface. It is connected to horizontal cable in the transition point. It can have category 3, 4, or 5.

Cable level – terminology introduced by Underwriters Laboratories (UL) together with Anixter Company in 1989 to classify twisted-pair cables. Depending on the upper frequency of transmitted signals the levels are divided as follows: level 1-is not specified, level 2- up to 2 mHz, level 3- up to 16 mHz, level 4- up to 20 mHz and level 5- up to 100 mHz. Currently, the category classification has substituted for Level Classification.

Cable of external layout – cable for subsystem creation of external SCS backbones. Its construction feature is high mechanical strength to stretching and squeezing stress, presence of special construction elements for rodent protection as well as water-resistance and broadened temperature range.

Cable of internal layout- cable that is used to create horizontal subsystems and subsystems of internal SCS backbones. It differs by it use of facilitated strengthening coatings, it has no protection elements from moisture and rodents, and it provides more flexibility and better mass and size parameters. Cables of this group are made with high demands for fire safety.

Cable sharing – principle of SCS horizontal cable usage included in simultaneous signal transmission of more than one application over different pairs. Allowed by standards ISO/IEC 11801 and EN 50173, requires usage of special Y-adapters or adapter insertions. The most effective for shielded constructions of STP and S-STP types and often are realized in small networks.

Cable termination – procedure of cable preparation to connector.

Category – throughput classification of separate elements, channels, and communication lines based on twisted-pair cables. Category number is connected with the maximum frequency to which the electrical characteristics are normalized: category 3- up to 16 mHz, category 4- up to 20 mHz, and category 5- up to 100 mHz. First introduced in 1991 in TIA/EIA TSB-36 Technical Bulletin and changed the classification by levels.

Centralized administration – see Single-point administration.

Channel – the entire path over the cable system from connector to connector among network equipment. It includes termination cords.

Chromatic dispersion – dispersion of electromagnetic radiation arising from non-zero spectrum width of optical carrier of transmitter output signal.

Circular topology- network topology in the form of a ring in whose breaks the network equipment is connected. When data are transmitted over such networks, they pass in series all or part of the stations.

Closed architecture – architecture compliant only with hardware and/or software of a single manufacturer.

Coaxial cable – cable for data transmission in local networks. It consists of an inner conductor surrounded by dielectric isolation and external conductor as a shield, which has a polymer protective sheath. It cannot be used in SCS by present and future standards revisions.

Color coding – see Color marking.

Color marking – is used to identify twisted pairs and their conductors in electrical cables as well as in fiber-optic waveguides. IEC 708-1 and TIA/EIA-598 standards determine the rules of color marking on the basis of twelve main marking colors.

Combined cable- cable containing two or more different types of cable elements simultaneously for signal transmission (for example, optical fibers and twisted pairs). These elements can be located under general sheath or knitted with each other by other ways.

Communications center – premise that is especially intended for location of automatic telephone exchanges and other network equipment that provides functioning of external telecommunications.

Component or communications line throughput rate – ability of electrical or optical component (communications line) to transmit signal on intended frequencies with intended quality.

Connecting block – construction element of 110-type patch panel that carries out electrical contact with cable conductors and access to them for connection.

Connecting module – unit for splicing two ends of horizontal cables. It allows users to lengthen 4-pair cables or to splice the ends of damaged cables without laying out new segments. It is used in repair work for rapid communications restoration and as an element for transition point organizing.

Connection bar to main grounding system – element of telecommunications grounding that is the grounding conductor connecting the main grounding plate with the main building grounding system.

Connector of the 110 type – connector for switching equipment. It consists of a plug and a ruler, which after installation of 110 C unit on it carries out function of outlet. Plug contacts are metallic plates perpendicular to the case plate that comes into the gap among the contacts at the ruler. Plugs of 110 connectors are for 1-, 2-, 3-, and 4-pairs.

Connector- optical or electrical plug&socket that makes it possible to carry out multiple connection and disconnection of optical fibers or twisted pairs. Connectors differ by size, form, and principles of plug fixation in outlets in operation situations, number of spliced conductors or optical waveguides. They provide several hundreds of cycles of connections and disconnections without decreasing the maintenance parameters

Connector outlet – connector section with a female in which plug is inserted.

Connector plug – connector part that is inserted into an outlet. In most constructions the connector cable part is carried out as a plug.

Consolidation point – analogue to transfer point in open office.

Contact, 110-type – variety of IDC contact with perpendicular to conductor axis location of cutting edges in which it is not allowed to connect more than one conductor to a single contact.

Contact, 66-type – variety of IDC contact with perpendicular to conductor axis location of cutting edges in which it is allowable to connect more than one conductor to a single contact.

Cord – cable section with terminal connector plugs.

Cordless panel – patch panel in which outlet contacts of modular connector pairs are connected one with another by short external jumpers made of current-carrying lanes of printed boards. Using cordless panels makes it possible to carry out port connection of different equipment without cable layout on IDC contacts.

Core eccentricity – core variation of optical waveguide relative to sheath center. It is one of the reasons for attenuation in optical connectors.

Core ellipticity – shift of optical waveguide core area from circular. It is one of the reasons for attenuation in optical connectors.

Core – optical transmitting section of fiber-optic waveguide with average value of refractive index that is greater than the refractive index of the reflecting sheath.

Crimping tools – hand tools intended for the installation of electrical connector plugs to cables by crimping method. Special types of crimping tools are also widely used when assembling the plugs of optical connectors.

Critical angle of total refraction – reference angle at which refracted ray begins to slide over boundary of two mediums without transition to optically more dense medium.

Cross panel – variety of switching equipment developed to use patch cords with plugs that are different from modular ones.

Cross-room – business premises to locate SCS switching equipment, network units, and other auxiliary elements.

Cross-room for external backbone – cross-room in which SCS backbone cables connecting it to the cross-rooms of the building. It can be combined with cross-rooms of the building.

Cross-room of a building – cross-room in which internal SCS backbone cables connecting it with the cross-rooms of floors. It can be combined with the cross-room of its floor and cross-rooms of external backbones if they are situated in one building.

Cross-room of floor (floor cross-room) – cross-room to which horizontal cables of working areas located on the same floor come. It can be combined with a cross-room of the building if they are located on the same floor.

Cross-tower – construction that is made out of vertically located 110-type switching units with separating organizers fastened on a unified metallic base (chassis). The typical number of pairs is equal to 300–500.

Cross-wire- see Wire for jumpers.

Cross-talk attenuation – level difference of signal being transmitted and cross-talk interference at adjacent pair that is made by it.

Cross-talk noise pickup – frequency-dependent influence of two adjacent pairs of conductors each with another. It is primarily caused by the presence of capacitive coupling between them. To decrease the value of cross-talk noise pickup in balanced cables one uses twisted pairs with different steps and individual shields of each pair.

Decibel – logarithmic ratio of two electrical (current, voltage, power) or optical (power) values. It is a dimensionless unit that is often used in engineering calculations and is abbreviated as dB.

Descrambler – functional block restoring the initial form of a scrambled signal. It is used in high-speed interfaces and installed in the receiver. It always operates in a pair with a scrambler that is installed in the transmitter.

Directed mode – mode propagating along the core of an optical waveguide and providing information transmission.

Dispersion losses – signal attenuation when propagating over optical waveguide based on light dispersion in the core material.

Dispersion of electromagnetic radiation – time dispersion of separate components of electromagnetic radiation. It leads to passband limitation of optical fibers and to pulse widening over optical cables. The idea of dispersion is widely used to describe radiation propagation processes in fiber-optic waveguides, and is made up of a number of components: intermodal, waveguide, and material.

Double cable – two similar or (rarely) different horizontal cable elements connected in a single whole with a narrow jumper or general sheath and which makes it easier to create two-port work areas during SCS creation.

Drain wire – thin, bare copper-tinned conductor or zinc-coated wire with diameter about 0.5 mm that is laid directly over a film shield to provide its electrical continuity for occasional film breaks during layout and maintenance. It is used in some types of electrical modular connectors to provide shield contact of the cable and outlet.

Duplex cable – see Double cable.

Duplex coupler – see. Y-adapter.

Duplex transmission – information transmission over the communication line in both directions. It is distinguished by half-duplex and duplex transmission.

Electromagnetic compatibility – ability of electronic unit, equipment, or other system to operate in the conditions of external electromagnetic radiation effects that are created by other units and systems and not to make electromagnetic interference.

Elementary beam – assembly of several cable elements of one type and category. Can have shielding sheath. Bunch of electrical multipair cable is carried out as elementary beam.

Equipment room – service room in which the network equipment of collective use is located (automatic telephone exchange, servers, hubs). Can be combined with cross-room of the building.

Ethernet – LAN with bit rates of 10, 100 and 1000 Mbps. First versions used bus topology and operated over coaxial cables (50 Ω): 10BASE-5 (up to 500 m) and 10Base-2 (185 m). All the following variants of Ethernet networks have star topology and operate over twisted pairs (100 Ω) or over optical fibers. In 10Base-T (10 Mbps) and 100Base-T4 modifications, cables of category 3 (2 and 4 pairs correspondingly) are used; in 100Base-TX (100 Mbps) two pairs of category 5 are used. The 1000Base-T (1 Gbps) networks, which will have four pairs of enhanced Category 5, Category 6, and older, are in the standardization stage.

External backbone – cable communication line between two buildings. It is created using cables of external layout. Connects cross-rooms of external backbones with cross-rooms of another building.

External protection sheath – external lay of isolating material protecting cable core from external damage.

Far-end cross talk – see FEXT.

Fastening along the profile – mounting frame fastening to install telecommunications or power outlets near ducting.

Fastening in profile – mounting frame fastening to install telecommunications or power outlets on ducting.

FC connector – optical connector that is used, in general, in single-mode systems. The outlets of this connector are manufactured in two variants: SF types with square flanges and fastening with two screws, and RF types with circular flanges and fastening with a nut.

FDDI (Fiber-distributed data interface) – LAN of circular topology with bit rate of 100 Mbps based on the principle of marker transmission. FDDI classic network uses multimode and single-mode fiber-optic cables as transmission media. There are also the varieties of this network that work over shielded and unshielded twisted pairs of category 5 (100 Ω).

FEXT (Far-end cross talk) – cross-talk attenuation, measured at the far end of adjacent pairs.

Fiber-optic cable- cable containing one or several optical fibers. Depending on construction variant, SCS fiber-optic cables are divided into cables for internal and external layouts as well as cables for cords.

Fiber-optic cable for internal layout – variant of fiber-optic cable. Contains optical waveguides in 0.9-mm buffer coverings that, together with Kevlar-reinforced fibers, are located in a general protection sheath. Differs by increased flexibility, hence it does not have protection elements from moisture and, in general, is intended for use inside buildings.

Fiber-optic waveguide – transparent dielectric optical waveguide to transmit optical radiation based on operation the phenomenon of total reflection. It consists of a cylindrical core surrounded with one or several reflecting sheaths with lower material indexes of refractive than the core.

Film shield – shield as a wound-to-cable element sputtered with aluminum thin film (foil). The sputtered side can be oriented both inside and outside. Foil edges are laid each other with overlapping, or are connected with longitudinal roofing-type joints. Film shields are used to shield separate pairs and are, in general, external shields. It protects cable from high-frequency interference very well.

FM-adapter (female–male) – optical adapter providing the plug connection of one type connector to the connector outlet of another type.

Form factor – presumptive parameter that characterizes form and geometrical sizes of sites for outlets of different types of connectors, including optical and electrical ones. Requirements for form factor equality are the necessary condition to provide SCS universality.

Frequency – oscillation number per unit of time, often per second, given in Hertz (Hz).

Full duplex transmission – simultaneous transmission and receipt of a signal over a communication line in two directions.

Global transfer interference- total transfer interference at both ends of cable.

Gradient optical waveguide – waveguide in which the index of refraction smoothly decreases with distance from the core axis by a law close to a quadratic parabola.

Graduated optical waveguide – optical waveguide in which the core refractive index stays constant along its radius.

Grounding plate – element of telecommunications grounding that is a metallic plate with screw connections located in cross- or equipment rooms. Backbone and interbackbone grounding bars as well as grounding conductors of network equipment are connected to the plate.

Grounding – potential balancing of different devices, systems, or units in order to protect personnel from electrical shock (protective grounding) and to provide effective signal transmission system functioning (telecommunications grounding).

Guarantee for application operation – ability of correctly mounted and installed SCS to support operation of various applications from a predetermined list or developed especially for operation over SCS of certain category that is guaranteed by manufacturer.

Guarantee for components – lack of manufacturing defects for all components of cable system and their operational integrity for a certain period of time from purchase that is guaranteed by the SCS manufacturer.

GXT (Global cross talk) – see Global transfer interference.

Half-duplex transmission – alternate signal transmission and receipt over communications line in two directions.

Halogen-free material – material meeting special requirements for the level of halogen-containing matter in gaseous burning byproducts with the aim of minimizing their corrosion and choking effects.

Harmonics – adapter splitting 25-pair cable into several groups of conductors with 2-, 3-, or 4-pairs. To connect multipair cables Telco connectors or six 4-pair 110-type blocks are used, output adaptor interface is realized by the outlets of modular connectors.

Hexagonal diagram – NEXT index diagram for 4-pair cable and connector as well as communication lines based on them. It is carried out as six axes (one for every possible combination of interacting pairs) along which the values of cross-talk attenuation that are given by the standard as well as the real ones, are measured. It is an obvious example of the quality of mounting fulfillment and possible reserves on cross-talk attenuation value.

Hierarchical star – variant of star-like topology, characterizing some "branches" radiating from the main center that are split at the second end.

HIGH Screen – see Two-layer screen.

Horizontal cable – cable of horizontal subsystem from floor cross-room to telecommunications outlet of work area.

Horizontal organizer – organizer to lay out horizontally located sections of cables and patch cords. It is often an integral part of patch panel construction.

Horizontal subsystem – SCS subsystem providing floor cross-room connection with work areas. Includes telecommunications outlets of work areas, transition points, switching equipment in the floor cross-rooms connecting their horizontal cables, as well as patch cords and/or jumpers.

Hybrid cable – see Combined cable.

Hydra – see Splitting cord.

Hydrophobic gel – jelly-like mass filling the internal hollows of cable cores for external layout that provides longitudinal cable hermeticity and separate optical waveguide protection from air and humidity.

IBM cable system – first universal cable system, developed by IBM Corporation in 1980s and intended for Token Ring networks, AS/400 servers, 3270 terminals, and another similar units. In addition, it included components for telephone signal transmission. Cable part specifications determined 9 different cable types, the most popular were the 1 and 6 types.

IDC (Insulation displacement connection) – Technology of connector connection to copper wires of twisted pairs based on cutting the cable dielectric sheath from opposite sides by sharp knife contacts and conductor fixation between them by spring reaction.

Immunity – difference between friendly signals and interference levels. In electrical systems it is necessarily indicated for a certain frequency.

Impact tool – tool for twisted-pair wire mounting to IDC contacts. It is carried out in single conductors and multipair variants. Clamping force is done by an internal spring-loaded striker. The tool has a part to cut the excess wire when mounting. Five-pair tool is also used during the installation of connective blocks of 110-type connectors.

Impedance – resistance that the electromagnetic waves meet when propagating along homogeneous lines without reflection, when mismatches at the ends of the line fail to affect the transmission process. It is numerically equal to the ratio of the transverse vector component of the electrical and magnetic fields and depends on wire material and construction as well as its isolation and also on screen characteristics.

Index of protection – letter-and-digit index of 19-in or any other element showing the protection effectiveness of the equipment installed in it from different external mechanical, climatic, chemical, and other actions. .

Information grounding loop – grounding loop that connects cross- and equipment rooms with each other and is connected to the main grounding loop. It is intended to increase the communications reliability due to additional ground potential balancing in different business premises.

In-object cables – see Cables of internal layout.

Insulation resistance – resistance of insulation material of conductor wires and cable sheaths to electric current. It depends, in general, on material features and, to a lesser degree, on its thickness.

Interbackbone grounding bar – element of telecommunications grounding that is an isolated grounding conductor connecting two or more grounding plates in the cross-rooms on one floor and is intended for their additional potential equalization.

Interconnection – connection of SCS subsystems to each other.

Intermediate coupling – unit in which splicing of cables for external layout of external backbone subsystems is carried out. Usually, it has a cylindrical form or that of a parallelepiped with round panes. It is supplied with restoring elements to strengthen the coating continuity and special organizers to lay out splices and technological excess lengths of conductors or optical waveguides. It is pressurized by mechanical, thermal, or another way. Depending on quantity and input location one finds straight, deadlock, split, and pass intermediate couplings.

Intermode dispersion – dispersion of electromagnetic radiation arising in multimode optical waveguides from a large number of modes.

Internal backbone – cable communications line inside the building. It is created using cables of internal layout. Connects cross-room of the building to the cross-rooms of floors.

Inverted terminal cord – terminal cord in which conductor connection to the contacts of second plug is made in reverse order with reference to the first plug.

ISDN (Integrated services digital network) – Network providing data transmission, telephone, and video signal transmission over a single channel.

I-adapter – Unit that has two modular outlets connected to each other that serve to increase the length of patch and terminal cords with modular connector plugs.

Jumper – twisted pair section (more often without external sheath) to switch in cross-rooms.

KATT contact – variety of IDC contact in which the cutting edges are located at angle of 45° to the conductor axis and are perpendicular to each other.

Krone contact – variety of IDC contact in which the cutting edges are located at angle of 45° to the conductor axis and are coplanar.

LAN – data, voice, and video transmission network that is located in a geographically limited territory.

Leaking mode – decaying wave in optical waveguides from power loss caused by refraction into optical waveguide sheaths. Appears when the angle from an external source to the side of the optical waveguide exceeds the angle of beam incidence.

Line – entire path over the cable system between two connectors of cross-equipment. Termination cords are not component parts of line.

Link – see Communication line.

LSA-Plus contact – name for Krone contact

Main grounding plate – element of telecommunications grounding mat. Represents metallic plate with screw connections to which the bar of the main building grounding system is connected; backbone bars of telecommunications grounding as well as network equipment are located near of it.

Manchester plate – see Mounting box.

Material dispersion – component of chromatic dispersion of fiber-optic waveguide caused by dependence of the refractive index of the core on the wavelength of light.

Maximum tensile strain – parameter giving the requirements for the conditions of cable layout. It is indicated taking into account the minimum bending radius during layout.

Mechanical splice – unit for nondetachable splicing of two fiber-optic waveguides without using welding. Constructively it is carried out as V-formed groove or capillary to center the fibers that are spliced. It allows the connection of single-mode and multimode optical waveguides with losses not more than 0.2 dB.

MIC connector – see MIC.

MIC (Medium interface connector) – Optical connector of FDDI networks and electrical connector of Token Ring networks.

Micromodular construction – variety of quasi-secondary protective coating of an optical waveguide. It is a tube with external diameter 0.9 mm and with a free internal laying of a single optical waveguide. The remaining internal free space is filled with hydrophobic gel. Cables of this construction can serve to organize external backbones of small length.

Mini cable – optical cable with one or two optical waveguides in 0.9-mm buffer coating; each is located in separate hose. It is intended for using in cords and implementation of layout in decorative protective ducts.

Minimum bending radius – the least radius at which cable bending fails to break its electrical or optical characteristics. This parameter is normalized separately for layout and maintenance conditions.

Mode – electromagnetic light wave arising in fiber-optic waveguide when inputting the radiation from external source to the input side of an optical waveguide.

Modular connector – SCS electrical connector. It consists of two parts: plug and outlet. These connectors differ by categories (3, 4 or 5) and can be both unshielded and shielded. It is allowable to use only 8-contact modular connectors in SCS. It is allowable to use 6-contact plugs only to connect network equipment on conditions that it is compliant with 8-contact outlets.

Module – variety of optical waveguide secondary protective coating that is used in backbone cables of external layout. It is a tube made out of plastics of different rigidity with diameter of 2–3 mm in which one or several optical waveguides are laid freely.

Mounting box – element of electrical, optical, or power outlet as well as other elements in cable ducting. It is a plastic case that is open from the front side with installation elements into internal ducting cavity, holes for cable input, and holes for fixing screws of outlet front panels.

Mounting closet – closed 19-in units in which the basic construction is a case with doors and mounting guides. It is carried out in wall and floor variants. It provides protection of installed equipment from unauthorized access and in some variants of construction it provides effective shielding.

Mounting cord – electrical cable or fiber-optic waveguide section (pigtail) with connector plug installed at one end.

Mounting frame – plastic base with installation elements to the ducting or to the flat surface and with a cut for an outlet module that is fixed in the frame using latches or screws and is closed with a decorative front plate.

Multielement cable – cable whose core is based on 2- or 4- pair elements that are analogous to horizontal cables and supplied by individual protective sheaths. Functionally, it is equal to a multipair cable, and usually exceeds its electrical characteristics and mechanical robustness, however, it has poorer weight and size characteristics. In general, it is used in SCS backbone subsystems.

Multimedia outlet – case with sites for electrical and optical outlets of different types. It is necessarily supplied with built-in optical waveguide organizer.

Multimode optical waveguide – optical waveguide whose core diameter is much greater than the wavelength of the optical carrier. The standard values of core diameter for SCS optical cables are 50 and 62.5 μm. There exist about 1500 directed modes in this optical waveguide.

Multipair cable – cable containing more than four twisted pairs. It is usually used in SCS backbone subsystems.

Multipoint administration – cable system administration created on the classical scheme of a hierarchy star. In general, to change the configuration it is necessary to reswitch more than in one cross-room.

Multiuser outlets – case with several outlets intended to serve groups of users in open offices.

Multiwire conductor – conductor consisting of several intertwisted thin wires. It is characterized by an increased stability to bending force effects and is used in cables for cords.

MUTO (Multi-user telecommunication outlet) –installed in open offices. Allows using terminal cords with lengths up to 20 m.

Near-end cross talk – see NEXT.

Network equipment- active transceiver equipment that makes it possible to exchange information of any kind (data, voice, video). It provides network operation of various protocols, for example, Token Ring, Ethernet, ATM, etc.

NEXT (Near-end cross talk) – Cross-talk attenuation measured at the near end of an adjacent pair.

Numerical aperture – sine of the aperture angle. Usage of this parameter makes it easier to make some theoretical and engineering calculations.

NVP (Nominal velocity of propagation) – Velocity of signal propagation over twisted pairs. Normalized relative to light velocity in vacuum and is indicated in percent or relative units.

OI coefficient – Minimum oxygen content, in percent, in nitrogen–oxygen mixture at which self-burning of material begins without external heat.

Open office – large-area work premises divided by independent sections with specialized furniture or with light nonpermanent partitions.

Open rack – open 19-in mounting equipment whose construction is based on a base and one or two vertical rows of mounting guides. It provides compact and convenient in-service location of different kinds of equipment in cross-rooms, equipment rooms, and other business premises with limited access of nonauthorized persons.

Optical waveguide organizer – construction element of optical switching equipment that stores the technological reserve of fiber length while keeping the minimum radius of bending. Constructively it can be carried out as a drum with side lobes, as a tray with edges bent up, or as separate split rings.

Optical waveguide sheath – optical waveguide core coating that is made out of glass with an index of refraction that is lower than that of the core. It provides optical isolation and mechanical protection of the core.

Optical waveguide – thin fiber of cylindrical form made out of quartz glass over which electromagnetic radiation of visible or near-infrared range of wavelengths is transmitted.

Organizer of nondetachable connectors – construction element of optical switching equipment to fasten the tubes of protection sleeves of welded fiber connections or mechanical splices.

Organizer – unit to lay out the excess cable length, mounting, and patch cords as well as the separate optical waveguides of optical cable, protection sleeves of welded connections, and mechanical splices. It provides fastening of listed elements in operating position, keeping the bending radius and pressing and pulling mechanical effects limits.

Outlet module-construction element consisting of an 8-contact female of modular connector and terminator. Terminators for twisted-pair conductor connection are usually realized as a set of IDC contacts.

Pair – two conductors in quad, twisted or diametrically opposite, serving for electrical signal transmission.

Passive patch panel – scheme of single-point administration with interconnection based on inputting the optical cable to the work area, at which the backbone cable is laid out to FD and further layout is done with horizontal cable that is connected to the backbone cable with simple patch cord. Maximum distance from telecommunications outlet to FD is 90 m. It makes possible to keep succession in relative to horizontal layout and provides easiness of return to standard two-level topology. Maximum channel length with interconnection is selected 300 m.

Patch cord – cord intended for switching at the switching equipment in the cross-rooms.

Patch panel – variety of switching equipment

Permanent connection – splicing or connection without using connectors, providing signal passage between two electrical conductors (welding, twisting, IDC contact) or optical fibers (welding, mechanical splicing).

Plenum cable – cable whose fire-safety characteristics make it possible use it in plenum cavities without using additional protective measures.

Plenum cavity – space between suspended ceiling and covering above it where air flow in a volume that is necessary to support burning is possible. In some cases, the space under a false floor is also considered a plenum cavity. In order to lay cables in plenum cavities, very strict requirements are placed relative to fire safety.

Plug cap – element of optical connector. It has a form of cylinder or parallelepiped with rounded panels. It serves to fix and position the terminal sections of fibers that are spliced. It is manufactured out of ceramic, metal, or plastic.

Power sum of induced transitions –sum of induced transitions power at near end from all adjacent pairs in the cable.

Power sum – see Total power of induced transitions.

Preterminated assembly – construction to organize short communication lines without mounting a connector at the object. It is a cable section with connector plugs and outlets mounted at the plant, protected with hermetic terminal armor of metallic or plastic tubes. Preterminated assemblies are often used to create optical communication lines.

Preterminated panel – construction of one or several 66-type or 110-type switching blocks with preliminarily prepared multipair cables at whose other end a plug or outlet of a Telco connector is installed. It reduces the time for SCS mounting.

Primary parameters of twisted pair – set of parameters of electrical model of twisted pair (conductor resistance and inductance, capacitance, and conductivity of isolation) that entirely characterize its electrical features. They allow designers to calculate the remaining parameters and are used when theoretically analyzing communication lines.

Primary protection coating – optical waveguide protective coating made from epoxy acrylate with external diameter of 250±15 μm that is applied to the lacquered surface of optical waveguide sheath.

Profile of refractive index – function of refractive index change of an optical waveguide core over the cross section from the axis to the edge.

Protective grounding – see Grounding.

Proximity effect – effect of alternating current displacement to the conductors of twisted pair turned to each other that leads to a decrease of effective conductor section and an increase in its active resistance. It appears more sharply with increasing frequency.

PS (power sum) – prefix meaning the normalization of given certain parameters of interference immunity using the model of total power, that is the influence of all pairs to one.

Public Network Interface – place of division for external public networks and SCS. In the great number of cases, it is the interface to the network equipment of the communications service carrier.

Quad, Star Quad – element of cable core consisting of four isolated conductors twisted together. To minimize the value of cross-talk signals there are two mutually opposing conductors in the pairs of single quad.

Quartz optical waveguide – waveguide whose core and sheath are made of high-purity quartz glass with the addition of different compounds to form a given refractive index profile.

Quasi-secondary coating – variety of secondary optical waveguide protective coating. It provides effective mechanical isolation of the external layer and fiber surface that greatly decreases losses from microbends. There are two variants of quasi-secondary coatings: micromodular construction and two-layer secondary protective coating.

Radiated mode – mode radiated by fiber-optic waveguide into ambient space or protective sheath.

Riser – vertical channel for layout of cables that serve floors located above each other.

Riser cable – cable whose fire safety characteristics make it possible to use it in vertical risers of buildings without additional protection.

S/STP (Shielded/shielded twisted pair) – cable with twisted pairs, each with individual shielding, and a general protection shield.

S/UTP (Shielded/unshielded twisted pair) – cable with general external shield around unshielded twisted pairs.

SC connector – optical connector that is determined by present review of standards as the main connector type for SCS use. It is available in single or double (duplex) variants.

Scrambler – functional block that handles formed code before its transmission for the purpose of decreasing the probability of the appearance of long sequences with similar signs and intermixing spectral signal components. Use of scrambling makes it possible to increase synchronization stability, mainly of high-speed interfaces. It is realized by software and hardware. It always operates together with a descrambler that is mounted in a receiver.

SCS administration – process of SCS monitoring, management, and documentation for maintenance. Both single-point and multipoint administration can be performed.

SCS component – any functionally complete element of a cable system providing signal transmission. The main SCS components are: cables, telecommunications outlets, switching equipment, patch cords, and jumpers.

SCS redundancy – at the technical design stage, the designed introduction in SCS composition of additional telecommunications outlets whose number and location is determined by the area and work premises topology, but not by specific plans of workstation location and office furniture arrangement. It allows to users to easily organize new work areas during maintenance as well as to carry out the workstation and equipment relocations.

SCS structurization – principle of cable system creation referring to its division into functional subsystems with standard interfaces to connect with each other and with network equipment.

SCS technical premises – business premises meeting the particular requirements for size, climatic and other conditions equipped with ventilation, power, and communications systems. It is intended for installation of switching and network equipment. It is chosen from cross-rooms of different types and equipment rooms.

SCS universality – cable system feature stating that initially a system is created not for some specific network application but on the principles of open architecture that are given and fixed in the standards for the main technical characteristics. The vast majority of modern network applications are being developed by such a way to support these standards.

Secondary buffer coat – secondary optical waveguide protective coat that is used in cables for internal layout. It has external diameter of 0.9 mm (tight buffer) and is laid without gaps over the first protective coating with diameter of 0.25 mm. It allows mounting the plugs of optical connectors without connector adapters.

Secondary electrical parameters of twisted pair – set of twisted pair parameters that can be calculated on the basis of the primary parameters of the electrical model of this pair.

Secondary protective coat – additional protective coat of an optical waveguide that is applied over the main one.

Self-interference – see Cross-talk noise pickup.

Server room – premise that is especially intended for central LAN equipment location of servers, switches, data storage, etc.

Shield – electrically conducting sheath surrounding an electroconducting medium. The use of a shield decreases electromagnetic effects at the adjacent conductors and increases cable stability to external electromagnetic fields. Shields are available in film-containing and armored forms.

Shield with armor – external shield carried out as armor of zinc-coated or tinned copper wire. It can be characterized by different density as is pointed out in its technical specifications. It protects cables from low-frequency interference very well and gives them additional mechanical robustness.

Shielded balanced cable – see STP, S/STP and S/UTP.

Shotgun cables – see Double cable.

Siamese cable – see Double cable.

Signal strength – logarithm of the ratio of power or voltage value in the point under consideration to the power or voltage value in the circuit point selected for comparison.

Simplex transmission – signal transmission in only one direction over communications lines.

Single-mode fiber – optical waveguide whose core diameter (7–10 μm) is commensurate with the wavelength of the optical carrier. There is only one directed mode in such a fiber. It has much better broadband characteristics and losses than multimode fibers , however, operation with it requires more complex and expensive equipment.

Single-point administration – administration of a cable system that has direct connections of all work area telecommunications outlets with a single cross-room. Changing the SCS configuration is done by reswitching only in this cross-room.

Skew – delay dispersion of signal propagation over twisted pairs due to the velocity dispersion of signal propagation and the difference in the physical length of twisted pairs because of different twisting steps.

Skin effect – effect of current displacement to the surface of twisted pair conductors with increasing frequency growth, which leads to a decrease in the effective cross section of the conductor and to an increase in its active resistance.

Slot – rectangular aperture with skirting in interfloor overlaps. It carries out the functions of a riser for backbone cables.

SMA connector – optical connector with metallic end in which plug-to-outlet fastening is carried out with hexagonal or (rarely) circular coupling nuts. Currently, it is considered to be old technology.

Solid conductor – electrical conductor containing only one wire. In contrast to multiwire conductors, it minimizes attenuation and is therefore used in horizontal and backbone cables.

Spectral window – fields of wavelengths within which the minimum attenuation of an optical signal for its propagation over an optical waveguide is reached.

Splice-plate – construction element of optical switching equipment in which optical waveguides and nondetachable connector organizers are integrated in one unit.

Splicing – nondetachable connection of twisted pairs or optical waveguides.

Splitter – see Y-adapter.

Splitting cord – special multipair termination cord consisting of a plug or outlet of a Telco connector on which six 4-pair cables are prepared with modular plugs installed at the second end.

SRL see Structural return loss –

ST connector – optical connector for fastening plug in outlets with spring bayonet elements. In SCS standards it is often called BFOC-type connector (bayonet fiber-optic connector).

STP – **Shielded Twisted Pair** – cable out of twisted pairs with individual shielding each of them.

Straight termination cord – termination cord with two plugs of modular connectors in which cable conductors are connected to similar plug contacts.

Structural return loss – signal reflection appearing from discontinuity of medium resistance. It depends on frequency, leads to distortions of the signal being transmitted, and is determined by standards.

Structured cabling system (SCS) – cable systems designed and installed according to requirements of ISO/IEC 11801, TIA/EIA-568-A standards, and others. The main SCS features are structurization, universality, and redundancy.

Subsystem of external backbones – SCS subsystems between buildings. It connects cross-room of external backbones with cross-rooms of the buildings.

Subsystem of internal backbones – SCS subsystem providing connections between cross-rooms of the building and cross-rooms of the floors.

Switching block – patch panel base construction element including means of conductor fastening and preparation at the plug&socket contacts in the patch panel composition.

Switching – connection switching between SCS subsystems during maintenance.

Switching equipment – passive equipment located in cross-rooms that is intended for hand connection of different SCS segments with each other and with network equipment by patch cords or jumpers

Switching shelf – variety of optical switching equipment intended for installation into 19-in mounting units by holding brackets. It has a flat construction with height not more than 3*U*. When a great number of optical waveguides are prepared on it, it can be combined with the shelves for storage of welded splices as well as mechanical splices.

System of network equipment connection – total aggregate of terminal cords and adapters to connect network equipment to SCS. In the simplex case it consists only of terminal cord. It is not standardized by ISO/IEC 11801 or TIA/EIA-568-A and is not a part of SCS.

T-adapter – adapter allowing changes in modular connector connection schemes, for example, from T568A to USOC.

Telco connector – connector for electrical cables with two parallel rows of 25 contacts that allows connection of a large number of contacts in a small space.

Telecommunications grounding – see Grounding.

Telecommunications outlet – outlet by which horizontal SCS subsystem is terminated. It is intended to connect the equipment of work areas.

Telecommunications – branch of technology connected with transmission and reception of any type of signal (symbols, signs, texts, video, and voice) over any communications lines (cables, radio communication, optical systems, etc.).

Termination cord – cord to connect network equipment to a cable system. It is used at work areas and in cross-rooms.

Termination point – elastic tube limiting cable bends where the cable enters the connector plug. It is often supplied with a cut system to increase flexibility. It is a standard element of optical connectors and is often used in electrical connectors. Sometimes, it additionally carries out color-coding of cords.

Through-pass adapter – see I-adapter.

Token Ring – LAN of circular topology developed by IBM Corporation. Bit rates of 4 or 16 Mbps; in classic variant it uses the IBM cable system, that is, shielded twisted pairs with impedance of 150 Ω. Later, versions for twisted pairs of Categories 3 and 4 (100 Ω) with bit rates of 4 and16 Mbps, respectively, were developed.

Topology – network configuration. Bar, circular, and star-like topologies are mostly used. One can define both physical and logical topologies. Single networks can have different kinds of physical and logical topology.

Total reflection – effect that is the base of optical signal transmission over an optical waveguide. It is possible for all reference angles that exceed the critical angle of total reflection. In this case, there is only reflection, and refracted waves are absent.

Transceiver – unit for signal transmission and reception between two physically different mediums of communications systems.

Transfer scheme – scheme of single-point administration without interconnection based on passing the optical cables from the cross-room of a building directly to the work area by transit through a floor cross-room. The maximum distance from a telecommunications outlet to the floor cross-room in the version under consideration is 90 m. Maximum channel length is selected as 300 m.

Transition Point, TP – unit of horizontal subsystem in which the connection of two different types of cables (for example, circular cable with flat cable) or splitting of multipair cables to several 4-pair cables is carried out. Network equipment connection and switching operations in transition points are prohibited.

Tubular contact – variety of IDC contact. It is made as split mold tube with cutting knives at the edges of the notch. It has limited use in field-mounted Category 3 outlets.

Twisted pair – cable element containing two isolated wires twisted together with a regular step.

Twisting step – distance at which two conductors of a twisted pair make a full turn around each other.

Two-layer secondary protective cover – variety of secondary protective covers of optical waveguides in which an internal layer with external diameter near 500 μm is made from a soft silicon-type material to decrease fiber deformation and reduce losses.

Two-layer shield – Shield of a cable core containing a general film shield added to an armored shield. The armor has higher mechanical robustness and when terminating the connector provides reliable circular contact with the shielding casing. Shielding film protects cable twisted pairs from high-frequency interference very well and armored shields protect from low-frequency interference. That is, a two-layer shield guarantees reliable shielding over the entire frequency range.

U (Unit) – unit for measuring the height of the workspace of 19-in mounting cabinet that equals 44.45 mm (1.75 in).

Unbalanced circuit – circuit in which one of conductors is grounded from one or two sides and signals are transmitted over the remaining conductors. It is very sensitive to electromagnetic radiation. It serves for low-frequency signal transmission over short distances.

Unbalanced transmission – see Unbalanced circuit.

Unshielded symmetrical cable – see UTP.

Upper cutoff frequency – for electrical routes it is the signal frequency at which signal protection from interferences is equal to10 dB. For optical routes it is numerically equal to the sine of the frequency modulation of the far-end signal at which the amplitude of output signal decreases by 1.41. One can define upper cutoff frequencies of separate elements and of transmission routes mounted from them.

UTP (Unshielded twisted pair) – cable out of unshielded twisted pairs and without general shielding around the core.

Vertical organizer – organizer to lay out vertical parts of patch cords and cables.

VOP – Velocity of propagation

Wall coupling – variety of optical switching equipment for wall mounting. It is made as a box with optical connector outlets on the side wall. Sometimes, it is supplied with a protective cover and an extra-solid case .

Waveguide dispersion – Chromatic dispersion component conditioned by the dependence of the mode propagation conditions from a particular construction of optical waveguide.

Wavelength of zero dispersion – wavelength of optical radiation at which zero dispersion of a single-mode optical waveguide is reached. It is usually equal to 1300–1310 nm.

Welding – technology of nondetachable splicing of fiber-optic waveguides based on melting of their ends with an electric arc with subsequent joining. It provides the best characteristics of connection on insertion losses and stability. It requires the use of precision equipment and is used for large volumes of work in SCS optical subsystems.

Wire for jumpers –one or several unshielded twisted pairs, usually of Category 3, without external protective sheaths. It is used on 66-type patch panels with 66-type contacts.

Work area – place where user network equipment is located (PC, telephone) in the office premises. Sometimes, equipment of common usage is installed at work areas (network printer, fax, etc.).

Y-adapter – adapter to split cable pairs connected to a single modular connector plug to two modular outlets.

Zipcord – two fiber-optic cables for cords whose protective hoses are connected by a narrow jumper

References

1. Smirnov IG: Structured Cabling Systems. Eco-Trends, Moscow (1998)
2. Volovodov AA: Open systems-certain terms (Preface to SCS glossary). Networks and communication systems 49: 29–37 (1999)
3. Volovodov AA: SCS glossary. Networks and communication systems, 50: 32–39 (1999)
4. Galperovich D: Glossary-handbook on structured cabling systems. LAN network solutions journal, 6: 77–85 (2000)
5. Alecseev ES, Myachev AA: English–Russian thesaurus on PC system techniques for users. Finance and Statistics, 1993, 256p, Moscow (1993)
6. Dison P: Dictionary on modern network technologies. Comizdat, Kiev (1997)
7. Semenov AB: Fiber optics in local and corporative communication networks. Computer Press, Moscow (1998)
8. The Siemon Company, 1996 Catalog
9. Semenov AB, Strijzakov SK, Samarsky PA: IT-SCS structured cabling system. IT Company, Moscow (1998)
10. Web site: www.microtest.com
11. EIA/TIA-569: Commercial Building Standard for Telecommunication Pathways and Spaces - Global Engineering Documents, Washington, October 1995
12. Alcatel Cabling Systems, Alcatel, Paris, (1998)
13. Flagman A: Evolution of cable standards. Networks and communication systems, 7 :2631 (2000)
14. TSB-36, Technical Systems Bulletin: Additional cable specifications for unshilded twisted pair cables. Global Engineering Documents, Washington, October (1995)
15. TIA/EIA TSB-40: Additional Transmission Specification for Unshilded Twisted Pair Connecting Hardware. – Global Engineering Documents, Washington, (1995) 18 p
16. Volovodov AA: SCS of Category 6: pro and con. Networks and communication systems, 8: 28–33 (1999)
17. ISO/IEC 14763-1:1999 (E). Information Technology – implementation and operation of customer premises cabling. 1. Administration.
18. Systimax Structured Cabling Systems, Product guide. Avaya, Basking Ridge, Spring 1996
19. Web site: www.siemon.com/references/tech
20. TIA/EIA TSB-75: Additional horizontal cabling practices for open offices, August 1996
21. TIA/EIA TSB-72: Centralized optical fiber cabling guidelines. October 1995
22. Local network using fiber optics. Corning data sheet, Corning incorporated, New York, 1999
23. Giyduk SK: What SCS from being offered to select? Vestnik Svyazy 5: 57–59 (2000)

24. Avduevsky AV: SCS from bird's-eye height. LAN Magazine, Russian issue 6: 55–66 (1998)
25. Grodnev II, Vernik SM, Kochanovsky LN: Communication lines. Radio and Communications, Moscow (1995) p. 488
26. Weidhaus W. Symmetrische Kabel fur zukunftssichere Datennetze. Renningen-Malmsheim: Expert Verlag, (1996) 124 S
27. GOST 27893-88 (CT CMEA, 1101-87): Communication cables. Testing methods. Standards Presshouse, Moscow (1989) p.24
28. International Standard ISO/IEC 11801: Information Technology, Generic Cabling for Customer Premises. (1995)
29. Deutsch B, Mohr S, Roller A, Rost H: Elektrische Nachrichtenkabel. Grundlagen, Kabeltechnik, Kabelanlagen. Publicis MCD Verlag, München (1998)
30. Drayer HE: Searching for new cable tester. LAN Magazine, Russian issue 4: 57–62 (1998)
31. Prudon D: Cable systems for speed data transmission. Networks and communication systems, 5: 44–48 (1998)
32. Kenny B: Returning to the fundamentals. LAN Journal of network solutions, 5: 59–68 (1999)
33. Green D, Volovodov AA: Cable-to-cable crosstalk. Networks and communication systems, 3 (53): 40–47, (2000)
34. Smirnov I.G. Signal propagation: passing delay and delay shifting//Vestnik svyazy, 8: 39–40 (1998)
35. Web site: www.molexpn.com.pl
36. The Siemon Company. Copper and Fiber Optic Cable Catalog. 1999. – 38 p
37. Grodnev A.G, Muradyan R.M, Sharafutdinov et alias. Fiber optical transmission systems and cables. Handbook, Moscow, Radio and communication, 1993, p.264
38. Neues von Gigabit Ethernet IEEE 802.3ab: www.lucent. com.netsys/systimax
39. Kemmler W: 1000 Mbit/s auf UTP-Kupferkabel. 2. Kodierung, Kabel und Kompensationen. Technik News, Februar 2000, Ausgabe 02/2000, pp. 64–67
40. Volovodov AA: From clock frequency to information backbone. Networks and communication systems, 9: 13–17 (1999)
41. Gandja D: How does FC-AL operate?. LAN Journal of network solutions, 6: 19–21 (2000)
42. Galperovich DY: Horizontal layout of Category 5. LAN Magazine, Russian issue 8: 2 (1996)
43. Uninet: Data Cables Satisfying Highest Demands. Datwyler Catalog, Datwyler Company, Berlin, December 1999 edn.
44. LAN Journal of network solutions, 4: 12–17 (1998)
45. Link Performance Definitions Cabling Vision, Alcatel Cabling Systems International Newsletter, September 1998, pp. 39–44
46. Data Cable LANmark Category 6 UTP, 4-Pair Horizontal Cable. Alcatel data sheet
47. NetConnect, Open Wiring System. AMP Catalog
48. Belden. Master Catalog (1997)
49. A Division of Cable Design Technologies. Master Catalog. Mohawk/CDT
50. Copper Data Cables, Pirelli data sheet, Milano (1998)
51. Fiber Optic Cables and Copper Data Cables, BICC Brand-Rex Catalog (1999)
52. Automation and Data Cables, NK Cables Catalog (1999)

53. Structured Cabling Systems 2000, Elgadphon Catalog (2000)
54. PCnet Network Cable, Pun Chang Wire & Cable Co. data sheet
55. Systimax SCS GigaSPEED Copper Cable. Lightspeed Performance Over Copper Cable, Lucent Technologies data sheet. – 5 p
56. Datacom Products (1998) . RiT Technologies. – 363 p
57. Belden Master Catalog,Belden data sheet, (1998) – 358 p
58. Dittrich J, Thienen U. Moderne Datenverkabelung. – Bonn: Datakom, (1998) . – 517 S
59. Web site: www.teldor.com/products/hi-giga.htm
60. Boothroyd D: Hybrid Cable lives up to a mixed reputation. Cabling Installation & Maintenance Europe, May/June 2000, pp. 9–10
61. Technical information and user's manual, Krone Catalog (1998)
62. Kish P: Hardware and category 6 backward-compatibility. Cabling Installation & Maintenance, 6, pp. 53–58 (1999)
63. International Standard IEC 603-7: Connectors for frequencies below 3 MHz for use with printed boards. 7. Detail specification for connectors, 8-way, including fixed and free connectors with common mating features, with assessed quality, 2nd edn, vol. 11 (1996)
64. Category 5e Networking & Communication Cabling Accessories 2000–2001. Hsing Chan Industrial Catalog (1999)
65. Product Guide, Spring/Summer, Molex Catalog (1999)
66. Avduevsky AV: New proposals on connections of Category 7. LAN Journal of network solutions, 5: 6 (1999)
67. The Siemon Company Catalog (1999)
68. Technical specification for structured cabling systems: UTP Cat 5/Class D. Reichle & De-Massari Company (1999)
69. Daten- und Telekommunikation,Telegartner Catalog, 1998/1999
70. FMT, FMT Fernmeldetechnik Produktions GmbH Catalog (1999)
71. V.D.I. Guide, Legrand Catalog
72. Produkt Katalog, Telesafe Catalog (1999)
73. Catalog 2000, Digital Data Communication Company (2000)
74. Gerschau L: Strukturierte Verkabelung. DATAKOM, Bergheim (1995)
75. Systemlosungen fur Netzwerke und Komponenten, Ackermann Catalog (1999)
76. Debiec T, Wagner M: Hybrid fiber-optic cable gets high marks on campus. Cabling Installation & Maintenance, 4: 27–32 (1997)
77. Volition, LWL-Installationkabel. 3M data sheet, December (1998)
78. Akshead D: Cable alternative. Networks, 5:65–67 (2000)
79. Fiber optical cables, Product information. Ericsson Cables AB Catalog (1998)
80. Ubaldullaev RR: Fiber optical networks. Eco-Trends, Moscow (1998) p.268
81. Malke G, Gessing P: Fiber optical cables. Izdatel, Novosibirsk (1997) p.264
82. Chomycz B: Fiber-optic installation. McGraw-Hill, New York (1996)
83. Optical connectors. Diamond data sheet (1998)
84. Optical fiber connectors LGC, LGA series. Honda Tsushiu Kogyo Catalog (1999)
85. Wrobel CP: Optische Übertragungstechnik in der Praxis. 2. Überarbeitete und Erweiterte Auflag. Huthig Verlag, Heidelberg (1998)
86. Haber L: Cable systems becoming speedy. Networks and communication systems, 55: 22–25 (2000)
87. Ivanov P: Renovated view of cable systems. Networks, 2: 44–47 (2000)
88. Shin'ichi Iwano: MU fiber-optic connector system. Lightwave, 10: 55–62 (1998)

89. Panduit Communication Product. Panduit Catalog (1998)
90. Kilmer JP: Connectors evolve for the premise market. Lightwave, 5: 39–42 (1998)
91. Telecommunication products: PD Telecom, Rev. 4. Hellermann-Tyton Catalog (1999)
92. Koshon GF: Modular components make easier installation of optical cable system. Networks and communication systems, 7: 16–20 (1999)
93. Berdichevsky M: Standard dimension-types of electronic equipment cases. CTA 1:96–106 (1997)
94. Bender D: LAN-Datenschranke. : VDE-Verlag, Berlin (1998)
95. MFB Products, MFB Catalog (1999)
96. Szafy 19-in Wiosna 99, Zaklad Produkcji Automatyki Sieciowej SA w Przygorzu Catalog, Warszava, (1999)
97. Rittal Handbuch 29, Rittal-WerkRudolf Loh GmbH & Co, Herborn (1999)
98. Electrical fitting and wiring accessories 97/98 catalogue, Legrand Catalog (1998)
99. Systems for electroinstallation, Rehau Catalog (1999) p.298
100. Rigglswort K, Volovodov AA: Effectiveness of cable box shielding. Networks and communication systems 4:50–57 (1998)
101. Flexilink 6000, 7000, 8000 Series. PINACL Products data sheet, PINACL Communication Systems Limited, London, (1999)
102. Reeves EA: Cable management systems. Oxford: Blackwell Scientific Publications (1992)
103. The First Name for Quality. The MK Electric Catalogue, Int. Edn. (1995–1997)
104. Product guide. Raychem Catalog (1998)
105. Complete optical fiber cabling solution. Panduit Catalog (1996)
106. General Catalogue. Unex Catalog (1998)
107. Full FTTO solution combines the best of fiber and copper installation. Cabling Vision, Alcatel Cabling Systems International Newsletter, September 1998, pp. 72–75
108. Semenov AB: LAN Installation units with fiber optical interface for SCS technique usage. LAN Magazine, Russian issue, 8: 25–30 (2000)
109. Chernobrovtsev A: DWDM air debute. Computerworld-Russia. 194: 33 (1999)
110. Rules of electric installation arrangement, vol. 6, Revised and updated. Glavgosenergonadzor, Moscow, p.608
111. Pelt M: Electromagnetic protection and grounding. Networks, 6: 52–53 (2000)
112. GOST 12.1.044-89: Fire and explosion danger substances and materials. IPK Standard Publishers, Moscow (1996) p.143
114. GOST 12176-89: Cables, wires and cords, Test methods for burning dispropagation. Standard Publishers, Moscow (1996) p.12
114. Maklaflin P: Deficit of fire resistance cables. Networks and communication cables, 3: 22–26 (1999)
115. Mair HJ: Kunstoffe in der Kabeltechnik. Renningen-Malmscheim: Expert Verlag, (1999) . – 272 S
116. Frose H. D. Brandschutz fur Kabel und Leitungen. Pflaum Verlag, München (1998)
117. GOST 34.601-90 Information technology. Standard complex on automated systems. Automated systems. Stages of creation. Standard Publishers, Moscow (1991) p.8
118. GOST 34.201-89. Information technology. Standard complex on automated systems. Types, completeness and document designation on automated system creation, Standard Publishers, Moscow, 1991, p.14.

119. GOST 34.602-89: Information technology. Standard complex on automated systems. Technical task for automatic system creation, Standard Publishers, Moscow (1991) p.16
120. RD 50-34.698-90: Information technology. Methodical guideline. Complex of Standards on automated systems. Automated systems. Requirements to document contents. Standard Publishers, Moscow (1991) p.39
121. Backbone for fiber optical cables. Dura-line Company data sheet, St. Louis, (1999) p.24
122. Cabling Installation & Maintenance, 6: 8–10 (1999).
123. Semenov AB: Fiber optical technique in LAN and SCS and its features. In: Dmitriev SA, Slepov NN (eds.) Fiber optical technique, history, achievements, future. «Connect» Publishers, Moscow (2000) pp.351–369
124. Maiorov AP: Cable systems for office buildings. 11. Administration. Networks and communication systems 7:30–38 (1998)
125. GOST 21.110: Rules of fulfillment of equipment, products and materials specifications. Moscow (1995) p.12
126. Avduevsky AV: Through fire and water. LAN Journal of network solutions 5:55–62 (1999)
127. Ivantsov I: Cable fastening. LAN Journal of network solutions, 5: 28–29 (1999)
128. Field testing of high performance premise cabling. Microtest, Münich (1998)
129. Hayes J, Resenberg P: Data, voice and video cabling. Delmar Thomson Learning New York(2000)
130. Flagman A: Future of SCS standards. Networks and communication systems, 3: 27–35 (1999)
131. Gratsiano K: Digital multimeters on guard of safety. Networks and communication systems 33:40–45 (1998)
132. DeSorbo MA: Testers of tomorrow available today. Cabling installation & maintenance 5, pp. 70–74 (1999)
133. Backlanov IG: Reflectometers for copper wire. Networks and communication systems, 11: 82–84 (1997)
134. Backlanov IG: Measuring technology in up-to-day telecommunications. Eco-Trades, Moscow (1998) p.140
135. Jensen Master Catalog (1997–1998)
136. Komarov MU: Controlling and measuring equipment for mounting and maintenance of fiber optical communication lines. Fiber optical technique 8 :19–24 (1999)
137. Ivanov AB: Comparative analysis of fiber optical communication line controlling and measuring equipment. Vestnik svyazy, 1: 52–56 (1998)
138. TIA/EIA-606: Administration standard for telecommunications infrastructure of commercial buildings Global Engineering Documents, Washington (1993)
139. Enterprise 1, Information resources management system. RiT Technologies Company data sheet (2000) p.4
140. Chetverikov V: Network infrastructure administration. LAN Journal of network solutions, 5: 65–74 (1999)
141. Electrical marking solution, Product guide. Brady Company (1997)
142. PAN-CODE, Identification and labeling solution products. Catalog of Panduit Company (1999)
143. Weidmüller, Die Interface Partner, Gesamtkatalog 98/99. Catalog of Weidmüller Company (1998)

144. Safety and facility identification products. Catalog of Panduit Company (1999)
145. Ivantsov I: Marking. LAN Journal of network solutions, 2: 28–29 (1997)
146. Smirnov IG: Should the cable systems be structured. Vestnik svyazy, 12: 66–69 (1998)
147. Your connection to the future, advanced electronic support products. AESPCompany data sheet (1999)
148. Shielded cable systems. AMP Company data sheet (1998) p. 8
149. AMP Netconnect structured cabling systems. Catalog of Moscow representatives of AMP Tyco Electronics (2000) p. 56
150. LAN Journal of network solutions, 1: 10–15 (2000)
151. Millenium Strukturierte Verkabelungssysteme, Produktkatalog. BICC Brand-Rex Company Catalog (1997)
152. Advanced connectivity system. IBM Product Catalog, Release 2.3, September (1997)
153. IT-SCS. IT Company Catalog (1998) p. 88
154. LANConnect network components. TRALE Ltd, Data sheet M (1998) p. 36
155. ISCS, Shielded cable systems. TRALE Ltd, Data sheet M (1998) p. 36
156. Products and order. Catalog of Krone Company (2000) p. 600
157. Technical information and user' manual, Catalog of Krone Company, p. 600
158. Krone Highband, Verteilfeld RJ45 STP, Technische Daten. – Krone Company data sheet (1998)
159. Fiber optic product catalog no. 1098A: Active & passive network solution. Molex Fiber Optics (1997)
160. Channel solutions for higher standards. Ortronics Company data sheet (1998)
161. European Systems Catalog, 3rd Qtr. Catalog of Ortronics Company (1998)
162. R&M Freenet. Catalog of Reichle & De-Massari Company (1998)
163. ICCS und FutureLink. Kabel und Komponenten fur die strukturierte Gebaudeverkabelung. Catalog of Siemens Company (1998)
164. GOST 24204-80: Decibel unit for measuring levels, attenuations and gains in wire communication technique. Standard Publishers Moscow, (1988) p. 4

Index